# DATE DUE

| MR 1 5 '99 | | | |
|---|---|---|---|
| ~~MR 17 '99~~ | | | |
| | | | |
| | | | |
| | | | |
| | | | |
| | | | |
| | | | |
| | | | |
| | | | |
| | | | |
| | | | |
| | | | |
| | | | |
| | | | |
| | | | |
| | | | |
| | | | |
| | | | |

DEMCO 38-296

# Fluid Physics in Geology

# Fluid Physics
# in Geology

## An Introduction to Fluid Motions on Earth's Surface and Within Its Crust

**DAVID JON FURBISH**

New York    Oxford
OXFORD UNIVERSITY PRESS
1997

...ersity Press

Oxford    New York
Athens    Auckland    Bangkok
Bogota    Bombay    Buenos Aires    Calcutta
Cape Town    Dar es Salaam    Delhi
Florence    Hong Kong    Istanbul    Karachi
Kuala Lumpur    Madras    Madrid    Melbourne
Mexico City    Nairobi    Paris    Singapore
Taipei    Tokyo    Toronto

and associated companies in
Berlin    Ibadan

Copyright ©1997 by Oxford University Press, Inc.

Published by Oxford University Press, Inc.,
198 Madison Avenue, New York, New York 10016

Oxford is a registered trademark of Oxford University Press

**Library of Congress Cataloging-in-Publication Data**
Furbish, David Jon.
Fluid physics in geology :
an introduction to fluid motions on Earth's
surface and within its crust /
David Jon Furbish.
p. cm.
Includes bibliographical references (p. – ) and index.
ISBN 0-19-507701-6 (cloth)
1. Fluid dynamics. 2. Geophysics.
I. Title.
QE517.5.F87   1996   550'.1'532—dc20   95-51658

1  3  5  7  9  8  6  4  2

Printed in the United States of America
on acid-free paper

*To Bill and Ruth ...*
*with immeasurable gratitude*

## Distilled Water

Some water
Deigned itself
To flowing!


      It
Sparkled
Clearly
With this caution:
Not for use
In washing clothing.

And this was not without just cause.

      It
Did not grasp at
Willow shrubbery,
Willows in full blossom. Quite

      Ignored
The streaming threads of algae
And fishes fat on damselflies.

      It
Did not frequent wavy places,
Nor did it wish to travel far.
Its life was most uncomplicated—
Clear,
Unadulterated
Water!

Leonid Martynov (1905–1980)
*Translated from the Russian by Dean Furbish*

# Contents

## 12. Viscous Flows,   261

## 13. Porous Media Flows,   299

## 14. Turbulent Flows,   347

# Preface

Geology is chock-full of the stuff of fluid physics. Geology has rivers and ground water, ocean surfs, glaciers, settling silt and saltating sand, gas bubbles and liquid drops, magma diapirs and lava flows, geysers, and volcanic jets. What better reasons are there for being a student of geological fluid physics?

Geologists typically take courses in mathematics or physics to obtain a formal training in fluid physics—which is laudable. Such courses and accompanying texts present the material in a way that favors those things that are traditionally of most interest to mathematicians and physicists. To fill in the geologically applied spaces, one goes elsewhere: to chemistry and physics texts for information on bubble growth; to hydrology texts for an explanation of Hubbert's potential; and to engineering texts for treatments of turbulent flows next to rough boundaries. For these reasons, I thought it would be useful to have a text that covers introductory fluid physics in a way that features the problems geologists traditionally ponder. In this vein, this text is intended to prepare geology students who might wish to pursue further course work in fluid physics, as well as to illustrate the relevancy of many problems from other fields to geological problems.

This text examines fluid behavior within a geologic context, rather than consisting of a collection of geological problems that happen to involve fluids. As such, it has the flavor of a conventional text in fluid mechanics. But it is unconventional in the sense that it offers a geological motivation for the material, and the Example Problems following each chapter are given lengthy treatment. My experience in teaching the material has been that students at this level are still exploring how the process of problem solving is actually undertaken, and respond well to studying specific examples. The examples are also intended to illustrate how a complete solution is not always essential; but rather, that significant insight to problems can be obtained from partial solutions.

This text is intended to provide material for a one-semester course. I have used the notes on which the text is based to teach a course aimed at providing students with the grounding necessary for my own advanced courses in hydrology and geomorphology. The material also is entirely suited for training students wishing to pursue more advanced topics involving fluid physics in areas such as igneous petrology and sedimentation. The text is written for well-prepared undergraduate and graduate students in geology. It is assumed that students have a minimum of two semesters of university physics and calculus. A third semester of calculus, with introduction to vectors, vector-valued functions, and ordinary and partial differential equations, would be beneficial, and I urge students to take such a course before undertaking the material in this text. Nonetheless, completion of such a course is not assumed; students without this training should be able to gain much from the material by carefully studying the example problems and undertaking supplemental reading.

The material is presented at an introductory level, motivated by the conviction that a solid grounding is essential for pursuing more difficult problems. At certain places, digressions are made to cover ideas and tools that are cornerstones of continuum (fluid) mechanics. Some of these are used throughout the text, and students who wish to pursue the study of fluids on their own ought to become very familiar with them. The use of Taylor expansion to approximate physical quantities in the vicinity of a control volume, or over a short interval of time, is one example.

Topics are presented in an order that is intended to facilitate teaching of the material. For example, the presentation of fluid strain leading to the rate-of-strain tensor is actually a kinematical topic. But rather than covering this material in Chapter 7 (Fluid Kinematics), ideas of fluid strain coherently follow the coverage of vorticity, inasmuch as the latter is helpful in understanding components of fluid strain. Some of the mathematical developments are relegated to appendixes, for example, those leading to equations of conservation of mass, energy, and momentum in non-Cartesian coordinates. This is to save space and does not reflect that these developments are unimportant.

The text is written to prepare students for creative work, and not just for manipulating formulae in patterned problems. Creative work is, of course, a very personal thing, and goes hand in hand with the idea that learning essentially is a random process. To paraphrase the remarks of W. R. LePage on this matter, an author cannot insist that events in a student's program of learning will occur in any predictable order. In reading a text of this nature, it is therefore important to recognize the interrelatedness of the material with a wealth of ideas to be found in many different sources. By referring to other texts, and by experimenting with one's own ideas as each topic is covered—referring back and forth to different chapters—one is likely to eventually cover the material in the text without ever having read it cover to cover.

This text, then, is written for students whose interests include, among other things, groundwater hydrology, surface water hydrology, geomorphology, sedimentation, and igneous petrology. Its coverage is therefore intended to encompass sufficient material to satisfy this general pedagogical purpose. Nevertheless, individuals may wish to concentrate on parts of the text that are most relevant to their own interests. For this reason, comments in the next section (Menu) regarding each chapter are offered as a guide to those wishing to organize a specific course of study. Certain chapters, as noted, should be examined fully by all, because they cover topics that bear, in one way or another, on most fluid physics problems in geology. In addition, some of the example problems are used to introduce material not explicitly covered in the preceding chapter, and it is therefore worthwhile to examine these.

On a more personal note, two essential ingredients for becoming comfortable with the material in this text are: (1) study of how others have solved problems; and (2) practice. It is important to realize that the insight and confidence to undertake a problem often come from mimicking related examples in the literature. A key is learning how to formulate a problem well—to ask concise questions, and to simplify a problem to a tractable form while retaining its original essence. This may be as simple as specifying coordinates in an unconventional way (Example

Problems 8.4.4 and 10.3.3), or as involved as undertaking a formal linearization of the equations governing fluid motion (Example Problems 15.8.3 and 16.5.1). On this note, I have tried to illustrate some of these ideas in a way that makes the material more accessible than might otherwise be to geology students. I learned much of this stuff on my own; and I cannot claim the wisdom to know whether it was the easy or the hard way. Have fun!

## MENU

Chapter 1 provides a general motivation for the contents of the text. It introduces several strategies that are used in the study of fluids, including using structural homologies to examine problems. The chapter concludes with an outline of the system of units and mathematical conventions adopted in the text, a statement regarding the scope of mathematics used, and with examples of structural homologies.

Chapter 2 examines why, and how, fluids and porous media are treated as continua. It begins with a molecular-scale view of fluids to motivate the necessary averaging process leading to the continuum concept. Similar concepts are then developed for porous media. The chapter concludes with a brief introduction of alternatives to the continuum concept, including an example in which soil pores are treated as possessing a fractal geometry. Example problems also examine recent ideas of fluid behavior at scales below the continuum scale, an important topic in studies of fluid flow within very small rock pores, and the geochemistry of fluid–solid boundaries at the pore scale.

Chapter 3 describes the basic mechanical properties of fluids and porous media that are essential for treating fluid behavior throughout the remainder of the text. Both Newtonian and non-Newtonian fluids are considered. Interpretations of fluid viscosity and the no-slip condition, based on molecular kinetics, are presented to illustrate the connection between the molecular and continuum scales. The coverage of porous media properties is restricted to items most relevant to flow problems. Examples are selected to illustrate how the definitions and ideas presented earlier in the chapter can immediately be used to solve several simple, but realistic, problems. These then serve to motivate material presented in later chapters, and as problems to be revised when certain ideas of fluid behavior are treated on a more formal basis. For example, simple one-dimensional flows are treated for later reference in chapters on dimensional analysis and viscous flows.

Chapter 4 describes basic thermodynamic properties of fluids, and is essentially a companion to Chapter 3. It introduces the idea of specific heat and Fourier's law for heat flow by conduction. The idea of a gradient as the derivative of a vector quantity is first formalized here; this should be studied carefully, because it is used several times later in the text. This chapter then introduces equations of state for gases and liquids, and examines how such equations vary from the ideal gas law for real fluids. The ideal law is then used together with the first law of thermodynamics to examine isobaric, isothermal, and adiabatic processes involving compressible fluids. The value of this chapter will rest chiefly with those wishing to examine problems where thermal effects are important, for example, magma dynamics, groundwater flow over large space and time scales, bubble dynamics, and geyser dynamics.

Chapter 5 introduces principles of dimensional analysis and similitude. Dimensional analysis is illustrated with a series of examples that build on each other, so these should be examined in order. Ideas of geometrical and dynamical similitude are likewise illustrated, and ideas of exact and statistical similitude are distinguished. The chapter concludes by introducing several important dimensionless quantities, for example, the Reynolds number, that frequently arise in fluid physics problems.

Chapter 6 provides an introduction to fluid statistics and buoyancy. This includes developments of the idea of static pressure, the equation of fluid statics, the hydrostatic equation, and hypsometric equations for both liquids and gases. The use of a Taylor expansion to approximate physical quantities in the vicinity of a control volume is first introduced here; because this basic procedure is used throughout the text, students should examine this section carefully. In addition, this chapter develops the idea of buoyancy as applied to both rigid objects and fluid elements, and considers the stability of a thermally stratified fluid.

Chapter 7 examines how motions of fluids are described. It begins with qualitative descriptions of fluid motion, including the ideas of a pathline, a streakline, and a streamline. The substantive derivative is then formally introduced in the context of the distinction between Lagrangian and Eulerian views of fluid motion. This leads to kinematical expressions for convective accelerations, which are incorporated later into the full dynamical equations of motion. In this regard, Chapters 7, 10, and 11 are mainly pedagogical steps toward treating the full dynamical equations of viscous flow discussed in Chapter 12.

Chapter 8 formalizes the idea of conservation of mass. Coverage includes unsaturated and saturated flows in porous media as well as purely fluid flows, and conservation of dissolved solids and gases in liquids. The idea of developing equations of continuity for large control volumes is also introduced; examples include the St. Venant equation of continuity for open channel flow, and the Dupuit approximation for flow in an unconfined aquifer.

Chapter 9 formalizes the idea of conservation of energy, and is a companion to Chapter 8. Coverage is limited in this chapter to inviscid flows. The special case of purely mechanical energy is considered, leading to an initial development of Euler's equations (Chapter 10). The development of the full equation describing conservation of energy is then completed in subsequent chapters once Euler's equations are reintroduced, then when viscous forces are covered. An introduction to Hubbert's concept of fluid potential as applied to porous media flow is presented, illustrating how this concept arises from conservation of energy.

Chapter 10 develops Euler's equations, the momentum equations for inviscid flow. An explicit treatment of forces inducing fluid acceleration is first provided here. This is a significant step toward the development of the Navier–Stokes equations for Newtonian flows in Chapter 12. Bernoulli's equation also is introduced, and is shown to be essentially a statement of conservation of mechanical energy.

Chapter 11 formalizes the ideas of vorticity and fluid strain. Vorticity is treated as a step toward describing fluid strain, and is particularly useful in illustrating the idea that simple shearing motion consists of pure shear and rotation, which, in turn, leads to the idea of separating strain components. The rate-of-strain tensor is then introduced as a necessary step toward material covered in Chapter 12, wherein rates of strain are formally related to the normal and shear stresses that produce them. Coverage of vorticity in this chapter also serves to illustrate certain ideas about turbulence (Chapters 14 and 15).

Chapter 12 forms the cornerstone of how real fluid flows are formally treated. This chapter begins with a general description of viscous forces, and then incorporates this description into the full momentum equations, the set of equations that describe Newton's second law for the motion of a viscous fluid. It then develops the constitutive equations for a Newtonian fluid, and incorporates these into the momentum equations to arrive at the Navier–Stokes equations. The full energy equation is derived, including the dissipation function, which characterizes the continuous transformation of mechanical energy to heat due to work performed against viscous friction. The chapter concludes with a brief examination of the constitutive equations for ice, as an example of a non-Newtonian fluid.

Chapter 13 presents the essential ideas involved in mechanical interpretations of Darcy's law, and develops the idea of describing hydraulic conductivity in tensor form. It then develops the equations governing groundwater motion for both saturated and unsaturated conditions. This chapter also examines the conventional view of mechanical dispersion, leading to the advection–dispersion equation. It concludes with models of effective thermal diffusivity and specific heat, and applies them in the energy equation for heat flow under saturated conditions.

Chapter 14 begins with a qualitative description of turbulence, including the general conditions under which turbulence is generated, and the classic view of the turbulence energy cascade. It introduces the procedure of defining time-averaged and fluctuating flow quantities, then applies this procedure to the Navier–Stokes equations for incompressible fluids to obtain expressions for the Reynolds stresses. The chapter concludes with a brief coverage of statistical interpretations of fluctuating fluid motions.

Chapter 15 introduces the two-dimensional, boundary-layer equations for turbulent flow as a basis for describing the development of a turbulent boundary layer. After examining the idea of a fully developed boundary layer, these equations are simplified for the case of a unidirectional shear flow. In this context, the statistical descriptions of turbulence introduced in Chapter 14 are used to clarify how the Reynolds stresses are distributed near a boundary. The Boussinesq hypothesis is introduced, which leads to Prandtl's mixing-length concept and the idea of eddy viscosity. These are used to obtain the classic logarithmic velocity law, concentrating on rough boundaries. The chapter concludes with a brief discussion of kinetic energy production and dissipation.

Chapter 16 examines the idea introduced in Chapter 6, that thermal stratification of a fluid may lead to mechanical instability and free convective motions. The chapter starts with a development of the Boussinesq approximation—a set of simplified versions of the equations of motion, energy, and state—which forms the cornerstone of formal descriptions of convection. It examines results of laboratory experiments, starting with Rayleigh–Bénard convection, then briefly considers these results within the context of convection in geological situations, including convection within porous media. The Example Problems are for deriving the critical Rayleigh number marking the onset of convection in purely fluid systems and in porous media.

## ACKNOWLEDGMENTS

I am very grateful to several individuals who helped me improve an early draft of the text. George DeVore helped me improve the presentation of thermodynamic topics, David Loper helped me clarify aspects of the chapter on viscous flows and provided insight to several of the Example Problems, and James Pizzuto and

Jeff Warburton provided comprehensive reviews of several chapters. I am particularly grateful to Ruby Krishnamurti, who contributed substantively to the chapter on thermally driven flows, and who has given generously of her time over the past five years to help me with flow problems. I also am grateful to Tracy Byrd and Stephen Thorne for helping me edit parts of the manuscript, and especially to Rosmarie Raymond for drafting the figures.

I am particularly grateful to Chelsea and Amelia for what I fondly recall as immeasurably pleasurable breaks from my computer screen, during which they concocted the most delightful of childish diversions. Most importantly, I am grateful to Anne Choquette for her unwavering encouragement, patience, and support.

Tallahassee, Florida                                                  D. J. F.
April 1996

# CHAPTER 1

# Introduction

Fluids are involved in virtually all geological processes. Obvious examples are phenomena occurring at Earth's surface in which fluid flow is a highlight: the flow of a lava stream, the play of a geyser, river flow and wind currents, the swash and backswash on a beach. Also obvious are phenomena that occur in the presence of fluid flows, such as sediment motion. Less obvious, but readily imaginable in terms of their behaviors, are fluid motions occurring within Earth's crust: flows of magma and ground water, and expulsion of brines from sediments during compaction. In addition, a bit of reflection will recall a host of phenomena in which fluid behavior, although not the highlight, may nonetheless take on a significant role: initiation of landslides, seismic activity, glacier movement, taphonomic organization, and fracture mechanics. With these should be considered instances in which the geological material containing a fluid can influence its fundamental behavior at a molecular scale. An example is flow through very small rock pores, where molecular forces interacting among fluid molecules and pore surfaces can lead to a structural arrangement of the fluid molecules such that their mechanical behavior is unlike that which occurs in large pores, where the bulk of the fluid is "far" from pore surfaces. It is thus understandable that to describe many geologic phenomena requires knowing how fluids work. It is also natural to begin by considering how fluids behave in a general way, then in turn, how they are involved in specific geological processes.

## 1.1 TOPICS AND STRATEGIES IN THE STUDY OF FLUID PHYSICS

There are several approaches for describing fluids and their motions, and the choice of one, or some combination, depends on the sort of insight desired as well as the specific problem being considered. *Fluid statics,* as the name implies, involves considering the properties of fluids that are at rest in some inertial frame of reference. Note that this frame of reference may actually be moving relative to the Earth frame of reference, so long as the fluid motion is like that of a rigid body. An important example of our use of fluid statics will be in developing the *hydrostatic equation,* which formalizes how fluid pressure varies with depth. This will serve as a base for describing the forces that act on submerged bodies, including buoyant forces, and for assessing the stability of fluids that are stratified, thermally or otherwise.

*Fluid kinematics* involves describing motions of fluids without explicit regard given to the forces causing the motions. Fluid kinematics is entirely analogous to the topic of kinematics covered in introductory physics, wherein expressions are obtained for the speed, velocity, and acceleration of a ballistic particle. Thus, we will use fluid kinematics to introduce the important idea of a *convective acceleration,* and obtain an expression for this acceleration that is analogous to that of a ballistic particle. In addition, we will make use of fluid kinematics to introduce the idea of examining fluid behavior at fixed positions in space, an *Eulerian* description, rather than tracking the behavior of individual fluid parcels, a *Lagrangian* description.

In contrast to fluid kinematics, *fluid dynamics* involves describing motions of fluids with due regard for the forces causing them. Herein consideration is given not only to forces inducing fluid motion, but also to those resisting motion, both of which depend on the geological setting and flow geometry, as well as the mechanical properties of the fluid. The *momentum equations*—the equations that express Newton's second law for the motion of fluids—form the cornerstone of fluid dynamics.

To these we add *fluid thermodynamics,* wherein fluid pressure, temperature and density, and the relations of these variables to heat and mechanical work, are considered. We will see that to properly account for the total energy of a fluid system requires considering energy in the form of work performed by the system on its surroundings, or by the surroundings on the system. An important phenomenon included in this subject is the *adiabatic cooling* of a fluid as it expands and thereby performs work on its surroundings; an example is the expansion of a gaseous, volcanic eruption plume as it ascends within the atmosphere. The study of fluids thus can involve a breadth of approaches; we will consider each in the development below of fluid physics in geology.

Several concepts are fundamental to the study of fluid physics. Foremost is the treatment of a fluid as a *continuum,* wherein we define the geometrical and mechanical conditions under which it is permissible to associate variable fluid quantities such as pressure, density, and velocity with each coordinate position in a region, treat these quantities in terms of continuous functions, and then make use of differential calculus to describe the variables. This is despite the fact that fluid is, at a molecular scale, discontinuous stuff. Also useful is the concept of an *ideal fluid* or *perfect fluid,* which, by definition, cannot sustain a tangential stress in the sense that it does not resist shearing motion. An ideal fluid therefore does not possess *viscosity,* and is said to be *inviscid.* An interesting outcome of this concept is that flow can "slip" past a solid boundary; the fluid adjacent to a boundary can have a nonzero velocity relative to the boundary. In contrast, a *real fluid* can sustain a tangential stress; it resists shearing motion, and therefore possesses viscosity. In this case, a *no-slip condition* exists between a solid boundary and adjacent fluid, and the relative velocity is zero. In addition, real fluids can exhibit varying behaviors under stress. We will consider some of these behaviors, but concentrate on *Newtonian fluids.* A Newtonian fluid is one whose rate of strain is linearly proportional to the shearing stress applied to it.

We also will consider behaviors of both liquids and gases. With exceptions, the physics and equations governing their motions are identical, so they can be treated interchangeably in many of the following sections. However, liquids are virtually *incompressible* in many geological situations, which allows their mathematical treatment to be simplified. An important example is groundwater flow near the surface of Earth. Nonetheless, problems exist where liquid compressibility must be treated explicitly. In fact, the compressible behavior of water is an essential reason that ground water can be extracted from an aquifer by pumping a well. Moreover, the finite compressibility of a magma is an essential reason that bubbles can pressurize the magma as they ascend within it. Likewise, although gases are *compressible,* and often must be treated as such, problems exist where a gas may be treated as incompressible. We will examine the conditions under which this simplification is permissible, and learn that it is appropriate, for example, when treating eolian transport in air, a compressible fluid.

Two important strategies, developed as themes throughout this text, are particularly useful for approaching problems in fluid physics. The first involves making use of *structural homologies* to investigate a problem. A structural homology exists between two systems when the forms of the equations governing each are the same, despite the fact that the physical meaning of the variables and coefficients may be quite different. In this situation, the results and techniques devised to investigate one of the systems immediately can be adopted and applied to the other. An important example of this involves use of the structural homologies that exist between many heat flow and groundwater flow problems. Recall Fourier's law for heat flow by conduction in one dimension:

$$q_T = -K_T \frac{\partial T}{\partial s} \tag{1.1}$$

Here $q_T$ is the rate of heat flow per unit cross-sectional area of conducting medium, $T$ is temperature, $s$ is a spatial coordinate along the medium, and $K_T$ is its *thermal conductivity* (Figure 1.1a). (Use of the partial notation $\partial/\partial s$ will be clarified in Section 1.3, and in Example Problem 1.3.1; for now think of $\partial T/\partial s$ as $dT/ds$.) For comparison, consider Darcy's law for groundwater flow in one dimension:

$$q_h = -K_h \frac{\partial h}{\partial s} \tag{1.2}$$

Here $q_h$ is the volumetric rate of water flow per unit cross-sectional area of porous medium, $h$ is *hydraulic head,* and $K_h$ is the *hydraulic conductivity* of the medium (Figure 1.1b). These laws have the same form: the rate of flow (heat or water) is proportional to a change in potential (temperature or hydraulic head) with respect to space $s$, where the constants of proportionality have the physical interpretation of conductivities (thermal or hydraulic). As we will see below, when these laws are substituted into equations describing conservation of heat and of mass, the resulting governing equations of heat and water flow also have similar forms. It should not be surprising, then, that hydrologists have solved many groundwater flow problems, not so much by independently developing the necessary mathematics, but rather by looking up solutions to the governing equations, with homologous initial and boundary conditions, in texts on heat flow.

In pondering the implications of structural homologies, however, one should give pause to an important caveat. The existence of a structural homology between two systems does not necessarily imply that the processes underlying the two systems are similar. Indeed, mechanisms of flow through a porous medium are quite different from mechanisms of heat conduction; one should, therefore, avoid inferring from a homology that some fundamental similarity exists in the two processes, but rather seek to understand each in its own context.

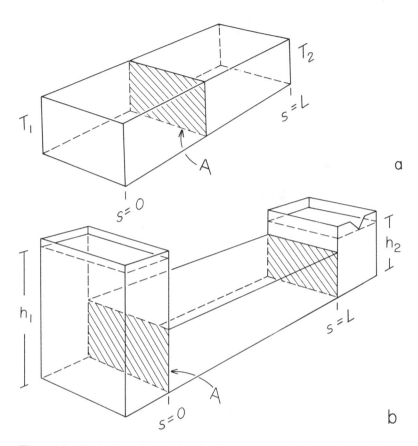

**Figure 1.1** Rock slabs of length $L$ and uniform cross-sectional area $A$: steady heat flow in the direction of positive $s$ is induced by a constant temperature difference $T_2 - T_1$ between inflow and outflow faces ($a$); steady water flow in the direction of positive $s$ is induced by a constant difference in water level $h_2 - h_1$ between reservoirs at inflow and outflow faces ($b$).

A second, closely related strategy involves making use of the geometrical similarities of different problems to systematically investigate them. This strategy will be formalized in developing ideas of geometrical and dynamical *similitude*. At a basic level, it should be intuitively clear that describing the motions of a settling spherical particle, an ascending bubble, and a falling drop, are similar problems involving different fluids. Likewise, describing water flow in a rock fracture with parallel walls, and magma flow in a dike or sill, are very similar problems. The treatment of one will closely follow the treatment of the other.

The text, as its title implies, is chiefly concerned with geological fluids occurring at Earth's surface and within its crust, as opposed to the fluids composing the atmosphere, ocean, mantle, and core. Most flows encountered in the surficial and crustal setting involve viscous, incompressible fluids moving at low speeds in the vicinity of boundaries. In addition, many flows on Earth's surface, and some within its crust, involve turbulence; so this topic also is developed, although a full treatment of turbulence is beyond the scope of the text.

## 1.2 UNITS AND MATHEMATICAL CONVENTIONS

The SI (Système International) system of measurements is used throughout the text. Fundamental quantities used in our treatment of fluid physics include mass, length, time, and temperature; other quantities are derived (Table 1.1). In addition, several symbolic conventions are adopted for clarity: The symbol $s$ is used to denote a general spatial coordinate when a coordinate system is not explicitly specified; $x$, $y$, and $z$ denote Cartesian coordinates; $r$, $\theta$, and $z$ denote cylindrical coordinates; $r$, $\theta$, and $\phi$ denote spherical coordinates; and $s$, $n$, and $z$ denote curvilinear coordinates. With the three systems involving $z$, this

**Table 1.1**   Quantities of Système International d'Unités Used in Text

| Quantity | Units | Unit Symbols |
|---|---|---|
| *Fundamental* | | |
| Length | Meter | m |
| Mass | Kilogram | kg |
| Temperature | Kelvin, degrees Celsius | K, °C |
| Time | Second | s |
| *Derived* | | |
| Acceleration | Meter per second squared | $\text{m s}^{-2}$ |
| Angular acceleration | Radian per second squared | $s^{-2}$ |
| Angular velocity | Radian per second | $s^{-1}$ |
| Area | Meter squared | $m^2$ |
| Bulk viscosity | Newton second per square meter | $\text{N s m}^{-2}$ |
| Dynamic viscosity | Newton second per squared meter | $\text{N s m}^{-2}$ |
| Energy (work) | Joule | J |
| Force | Newton | N |
| Hydraulic conductivity | Meter per second | $\text{m s}^{-1}$ |
| Hydraulic head | Meter | m |
| Intrinsic permeability | Meter squared | $m^{-2}$ |
| Kinematic viscosity | Squared meter per second | $\text{m}^2\,\text{s}^{-1}$ |
| Mass concentration | Kilogram per cubic meter | $\text{kg m}^{-3}$ |
| Mass density | Kilogram per cubic meter | $\text{kg m}^{-3}$ |
| Molar specific head | Joule per mole per degree | $\text{J n}^{-1}\,°\text{C}^{-1}$ |
| Molecular density | Number per cubic meter | $m^{-3}$ |
| Power | Watt | W |
| Pressure | Pascal | Pa |
| Specific heat capacity | Joule per kilogram per degree | $\text{J kg}^{-1}\,°\text{C}^{-1}$ |
| Stress | Pascal | Pa |
| Surface tension | Newton per meter | $\text{N m}^{-1}$ |
| Thermal conductivity | Watt per meter per degree | $\text{W m}^{-1}\,°\text{C}^{-1}$ |
| Velocity | Meter per second | $\text{m s}^{-1}$ |
| Volume | Cubic meter | $m^3$ |
| Volumetric discharge | Cubic meter per second | $\text{m}^3\,\text{s}^{-1}$ |
| Vorticity | Radian per second | $s^{-1}$ |
| Weight | Newton | N |

coordinate usually is set to coincide with the vertical, or near vertical, coordinate axis in the Earth reference frame. Fluid velocities in the directions of coordinate axes are denoted by $u$, $v$, and $w$ in an $xyz$ system, $u_r$, $v_\theta$, and $w_z$ in an $r\theta z$ system, $u_r$, $v_\theta$, and $w_\phi$ in an $r\theta\phi$ system, and $u_s$, $v_n$, and $w_z$ in an $snz$ system (Figure 1.2). Specific discharges normally are denoted by $q_h$ to differentiate them from velocities, and are also subscripted according to the coordinate system used. For example, values of specific discharge in the directions of coordinate axes are denoted by $q_{hx}$, $q_{hy}$, and $q_{hz}$ in an $xyz$ system, and similarly for $r\theta z$ and $snz$ systems. Heat flux densities are denoted by $q_T$, where the subscript $T$ is used to distinguish them from specific discharges. Similarly, mass flux densities of solids or gases dissolved in a liquid are denoted by $q_c$.

Students no doubt are familiar with the convention in mathematics texts of using the symbolism $y = f(x)$ to denote that the variable quantity $y$ is some function $f(x)$ of the independent variable or coordinate $x$. It is customary in treatments of fluid physics, however, to use the variable symbol to denote its functional relation to one or more independent variables or coordinates. For example, $u = u(x)$ denotes that the velocity component $u$ is a function of the coordinate position $x$. Similarly, $u = u(x,y,z,t)$ denotes that the velocity component $u$ is a function of the three coordinate positions $x$, $y$, and $z$, and time $t$. In addition, it is sometimes useful to include one or more parameters within this functional notation to emphasize the importance of the parameters. For example, $T(x,t;b)$ denotes that temperature $T$ is a function of the coordinate position $x$ and time $t$, where $b$, placed after a semicolon, is a specified parameter. Here $b$ might denote a characteristic dimension of the geometry of the flow problem, such as the width of a dike through which magma is flowing.

Operators are printed in a standard Roman font so that they are readily distinguishable from symbols for variables and coefficients. For example, the ordinary differential operator with respect to time $t$ is written d/d$t$, as opposed to the italicized form $d/dt$, which often is used in mathematics texts. Likewise, dimensions and units of physical quantities are printed in a standard Roman font.

Vectors and tensors are written in bold Roman. Then, using the functional notation from above, $\mathbf{u}(x,y)$ denotes, for example, that the velocity vector $\mathbf{u}$ is a function of the coordinate positions $x$ and $y$. When written as an ordered set of components, a vector is denoted using bent brackets. For example, the velocity vector $\mathbf{u} = \langle u, v, w \rangle$. It also will be convenient to sometimes consider the matrix version of a vector, symbolized using square brackets. For example, we may express $\mathbf{u}$ as a row matrix or column matrix (or as a row vector or column vector):

$$\mathbf{u} = \langle u, v, w \rangle \leftrightarrow [u, v, w] \leftrightarrow \begin{bmatrix} u \\ v \\ w \end{bmatrix} \tag{1.3}$$

where the arrows mean "corresponds to." Similarly, a tensor can be written in matrix form. For example, we will later examine the stress tensor $\boldsymbol{\tau}$, whose nine elements can be written as a $3 \times 3$ matrix:

$$\boldsymbol{\tau} = \begin{bmatrix} \sigma_x & \tau_{xy} & \tau_{xz} \\ \tau_{yx} & \sigma_y & \tau_{yz} \\ \tau_{zx} & \tau_{zy} & \sigma_z \end{bmatrix} \tag{1.4}$$

(The notation of [1.4] is fully explained in Chapter 12.)

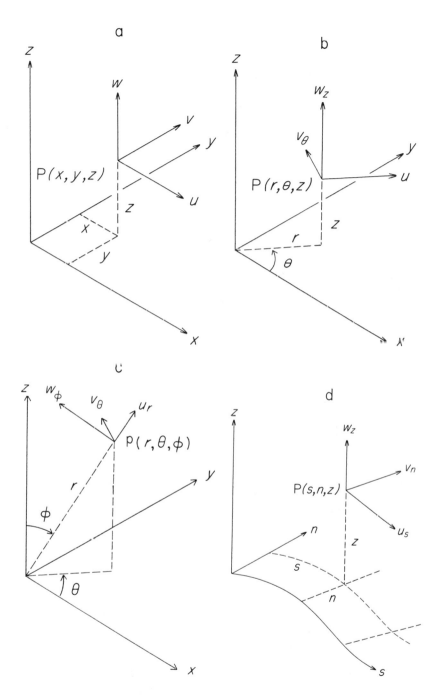

**Figure 1.2**   Coordinate systems and fluid velocity components used in text: Cartesian (*a*); cylindrical (*b*); spherical (*c*); and curvilinear (*d*).

Often the coordinate system used in a flow problem is positioned such that the $z$-axis is vertical and positive upward relative to the gravitational field of Earth. In this case it is sometimes sufficient to use scalar quantities instead of vectors in the context of a one-dimensional problem. For example, the velocity of a particle settling within a fluid under the influence of gravity usually may be denoted simply by the component $w$, as opposed to the more formal description of its motion as specified by the vector $\mathbf{u} = \langle u, v, w \rangle$, where $u = v = 0$. Similarly, the force acting on a particle of mass $m$ due to gravity is in vector form equal to $m\mathbf{g}$. But often this may be expressed in terms of a scalar, $-mg$, where the negative sign indicates that this force (the weight of the particle) acts downward in the Earth coordinate system. To clarify this, the gravitational acceleration $\mathbf{g}$ acting on the particle may be formally written as

$$\mathbf{g} = -g \left\langle \frac{\partial h}{\partial x}, \frac{\partial h}{\partial y}, \frac{\partial h}{\partial z} \right\rangle \tag{1.5}$$

where $h$ denotes the height of the particle relative to an arbitrary horizontal datum. Since $h$ is merely a vertical coordinate, observe that with vertical $z$-axis, $\partial h/\partial x = \partial h/\partial y = 0$ and $\partial h/\partial z = 1$. Therefore (1.5) becomes

$$\mathbf{g} = -\langle 0, 0, g \rangle \tag{1.6}$$

which indicates that it is sufficient in this case to work with the scalar quantity, $-mg$, to describe the weight of the particle.

We will frequently express the flux of mass, energy, momentum, etc. as the product of a coefficient and a potential gradient, as in Fourier's law and Darcy's law above. In these examples, the thermal and hydraulic conductivities are, in the language of thermodynamics, *phenomenological coefficients*. Note that a gradient of a quantity, for example, $\partial T/\partial s$, is treated as a negative value when the quantity decreases in the direction of positive $s$. Thus, a positive value of heat flow $q_T$ in (1.1) indicates flow in the direction of positive $s$ arising from a negative temperature gradient. This conscious choice, although not a universal convention, is normally adopted in physics-based treatments of fluids. In addition, we will normally adopt this convention for expressions relating the rate of strain of a fluid to the stress applied to it, a topic to be covered in Chapters 3 and 12. Consider, for example, the similarity between Fourier's law and the definition of a Newtonian fluid (Chapter 3):

$$\tau = -\mu \frac{\partial u}{\partial y} \tag{1.7}$$

Here $\tau$ is a shear stress; comparing with (1.1), $\tau$ is mathematically homologous to the heat flux $q_T$. The quantity $\partial u/\partial y$ denotes a rate of change in the fluid velocity component $u$ with respect to position $y$, and is homologous to the temperature gradient $\partial T/\partial s$. The phenomenological coefficient $\mu$ is the dynamic viscosity, and is homologous to the thermal conductivity $K_T$. The choice of a leading negative sign, however, departs from many treatments of fluid mechanics, but it is the convention adopted in the thermodynamics literature. In fact, we will see below that the stress $\tau$ can be interpreted as a momentum flux, at least in the case of gases.

Our description of functional notation above hints at an important idea that recurs throughout the text; namely, that descriptions of the physical properties and motions of fluids often involve the idea of a mathematical *field*. When we state that temperature $T$ is a function $T(x, y, z)$ of coordinate position, for example, we are associating with each coordinate position in some region a specific value of $T$. The totality of such temperature values is a *temperature field*. Because in this example the scalar function $T(x, y, z)$ does not depend on time, we are considering a *steady* temperature field. We will later examine the idea of associating with each point in a region a specific vector quantity, the totality of which defines a *vector field*. In the example in which the vector quantity is the fluid velocity $\mathbf{u} = \langle u, v, w \rangle$, the vector field is then a *velocity field*. By denoting $\mathbf{u} = \mathbf{u}(x, y, z, t)$, we are considering the possibility that this velocity field is *unsteady*. It is important to remember that the quantity (for example, temperature or velocity) is associated with *every* coordinate position in the region of interest.

## 1.3 SCOPE OF MATHEMATICS USED IN THIS TEXT

### 1.3.1 Analytical solutions, vectors, and tensors

One goal in our study of fluid physics is to develop equations that express Newton's second law for the motion of a fluid. These equations then can be solved, in principle, to obtain a function that describes how fluid velocity varies as a function of position and time over the domain of a specified flow geometry. We will see, however, that these equations of fluid motion are nonlinear because they contain terms of the form $u\,\partial u/\partial x$. (This term describes a convective acceleration, as we will see in Chapter 7.) This means that only a few flow geometries exist for which analytical (exact) solutions are possible. These pertain chiefly to situations—unidirectional flows, for example—in which the equations of motion can be simplified. The emphasis of the text, then, will be to examine some of these exact solutions in the context of geological problems. In addition, we will briefly consider certain numerical solutions, and techniques for recasting the nonlinear equations of motion into forms that can be solved. But this topic is not emphasized. Thus, although the class of flow geometries that we will examine is limited, there is a very bright side: We will discover that a surprisingly large number of geological flow situations exist in which simplified solutions reveal much about them. In addition, many of the topics covered in the text, fluid statics, for example, require only elementary functions with which students having two semesters of calculus should be familiar.

Our manipulation of vectors will include the elementary operations of vector addition, multiplication of a scalar and vector, the inner product of vectors, and the cross product of two vectors. We also will examine the ideas of the gradient, the divergence, and the curl of a vector. Our manipulation of tensors will be restricted to addition and multiplication, in keeping with the introductory treatment of the text. Nonetheless, tensor notation is introduced in appropriate places for the value of illustrating the compact forms of equations that arise from using this notation. Each vector and tensor notation used in the text is fully explained when it is first introduced. Then, successive chapters progressively rely more on this notation, assuming students have become familiar with it. Several formulae from vector analysis, which we will use at various points in the text, are provided in Appendix 17.1.

## 1.3.2 Partial differential notation

For those who are not entirely familiar with the partial differential notation introduced above, let us examine its significance by briefly considering several examples. A partial derivative serves the same purpose as an ordinary derivative, but is used when two or more independent variables or coordinates are involved.

Consider the function $h = h(x, y)$. The partial derivative $\partial h/\partial x$ describes the rate at which $h$ changes with respect to $x$, and $\partial h/\partial y$ describes the rate at which $h$ changes with respect to $y$. These have a simple interpretation for the case in which $h$ represents the elevation of the land surface; then $h(x, y)$ is a description of a topographic (contour) map, giving $h$ at each coordinate pair $(x, y)$, where $x$ and $y$ are essentially coordinates of longitude and latitude. The partial derivative $\partial h/\partial x$ is the local slope of the land surface at any specified coordinate pair $(x, y)$ measured in a direction parallel to the $x$-axis; and $\partial h/\partial y$ is the local slope parallel to the $y$-axis. This idea can be generalized to three dimensions. For example, the function $T(x, y, z)$ might describe how temperature $T$ varies with coordinate position $(x, y, z)$ within a fluid. Then the partial derivatives $\partial T/\partial x$, $\partial T/\partial y$, and $\partial T/\partial z$ denote the local, spatial rates at which $T$ changes in directions parallel to the $x$, $y$, and $z$ axes.

Now consider the function

$$u = u(y, t) = \frac{U}{b} y \left( 1 - e^{-t/t_0} \right) \qquad 0 \le y \le b \qquad (1.8)$$

We do not need to know the specific flow problem for which this equation is relevant, but merely note that it describes how the velocity $u$ varies as a function of the coordinate position $y$ and time $t$, where $U$, $b$, and $t_0$ are constants (parameters). Evidently when $t = 0$, $u = 0$ for all values of $y$; and when $t$ becomes very large, $u$ approaches $(U/b)y$. Further, when $y = 0$, $u = 0$ for all values of $t$; and when $y = b$, $u = U[1 - \exp(-t/t_0)]$. The partial derivatives of (1.8),

$$\frac{\partial u}{\partial y} = \frac{U}{b} \left( 1 - e^{-t/t_0} \right) \qquad (1.9)$$

$$\frac{\partial u}{\partial t} = \frac{U}{bt_0} y e^{-t/t_0} \qquad (1.10)$$

then indicate that the spatial rate at which $u$ changes with respect to $y$, $\partial u/\partial y$, is independent of $y$, but varies with time $t$ such that, as $t$ becomes very large, $\partial u/\partial y$ approaches the constant $U/b$. The rate at which $u$ changes with respect to time $t$, $\partial u/\partial t$, in contrast, depends on both $y$ and $t$ such that, as $t$ becomes very large, $\partial u/\partial t$ approaches zero for all $y$.

We will see later that the pressure field $p(x, y, z)$ of a static, incompressible liquid with constant density $\rho$ is given by $p(x, y, z) = p_0 + \rho g z$, where $p_0$ is atmospheric pressure and $z$ is depth below the liquid surface. Taking partial derivatives, $\partial p/\partial x = \partial p/\partial y = 0$, and $\partial p/\partial z = \rho g$. Thus pressure $p$ does not vary in the horizontal directions of $x$ and $y$, and it changes at a constant rate with depth $z$. The functional notation for the pressure field, therefore, may be simplified from $p(x, y, z)$ to $p(z)$. Further, since pressure varies only with the vertical coordinate $z$, we may write $dp/dz = \rho g$. We will see later that this is one form of the equation of hydrostatics (Chapter 6).

The use of partial differential notation also provides a way to denote an infinitesimal change in a variable quantity or coordinate, such that this change is readily distinguishable from such quantities as $dx$, $dy$, and $dz$, which we frequently will use to denote infinitesimal distances, and $dt$, which we will use to denote an infinitesimal interval. Consider the function $p(x)$, which denotes that pressure $p$ varies with coordinate direction $x$. Then consider an equation of the form

$$p(a + dx) = p(a) + \frac{\partial p}{\partial x}dx \qquad (1.11)$$

which gives the pressure at a position an infinitesimal distance $dx$ away from the position $x = a$. Observe that this is an equation of a straight line: $p(a)$ is the intercept, $dx$ is a variable distance, and $\partial p/\partial x$ is the slope of the line defining the spatial rate at which $p$ changes with $x$. Thus $p(a + dx)$ equals $p(a)$ when $dx$ goes to zero, or when pressure does not change with respect to $x$ such that $\partial p/\partial x = 0$. We will examine this equation again in Chapter 6 when we introduce the idea of using a Taylor expansion to approximate the value of a physical quantity in the vicinity of a rectangular control volume with one edge having length $dx$.

## 1.4 EXAMPLE PROBLEMS

### 1.4.1 Steady one-dimensional heat and water flows: a structural homology

Consider the two systems involving heat flow by conduction through an insulated rock slab, and water flow through a porous rock that is open to flow only across the two faces connected with the water reservoirs, as illustrated in Figure 1.1. Assume that the thermal and hydraulic conductivities $K_T$ and $K_h$ are constants, and that the cross-sectional area $A$ is constant along each medium. Further suppose that the temperatures $T_1$ and $T_2$ at the ends of the slab, and the heights $h_1$ and $h_2$ of the water in the reservoirs, are maintained at fixed values. Under these circumstances, each system involves steady one-dimensional flow in the direction of the coordinate $s$.

Let us now rewrite (1.1) as

$$\frac{dT}{ds} = -\frac{Q_T}{K_T A} \qquad (1.12)$$

where $Q_T = q_T A$ and the total derivative $dT/ds$ is used instead of the partial derivative $\partial T/\partial s$ because flow occurs only in the one coordinate direction $s$. Like (1.1), (1.12) is the governing law of heat flow. We may thus think of (1.12) as a condition that must be satisfied at each position $s$ along the slab, and we may anticipate that there exists a specific distribution of temperature along the slab where this condition (1.12) is everywhere satisfied, *and* which matches the values $T_1$ and $T_2$ at the boundaries. Such a distribution of temperature constitutes a solution to (1.12), and we immediately can infer that this solution is a linear function $T(s)$ of position $s$ because the derivative of $T(s)$, that is $dT/ds$, is a constant value equal to the right side of (1.12). The solution $T(s)$ may formally be determined as follows:

Rearranging (1.12) and integrating

$$dT = -\frac{Q_T}{K_T A} ds \tag{1.13}$$

$$\int dT = -\frac{Q_T}{K_T A} \int ds \tag{1.14}$$

$$T = -\frac{Q_T}{K_T A} s + C \tag{1.15}$$

The constant of integration $C$ can be determined by either boundary condition: $T = T_1$ at $s = 0$ or $T = T_2$ at $s = L$, whence

$$T = -\frac{Q_T}{K_T A} s + T_1; \qquad T = \frac{Q_T}{K_T A}(L - s) + T_2 \tag{1.16}$$

Thus, the distribution of temperature, as expected, is a linear function of $s$, expressed in terms of either boundary condition $T_1$ or $T_2$. This is known as a *boundary-value problem,* thus designated because the solutions (1.16) can be deduced from the known boundary values $T_1$ and $T_2$. To see that these indeed are solutions, one may take the derivative $dT/ds$ of each to retrieve (1.12). Thus, as required, (1.16) are solutions in the sense that they satisfy the condition (1.12) and are compatible with the values $T_1$ and $T_2$ at the boundaries. It is left to the reader to manipulate Equations (1.16) to obtain

$$Q_T = K_T A \frac{T_1 - T_2}{L} \tag{1.17}$$

which should be recognized as an expression for total heat flow by conduction (with dimensions of power), as typically presented in introductory physics texts.

Let us now consider the structural homology that exists between (1.1) and (1.2). From Figure 1.1 and the preceding description, we could proceed as we did for the heat flow system to determine a solution decribing how the hydraulic head $h$ varies along the porous rock system. Alternatively, knowing in advance that the mathematical manipulation will be identical to that for the heat flow problem, we may take note of homologous variables, coefficients, and boundary conditions and immediately write

$$h = -\frac{Q_h}{K_h A} s + h_1; \qquad h = \frac{Q_h}{K_h A}(L - s) + h_2 \tag{1.18}$$

and

$$Q_h = K_h A \frac{h_1 - h_2}{L} \tag{1.19}$$

where $Q_h = q_h A$. Thus, the distribution of hydraulic head is a linear function of $s$, expressed in terms of either boundary condition $h_1$ or $h_2$. The last equation should be recognized as an expression for total water flow through a porous medium (with dimensions $L^3 t^{-1}$), as typically presented in introductory geology texts.

The practical value of a structural homology is thus apparent, inasmuch as solutions to problems involving porous media flow often can be readily determined from homologous cases involving heat flow. Although this is an elementary example, numerous more complicated examples are available, including a description by Cooper et al. (1967) for a "slug test" in well hydraulics (see Reading list). We will also examine several other examples throughout the text. Note, however, that such borrowed solutions reveal little about the details of flow in porous media.

Now, suppose that the hydraulic conductivity $K_h$ of the porous rock varies as a simple linear function of $s$; that is $K_h = K_h(s) = b + as$. Here $b$ is a constant and $a$ is the rate of change in $K_h$ with $s$. Thus, $K_h = K_h(0) = b$ at the inflow face, and $K_h = K_h(L) = b + aL$ at the outflow face. Such a situation simply describes the variation in hydraulic conductivity that could accompany a systematic variation in the grain size of a rock over space. It is left as an exercise to determine the distribution of $h$ along $s$; this also is a boundary-value problem. To begin, set the problem up as above, then substitute $K_h(s) = b + as$ for $K_h$.

### 1.4.2 Sediment particles, vesicles, and olivine crystals: a case of geometric similitude

Vesicles in a basalt flow indicate that it solidified before the gas bubbles could escape (or disappear for other reasons). An interesting problem is to consider how slowly a flow would have to cool to maintain sufficient fluidity for gas bubbles of given size and proximity near the surface to escape. The complete answer is beyond the scope of this example. Nonetheless, an essential ingredient is to know how rapidly a bubble of given size is likely to rise within its molten surroundings. One can imagine, however, that empirical data concerning ascents of bubbles in lava are difficult to obtain. So an obvious possibility is to consider analogues, for example, the ascents of carbon dioxide bubbles in beer. Perhaps less obvious is the relevance of a settling sediment particle in a liquid, an analogue that has the advantage of a wealth of available information.

Consider for simplicity a spherical particle of mass $m$ settling with velocity $w_s$ through a fluid. The force producing this settling motion of the particle is its buoyant weight $W$ given by Archimedes's principle (Chapter 6):

$$W = -\frac{4}{3}\pi(\rho_s - \rho)gR^3 \qquad (1.20)$$

where $\rho$ is the density of the fluid, $\rho_s$ is the density of the particle, and $R$ is its radius. The leading negative sign indicates that the force $W$ is directed downward. Countering this is a drag force $F_D$ exerted on the particle by the fluid. According to Newton's second law, the net force acting on the particle—the sum of its buoyant weight $W$ and the drag $F_D$—must equal the product of its mass and acceleration. Thus

$$W + F_D = m\frac{dw_s}{dt} \qquad (1.21)$$

The drag on a slowly settling rigid sphere, in the case in which flow around the sphere is entirely laminar and steady, is given by the well-known formula obtained by G. G. Stokes in 1851:

$$F_D = -6\pi\mu R w_s \qquad (1.22)$$

where $\mu$ is the dynamic viscosity of the fluid, a measure of its resistance to deformation during flow (Chapter 3). The negative sign arises because we wish to associate a positive value of drag (directed upward) with a negative value of velocity (directed downward).

Here arises the temptation to substitute (1.22) into (1.21) to obtain a differential equation that can be solved for the velocity $w_s$. Recall, however, that (1.22) pertains to steady flow; it is relevant only when the particle motion has attained a dynamic equilibrium such that its acceleration $dw_s/dt$ is zero. An expression for the drag force is more complicated than (1.22) for the general case of an accelerating particle. This is because both the particle and fluid are being accelerated, say, when a particle starts its motion from rest. A general formulation of this problem, therefore, must involve the equations of motion for the fluid.

Let us assume that the particle motion is steady; then (1.21) becomes

$$W + F_D = 0 \tag{1.23}$$

Now substituting (1.22) and rearranging,

$$w_s = -\frac{2g}{9\mu}(\rho_s - \rho)R^2 \tag{1.24}$$

This should be recognized from introductory sedimentology as Stokes's law of settling. The leading negative sign indicates that $w_s$ is directed downward.

Consider now a spherical bubble ascending within molten lava. The force producing this ascending motion of the bubble, as we shall see later, is its buoyant weight $W$, again given by Archimedes's principle. We have only to note in this case that $\rho_s$ denotes the density of the bubble, and $R$ is its radius, in equation (1.20). Note that $\rho_s$ will generally be less than $\rho$ leading to a positive value for $W$, which indicates that $W$ is directed upward. Let us for the moment treat the bubble as rigid (see Example Problem 6.7.3). Countering the buoyant force $W$ is the drag $F_D$ exerted on the bubble by the fluid. Assuming for the moment that the lava behaves approximately like a Newtonian fluid, the drag $F_D$ is given by (1.22), in which we note that $R$ denotes the bubble radius. As with the settling particle, therefore, the sum of drag and buoyant forces must equal zero in absence of acceleration. Summing forces and solving for the velocity $w_s$ again leads to (1.24). Noting that $\rho_s$ is less than $\rho$, the resulting positive value of $w_s$ now indicates that $w_s$ is directed upward.

Thus, recognizing the geometric similarity between a rising gas bubble and a settling sphere, we immediately can take principles that have been thoroughly examined for the latter and apply them to the bubble problem. From (1.24) we see that the ascending velocity is proportional to the square of the bubble radius, and inversely proportional to the fluid viscosity. Thus big bubbles have a better chance than little ones of escaping, and this chance increases in proportion to the cross-sectional area of the bubble. The dynamic viscosity of pure water at 20 °C is $1.0 \times 10^{-3}$ (N s m$^{-2}$). In comparison, the viscosity of a basaltic lava (olivine tholeiite) at 100 kPa varies from about 16 (N s m$^{-2}$) at 1,300 °C to 3.5 (N s m$^{-2}$) at 1,400 °C; and at 500 kPa the viscosity varies from about 6.3 (N s m$^{-2}$) at 1,300 °C to 3.3 (N s m$^{-2}$) at 1,400 °C. We may infer from (1.24) that the ascending velocity of a bubble in water is about 3,000 to 16,000 times that for a bubble of equal size in lava! Here our comparison should end; the problem is actually more complicated due to exchange of gas between the bubble and lava, and due to possible changes in the size and shape of the bubble while it ascends. We will examine this further below, and make use of the bubbles-in-beer analogue (Example Problem 4.10.5). In addition, a basaltic lava may behave more like a Bingham plastic than a Newtonian fluid, depending on pressure, temperature, chemical composition, water content, and crystal concentration. We will examine this also in a later chapter.

Now consider a basaltic magma that is moving slowly upward with constant velocity $w_m$ within the crust. Suppose that approximately spherical olivine crystals of radius $R$ are suspended within the magma. It is left as an exercise to determine the velocity $w_s$ of the olivine crystals as a function of $w_m$ and $R$, and to surmise whether settling of olivine crystals during magma ascent could lead to a significant compositional variation between lava that is extruded early in an eruption relative to lava that reaches the surface later. Assume that the density of the magma ($\rho \approx 2.7 \times 10^3$ kg m$^{-3}$), and the size and density of the olivine crystals ($\rho_s \approx 3.3 \times 10^3$ kg m$^{-3}$), do not vary significantly with position in the crust.

How might one infer the vertical velocity of a magma from knowledge of settling velocities of olivine crystals? How will this problem change if one considers pooling of the magma within a chamber prior to eruption? Suppose that convective motions occur during pooling. What conditions determine whether convection is likely to occur? How will this problem change when one considers emplacement of a horizontal dike or sill? We will return to some of these questions later in the text.

### 1.4.3 Series, averages, covariances, and autocovariances of physical quantities

By describing properties of fluids and porous media in terms of scalar and vector fields, we are by definition concerned with how these properties vary over the spatial coordinates $x$, $y$, and $z$, and the time coordinate $t$. At several places in the text we will need to describe these variations statistically. This includes describing how two or more physical quantities vary jointly over space or time. We therefore require a set of rules for obtaining average quantities over time and space. These rules, described next, lead to definitions of the important statistical ideas of *covariance* and *correlation*, and *autocovariance* and *autocorrelation*. We will first use these definitions in Example Problem 2.4.5 at the end of the next chapter, then again when we consider the topics of solute dispersion in porous media flows, turbulence, and turbulent boundary layers.

Let $s$ denote any one of the independent coordinates $x$, $y$, $z$, or $t$, and let $f$ denote a physical quantity that varies as a continuous function $f(s)$ of $s$. The (infinite) set of sequential values of $f(s)$ is regarded as a *continuous series*. When $s$ denotes time, $f(s)$ is a *time series*; and when $s$ denotes $x$, $y$, or $z$, $f(s)$ is a *spatial series*. In practice, we usually describe $f(s)$ based on a sample of $N$ values observed at discrete values of $s$. The set of $N$ sequential values of $f$ is then referred to as a *discrete series*, and is denoted by $f_s$. The notation $f_s$ is a counterpart to the notation for a continuous function, $f(s)$. If the sampling interval is constant, it is understood by convention that the subscript $s$ may take on integer values such that a discrete series may be denoted by, say, $f_1$, $f_2$, $f_3$, ..., $f_s$, ..., $f_N$.

When $f(s)$ or $f_s$ is expressed as an analytical function such that future values of $f$ are explicitly given in terms of $s$, then $f$ is regarded as a *deterministic* series. When future values can be obtained only in probabilistic terms, however, then $f$ is nondeterministic, despite the fact that the underlying process that produces $f$ may in principle be deterministic. A process $f$ that varies with $s$ according to statistical (rather than deterministic) rules is a *stochastic process*. Any given series is then one *realization* of an underlying stochastic process.

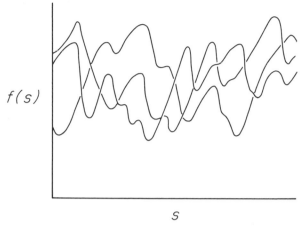

**Figure 1.3** Sample of three time (or space) series $f(s)$ obtained from an infinity of possible series associated with the same underlying process.

$f(s)$

$s$

To clarify these ideas, envision a process from which an infinity of possible series can be obtained (Figure 1.3). An example is when one measures the instantaneous velocity at a fixed point in a turbulent flow; fluctuations in the velocity are associated with the complex motions of turbulent eddies (Chapter 14). Successive experiments of specified duration will each provide a unique, continuous time series (realization). The underlying process is deterministic in the sense that each velocity series arises from the Newtonian mechanics governing fluid flow. But although we can write the equations of motion from which all such realizations arise, we cannot write an analytical equation that gives the fluctuating velocity explicitly in terms of time. The time series obtained therefore may be treated as stochastic series.

A *stationary* process is a particularly important type of stochastic process. Envision obtaining values from a large number of realizations, where the values obtained are associated with a specific time $t$ after the onset of measurement ($t = 0$) of each realization. Such values would be characterized by a particular statistical distribution, for example, a normal distribution (more on this below). A stationary process is one for which all properties of this distribution at any time $t$ are the same as the properties associated with the distribution defined at time $t + \tau$, for arbitrary $\tau$. A stationary process then has constant mean $\mu$ and variance $\sigma^2$. Stationarity requires in addition that the *covariance* (as described below) between values separated by an amount $\tau$ is constant for arbitrary translation of the time axis. Similar concepts apply to spatial series; the analogous process is referred to as being *homogeneous*.

In practice an estimate of the mean $\mu$ of a stationary (or homogeneous) process can be obtained from the mean of a discrete series involving $N$ observations,

$$\overline{f} = \frac{1}{N} \sum_{s=1}^{N} f_s \qquad (1.25)$$

where the overbar notation on the left side is used to denote the operation embodied on the right side. Notice the similarity of (1.25) to a Riemann sum; indeed, in the limit as the sampling interval goes to zero, the mean $\overline{f(s)}$ of the underlying continuous realization $f(s)$ is exactly given by

$$\overline{f} = \overline{f(s)} = \frac{1}{S} \int_0^S f(s)\,ds \qquad (1.26)$$

where $S$ is the total measurement interval. Hereafter we will not distinguish notationally between means obtained from a realization using (1.25) or (1.26).

Often it is useful to describe a series $f$ in terms of its mean value and fluctuations $f'$ about the mean. (The prime on $f'$ should not be confused with similar notation used to represent a derivative.) In the discrete case,

$$f_s = \overline{f} + f_s' \qquad (1.27)$$

and in the continuous case,

$$f(s) = \overline{f} + f'(s) \qquad (1.28)$$

For a stationary process the measurement interval $S$ in (1.26) must be sufficiently long that the average is independent of $s$, as suggested by the notation in (1.27) and (1.28). This means that $S$ must be much larger than the typical durations of the fluctuating parts. It is, however, sometimes necessary to allow for the possibility of variations in mean quantities, in which case $S$ cannot be "too" large. We will examine this further below.

By definition, means of fluctuating quantities must equal zero. For example, substituting $f + f'$ for $f$ in (1.26),

$$\frac{1}{S} \int_0^S f(s)\,ds = \frac{1}{S} \int_0^S \overline{f}\,ds + \frac{1}{S} \int_0^S f'(s)\,ds = \overline{f} \qquad (1.29)$$

The integral involving $\overline{f}$ equals the "area" beneath the constant value $f$, and the integral involving $f'$ equals the "area" between the curve of $f(s)$ and $\overline{f}$ (Figure 1.4). Because the magnitude and duration for which $f'$ is positive is on average equal to the magnitude and duration for which it is negative, the latter integral equals zero. Thus,

$$\overline{f_s'} = 0; \qquad \overline{f'(s)} = 0 \qquad (1.30)$$

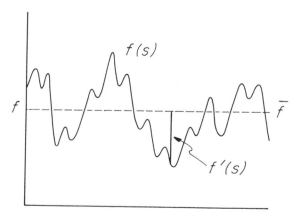

**Figure 1.4** Geometrical interpretation of the mean $\overline{f}$ and fluctuating values $f'(s)$ of a time (or space) series.

It also is necessary to obtain averages involving derivatives, integrals, sums, and products of one or more quantities. For this reason, let us list several rules. Let $f$ and $g$ denote any two dependent quantities for which an average is to be formed. Then

$$\overline{\overline{f}} = \overline{f}; \qquad \overline{f + g} = \overline{f} + \overline{g}$$
$$\overline{\overline{f} \cdot g} = \overline{f} \cdot \overline{g}; \qquad \overline{\overline{f} \cdot \overline{g}} = \overline{f} \cdot \overline{g} \qquad (1.31)$$
$$\overline{\frac{\partial f}{\partial s}} = \frac{\partial \overline{f}}{\partial s}; \qquad \overline{\int f \, ds} = \int \overline{f} \, ds$$

The first of these is represented by the second integral operation in (1.29). The second indicates that the average of the sum of two dependent quantities is equal to the sum of their averages. Similar remarks pertain to the remainder of these rules.

Consider measuring two quantities $f$ and $g$ at regular intervals of $s$. Estimates of the variances, $\sigma_f^2$ and $\sigma_g^2$, of stationary processes $f$ and $g$ are obtained from the variances of the discrete series:

$$s_f^2 = \overline{(f_s - \overline{f})^2} = \frac{1}{N} \sum_{s=1}^{N} (f_s - \overline{f})^2 \qquad (1.32)$$

$$s_g^2 = \overline{(g_s - \overline{g})^2} = \frac{1}{N} \sum_{s=1}^{N} (g_s - \overline{g})^2 \qquad (1.33)$$

Here $N$ is assumed to be large. An estimate of the *covariance* $\gamma_{fg}$ of $f$ and $g$ is obtained from the covariance of the discrete series:

$$c_{fg} = \overline{(f_s - \overline{f})(g_s - \overline{g})} = \frac{1}{N} \sum_{s=1}^{N} (f_s - \overline{f})(g_s - \overline{g}) \qquad (1.34)$$

Using (1.27),

$$s_f^2 = \overline{f_s'^2}; \qquad s_g^2 = \overline{g_s'^2}; \qquad c_{fg} = \overline{f_s' g_s'} \qquad (1.35)$$

Those familiar with correlation analysis will know that we may now define a *correlation coefficient* $r_{fg}$ by

$$r_{fg} = \frac{\overline{f_s' g_s'}}{\sqrt{\overline{f_s'^2} \, \overline{g_s'^2}}} = \frac{c_{fg}}{\sqrt{s_f^2 s_g^2}} \qquad (1.36)$$

The value of $r_{fg}$ may vary such that $-1 \le r_{fg} \le 1$; it characterizes how closely $f$ covaries with $g$ (or vice versa). We will use this several ways; let us start by examining the idea of *autocorrelation*.

Suppose for example that $f$ is the velocity component $u_t$ measured at a fixed position at regular times $t_1, t_2, \ldots, t_N$. Let $g$ denote the same velocity component $u_{t+k}$ (at the same fixed position) measured $k$ intervals of time later. An estimate $c_k$ of the *autocovariance* $\gamma_k$ between values $u_t$ and values $u_{t+k}$ separated by $k$ intervals is then

$$c_k = \overline{u_t' u_{t+k}'} \qquad (1.37)$$

When $k = 0$, this reduces, according to (1.35), to an estimate $s_u^2$ of the variance $\sigma_u^2$ of $u_t$, where $N$ is assumed to be large. Because $u_t$ and $u_{t+k}$ are obtained from the same time series, $\overline{u_t'^2} = \overline{u_{t+k}'^2}$, so $(\overline{u_t'^2 u_{t+k}'^2})^{1/2} = s_u^2$. Substitution of these quantities into (1.36) then leads to an estimate $r_k$ of the *autocorrelation* $\sigma_k$ of $u_t$ associated with lag $k$

$$r_k = \frac{c_k}{c_0} = \frac{c_k}{s_u^2} \tag{1.38}$$

Like the correlation coefficient (1.36), $r_k$ can be positive or negative, and characterizes how closely $u_t$ covaries with $u_{t+k}$. Qualitatively, if one measures $u$ at any instant, then again a short interval later, the chance is good that $u$ has not changed much. That is, values of $u_t$ separated by a small interval (small $k$) are on average strongly correlated. Indeed, when the interval goes to zero ($k = 0$), the correlation $r_0 = 1$ since, in the limit, each value of $u_t$ is being compared with itself. As the interval between measurements increases, the chance is greater that $u$ will change significantly during the interval; that is, values of $u_t$ are not as well correlated. Indeed, if the interval between measurements is sufficiently large, no correlation between measured values exists; observations of $u$ are independent and $r_k \to 0$ as $k \to \infty$. A plot of $r_k$ against $k$ (Figure 1.5) is a *discrete autocorrelation function*.

Expressions (1.32) through (1.34) have continuous counterparts:

$$s_f^2 = \frac{1}{S}\int_0^S [f(s) - \overline{f}]^2 = \overline{[f(s) - \overline{f}]^2} \tag{1.39}$$

$$s_g^2 = \frac{1}{S}\int_0^S [g(s) - \overline{g}]^2 = \overline{[g(s) - \overline{g}]^2} \tag{1.40}$$

$$c_{fg} = \frac{1}{S}\int_0^S [f(s) - \overline{f}][g(s) - \overline{g}] = \overline{[f(s) - \overline{f}][g(s) - \overline{g}]} \tag{1.41}$$

Using (1.28),

$$s_f^2 = \overline{[f'(s)]^2}; \qquad s_g^2 = \overline{[g'(s)]^2}; \qquad c_{fg} = \overline{f'(s)g'(s)} \tag{1.42}$$

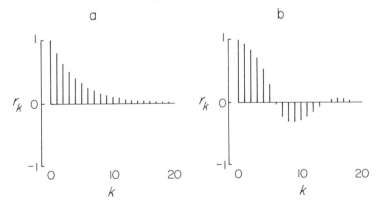

**Figure 1.5** Examples of discrete autocorrelation functions $r_k$: exponential (*a*) and damped oscillatory (*b*).

Following the example above, where $f$ now denotes a continuous velocity component $u(t)$ associated with a fixed position, the autocovariance is

$$c(\tau) = \overline{u'(t)u'(t + \tau)} \tag{1.43}$$

where $\tau$ is the interval between values of $u$; it is analogous to the discrete lag $k$.
When $\tau = 0$, (1.43) reduces to the variance of $u$. The autocorrelation function is
then

$$r(\tau) = \frac{c(\tau)}{c(0)} = \frac{c(\tau)}{s_u^2} \tag{1.44}$$

The autocorrelation function $r(\tau)$ has several important properties. Like $r_k$, $r(\tau)$
in general converges to zero with $\tau \to \infty$. If the rate of convergence is sufficiently
rapid (which is the case for a stationary or homogeneous process), the integral

$$\int_0^\infty r(\tau)\,\mathrm{d}\tau = \frac{1}{s_u^2}\int_0^\infty \overline{u'(t)u'(t + \tau)}\,\mathrm{d}\tau = T_0 \tag{1.45}$$

equals a finite quantity $T_0$. This quantity is referred to as a *correlation interval,*
and is a measure of the interval beyond which observations of $u$ are statistically
independent. Therefore $T_0$ is a measure of the "memory" in the velocity signal $u$.
This suggests that $T_0$ is on the order of the typical durations of the fluctuating part
$u'$ of $u$. Moreover, the minimum duration of $T$ that must be used in (1.26) (where
$S = T$ and $f = u$) to ensure that $\bar{u}$ is a constant value must be such that $T \gg T_0$.
Similar remarks apply to a spatial series, for which one may obtain a correlation
length $S_0$ from the autocorrelation function $r(s)$.

Also of interest are *nonstationary* or *nonhomogeneous* series. Such series describe a condition that we wish to allow for: the possibility that mean quantities
vary with $s$. In this regard, $S$ in (1.26) must be defined such that stationarity (or homogeneity) is satisfied for any interval of length $S$ taken individually, but such that
statistical properties (notably the mean) may change from one interval of length $S$
to the next. A nonstationary process that satisfies this condition is referred to as a
*quasi-steady* process.

The developments above involving series will be sufficient for our purposes.
Nonetheless, the idea of a stochastic process can be generalized to the idea of a
*random field*. For example, $f(x, y)$ might refer to a physical quantity that varies
as a stochastic process over an $xy$-coordinate system. In fact, this idea forms the
cornerstone of certain theoretical descriptions of how hydraulic properties (for example, the hydraulic conductivity $K_h$) vary over a porous geological unit, due to
variations in pore structure and arrangement in relation to bedding and texture. This
topic is beyond the scope of this text, so interested students are urged to examine
relevant references listed below.

It is important to be familiar with two additional averaging procedures that are
closely related to those above. The first involves obtaining an *ensemble average*
of a quantity. An ensemble refers to a collection of items or systems. Suppose, for
example, that we wish to obtain the average of the velocities of a large number
(ensemble) of molecules. This operation is like (1.25) above, where then $N$ refers
to the number of molecules. Similarly, consider an ensemble of identical systems,
each consisting of a sphere that moves with specified relative velocity in a fluid.
Envision obtaining measurements of the fluid velocity component $u$ at geometrically similar positions in each of the systems at the same instant. The average of

these measurements is an ensemble average (for the specified geometrical position). Formally, such velocity components would be characterized by a *probability density function* $p(u)$. The ensemble average $\mu_u$ is then obtained by

$$\mu_u = \int_{-\infty}^{\infty} u\, p(u)\, du \tag{1.46}$$

For students not familiar with continuous distributions, the probability density function $p(u)$ may be envisioned as a continuous histogram of $u$, where each of the histogram bars represents a relative frequency (the number of observations in the interval of $u$ represented by a bar, divided by the total number of observations). In the limit as the interval of $u$ represented by each bar goes to zero, the histogram tends to the function $p(u)$, which satisfies the condition:

$$\int_{-\infty}^{\infty} p(u)\, du = 1 \tag{1.47}$$

This states that the total probability beneath $p(u)$ must equal one, just as the total relative frequency represented by all bars of a histogram must sum to one. Often it is assumed that measurements obtained over time at a specified position in one system will represent the distribution of values that otherwise would be obtained at geometrically similar positions in the ensemble of systems at one instant. This is referred to as the *ergodic hypothesis*. In practice, therefore, (1.25) again provides an estimate of the average $\mu_u$, based on a time series $u_t$. By similar reasoning, the set of possible realizations of a stochastic process, in general, constitutes an ensemble. As described above, (1.25) provides an estimate of the ensemble average $\mu$ of such a process if it is stationary (or homogeneous).

Throughout the text, the appropriate averaging procedure will be made clear within the context of the problem or example examined. In this regard it is important to note that the rules of (1.31) apply to ensemble averages. For example, suppose that $f$ denotes a molecular velocity component that varies with time $t$. Then, the average of an integral of $f$ (for many molecules) is equal to the integral of the ensemble average of $f$ according to the last rule in (1.31).

The second procedure involves obtaining a *spatial average*. Let us start with a familiar example: If $Q$ denotes the volumetric discharge ($L^3 t^{-1}$) within a channel that has cross-sectional area $A$, the (spatial) average velocity $u_0$ through the cross section is $u_0 = Q/A$. Viewed another way, suppose that the streamwise velocity $u$ is a function $u(y, z)$ of the local coordinate position along axes normal to the streamwise flow. Then the discharge $Q$ is

$$Q = \int\int_A u(y, z)\, dy\, dz \tag{1.48}$$

so the average velocity $u_0$ is

$$u_0 = \frac{1}{A} \int\int_A u(y, z)\, dy\, dz \tag{1.49}$$

This is, in fact, a two-dimensional version of (1.26). For example, suppose that the component velocity $u$ of a fluid varies as a function $u(y)$ over the domain

$y_1 \leq y \leq y_2$. The spatial average of $u$ over this domain is obtained according to (1.26), where $S = y_2 - y_1$. Similarly, the volumetric average $f_0$ of a quantity $f(x, y, z)$ is

$$f_0 = \frac{1}{V} \int \int \int_V f(x, y, z)\, dx\, dy\, dz \qquad (1.50)$$

where $V$ denotes the (volume) domain of interest. A spatial average has qualities of an ensemble average. By associating a physical quantity with each coordinate position in a specified region, then taking the average of this quantity over the region, we are in effect obtaining an ensemble average if we think of the ensemble as consisting of the (infinite) set of coordinate positions within the region.

## 1.5 READING

Box, G. E. P. and Jenkins, G. M. 1976. *Time series analysis: forecasting and control.* San Francisco: Holden-Day, 575 pp. This text has a nice introductory treatment of the ideas of time series, stationarity, and autocovariance. It is a "classic" treatise in the sense that the models presented in the text for describing time series have become known as Box–Jenkins models of stochastic series.

Cooper, H. H., Jr., Bredehoeft, J. D., and Papadopulos, I. S. 1967. Response of a finite diameter well to an instantaneous charge of water. *Water Resources Research* 3:263–69. This paper is an example in which a problem involving radial flow associated with a "slug test" in a water well was solved by making use of well-known solutions to the homologous heat flow problem.

Cox, K. G., Bell, J. D., and Pankhurst, R. J. 1979. *The interpretation of igneous rocks.* London: Allen & Unwin, 450 pp. This text includes a discussion (pp. 169–73, 272–82) of compositional zoning in magma bodies due to crystal settling.

Goldstein, H. 1980. *Classical mechanics,* 2nd ed. Reading, Mass.: Addison-Wesley, 672 pp. This text includes a brief discussion (p. 53) of the use of structural homologies to examine different mechanical systems.

Huppert, H. E. 1986. The intrusion of fluid mechanics into geology. *Journal of Fluid Mechanics* 173:557–94. This paper provides an overview of how theoretical, experimental, and field-based work involving fluid mechanics is being applied to problems in geology. It emphasizes magmatic and volcanic processes, and includes a set of color photographs illustrating a variety of laboratory and field phenomena. The paper provides a comprehensive list of references.

Kushiro, I. 1980. Viscosity, density, and structure of silicate melts at high pressures, and their petrological applications. In Hargraves, R. B., ed., *Physics of magmatic processes.* Princeton, N. J.: Princeton University Press, pp. 93–120. This paper summarizes laboratory measurements of viscosity for various melts, and discusses how viscosity varies with pressure, temperature, and melt composition.

Priestley, M. B. 1981. *Spectral analysis and time series.* London: Academic Press, 940 pp. This is a comprehensive treatment of time series, spatial series, and random fields, written at an advanced level.

Shaw, H. R., Wright, T. L., Peck, D. L., and Okamura, R. 1968. The viscosity of basaltic magma: analysis of field measurements in Makaopuhi Lava Lake, Hawaii. *American Journal of Science* 266:225–64. This article discusses early field attempts to measure the viscosity of a basaltic lava, and its behavior as a Bingham plastic.

Spera, F. J. 1980. Aspects of magma transport. In Hargraves, R. B., ed., *Physics of magmatic processes.* Princeton, N. J.: Princeton University Press, pp. 265–323. This paper includes a clear discussion of the rudiments of magma flow and intrusion, and crystal and xenolith settling, using basic fluid mechanics.

Stokes, G. G. 1851. On the effect of internal friction of fluids on the motion of pendulums. *Transactions of the Cambridge Philosophical Society* 9(Part II):8–106. This is the original paper in which Stokes derived his well-known expression for the drag force on a slowly moving rigid sphere.

Wang, H. F. and Anderson, M. P. 1982. *Introduction to groundwater modeling.* San Francisco: W. H. Freeman, 237 pp. Appendix D of this text (pp. 219–24) is a delightful, technical, and historical summary of how physical analogues (structural homologies) involving electricity, heat flow, and structural mechanics have provided a significant role in the development of mathematical modeling of groundwater flow.

# CHAPTER 2

# Fluids and Porous Media
# as Continua

Let us anticipate that we wish to treat fluids as continuous substances at a *microscopic scale*. To see the motivation for this, it is instructive to consider the possibility of describing fluid behavior at a *molecular scale* by making use of Lagrangian mechanics to track the behavior of each molecule, just as we would describe the ballistics of a moving, rigid body. Consider describing the state of a simple diatomic molecule at some instant; to do this, we must decide what minimum set of coordinates completely specifies the position and configuration of the molecule. For example, we must specify its position within an inertial reference frame, which requires the three Cartesian coordinates $x$, $y$, and $z$. We also must specify its velocity with respect to this coordinate system, which requires the three corresponding components of velocity $u_m$, $v_m$, and $w_m$ (Figure 2.1). The molecule may be spinning; to describe this, we must assign to the molecule three local coordinate axes to specify three angular coordinates that give its orientation within the inertial reference frame. Because the axis of rotation may not coincide with one of the local axes, we also must specify two angular coordinates that give the orientation of the axis of rotation within the local coordinate system (Figure 2.1). Finally, we must specify the angular velocity about this axis of rotation.

Thus, in addition to specifying the mass of a molecule, we need twelve variables or *generalized coordinates* to specify its state. Moreover, we must know initial values of these twelve coordinates, just as we need to know the initial position and velocity of a ballistic body to track its course. To track the behavior of $N$ molecules, we therefore must know $N$ masses plus $12N$ initial position, velocity, and orientation coordinates. It becomes clear that to adopt this approach to describe the behavior of a fluid constitutes a formidable task! One has no choice but to abandon a molecular-scale treatment and adopt a view involving the microscopic scale, where behaviors of individual molecules are ignored, and instead, the collective behavior of a suitably defined ensemble of molecules is treated in a statistical (average) sense in terms of bulk properties such as fluid density, temperature, and viscosity. (This is essentially the problem that motivated Josiah Willard Gibbs to formally introduce principles of statistical mechanics in classical physics.)

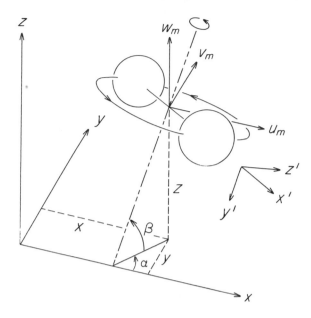

**Figure 2.1** Diatomic molecule with local $x'y'z'$-coordinate system, and axis of rotation passing through its center of mass, with Cartesian coordinates $x$, $y$, and $z$, and component velocities $u_m$, $v_m$, and $w_m$.

For our purposes, we will consider a specific volume of fluid: It must be large enough to contain a sufficient number of molecules that their collective behavior, as measured by bulk properties such as density, does not discernibly fluctuate from one instant to the next due to a disproportionate influence of the random behaviors of a few individual molecules. Similarly, it must be sufficiently small that the collective behavior of the molecules is homogeneous over the volume of fluid. Note that in choosing to work with bulk properties, we assume that these properties ultimately have a molecular-scale interpretation. This idea will be clarified in our treatment of fluid density below. In addition, we will consider an explanation of viscosity using a surprisingly simple argument involving molecular kinetics (Chapter 3), which stands as a triumph of this approach. But before doing this, it is useful to consider what defines the right scale to perform our averaging procedure; and it seems natural to consider the characteristic distances involved in the motions of molecules as they race around, continually colliding with each other, due to their thermal agitation.

## 2.1 MEAN FREE PATH

The *mean free path* $l$ is the average distance traveled by a molecule between collisions. The value of $l$, in theory, can be estimated from a sample of the distances traveled between many successive collisions of an individual molecule, or from a sample of such distances traveled by many molecules, taken at one instant. For an ideal gas consisting of identical molecules having radius $r$, kinetic theory suggests that

$$l = \frac{1}{4\pi \sqrt{2}\, r^2} \frac{V}{N} \tag{2.1}$$

This indicates that $l$ is inversely proportional to the silhouette area $\pi r^2$ of a molecule. Furthermore, $l$ is inversely proportional to the number of molecules per unit volume $N/V$, which is the molecular density $\rho_N$ of the fluid. Note that molecular speed does not appear in this equation.

Making use of the ideal gas law (Chapter 4), Equation (2.1) takes on a slightly different appearance

$$l = \frac{k}{4\pi \sqrt{2}\, r^2} \frac{T}{p} \tag{2.2}$$

where $k$ is Boltzmann's constant ($1.38 \times 10^{-23}$ J K$^{-1}$). This indicates that for a given pressure $p$, $l$ is proportional to temperature $T$. We may thus infer from these equations that the mean free path in general varies with the size of molecule, the molecular density, and thermal agitation.

Equation (2.1) implies that when the mean free path $l$ is large, molecules are on average farther apart than when it is small. This in turn implies that with increasing $l$ we must consider larger fluid volumes to consistently observe over time the same value of a bulk property such as molecular density. For example, consider a specific volume of fluid $V_0$. During a brief interval of time, some molecules travel into this specified volume, and some travel out, such that a momentary net increase or decrease in the number of molecules within $V_0$ may occur. Over many such intervals, the net change in the number of molecules will be zero. Nevertheless, if $V_0$ is on the order of $l^3$, it will on average contain only a few molecules, and a net increase or decrease of two or three molecules during any given interval could lead to sizable fluctuations in molecular density from one instant to the next. However, if $V_0$ is much larger than $l^3$, on average containing a hundred or more molecules, a net increase or decrease of a few molecules would not alter the molecular density by much. Thus, a lower limit on the size of $V_0$ must exist, below which it is inappropriate to describe a fluid in terms of bulk properties.

Now consider an inhomogeneous fluid. For example, this might involve a change in the molecular density of the fluid with respect to space, which from (2.1) would imply a concomitant change in the mean free path $l$. A volume $V_0$ with linear measure $V_0^{1/3}$ on the order of $100l$ or more would very likely involve a discernible change in molecular density over its extent. However, using this volume to estimate the molecular density would have the effect of averaging this variation, thereby giving the impression that it did not exist. Yet such variations in density are of interest with regard to fluid behavior at a larger scale. So although $V_0$ might be suitable because it contains a sufficient number of molecules to exhibit steady bulk properties with respect to time, it would be too large in the sense that useful information regarding how these properties vary over space is lost. Let us now formalize these ideas further by introducing the idea of a characteristic volume.

## 2.2 MATHEMATICAL AND PHYSICAL POINTS

Suppose that we could measure instantaneous values of molecular density $\rho_N$ associated with a sample of points $P_1 = (x_1, y_1, z_1)$, $P_2 = (x_2, y_2, z_2)$, .... In the case where we used a sampling volume $V$ on the order of $l^3$, some of the measured

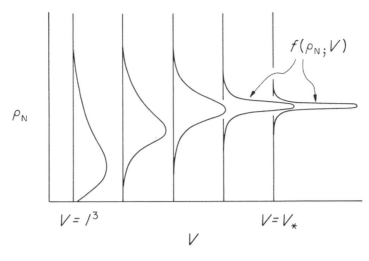

**Figure 2.2**  Schematic of how the distribution $f(\rho_N; V)$ of molecular density $\rho_N$ varies with sampling volume $V$, up to the characteristic volume $V_*$.

values of $\rho_N$ would be very small, because, based on the preceding discussion, only one or two molecules might be within the volume at the instant of measurement. Similarly, some values might be quite large if we happened to take a measurement, by chance, when the volume contained many molecules. Thus, we would anticipate observing a distribution $f(\rho_N; V)$ of molecular densities associated with a small sampling volume $V$ (Figure 2.2), where the notation indicates that the form of this distribution varies with the value of $V$. Suppose that we performed this experiment again, this time increasing the sampling volume. Based on the preceding discussion, we would again anticipate observing a distribution of measured densities, but with less variation because the larger sample volume would, on average, contain more molecules, and be less subject to large fluctuations in the numbers of molecules within it. Eventually we would reach a sampling volume $V_*$ where the variation in measured densities becomes negligible, and the molecular density would be independent of sampling volume. A similar outcome would emerge if, instead of measuring densities associated with a sample of points $P_i$ ($i = 1, 2, 3, \ldots$), we measured densities associated with one point, say $P_1$, at successive times $t_j$ ($j = 1, 2, 3, \ldots$).

The volume $V_*$ is considered to be a characteristic volume of the fluid; the molecular density associated with a *mathematical point* $P_i$ now can be defined by the limit

$$\rho_N(P_i) = \lim_{V \to V_*} \rho_N(V) = \lim_{V \to V_*} \frac{N(V)}{V} \tag{2.3}$$

Moreover, we assert that the limit

$$\lim_{\Delta s \to 0} \frac{\rho_N(s + \Delta s) - \rho_N(s)}{\Delta s} \tag{2.4}$$

exists, where $s$ denotes an arbitrary coordinate direction. This limit should be recognized as the first derivative of $\rho_N$ with respect to $s$. Equation (2.4) thus implies that molecular density, associated with an infinite set of mathematical points $P_i$, is a continuous function over space.

The volume $V_*$ also is referred to as a *physical point*. A physical point is distinguished from a mathematical point, not only because of the obvious difference in dimensions, but also in the following sense: Two neighboring physical points are not independent, because they may overlap, despite the fact that each is used to characterize the molecular density at distinct mathematical points. Now, imagine replacing the molecules with a continuous set of physical points; the result is a physical abstraction—a *fluid continuum*—in which we may specify the density (or some other property) at each coordinate position. Because of this continuity, we may immediately make use of the differential calculus to describe fluid properties. Note that by defining a physical point based on a volume less than $V_*$, associated properties would reflect individual molecular effects such that our physical abstraction would not be a continuous one (Figure 2.2).

We may now define characteristic lengths $L_x$, $L_y$, and $L_z$ for a fluid continuum in the coordinate directions $x$, $y$, and $z$:

$$L_x = \frac{\rho_N}{\partial \rho_N / \partial x}; \qquad L_y = \frac{\rho_N}{\partial \rho_N / \partial y}; \qquad L_z = \frac{\rho_N}{\partial \rho_N / \partial z} \qquad (2.5)$$

To see the physical interpretation of these lengths, first note that by using the chain rule, a change in density $\rho_N$ with respect to space $s$ can be written

$$\frac{\partial \rho_N}{\partial s} = \frac{\partial}{\partial s}\left(\frac{N}{V_*}\right) = \frac{1}{V_*}\frac{\partial N}{\partial s} + N\frac{\partial}{\partial s}\left(\frac{1}{V_*}\right) \qquad (2.6)$$

Because molecular density is now defined using the constant characteristic volume $V_*$, $\partial/\partial s \,(1/V_*)$ equals zero, and the lengths $L_x$, $L_y$, and $L_z$ become

$$L_x = \frac{N}{\partial N / \partial x}; \qquad L_y = \frac{N}{\partial N / \partial y}; \qquad L_z = \frac{N}{\partial N / \partial z} \qquad (2.7)$$

Thus, each of these characteristic lengths is a measure of the average number of molecules in the vicinity of a coordinate position relative to the rate of change in the number of molecules as one moves away from the coordinate position. With large $N$, or with a small change in $N$ over space for given $N$, the *proportional* variation in $N$ is not large, and the resulting large value of length $L$ indicates uniformity in $N$ over a large distance. With small $N$, or with a large gradient of $N$, the proportional variation in $N$ over space may be large, and the resulting shorter length $L$ indicates uniformity over a shorter distance. The volume $L_x L_y L_z$ may be taken as an upper limit for which $\rho_N$ is independent of $V$; note that with a homogeneous fluid, $\partial N / \partial s$ is zero, and $L_x L_y L_z$, in principle, is infinite.

It now should be intuitively clear that a fluid ought to exhibit uniformity over distances larger than the mean free path. In this regard, the dimensionless *Knudsen number Kn* is defined by the ratio $l/L$. When $Kn < 0.01$, the mean free path is sufficiently small relative to gradients in $N$ over space that the fluid behaves as a continuum.

## 2.3 REPRESENTATIVE ELEMENTARY VOLUME

Fluid flow through the interconnected pores of an unconsolidated sediment or rock involves a myriad of complex paths traced out by "parcels" of fluid as they move from one pore to the next (Figure 2.3). But rather than treating these details of fluid flow at the microscopic scale of the pores, it is appealing to consider the possibility of choosing a larger *macroscopic scale* at which we can ignore details of the flow, and instead, treat the average behavior of the fluid within the medium—in a manner similar to the development above for replacing detailed molecular behavior with a fictitious fluid continuum. The motivation for this will be explored further when we consider Darcy's law for flow through porous media. Meanwhile, let us anticipate that we wish to describe sediment and rock media as continuous materials with respect to flow, despite the fact that, being porous, they are decidedly discontinuous. A property that is particularly convenient to use in developing the idea of a porous medium as a continuum is its porosity $n$, although other properties could be used. In many ways the development follows the treatment of fluid as a continuum, but with several important differences.

Consider a sample of points $P_i$ ($i = 1, 2, 3, \ldots$) randomly located within a porous medium. Suppose that we measure the volume of pore space $V_p$ within the sampling volume $V$, centered about each point. The ratio $V_p/V$ is the porosity $n$ associated with the sampling volume $V$. In the case in which the sampling volume $V$ is smaller than the volume of individual pores, some of the measured values of $n$ likely would equal one because some of the sampling points, by chance, would be located entirely within pores. Also, some of the values of $n$ likely would equal zero where sampling points fell entirely within the solid medium. Still others would include both pore and solid. Thus, we would anticipate observing a distribution

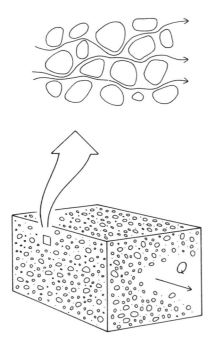

**Figure 2.3** Schematic of how macroscopic flow quantity $Q$ is envisioned as an integrated effect of detailed flow structure at microscopic (pore) scale.

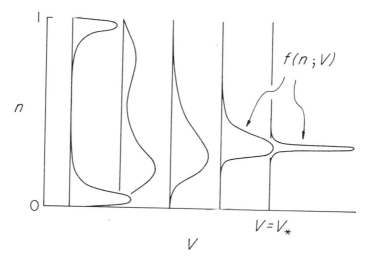

**Figure 2.4** Schematic of how the distribution $f(n; V)$ of porosity $n$ varies with sampling volume $V$, up to the representative elementary volume $V_*$.

$f(n; V)$ of porosities associated with a small sampling volume $V$ (Figure 2.4). Suppose that we performed this experiment again, this time increasing the sampling volume. The larger volume would more likely include both pore and solid. The case in which the sampling volume was entirely pore ($n = 1$) would occur infrequently, and involve the occasional, rare large pore. Likewise, the case in which the sampling volume was filled entirely with solid ($n = 0$) would occur infrequently. Thus, we would again anticipate observing a distribution of porosities, but with less variation (Figure 2.4). Eventually we would reach a sampling volume $V_*$, where the variation in measured porosities becomes negligible, and the porosity would be independent of sampling volume.

The volume $V_*$ is considered the *representative elementary volume* (REV) of the porous medium. The porosity $n$ associated with a mathematical point $P_i$ now is defined by the limit

$$n(P_i) = \lim_{V \to V_*} n(V) = \lim_{V \to V_*} \frac{V_p}{V} \tag{2.8}$$

Moreover, as with molecular density for a fluid, we assert that the limit

$$\lim_{\Delta s \to 0} \frac{n(s + \Delta s) - n(s)}{\Delta s} \tag{2.9}$$

exists, where $s$ denotes an arbitrary coordinate direction. This limit is the first derivative of $n$ with respect to $s$. Equation (2.9) thus implies that porosity, associated with an infinite set of mathematical points $P_i$, is a continuous function over space.

The volume $V_*$ is a physical point, again distinguishable from a mathematical point in the sense that two neighboring physical points are not independent because they may overlap, despite the fact that each is used to characterize the porosity at a distinct mathematical point. This is formally stated as

$$P[n_1 \le n(s) \le n_2] \ne P[n_1 \le n(s) \le n_2 \mid n(s + \Delta s)] \tag{2.10}$$

where $P[n_1 \leq n(s) \leq n_2]$ denotes the probability of observing a value of $n$ within the small range $n_1$ to $n_2$, and $P[n_1 \leq n(s) \leq n_2 \mid n(s + \Delta s)]$ is the probability of observing a value of $n$ within $n_1$ to $n_2$, given knowledge of the value of $n(s + \Delta s)$. If one randomly selects a volume $V_*$ at a point $P$, there is a definite probability that $n$ falls between $n_1$ and $n_2$, given by the left side of (2.10). However, if one selects a volume $V_*$ at the same point $P$, knowing the value of $n$ a short distance $\Delta s$ away, this information influences the likelihood (probability) that the value of $n$ at $P$ falls between $n_1$ and $n_2$, because the two values of $n$—that at $P$ and the neighboring known value—ought to be similar due to their proximity.

Now, imagine replacing the porous medium with a continuous set of physical points; the result is a physical abstraction—a continuum—in which we may specify the porosity (or some other property) at each coordinate position. As with a fluid continuum, we also may define characteristic lengths $L_x$, $L_y$, and $L_z$ in terms of porosity, analogous to (2.5). When a medium is homogeneous with respect to $n$, the volume $L_x L_y L_z$, in principle, is infinite.

It also is possible to characterize the porosity of a medium without directly using the idea of an REV. Consider a function $f(s)$ defined along a straight coordinate $s$ embedded within a homogeneous porous medium such that

$$f(s) = \begin{cases} 0 & \text{within solid phase} \\ 1 & \text{within pore space} \end{cases} \qquad (2.11)$$

Such a function would appear as a series of steps as the coordinate line crossed alternately between pore and solid portions of a medium (Figure 2.5). We then define an *ensemble average* of the function $f$:

$$\overline{f} = \lim_{s \to \infty} \frac{\displaystyle\int_{-s}^{s} f(s')\,ds'}{\displaystyle\int_{-s}^{s} ds'} \qquad (2.12)$$

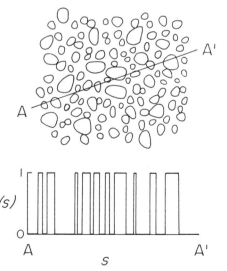

$f(s)$

**Figure 2.5** Example of piecewise continuous function $f(s)$, where $f(s)$ equals one when $s$ crosses pore space or zero when $s$ crosses solid.

where $s'$ denotes a variable of integration. The numerator equals the length of the coordinate line $s$ falling within pore. The denominator simply equals the length of the coordinate line over which the function $f(s)$ is considered. The quantity $\overline{f}$ is therefore the proportion of coordinate line represented by pores, and it follows that $\overline{f}$ is a measure of the *linear porosity*. Moreover, it can be demonstrated that the linear porosity coincides with the volumetric porosity $n$. This opens the possibility of estimating $n$ from a thin section of a porous medium. In practice, one uses a finite sample distance $S$ such that (2.12) becomes

$$\overline{f} = \frac{1}{S} \int_0^S f(s) \, \mathrm{d}s \qquad (2.13)$$

Note that characteristic linear and areal measures can be defined in a manner similar to that used to define an REV. These measures are referred to as the *representative elementary length* and the *representative elementary area* of a porous medium. Interested students are urged to examine Bear (1988). In addition, the classic concept of an REV as developed here is not entirely satisfactory. An extreme example is a medium whose pores consist of fractures as well as intergranular voids. (We will examine this in Example Problem 8.4.6.) The REV of such a medium can be so large that the idea of a property such as hydraulic conductivity being a continuous mathematical function is not physically sound. Students familiar with the theory of random variables and functions are urged to examine the text by de Marsily for an alternative development of the averaging proccess that is necessary for the idea of a continuum.

## 2.4 EXAMPLE PROBLEMS

### 2.4.1 Shear of thin liquid films: behavior below the continuum scale

Consider a film of liquid between two smooth, parallel surfaces. When a steady tangential force is applied to one of the surfaces, the liquid film deforms with a shearing motion. At sufficiently low rates of shear, the behavior of the liquid is described by a law of the form

$$\tau = \mu \frac{\mathrm{d}\varepsilon}{\mathrm{d}t} = \mu \dot{\varepsilon} \qquad (2.14)$$

Here $\tau$ is a shear stress that is equal to the applied tangential force divided by the area of the liquid film in contact with the surfaces, $\varepsilon$ is the shear strain so $\mathrm{d}\varepsilon/\mathrm{d}t = \dot{\varepsilon}$ is the rate of strain (or shear rate), and $\mu$ is the dynamic viscosity, a measure of the resistence offered by the liquid film to the shearing motion. We will see in Chapter 3 that (2.14) describes the behavior of a Newtonian fluid, and is therefore another way of expressing the relation (1.7).

Now suppose that the two smooth surfaces are brought very close together such that the distance separating them is on the order of, say, several tens of angstroms or less. The behavior of the liquid film in this situation may be very different from the behavior of the same liquid in bulk, when the distance separating the surfaces is much larger than the mean free path of the liquid molecules. Characterizing the behavior of thin liquid films is a problem that bears on fluid motions within very small rock pores, as well as the interactions that occur among fluid molecules and mineral surfaces. Let us therefore briefly examine this topic following the discussion of S. Granick.

Consider a hypothetical experiment in which a liquid drop is squeezed between a falling sphere and a flat surface. The sphere and surface are smooth relative to the molecular size of the liquid. The liquid initially is readily displaced, but as the sphere continues to approach the surface, the liquid is displaced at a slower rate. Eventually the liquid stabilizes as a thin film whose finite thickness is on the order of a few molecular diameters, and very high pressures are required thereafter to squeeze out the remaining liquid. At this point the liquid film is acting like a rigid solid because it supports the weight of the sphere.

The reason for this behavior is that a liquid tends to organize into strata parallel to a smooth solid surface. This is apparent in the distribution of the local molecular density between two solid surfaces (Figure 2.6). The local density exhibits an oscillatory structure away from each surface. A maximum density occurs at a distance equal to one molecular radius, then a minimum occurs at a distance equal to two radii, and so on. If the surfaces are sufficiently close, interference between these structured interfacial regions occurs, and this thin region then behaves more like a solid than a liquid.

The solid surfaces used in experiments to examine the behavior of thin liquid films typically are those of mica (muscovite) crystals. It is possible in such experiments to produce liquid films whose thicknesses are defined with a resolution of an angstrom—less than the thickness of one liquid molecule. If the mica surfaces are made to oscillate parallel to each other, an oscillatory shearing motion of the fluid occurs. With oscillations of large amplitude, this motion is similar to a steady shear of the fluid, at least for one-half the period of each oscillation. (We will see in Chapter 3 that this configuration is like that of a *Couette flow*.)

Shearing behavior varies with film thickness and shear rate. At low shear rates, the effective viscosity calculated from (2.14) increases with decreasing film thickness (Figure 2.7). The region of inhomogeneous density between the surfaces (Figure 2.6) occupies an increasing proportion of the film thickness, and contributes to a greater resistance to shearing motion than the liquid in bulk. For a given film thickness, in contrast, the effective viscosity is constant for small shear rates, but then steadily decreases with increasing shear rate (Figure 2.8). This nonlinear response occurs when the experiment distorts the structure of the liquid film. This happens when the oscillations are faster than a characteristic time scale of the Brownian motions of the liquid molecules. The relaxation time associated with molecular

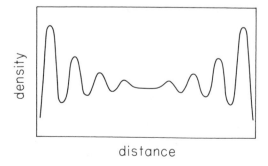

**Figure 2.6** Schematic diagram of the distribution of local molecular density between two solid surfaces; adapted with permission from Granick (1991), copyright 1991 American Association for the Advancement of Science.

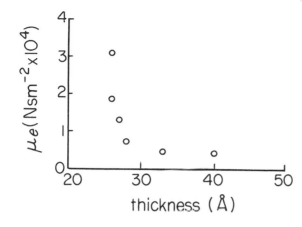

**Figure 2.7** Plot of effective viscosity $\mu_e$ of dodecane films at low shear rates versus film thickness; adapted with permission from Granick (1991), copyright 1991 American Association for the Advancement of Science.

motions is much longer than that associated with fluid molecules in bulk, and the structured arrangement of molecules next to the surfaces is less ordered during shear. Qualitatively, the shearing motion induces disorder in the molecular motions. This disordering occurs faster than the rate at which the molecules otherwise tend to reorganize themselves (relax) back into a structured arrangement

In addition, when a liquid film is sufficiently thin—less than four molecular diameters—its response to a tangential shear is more like that of a solid than a liquid, and its behavior is therefore more properly characterized in terms of properties normally associated with solids. That is, such films solidify in the sense that shear does not occur until a critical stress (yield stress) is exceeded, and at stresses below this yield stress a film exhibits an elastic rigidity. Films whose thicknesses are four to 10 molecular diameters exhibit a combination of viscous and elastic behaviors (Chapter 3). Moreover, solidification is suppressed by continuous shear, analogous to the behavior exhibited by solid materials undergoing shear, wherein the dynamic friction is less than the static friction. The mechanisms leading to this behavior of thin films are not entirely understood, and are being examined, in part, using computer simulations of the solidification process. Students are urged to

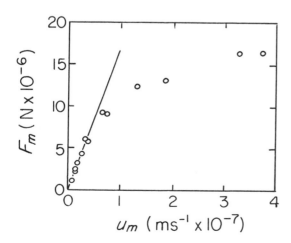

**Figure 2.8** Plot of maximum viscous force $F_m$ versus maximum velocity $u_m$ during cycle of oscillation; data are for dodecane films involving varying periods and amplitudes of oscillation; straight line represents extrapolation of domain of linear behavior; adapted with permission from Granick (1991), copyright 1991 American Association for the Advancement of Science.

examine Granick (1991) regarding these points, and the related paper by Drake and Klafter (1990) regarding the dynamics of fluid molecules in very small pores.

### 2.4.2 Fractal pore geometry

Consider an idealized porous medium consisting of a cube that is divided by planar pores into smaller cubes in the following special way: The largest cube is divided into eight smaller cubes, separated by three orthogonal planar pores that bisect the original cube; the apertures of these pores are a specified proportion $k$ of the smaller cube dimension (Figure 2.9). Each of these eight cubes is similarly divided into eight smaller cubes, again separated by three planar pores whose apertures equal the same specified proportion $k$ of the smallest cube dimension. Then, each of these 64 cubes is similarly divided into eight smaller cubes, and so on. In a mathematical sense we could continue this process of division indefinitely. Good geological sense, however, tells us that this process of division is not unbounded if our geometrical artifice is to resemble real porous media. Rieu and Sposito (1991a) have, in fact, applied this type of geometrical description to structured soils—where individual grains are clumped into aggregates or "peds"—as opposed to unstructured media like cohesionless sand. The smallest division must be limited by the size of indivisible solid grains, and the largest cube to which this process of division can be applied coincides with the largest peds. Over an intermediate range of sizes, this idealized porous medium exhibits *self-similarity* in the sense that its geometrical appearance is independent of how close or far away one is when observing it. Readers may recognize that this idealized medium possesses properties of a self-similar *fractal*.

Let us examine one part of this idea of fractal porous media. Consider a porous material whose porosity is composed of a distribution of pore sizes that decrease from diameter $p_0$ to $p_{m-1}$ ($m \geq 1$). Thus $p_0$ represents the largest pore size, $p_{m-1}$ represents the smallest pore size, and the quantity $m$ is associated with the smallest indivisible part of the medium. Suppose that a bulk sample of volume $V_0$ contains all pore sizes $p_i$ and possesses porosity $n$. We may then divide the pore–volume distribution of $V_0$ into $m$ virtual pore-size fractions, where the $i$th *virtual size fraction* is defined by

$$P_i = V_i - V_{i+1} \qquad i = 0, 1, \ldots, m-1 \tag{2.15}$$

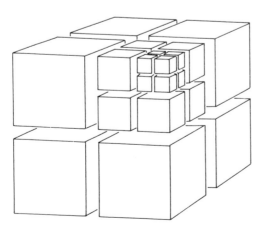

**Figure 2.9** Conceptual model of fragmented, fractal porous medium consisting of self-similar cubic elements; fragmentation is incomplete to illustrate division of elements; adapted from Rieu and Sposito (1991).

Here $P_i$ is the volume of pores solely of size $p_i$ contained in $V_i$, and the $i$th partial volume $V_i$ contains all pores of size $p_i$ or less. Now, define a *linear similarity ratio* $r$, which relates the successive sizes $p_i$:

$$p_{i+1} = rp_i \qquad r < 1 \qquad (2.16)$$

This states that each pore size ($p_{i+1}$) is a fixed proportion $r$ of the next larger pore size ($p_i$). The pore volume increment $P_i$ is proportional to the cube of $p_i$; likewise $P_{i+1}$ is proportional to the cube of $p_{i+1}$. It is left as an exercise to use this fact together with (2.15) and (2.16) to demonstrate that the pore volume increments and partial volumes are related by

$$P_{i+1} = r^3 P_i \qquad (2.17)$$

$$V_{i+1} = r^3 V_i \qquad (2.18)$$

This in turn implies that

$$\frac{P_i}{V_i} = \Gamma = \text{constant} \qquad i = 0, 1, \ldots, m-1 \qquad (2.19)$$

A fractal porous medium possesses the property that each volume $V_i$ contains a constant $N$ smaller volumes $V_{i+1}$ and one associated pore volume $P_i$. Thus,

$$V_i = NV_{i+1} + P_i \qquad i = 0, 1, \ldots, m-1 \qquad (2.20)$$

In turn, each volume $V_{i+1}$ contains $N$ volumes $V_{i+2}$ and one associated pore volume $P_{i+1}$; and so on for $i = 2, 3, \ldots$. For example, $N = 8$ for the medium described above (Figure 2.9).

Let us now introduce the *fractal dimension D* of a porous medium:

$$D = \log N - \log \frac{1}{r} \qquad D > 0 \qquad (2.21)$$

Although we will not show the development here (students are referred to Rieu and Sposito [1991a; 1991b]), it is straightforward to demonstrate that

$$\Gamma = 1 - r^{3-D} \qquad (2.22)$$

Further, the porosity $n$ is given by

$$n = 1 - (r^{3-D})^m \qquad (2.23)$$

where $m$ has a finite value. As a point of reference, the idealized fractal medium described above (Figure 2.9) has a fractal dimension $D = 2.8$ for $N = 8$ and $r = 0.476$. Rieu and Sposito (1991b) report fractal dimensions that vary from 2.88 to 2.95 for loam, silt-loam, and clay soils. Thus the fractal dimensions of these media are less than that associated with a three-dimensional Cartesian object. Rieu and Sposito (1991b) further used these relations to develop formulae describing bulk density, aggregate size distributions, and the relation between moisture content and hydraulic conductivity (Chapter 13).

This fractal description of a porous medium does not appeal to the concept of a representative elementary volume (REV) to define the porosity in (2.23). Also noteworthy is the conclusion that this description is not restricted in terms of the shapes of the solid matrix elements or pores, or how these are interconnected. Such fractal descriptions therefore are being used to describe a variety of properties of natural porous media, including pore and particle-aggregate sizes, motivated by the observation that these properties sometimes exhibit scaling relations and self-similarity over certain size domains. We will briefly revisit this idea in Chapter 13.

### 2.4.3 Porosity estimates of regularly and randomly packed particles

Consider a porous medium composed of ideally packed spheres, each having radius $R = 1$. The choice of packing arrangment is not important; results similar to those presented below hold for each possible arrangement. So, for simplicity, let us choose a cubic packing. To further simplify the mathematics, let us treat the medium as two-dimensional (Figure 2.10), anticipating that the three-dimensional case also will yield similar results. We may note, in fact, that this two-dimensional case corresponds to one of an infinite set of planes through the three-dimensional medium, and therefore represents one view of it in thin-section. We immediately can calculate the *areal porosity* of this two-dimensional medium as the ratio of pore area to total area in a unit cell: $n = 1 - \pi/4 \approx 0.215$.

Suppose that we alternatively wish to obtain a linear estimate of $n$ using (2.13). Of eight selected sample lines (Figure 2.10) whose origins ($s = 0$) are set at the Cartesian origin $(x, y) = (0, 0)$, two of the corresponding functions $f_i(s)$ are given by $f_1(s) = f_5(s) = 0$; one function is given by $f_8(s) = 1$; and the others appear as series of steps (Figure 2.11). For example, $f_3(s)$ is defined by

$$f_3(s) = \begin{cases} 0 & 0 < s < 1.789 \\ 1 & 1.789 < s < 2.683 \\ 0 & 2.683 < s < 6.258 \\ 1 & 6.258 < s < 7.152 \\ 0 & 7.152 < s < 10.73 \\ & etc. \end{cases} \tag{2.24}$$

Integrating these functions and dividing by the sample line length $S$ to estimate $n$ for each line, we obtain two interesting results.

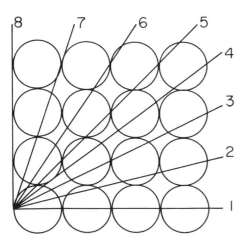

**Figure 2.10** Sample lines used to estimate linear porosity of ideal porous medium.

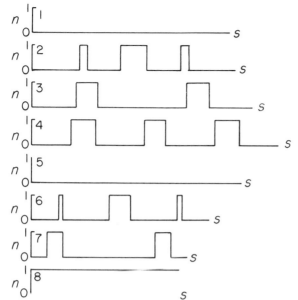

**Figure 2.11** Functions $f_i(s)$ associated with sample lines in Figure 2.10

First, each estimate of $n$, in general, does not converge to the value of $1 - \pi/4$, regardless of the length $S$. In fact, one can obtain a reasonable estimate of the areal porosity only by taking the mean of many such sample line values. The mean of the eight sample line values is about 0.256; one would expect this to be closer to the true value with a larger sample of lines. Second, evidently the medium is not isotropic with respect to the mathematical point $(0, 0)$, because the value of $n$ varies with sample line orientation. Yet, if we considered porosity associated with a set of physical points (each having the size of an REV) arranged along lines of arbitrary orientation, we would discover that the medium is indeed isotropic with respect to porosity. To obtain a reasonable estimate of $n$ using (2.13), one would need to use the average of a set of lines having different orientations *and different origins*.

These results evidently indicate that some degree of randomness in the arrangement of particles must exist to obtain a value of the linear porosity that approximates the areal (and therefore volumetric) porosity. In contrast to the ideal particle arrangement above, consider a thin section of an uncemented sandstone (Figure 2.12). One obtains a porosity of 0.18 from four randomly positioned sample lines, which is close to the value of 0.21 obtained from a point-count over the entire thin section.

## 2.4.4 Covariance of porosity

Let us examine several implications of the idea embodied in (2.10), that porosities measured at neighboring points are not independent. Consider for illustration a uniform sand. Suppose that we randomly selected $N$ points over the extent of the sand, then determined the porosities associated with these points using an REV and (2.8). One could infer from this sample of points that the porosity $n$ possesses a specific probability density function $f_n(n)$ (see Example Problem 1.4.3). The variance $\sigma_n^2$

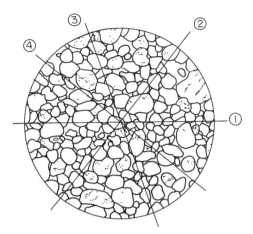

**Figure 2.12**   Sample lines used to estimate porosity from thin section of sandstone.

of the function $f_n(n)$ would reflect presence of local variations in grain size and grain packing, ultimately related to variations in local depositional environment. Moreover, $f_n(n)$ would characterize the sand as a whole.

Now suppose that we randomly selected another point and asked the question: What is the probability P that the porosity $n$ associated with this new point falls within the range $n_1$ to $n_2$? Based on knowledge of $f_n(n)$ alone, this probability is given by

$$P[n_1 \leq n \leq n_2] = \int_{n_1}^{n_2} f_n(n)\,dn \qquad (2.25)$$

According to (1.47), the total probability beneath $f_n(n)$ equals one. This is the same as saying that there is a 100 percent chance (P = 1) that a specified value of $n$ falls between the lower and upper limits of $f_n(n)$, which are zero and one in this problem. Since (2.25) represents the "area" beneath $f_n(n)$ between $n_1$ and $n_2$, this quantity is equal to a fraction of the total probability beneath $f_n(n)$ falling between $n_1$ and $n_2$.

Suppose, however, that we selected a point that is close to another point at which the porosity is known, and suppose that this known porosity is very different from the range $n_1$ to $n_2$. In the idealized case in which neighboring porosities are truly independent, (2.25) would provide the best guess of the likelihood that the porosity at the unknown point would fall within the range $n_1$ to $n_2$. Geological sense, however, suggests that for physical reasons the unknown porosity is likely to be similar to the known value due to their proximity, and that the formula above could be a bad guess in view of this information. It is useful to define a measure of the similarity in values of porosity based on proximity of sampling points.

Envision measuring the porosity $n_s$ at $N$ equally spaced positions ($s = 1, 2, 3, \ldots$) along the coordinate axis $s$. This set of measurements constitutes a discrete spatial series $n_1, n_2, n_3, \ldots, n_N$. Assuming large $N$, an estimate of the mean $\mu_n$ of the distribution $f_n(n)$ is obtained from the mean of the series using (1.25):

$$\bar{n} = \frac{1}{N} \sum_{s=1}^{N} n_s \qquad (2.26)$$

An estimate of the variance $\sigma_n^2$ is obtained from the variance of the series using (1.32):

$$s_n^2 = \frac{1}{N} \sum_{s=1}^{N} (n_s - \bar{n})^2 \tag{2.27}$$

We also may compute the autocovariance $c_k$ between values $n_s$ and values $n_{s+k}$ separated by $k$ measurement intervals using (1.34):

$$c_k = \frac{1}{N} \sum_{s-1}^{N-k} (n_s - \bar{n})(n_{s+k} - \bar{n}) \tag{2.28}$$

Notice that, as $k$ approaches zero, $c_k$ reduces to the ordinary variance (2.27); thus $c_0 = s_n^2$. Finally, we can compute the autocorrelation $r_k$ using (1.38):

$$r_k = \frac{c_k}{c_0} = \frac{c_k}{s_n^2} \tag{2.29}$$

The value of $r_k$ can be positive or negative, and is a measure of the statistical correlation between values of $n_s$ separated by the distance $k$. Observe that $r_k = 1$ when $k = 0$; that is, values of $n_s$, when compared with themselves, are perfectly correlated. In general, as $k$ increases, the correlation between values of $n_s$ decreases; and when $k$ is sufficiently large, $r_k$ approaches zero, reflecting that values of $n_s$ separated by large distances are statistically independent of each other.

Porosity data for unconsolidated sands typically possess a discrete autocorrelation function of the form $r_k = \phi^k$. This has a continuous counterpart $r(\tau)$ with the exponential form,

$$r(\tau) = e^{-\tau/S} \tag{2.30}$$

The quantity $S$ is a characteristic length; it is a measure of the distance over which porosities are statistically correlated. As $S$ decreases, values of porosity at neighboring positions along $s$ are less correlated. The relation between $r_k$ and $r(\tau)$ is given by $\phi = \exp(-1/S)$, where the continuous interval $\tau$ replaces the discrete lag $k$. The characteristic length $S$ is in general defined by (1.45):

$$S = \int_0^{\infty} r(\tau) \, d\tau \tag{2.31}$$

Note that measurements of $n$ associated with two closely spaced points are not independent because the REV in each case involves virtually the same set of pores. Thus $S$ generally cannot equal zero.

Let us now return to the problem of estimating the unknown porosity $n(\tau)$ associated with a point that is a distance $\tau$ from a position where the porosity $n_0$ is known. In the case where $r(\tau)$ has an exponential form, a good guess of the value of $n(\tau)$ is given by

$$n(\tau) = n_0 \phi^\tau + \bar{n} \tag{2.32}$$

Notice that as the interval $\tau$ becomes large such that $\phi^\tau$ approaches zero, the best estimate of the value of $n$ is the mean value $\bar{n}$. It also is possible to assign a statistical confidence to this estimate; without proof the variance $\text{var}(\tau)$ associated with the estimate (2.32) at a distance $\tau$ is

$$\text{var}(\tau) = s_n^2(1 - \phi^{2\tau}) \tag{2.33}$$

This indicates that the variance $\text{var}(\tau)$ is small when $\tau$ is small, and as $\tau$ becomes very large, the variance associated with the estimate (2.32) of $n$ approaches the overall variance $s_n^2$ about the mean $\bar{n}$. Students interested in this topic should consult Box and Jenkins (1976) regarding forecasting unknown values.

The ideas outlined above can be extended to three dimensions, and to other physical quantities such as the hydraulic conductivity. This is the basis for treating properties of porous media as *random fields,* a topic that is currently receiving considerable attention in the hydrological literature.

## 2.5 READING

Bear, J. 1988. *Dynamics of fluids in porous media.* New York: Dover, 764 pp. This text, a classic treatise on porous media flow, provides a concise summary (pp. 15–22) of the continuum concept applied to porous media, including treatments of the ideas of a representative elementary volume, area, and length.

Box, G. E. P. and Jenkins, G. M. 1976. *Time series analysis: forecasting and control.* San Francisco: Holden-Day, 575 pp. This text, also referred to in Example Problem 1.4.3, examines procedures for estimating autocorrelation functions, and forecasting of unknown values based on an autocorrelation function.

Drake, J. M. and Klafter, J. 1990. Dynamics of confined molecular systems. *Physics Today* 43(5):46–55. This paper examines the dynamical and thermodynamical behavior of fluid molecules in small spaces, drawing largely on experimental work with porous silica glasses.

Granick, S. 1991. Motions and relaxations of confined liquids. *Science* 253:1374–79. This paper provides a review of what is currently known about the rheological behavior of thin liquid films, and is the basis of Example Problem 2.5.1.

de Marsily, G. 1986. *Quantitative hydrogeology.* London: Academic, 440 pp. This text provides a clear theoretical description of the REV concept, its limitations, and an alternative derivation of the averaging process leading to the idea of a continuum for porous media (pp. 15–19).

Rieu, M. and Sposito, G. 1991. Fractal fragmentation, soil porosity, and soil water properties: I. Theory; and II. Applications. *Soil Science Society of America Journal* 55:1231–44. This pair of papers provides a clear description and development of the idea of fractal porous media as applied to natural soils, and is the basis of Example Problem 2.5.2.

Tolman, R. C. 1979. *The principles of statistical mechanics.* New York: Dover, 661 pp. This text, a classic treatise at the advanced level, provides a clear description (pp. 17–19) of the specification of generalized coordinates and velocities for examining dynamics of complex, as well as simple, mechanical systems.

Young, H. D. 1964. *Fundamentals of mechanics and heat.* New York: McGraw-Hill, 638 pp. This text provides a clear elementary introduction (Chapter 16) to the kinetic theory of gases, including a straightforward treatment of the molecular-scale model for an ideal gas, and the derivation of the mean free path.

# CHAPTER 3

# Mechanical Properties of Fluids and Porous Media

Fluids possess characteristic physical properties that govern how they behave when forces are applied to them. Of particular interest are properties that govern fluid responses to ordinary mechanical forces; thus we will not consider electrical and magnetic phenomena, and we will defer treating thermodynamical properties of fluids to the next chapter. Further, we will adopt a classic treatment of bulk behavior, as opposed to fluid behavior below the continuum scale. For example, we will examine how a Newtonian fluid deforms (in bulk) in a systematic way when it is subjected to shear stress, such that the rate of strain is proportional to the stress. In contrast, the same nominal fluid very close to a rock surface within a small pore may exhibit a much more complicated rheological behavior, from which we can infer that surface phenomena fundamentally alter the behavior of the fluid from that observed in bulk.

We also will see how different types of real fluids respond differently to an applied shear stress, including how the rate of strain of certain fluids varies with the duration of stress as well as other factors. In this regard we will distinguish between time-independent and time-dependent behaviors; in the latter case the behavior of a fluid depends on its history of strain. These basic ideas of fluid behavior are a foundation for most of the material covered in the remainder of the text. Nonetheless, this initial treatment of fluid properties will provide sufficient insight to fluid behavior to begin using simple mathematical analyses to examine several important fluid-flow problems. Likewise, the description here of mechanical properties of porous media also is a foundation for later use. Coverage of this topic, however, is restricted to items that will be useful to understanding fluid behavior.

## 3.1 BODY AND SURFACE FORCES

Certain mechanical properties of fluids are defined in terms of the forces acting on them. In this context, it is useful to distinguish two types of forces: *body forces* and *surface forces*. A body force is one whose magnitude is proportional to the volume or mass of a fluid parcel. A body force is distributed throughout a fluid parcel in the sense that each part of the parcel is equally subjected to the force. For example, the force on a fluid parcel imposed by the gravitational field of Earth—the weight of the parcel—is a body force. Similarly, a centrifugal force imposed on a fluid parcel due to angular acceleration of the fluid is a body force. In each of these cases, the magnitude of the force is proportional to the mass (or volume) of the fluid parcel because it equals the sum of the forces acting on all its parts. In contrast, a surface force is independent of fluid mass. A surface force acts on a fluid surface, which may be real or specified. For example, tangential (shear) forces and normal forces impressed on the faces of a specified fluid control volume are surface forces. Similarly, the shear force impressed on the real boundary between two dissimilar fluids or at the boundary between a fluid and solid, when there is relative motion between them, is a surface force.

## 3.2 IDEAL VERSUS REAL FLUIDS

It also is useful to distinguish two types of fluids: ideal, or perfect, fluids and real fluids. An ideal fluid is incompressible, and flow involves no friction. During flow, juxtaposed fluid "layers" experience no tangential stresses, only normal (pressure) stresses. As a consequence, there is no friction between layers and thus no internal resistance to shearing motion; an ideal fluid is inviscid. Although an ideal fluid is a physical abstraction, the idea is, nonetheless, useful in certain problems because fluids having small viscosities (including air and water), or far from real boundaries, can behave as if they are inviscid. In such cases the mathematics of a problem can be greatly simplified. As a point of reference, treatments of fluid flow in introductory physics courses typically involve ideal fluids when introducing Bernoulli's equation (Chapter 10).

There are many kinds of real fluids. A description of these is deferred until fluid viscosity is defined (Section 3.4). Meanwhile, note that the essential quality of a real fluid, in contrast to an ideal one, is that juxtaposed fluid layers experience tangential stresses during flow. As a consequence, there is friction between layers and thus internal resistance to shearing motion; a real fluid is viscous.

An important manifestation of these differences between ideal and real fluids involves how flow occurs next to a real boundary. As mentioned in Chapter 1, an ideal fluid exhibits "slip flow" at a boundary such that there is relative motion between the boundary and adjacent fluid. A real fluid exhibits "no-slip flow" such that no relative motion between the boundary and adjacent fluid exists. The velocity of the fluid must therefore approach zero at the boundary. The ideas of slip flow and no-slip flow, like fluid viscosity, ultimately have a molecular-scale explanation. An explanation involving molecular kinetics is presented below.

As we will see later, the no-slip condition often is very useful in solving flow problems involving viscous fluids near real boundaries. In addition, we will see the curious result that certain viscous flow problems involve mathematical descriptions that are entirely consistent with inviscid flow. To anticipate this, consider the case of flow in a porous medium. Despite the fact that such flow involves viscous forces at the microscopic scale of the pores, the flow may behave, in a mathematical sense, like an ideal fluid at a macroscopic scale. For example, consider how the specific discharge $q_{hx}$ in one-dimensional flow near a confined aquifer boundary is uniform

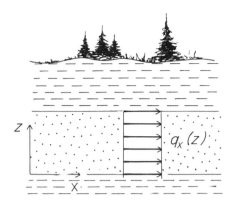

**Figure 3.1** One-dimensional flow within a confined aquifer; specific discharge $q_{hx}(z)$ is uniform over vertical coordinate $z$.

and nonzero (Figure 3.1). (Note that $q_{hx}$ has the dimensions of velocity since it is a volume per unit area per unit time.) In this *macroscopic* view, fluid next to the aquifer boundary "slips" past it rather than going to zero.

## 3.3 DENSITY

### 3.3.1 Fluids

The *mass density* of a fluid $\rho$ is defined as the product of molecular mass and molecular density. As such, mass density is subject to the constraints of the continuum concept introduced in Chapter 2. Because the product of molecular mass $M$ and the number of molecules $N$ within a specified volume $V$ equals the fluid mass $m$ associated with that volume $V$, in practice the mass density is simply defined as $m/V$. Further, we usually refer to mass density simply as density, since this usually will be clear in the context of its use.

Related to density is the specific weight $\gamma$ of a fluid defined by the product $\rho g$, where $g$ is the acceleration associated with a gravitational field. In addition we define the specific gravity $G$ of a fluid as the ratio of its density to that of pure water at 4°C. With this definition, we can measure the specific gravity, and therefore the density, of a liquid using a hydrometer. Such measurements typically are temperature dependent. In the case of gases, the density as a function of temperature and pressure can be inferred directly from thermodynamic considerations, making use of the ideal gas law (Chapter 4). Unfortunately, liquids are more complicated than gases, and a comparable, universal relation between pressure, temperature, and density has not been deduced. Nonetheless, it is sometimes possible to infer on empirical grounds a suitable relation of this sort for certain liquids.

### 3.3.2 Porous media

Because a porous medium is in general a multiphase system, it is useful to distinguish between the *density of solids* $\rho_s$ and the *bulk density* $\rho_b$. Let $V_T$ denote a total sample volume occupied by solid and fluid. This volume is subject to the constraints of the continuum and REV concepts. The solid within $V_T$ may consist of distinct phases, for example, grains having different mineralogies. Nonetheless, these collectively occupy a volume denoted by $V_s$. Likewise the fluid within $V_T$ may consist of different phases of the same substance, or different substances altogether. But instead of treating these individually, it is easier to work with the volume of pores $V_p$, which represents the part of $V_T$ that in principle *could* be occupied by fluid. Then $V_T = V_s + V_p$. Letting $m_s$ denote the mass of solids occupying $V_s$,

$$\rho_s = \frac{m_s}{V_s}; \qquad \rho_b = \frac{m_s}{V_T} \tag{3.1}$$

These definitions have the important feature of describing density properties of a porous medium, independent of any fluids present. This, of course, does not imply that values of densities based on (3.1) are constants. Both $\rho_s$ and $\rho_b$ may vary due to fluid–solid interactions. For example, chemical weathering over a long period can alter both $\rho_s$ and $\rho_b$ of media whose solid part consists of rock materials. Changes in phase between ice, meltwater, and water vapor over short periods can alter the density of porous snow. The bulk density of saturated porous media, in particular, also varies with fluid pressure and with other stresses, as we will see below.

It is left as an exercise to show that $\rho_s$, $\rho_b$, and porosity $n$ are related by the expression

$$n = 1 - \frac{\rho_b}{\rho_s} \tag{3.2}$$

which is useful for computing $n$ from measurements of $\rho_s$ and $\rho_b$.

## 3.4  FLUID RHEOLOGY AND SHEAR VISCOSITY

### 3.4.1  Newtonian fluids

Consider a Couette experiment: Two vertical, concentric right-circular cylinders have a homogeneous fluid between them (Figure 3.2). Both cylinders may rotate. Suppose that the outer cylinder is forced to rotate at a steady angular velocity. Although the inner cylinder initially is motionless, it soon begins to rotate slowly, then its velocity increases, eventually reaching a steady value. We can infer from

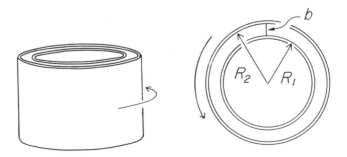

**Figure 3.2**  Couette experiment involving viscous fluid between vertical right-circular cylinders with radii $R_1$ and $R_2$; cylinders may rotate with different angular velocities.

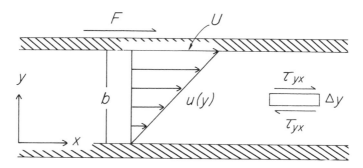

**Figure 3.3**  Shear of viscous fluid between parallel plates with aperture $b$.

this that a tangential stress only, due to the concentric geometry of the problem, is transmitted from the outer rotating cylinder through the fluid to the inner cylinder. (An interesting question arises as to whether the steady angular velocity of the inner cylinder equals that of the outer cylinder. The answer can be inferred several ways, and is left to the reader.)

Now consider a similar experiment involving parallel plates separated by the distance $b$, where one is stationary and the other has a steady velocity $U$ due to a steady tangential force $F$ applied to it (Figure 3.3). Note that this essentially is a small portion of the Couette experiment in the case in which $b$ is much smaller than the cylinder radii. Assume that the no-slip condition exists at the fluid–plate boundaries so that there is a zero streamwise fluid velocity infinitesimally close to the stationary plate, and a maximum fluid velocity equal to $U$ infinitesimally close to the moving plate. Note that $U$ may be considered a rate of strain of the fluid because it equals the maximum linear displacement (deformation) per unit time. Experience suggests that a linear relation exists between the strain rate $U$ and the stress $\tau_{yx} = F/A$ applied to a plate having area $A$. This relation has the form

$$\tau_{yx} = \mu \frac{U}{b} \tag{3.3}$$

As we will see below, the distribution of streamwise velocity $u$ is a linear function of $y$:

$$u = u(y) = \frac{U}{b} y \tag{3.4}$$

Equation (3.3) therefore holds for any small distance $\Delta y$ within the fluid (Figure 3.3). That is,

$$\tau_{yx} = \mu \frac{u(y + \Delta y) - u(y)}{\Delta y} \tag{3.5}$$

Taking the limit as $\Delta y$ goes to zero, we may write

$$\tau_{yx} = \mu \frac{\partial u}{\partial y} = \lim_{\Delta y \to 0} \mu \frac{u(y + \Delta y) - u(y)}{\Delta y} \tag{3.6}$$

In (3.3), (3.5), and (3.6), the constant of proportionality $\mu$ is the *dynamic viscosity*, a measure of the resistance to shearing motion. The subscripts $y$ and $x$, by convention, denote that the stress $\tau$ is acting on fluid faces normal to the $y$ axis, and in the (streamwise) direction of the $x$ axis. In addition, we define a *kinematic viscosity* $\nu$ as the ratio $\mu/\rho$, so that (3.6) becomes

$$\tau_{yx} = \rho\nu\,\frac{\partial u}{\partial y} \tag{3.7}$$

A fluid that behaves according to (3.6) or (3.7) is a Newtonian fluid.

For reasons that will become clear below, and in Chapter 12, it is useful to note that (3.6) and (3.7) also may be written as

$$\tau_{yx} = -\mu\,\frac{\partial u}{\partial y} \tag{3.8}$$

$$\tau_{yx} = -\rho\nu\,\frac{\partial u}{\partial y} \tag{3.9}$$

Now, whereas (3.6) and (3.7) are to be interpreted as the shear stress exerted on a fluid surface at position $y$ by fluid from above (at greater $y$), the latter expressions involving a negative sign are to be interpreted as the shear stress exerted on a fluid surface at position $y$ by fluid from below (at lesser $y$) (Figure 3.3). We will see in the Example Problems 3.7.1 and 3.7.2 that the choice of a positive or negative sign for a given problem depends on whether the velocity component $u$ is increasing or decreasing relative to the positive $y$-coordinate direction. (We will then adopt a specific convention in Chapter 12 when we examine viscous forces.)

Fluid viscosity, and the tangential fluid stresses evident from the Couette experiment, have a simple, molecular kinetic interpretation, at least for gases. Reconsider the Couette experiment where the separation $b$ between cylinders is much smaller than the cylinder radii. Assume that $u = u(y)$, and that fluid velocity components $v$ and $w$ parallel to the $y$-axis and $z$-axis equal zero. Consider an arbitrary, finite plane with area $A$ normal to the $y$-axis within the flow (Figure 3.4). Let $\mathbf{u}_m = \langle u_m, v_m, w_m \rangle$ denote the velocity of an individual molecule with component velocities $u_m$, $v_m$, and $w_m$ in the coordinate directions $x$, $y$, and $z$. Now,

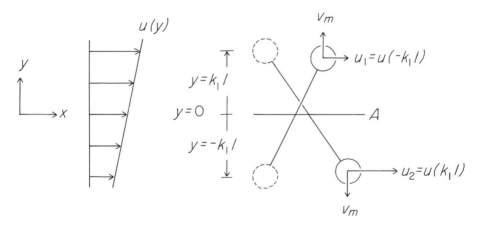

**Figure 3.4**  Arbitrary plane of area $A$ normal to $y$-coordinate axis within Couette flow; molecular momentum is transported upward and downward across plane.

$$u_m = \overline{u_m} + u'_m; \qquad v_m = \overline{v_m} + v'_m; \qquad w_m = \overline{w_m} + w'_m \qquad (3.10)$$

where $\overline{u_m}$, $\overline{v_m}$, and $\overline{w_m}$ are means of the component velocities taken for a large number of molecules, and $u'_m$, $v'_m$, and $w'_m$ are random fluctuations, positive and negative, about mean values. Based on vector addition

$$\mathbf{u}_m = \langle u_m, v_m, w_m \rangle = \langle \overline{u_m} + u'_m, \overline{v_m} + v'_m, \overline{w_m} + w'_m \rangle$$

$$= \langle \overline{u_m}, \overline{v_m}, \overline{w_m} \rangle + \langle u'_m, v'_m, w'_m \rangle \qquad (3.11)$$

The term $\langle \overline{u_m}, \overline{v_m}, \overline{w_m} \rangle$ equals the mean molecular velocity $\overline{\mathbf{u}_m}$; and letting $\mathbf{u}'_m$ denote the vector $\langle u'_m, v'_m, w'_m \rangle$, we may write

$$\mathbf{u}_m = \overline{\mathbf{u}_m} + \mathbf{u}'_m \qquad (3.12)$$

Because the fluid velocity components $v = w = 0$, no *net* transport of mass associated with molecular motions occurs in the coordinate directions $y$ and $z$. This implies that $\overline{v_m} = \overline{w_m} = 0$. Moreover, since $u$ is the only nonzero component of fluid velocity, the $x$-coordinate direction is the only one in which a net transport of mass associated with molecular motions occurs. It follows that the mean molecular velocity $\overline{\mathbf{u}_m}(y)$ at a position $y$ must equal $u(y)$, and we note that

$$\overline{\mathbf{u}_m}(y) = \langle u(y), 0, 0 \rangle \qquad (3.13)$$

In fact, this is the kinetic interpretation of the fluid velocity $u$.

Now, we want a measure of the average velocity fluctuation. By definition $\mathbf{u}'_m = 0$, so this cannot be used. Instead we choose the *root mean square c* of the fluctuations

$$c = \sqrt{\overline{u'^2_m} + \overline{v'^2_m} + \overline{w'^2_m}} \qquad (3.14)$$

Note that $c$ is a positive scalar quantity and is therefore a speed rather than a velocity.

It is plausible that, due to random molecular collisions, the number of molecules $N$ reaching the area $A$ from below ($y < 0$; Figure 3.4) during a small interval of time $dt$ is proportional to molecular density and speed, and we write

$$N = kc\rho_N A \, dt \qquad (3.15)$$

where $k$ is a dimensionless constant. This is intuitively appealing because a high molecular density $\rho_N$ ensures that many molecules will at any instant be in the vicinity of the area $A$, and because a high average molecular speed $c$ increases the likelihood that molecules will cross the area $A$ during the brief interval $dt$. Before arriving at $A$ from below, the $N$ molecules had their last collisions, on average, at a position $y = -k_1 l$, where $l$ is the mean free path, and $k_1$ is a dimensionless constant such that $0 < k_1 < 1$ (Figure 3.4). Assume that the mean molecular velocity $\overline{\mathbf{u}_m}(y) = \langle u(y), 0, 0 \rangle$ is determined by the average position of last collisions. The mean streamwise velocity $\overline{u_{m1}}$ of molecules reaching $A$ from below ($y < 0$) is therefore given by $\overline{u_{m1}} = u_1 = u(-k_1 l)$. Similarly, $N$ molecules reach $A$ from above ($y > 0$) during $dt$ with mean streamwise velocity $u_2 = u(k_1 l)$.

The export of streamwise ($x$-component) momentum across $A$ by an individual molecule moving upward is on average equal to $Mu_1$. The total export of streamwise momentum is $NMu_1$. Similarly, the total export of streamwise momentum across $A$ by molecules moving downward is $NMu_2$. The *net* rate of export of streamwise momentum upward is

$$\frac{NMu_1 - NMu_2}{dt} = \frac{NM(u_1 - u_2)}{dt} = F \qquad (3.16)$$

Note that the quantity $F$ has the dimensions of a force ($\text{M L t}^{-2}$). Substituting (3.15) for $N$

$$kc\rho_N MA(u_1 - u_2) = kc\rho A(u_1 - u_2) = F \qquad (3.17)$$

Here it is useful to observe that

$$\lim_{l \to 0} \frac{u_2 - u_1}{2k_1 l} = -\frac{du}{dy} \qquad (3.18)$$

at $y = 0$ (Figure 3.4). Solving (3.18) for $(u_1 - u_2)$, then substituting the result into (3.17) and dividing by $A$,

$$\tau_{yx} = \frac{F}{A} = -k_2 c\rho l \frac{du}{dy} \qquad (3.19)$$

which is a tangential stress acting on $A$ in the direction of $x$. Thus, $\tau_{yx}$ equals the net flow of momentum in the $x$ direction across a plane located at constant position $y$.

Comparing this with (3.8), evidently

$$\mu = k_2 c\rho l \qquad (3.20)$$

Thus, the dynamic viscosity $\mu$ is proportional to molecular speed $c$, fluid mass density $\rho$, and the mean free path $l$. Substituting the expression (2.1) for $l$,

$$\mu = \frac{k_2 cM}{4\pi \sqrt{2} r^2} \qquad (3.21)$$

This indicates that the dynamic viscosity is proportional to molecular speed $c$ and mass $M$, and inversely proportional to molecular silhouette area $\pi r^2$. Substituting the expression (2.2) for $l$ in (3.20),

$$\mu = \frac{k_2 c\rho}{4\pi \sqrt{2} r^2} \frac{T}{P} \qquad (3.22)$$

This indicates that, for an ideal gas, the dynamic viscosity is proportional to temperature $T$ and inversely proportional to pressure $p$. Due to the complexity of molecular interactions in liquids, kinetic theory unfortunately does not provide analogous insight regarding viscosities of liquids. Whereas $\mu$ generally increases with $T$ for gases, experiments indicate that $\mu$ typically decreases with $T$ for liquids.

A physical explanation of the no-slip condition follows from the kinetic treatment of shear stress and viscosity. During flow, molecules experience collisions with the solid boundary and then are "reflected" back into the fluid. This includes *specular reflection,* in which molecules leaving the boundary after collision have the same mean streamwise velocity as before collision. This essentially describes the case of ideal elastic collisions. Reflection following collision also includes *diffusion reflection,* in which molecules are captured by the electrostatic field of the boundary, then are "kicked off" in random directions by thermal oscillations of the boundary solid. Molecules whose motions are renewed from this state therefore have, on average, zero streamwise momentum as they move back into the fluid. Real reflection is a combination of these, and involves inelastic collisions. The effect, particularly due to diffusion reflection, is that the mean streamwise velocity approaches zero very near the boundary. In addition, we may interpret the shear stress acting between a solid boundary and fluid as the extraction of streamwise momentum from fluid molecules by the boundary during collision.

### 3.4.2 Non-Newtonian fluids

In contrast to a Newtonian fluid, the rheology of a non-Newtonian fluid departs from a simple linear relation between shear stress and strain rate. Let $ds$ denote the small linear displacement associated with a thin layer $dy$ of fluid undergoing strain, such that $u = ds/dt$ (Figure 3.5). Then

$$\frac{du}{dy} = \frac{d}{dy}\left(\frac{ds}{dt}\right) = \frac{d}{dt}\left(\frac{ds}{dy}\right) = \frac{d\varepsilon}{dt} \tag{3.23}$$

Thus, whereas $\varepsilon = ds/dy$ is strain associated with the shear stress $\tau$, $\dot{\varepsilon} = d\varepsilon/dt = du/dy$ is the strain rate, where the subscript $yx$ on $\tau$ has been omitted for simplicity.

Let us now list five, ideal rheological formulae:

$$\varepsilon = \frac{1}{S_m}\tau \tag{3.24}$$

$$\tau = \mu\frac{d\varepsilon}{dt} \tag{3.25}$$

$$\frac{d\varepsilon}{dt} = \frac{1}{S_m}\frac{d\tau}{dt} + \frac{1}{\mu}\tau \tag{3.26}$$

$$\tau = S_m\varepsilon + \mu\frac{d\varepsilon}{dt} \tag{3.27}$$

$$\tau = \tau_0 + \mu\frac{d\varepsilon}{dt}; \qquad \tau > \tau_0 \tag{3.28}$$

Here $S_m$ is an elastic *shear modulus* or *rigidity modulus*, and $\tau_0$ is a *yield stress*. The first of these should be recognized as *Hooke's law* for an elastic substance, where the shear strain is proportional to the shear stress applied. Recall that elastic strain is fully recovered when the stress is removed. The second of these formulae, as we have seen above, defines a viscous Newtonian fluid. The third and fourth, (3.26) and (3.27), thus are linear combinations of elastic and viscous terms. The

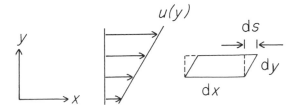

**Figure 3.5** Shear of a thin fluid layer used to define strain rate $\dot{\varepsilon}$.

fifth involves a viscous term, and a yield stress $\tau_0$ below which strain does not occur.

A sense of the differences in the behaviors of substances defined by these formulae can be gained by solving each, assuming that the applied load $\tau$ is some function $\tau(t)$ of time $t$. Let us begin with a simple example. Note that (3.25) may be rearranged to

$$d\varepsilon = \frac{\tau}{\mu} dt \qquad (3.29)$$

A solution to (3.29)—an equation that expresses the amount of strain $\varepsilon$ as a function of time $t$—can be found by integrating both sides:

$$\int d\varepsilon = \frac{1}{\mu} \int \tau(t) \, dt \qquad (3.30)$$

Assuming the load $\tau(t)$ is constant, that is $\tau(t) = \tau_c$, (3.30) then reduces to

$$\int d\varepsilon = \frac{\tau_c}{\mu} \int dt \qquad (3.31)$$

$$\varepsilon = \frac{\tau_c}{\mu} t + C \qquad (3.32)$$

The constant of integration $C$ is determined from the initial condition that when $t = 0$, strain $\varepsilon$ must equal zero. Therefore $C = 0$, and

$$\varepsilon = \frac{\tau_c}{\mu} t \qquad (3.33)$$

Thus, for a constant stress $\tau_c$, the amount of strain of a Newtonian fluid at any time $t$ is proportional to the stress, and is inversely proportional to its viscosity (Figure 3.6).

Now consider the case of a constant stress $\tau_c$ applied to a substance that behaves according to (3.26). We may immediately observe that the derivative $d\tau/dt$ of a constant $\tau_c$ equals zero, so the second term drops out of the formula, which is then identical to that of a Newtonian fluid. It follows that a material that behaves according to (3.26) acts like a Newtonian fluid when the applied stress is constant. As a more interesting example, consider the case in which the stress steadily in-

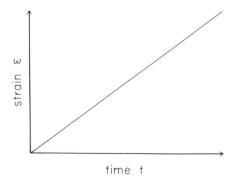

**Figure 3.6**   Strain-time diagram of Newtonian fluid for constant shear stress $\tau_c$.

creases or decreases with time. For example, let $\tau(t) = at + b$, where $a$ is the rate of change of the function $\tau(t)$, that is $a = d\tau/dt$, and $b$ is the initial stress, $b = \tau(0)$, at time $t = 0$. When $a > 0$ the stress $\tau$ increases, and when $a < 0$ it decreases. Substituting this expression into (3.26) and rearranging,

$$d\varepsilon = \left( \frac{a}{S_m} + \frac{b}{\mu} + \frac{a}{\mu}t \right) dt \qquad (3.34)$$

Integrating, then evaluating the constant of integration from the initial condition that strain $\varepsilon$ must equal zero when $t = 0$,

$$\varepsilon = \left( \frac{a}{S_m} + \frac{b}{\mu} \right) t + \frac{a}{2\mu}t^2 \qquad (3.35)$$

Note that at small times $t$, specifically $t < 1$, the effect of squaring $t$ in the last viscous term is to diminish its magnitude relative to the term in $t$. Therefore the strain $\varepsilon$ initially is governed more by the elasticoviscous term in $t$. Further, when $b = 0$, this term in $t$ is a purely elastic one, such that strain initially is chiefly elastic. With increasing time ($t > 1$), however, the magnitude of the last term increases proportionally more than the term in $t$. Therefore the strain $\varepsilon$ is increasingly governed by the viscous term in $t^2$. Note further that when $a > 0$, the viscous term is positive and the fluid deforms at an increasingly positive rate; it becomes easier to strain. When $a < 0$, the viscous term is negative and the fluid deforms at an increasingly negative rate; it becomes more difficult to strain (Figure 3.7). This describes an *elasticoviscous* or *Maxwell substance*.

Consider a substance that behaves according to (3.27). Readers familiar with linear differential equations will recognize this as a first-order equation. Suppose that the applied stress $\tau(t)$ is defined by a step function:

$$\tau(t) = 0, \ t < 0; \qquad \tau(t) = \tau_c, \ t \geq 0 \qquad (3.36)$$

We will note without performing the necessary manipulations that a solution to (3.26) is then given by

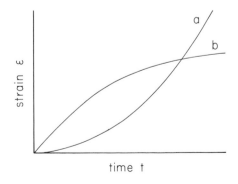

strain $\varepsilon$

time $t$

**Figure 3.7** Strain-time diagram of elasticoviscous substance for steadily increasing (a) and steadily decreasing (b) shear stress $\tau(t)$.

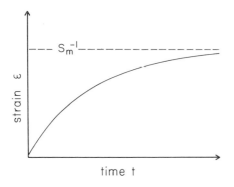

**Figure 3.8**  Strain-time diagram of firmoviscous substance for constant shear stress $\tau_c$.

$$\varepsilon = \frac{1}{S_m}\tau_c - \frac{1}{S_m}\tau_c e^{-S_m t/\mu} \tag{3.37}$$

Observe that strain $\varepsilon$ equals zero when $t = 0$. At large times $t$ the last term approaches zero. Strain $\varepsilon$ thus asymptotically approaches a limiting value equal to $\tau_c/S_m$ (Figure 3.8). Note that this limit is an elastic strain; it is reached via a viscous-like creep whose rate is governed by the second term in (3.37). This describes a *firmoviscous* or *Kelvin substance*.

Finally, consider a substance that behaves according to (3.28). This formula can be rearranged to

$$d\varepsilon = \frac{1}{\mu}(\tau - \tau_0)\,dt \tag{3.38}$$

Integrating, evaluating the constant of integration, and assuming that the applied stress $\tau(t)$ equals a constant $\tau_c$,

$$\varepsilon = \frac{1}{\mu}(\tau_c - \tau_0)t; \qquad \tau_c > \tau_0 \tag{3.39}$$

This has the same form as the solution above for a Newtonian fluid, where strain occurs only when $\tau_c$ is greater than the yield stress $\tau_0$. This describes the behavior of a *plasticoviscous substance,* often referred to as a *Bingham plastic*. Closely related to an ideal Bingham plastic is a *pseudoplastic* whose apparent viscosity decreases continuously with strain rate with no apparent yield stress (Figure 3.9).

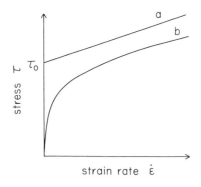

**Figure 3.9**  Stress versus strain-rate diagrams of Bingham plastic (a) and pseudoplastic (b).

Newtonian fluids and plastics exhibit time-independent behavior in the sense that their rate of strain is a single-valued function of the applied stress. The apparent viscosity—the slope of the curve relating stress to strain rate—does not vary with the duration of strain. *Dilatant* fluids, whose apparent viscosities increase with strain rate, also exhibit time-independent behavior. In contrast, certain fluids exhibit time-dependent behavior in the sense that their apparent viscosities depend on the duration of strain as well as the strain rate. *Strain hardening* (rheopectic) behavior characterizes fluids that become increasingly difficult to strain with time under conditions of constant stress. Such behavior is manifest as a decreasing slope of the strain-time diagram of the fluid under conditions of constant stress. In contrast to the simple linear behavior of a Newtonian fluid, the apparent viscosity of the fluid increases. *Strain softening* (thixotropic) behavior characterizes fluids that become easier to strain with time. Such behavior is manifest as an increasing slope of the strain-time diagram of the fluid under conditions of constant stress; the apparent viscosity decreases. The mechanical reasons for these two behaviors vary with the fluid; we will consider below possible explanations for both strain hardening and softening exhibited by ice.

The five rheological formulae listed above have linear forms. That is, strain $\varepsilon$ and its time derivative have only exponents of one, and no terms involve products, in any combination, of strain and its derivative. Other non-Newtonian fluids, however, can exhibit nonlinear rheologies. Let us consider ice, for example, as described by *Glen's law*.

$$\frac{d\varepsilon}{dt} = A\tau^m \qquad (3.40)$$

Here $m$ is a coefficient that can vary with the magnitude of stress; values of $m$ vary experimentally from 1.5 to 4.2, but an average value of 3 normally is adopted in glacier mechanics for the case of polycrystalline ice. The parameter $A$ depends on ice temperature, crystal size and orientation, and concentration of impurities (including gas bubbles and rock debris); values of $A$ vary with the type of experiment used to estimate it. Based on temperature alone (assuming $m = 3$), $A$ varies from about $5.3 \times 10^{-15}(s^{-1} \text{ kPa}^{-3})$ at $0\ °C$ to $3.8 \times 10^{-18}(s^{-1} \text{ kPa}^{-3})$ at $-50\ °C$. Taken in isolation, (3.40) decribes a time-independent behavior. However, the actual rheological behavior of ice under a sustained stress can be complex, depending on stress level as well as strain rate, and depending on whether strain involves individual crystals or polycrystalline ice.

Constant stress tests involving polycrystalline ice exhibit a characteristic strain-time behavior (Figure 3.10). Initial behavior is elastic followed by a transient period of strain hardening (primary creep) where the strain rate decreases. This normally is followed by a period where the strain rate is roughly constant (secondary creep), and then a period of strain softening (tertiary creep) where the strain rate increases. At low values of stress, behavior may remain in the transient, hardening phase for the duration of an experiment. These periods of distinct behavior after the onset of strain vary with the level of stress.

Explanations of this behavior involve *dislocation theory*. Dislocations are defects in a crystal structure that allow local deformation to occur more easily than in a perfect crystal. A dislocation may be regarded as a local plane separating two

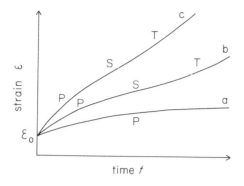

**Figure 3.10**   Strain-time diagram of polycrystalline ice for low (a), moderate (b), and high (c) values of shear stress $\tau$; all cases undergo initial elastic strain $\varepsilon_0$, then exhibit primary P, secondary S, and tertiary T creep stages; adapted from Paterson (1981), by permission of Butterworth-Heinemann Ltd.

parts of a crystal that are slipping tangentially to each other. Strain involves the production and migration of dislocations within crystals. A dislocation may interfere with the migration of others such that dislocations become "concentrated"; the effect is to locally harden the ice. The strain behavior of polycrystalline ice also involves crystal growth and recrystallization. For example, primary hardening is thought to be caused by dislocation hardening and by interference between crystals having different orientations. Tertiary softening involves recrystallization of crystals in orientations favorable to deformation. Secondary creep, then, represents a rough balance between softening in some parts of the ice and hardening in others. The value of $m$ in Glen's law (3.40) is taken to apply to secondary creep. Thus, as seen with this example of ice, the behavior of real substances can be much richer than the simple linear rheologies introduced above.

The rheological behaviors described so far can be observed in laboratory experiments, and therefore are known to be relevant to the behavior of real fluids undergoing strain at the spatial and temporal scales of laboratory experiments. However, many real geological substances, solid and fluid, may undergo strain at much larger spatial scales, and for periods considerably longer than the duration of a typical laboratory experiment. In the context of deformation of crustal materials, S. W. Carey (see Turner and Weiss, 1963) suggests the interesting possibility that these materials may behave essentially like a viscous fluid if they are loaded for a sufficiently long period. Assuming a constant shear stress $\tau_c$, strain is postulated to be described by a formula of the form

$$\varepsilon = \frac{\tau_c}{S_m} + f(\tau_c) + \beta t^{1/3} + \frac{\tau_c}{\mu} t \tag{3.41}$$

The four terms on the right correspond respectively to elastic, plastic, firmoviscous, and viscous behaviors. With long durations of strain, the first three terms become negligible relative to the viscous term, so only viscous behavior remains in the steady state. The time period necessary for this to occur is called the *rheidity* of the material and equals the time required for the viscous term to become a thousand times the elastic term, or about $1000\mu/S_m$ seconds.

## 3.5 COMPRESSIBILITY

### 3.5.1 Fluids

The compressibility $\beta$ of a fluid is a measure of how much its volume or density changes when it is subjected to a change in normal pressure $p$. For a fixed fluid mass associated with volume $V$, the compressibility is defined for isothermal conditions by

$$\beta = -\frac{1}{V}\frac{dV}{dp} = \frac{1}{\rho}\frac{d\rho}{dp} \tag{3.42}$$

The condition $d\rho/dp = 0$ implies that the fluid is incompressible; this also means that $\beta = 0$ and $\rho$ is constant. The reciprocal $(1/\beta)$ of compressibility is the modulus of elasticity $E$:

$$E = -\frac{dp}{dV/V} = \frac{dp}{d\rho/\rho} \tag{3.43}$$

If $\beta$ is independent of pressure, (3.42) can be rearranged to

$$\frac{1}{\rho}d\rho = \beta dp \tag{3.44}$$

In solving this to obtain an equation that expresses $\rho$ as a function of $p$, it is useful to specify the pressure $p$ in terms of a reference pressure $p_0$ and associated density $\rho_0$. These quantities therefore become limits of integration:

$$\int_{\rho_0}^{\rho}\frac{1}{\rho'}d\rho' = \beta\int_{p_0}^{p}dp' \tag{3.45}$$

$$\ln\rho - \ln\rho_0 = \ln\frac{\rho}{\rho_0} = \beta(p - p_0) \tag{3.46}$$

The primes on $\rho$ and $p$ indicate that these are variables of integration in the first step above. Recall that we can eliminate the ln operator by performing the inverse operation:

$$\exp\left(\ln\frac{\rho}{\rho_0}\right) = \exp[\beta(p - p_0)] \tag{3.47}$$

$$\rho = \rho_0 e^{\beta(p-p_0)} \tag{3.48}$$

This suggests that, for isothermal conditions, fluid density exponentially increases above the reference density $\rho_0$ with increasing pressure $(p > p_0)$, and exponentially decreases with decreasing pressure $(p < p_0)$. Note that when $p = p_0$, $\rho = \rho_0$. This *equation of state* describes the behavior of liquids that do not contain significant quantities of gas. We will reexamine this equation in Chapter 4, and note here that it is not an entirely satisfactory expression for the relation between the density and pressure of a liquid.

Under normal circumstances, it is difficult to maintain strict isothermal conditions when the pressure of a fluid is changing, and its density will also depend on associated changes in temperature. On this note, the compressible behaviors of gases in relation to pressure and temperature can be inferred directly from thermo-dynamic considerations. Comparable semiempirical, or wholly empirical, formulae can be obtained for liquids if consideration is restricted to small domains of pressure and temperature. These topics are taken up in the next chapter. Meanwhile, let us consider the assertion made in Chapter 1, that compressibility can be neglected under certain circumstances when treating the flow of gases.

The question to be addressed is whether changes in the density of a gas induced by flow are sufficiently large that the condition which defines the incompressible condition, $d\rho/dp = 0$, is significantly violated. In particular, the flow of a gas can be treated as incompressible if the relative change in density is small, that is $d\rho/\rho \ll 1$. Recall Bernoulli's equation from introductory physics: $p + 1/2\rho u^2 =$ constant (Chapter 10). The change in pressure $dp$ induced by flow is of the order of the *dynamic pressure* $1/2\rho u^2$. Substituting this into (3.43) and rearranging

$$\frac{d\rho}{\rho} \approx \frac{\rho u^2}{2E} \tag{3.49}$$

The speed of sound $c_s = \sqrt{E/\rho}$, so we may rewrite (3.49) as

$$\frac{d\rho}{\rho} \approx \frac{\rho u^2}{2E} \approx \frac{1}{2}\left(\frac{u}{c_s}\right)^2 \tag{3.50}$$

The ratio of the velocity of flow to the speed of sound, $u/c_s$, is the *Mach number M* (Chapter 5). Based on the condition that $d\rho/\rho \ll 1$, it follows that compressibility can be neglected when

$$\frac{1}{2}M^2 \ll 1 \tag{3.51}$$

As a point of reference, the speed of sound in air is about 335 m s$^{-1}$. The relative change in density $d\rho/\rho$ for a flow velocity of 50 m s$^{-1}$ is therefore about 0.01, which is a relatively small value. This implies that many gaseous flows, for example those involving aeolian transport near Earth's surface, can be treated as incompressible.

### 3.5.2 Porous media

The compressibility of a porous medium is defined in a similar way. However, the development is more detailed than that for fluids because deformation of a finite volume of porous medium involves the compressible behaviors of both the solid matrix and the pores, and because changes in volume can be induced by changes in the internal fluid pressure $p$ or an externally imposed stress $\sigma$. This external load can include, for example, the overburden weight of rock and water, or tectonic and seismic loading.

Of particular relevance to flow problems are changes in pore volume and related changes in porosity, because these define the space available for fluids. Changes in pore volume involve interparticle stresses that can elastically deform the particles and intervening pore spaces of the matrix, as well as irreversibly rearrange the particles. These ideas are embodied in the idea of *effective stress,* a concept introduced by Karl Terzaghi in 1925, which has the following simplified explanation.

Consider a horizontal plane of area $A$ embedded within a saturated porous medium, and assume that any changes in the load on $A$ are essentially one-dimensional and vertical. Further assume for simplicity that the medium consists of noncohesive particles whose sizes are sufficiently large that molecular and related interparticle forces may be neglected. The total load on $A$ consists of the weight $W$ of rock, soil, water, and atmosphere overlying this plane. The total stress $\sigma$ on $A$ is therefore equal to $W/A$. (We will obtain an explicit expression for $\sigma$ in terms of $W$ in later chapters.) The plane $A$ intersects both solid and pore regions. In the static case the total stress must be balanced by interparticle stresses and fluid pressure. The total stress $\sigma$ therefore is

$$\sigma = (1 - m)p + m\sigma_s \tag{3.52}$$

where $m = a_s/A$ is the ratio of the vertically projected area $a_s$ of solid–solid contacts between grains to the total area $A$, $(1 - m) = a_p/A$ equals the ratio of the projected area $a_p$ of solid–pore contacts to total area $A$, and $\sigma_s$ is the stress over the solid–solid contact area $a_s$. The value of $m$ generally is very small; that is $m \ll 1$, so $(1 - m)p \approx p$. The product $m\sigma_s$ is referred to as the effective stress and is denoted by $\sigma_e$. Then (3.52) becomes

$$\sigma = \sigma_e + p \tag{3.53}$$

Thus the total stress is the sum of fluid pressure $p$ and the effective stress $\sigma_e$. Then, changes in the effective stress, which lead to changes in porosity, can be induced by altering either the total stress or the fluid pressure, or both. This may be denoted by

$$d\sigma = d\sigma_e + dp \tag{3.54}$$

Suppose that we are concerned with possible changes in the total stress on a volume of porous medium over a period much shorter than a human lifetime. Possible mechanisms that could alter the total load might include, for example, seismic loading, fluctuations in atmospheric pressure, and periodic loading related to Earth and ocean tides. In fact, it is well documented that these factors can produce measurable variations in both fluid pressure and effective stress within aquifers. On a longer time scale, changes in total stress, and therefore fluid pressure and effective stress, can arise from such phenomena as tectonic loading, the addition of overburden during filling of a sedimentary basin, and unloading associated with geomorphic denudation. But let us return to the shorter time scale. Aside from the transient factors listed above, the total load $\sigma$ on a volume of porous medium often can be considered constant. Then, any changes in effective stress arise from changes in fluid pressure. In particular, an increase in fluid pressure, according to (3.54), leads to a decrease in effective stress. Conversely, a decrease in fluid pressure leads to an increase in effective stress such that the matrix of the porous medium must support more of the total load, including its own weight. A vivid example of this is when the fluid pressure within an aquifer is decreased over a period due to pumping at a rate that is greater than the rate of fluid recharge to the aquifer necessary to maintain a steady fluid pressure. An effect of this can be subsidence of the land surface, where the increased effective stress is sufficient to lead to irreversible consolidation and compaction of the aquifer matrix.

In this context, the compressibility $\alpha$ of a porous medium is a measure of how much its volume changes when it is subjected to a change in effective stress $\sigma_e$. The compressibility of a fixed mass of porous medium associated with volume $V$ is defined by

$$\alpha = -\frac{1}{V}\frac{dV}{d\sigma_e} \tag{3.55}$$

Equation (3.55) will be sufficient for our treatments of porous media loading and deformation related to fluid motion. Nonetheless, students should be aware that the problem is actually more complicated in the general case of three-dimensional loading; see Bear (1988). In this regard, the general topic of how geological porous media respond to loading is well-developed. And although this topic is beyond the scope of this text, several general results are worth noting.

Deformation of consolidated sediments and rocks due to changes in effective stress $\sigma_e$ is essentially elastic, particularly if variations in $\sigma_e$ are not great. With unconsolidated materials, however, deformation involves microscopic rearrangements of particles in response to the detailed stress field that develops among them. Deformation therefore is largely inelastic, although some deformation can be elastic and is recoverable during unloading. With unloading, particles may undergo new, small adjustments in response to the new stress field, but their movement is now from positions acquired during initial loading, so they generally do not return to their original positions. Hence, deformation is not only irreversible, but also depends on loading history. In addition, an increase in total load $\sigma$ can in some circumstances lead to an immediate response manifest as an increase in fluid pressure $p$, while $\sigma_e$ initially remains virtually unchanged. This occurs because pressurization of the fluid is essentially immediate, whereas changes in pore geometry may involve a lag time as stress is redistributed among grain contacts. If the fluid pressure then recovers, for example, due to drainage of fluid, $\sigma_e$ then increases, leading to consolidation. In this regard it is important to remember that consolidation is a response to changes in $\sigma_e$ only.

## 3.6  SURFACE TENSION

### 3.6.1  Surface tension as a potential energy

Surface tension arises from intermolecular forces acting at the interface between two or more substances having markedly different physical properties. Here it is useful to recall the macroscopic concepts of *cohesion* and *adhesion:* Cohesion refers to intermolecular forces acting between like molecules of a substance, manifest by the tendency for a substance to hold itself together; adhesion refers to intermolecular forces acting between unlike substances, manifest as the tendency for one substance to cling to the other.

Consider the interface between a liquid and gas, for example, water and air, and the intermolecular forces acting on molecules within the liquid near the interface. These molecules are, of course, thermally agitated and therefore are undergoing rapid changes in their positions within the liquid; nonetheless, imagine that this process is slowed in such a way that we may focus on individual molecules for a brief period. Molecules that are deep (several mean free paths) within the liquid are on average uniformly attracted to surrounding molecules, so there is a zero net balance of cohesive forces (Figure 3.11). In contrast, a molecule "at" the interface is subjected to both cohesive forces from below and adhesive forces from above. In the specific case of water and air, the cohesive forces are much stronger than the

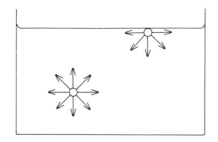

**Figure 3.11** Schematic of balanced, isotropic cohesive forces acting on a liquid molecule "deep" within the liquid, and net imbalance of cohesive and adhesive forces acting on a liquid molecule at the liquid surface; after Miller (1977).

adhesive forces between the two fluids. As a consequence, a net imbalance of forces acting on the molecule occurs (Figure 3.11), suggesting that, in bringing a liquid molecule upward toward the interface, work must be performed in overcoming this net downward force that develops on the molecule, much like the work performed in stretching a small spring. Moreover, like a spring, this work is equivalent to imparting a potential energy to the molecule.

We should be careful not to proceed too far with this simplistic model. Nevertheless, it provides the essential idea that work must be performed in "creating" interfacial area. Consider a line of length d$s$ on the liquid surface (Figure 3.12). Suppose this line is pulled a distance d$l$ from its original position by a force $\mathbf{F}$ and, in doing so, new liquid surface of area d$s$d$l$ is created. Because this involves bringing molecules from within the liquid to the surface, work is performed in overcoming the imbalance of cohesion–adhesion forces. This work— or the potential energy imparted to the liquid molecules in creating this area—is $\mathbf{F}$d$l$. A *coefficient of surface tension* $\sigma$ is then defined as the potential energy per unit area, $\sigma = \mathbf{F}\mathrm{d}l/\mathrm{d}s\mathrm{d}l = \mathbf{F}/\mathrm{d}s$, which equivalently may be considered a contractile force per unit length. (The symbol $\sigma$ for the coefficient of surface tension normally will be used in context so that it should not be confused with "total load" in the preceding section.) Note, then, that the product $\sigma$ d$s$ is a force acting over the length d$s$, normal to d$s$ and parallel to the plane of the surface containing d$s$ (Figure 3.12). The value of $\sigma$ is determined experimentally for different interfaces.

### 3.6.2 Bubble pressure and capillarity

Two interesting phenomena—bubble pressure and capillarity—can be examined using this idea of surface tension. Let us begin by considering a curved interface between two fluids, air and water. In particular, consider a small, interfacial surface segment d$S$ whose orthogonal edges have arc lengths d$s_1$ and d$s_2$ (Figure 3.13). Assume that the segment is sufficiently small that the lengths d$s_1$ and d$s_2$ are essentially circular arcs with constant radii of curvature $R_1$ and $R_2$ over the small

**Figure 3.12** Definition diagram for work performed in creating liquid surface area d$s$d$l$.

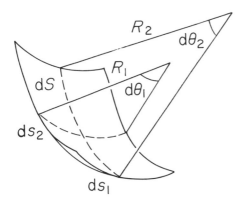

**Figure 3.13** Curved interfacial surface segment $dS$ with orthogonal edges $ds_1$ and $ds_2$ and radii of curvature $R_1$ and $R_2$.

angles $d\theta_1$ and $d\theta_2$. In general, $R_1$ and $R_2$ may be unequal and of either positive or negative sign; here we shall assume that both are positive, so the surface segment $dS$ is everywhere concave toward the air phase. Consider now a view normal to $ds_1$, and observe that a force associated with surface tension, equal to $\sigma ds_2$, acts over each of the edges of arc length $ds_2$, tangential to the arc $ds_1$ (Figure 3.14). These forces can be resolved into horizontal components equal to $\sigma\cos(d\theta_1/2)ds_2$, and vertical components equal to $\sigma\sin(d\theta_1/2)ds_2$. Similarly, in a view normal to $ds_2$, a force equal to $\sigma ds_1$ acts over each of the edges of arc length $ds_1$, tangential to the arc $ds_2$. These forces can be resolved into horizontal components equal to $\sigma\cos(d\theta_2/2)ds_1$, and vertical components equal to $\sigma\sin(d\theta_2/2)ds_1$.

The two horizontal components $\sigma\cos(d\theta_1/2)ds_2$ are equal in magnitude and act in opposite directions, and therefore induce no sense of motion of the interface. Likewise, the two horizontal components $\sigma\cos(d\theta_2/2)ds_1$ are balanced. The vertical components $\sigma\sin(d\theta_1/2)ds_2$ and $\sigma\sin(d\theta_2/2)ds_1$, however, sum to produce a net upward force. In absence of motion of the interface, these vertical forces must be balanced by a force equal in magnitude to their sum and directed downward. This downward force is supplied by a pressure difference across the interface, and is equal to $p_1 ds_1 ds_2 - p_2 ds_1 ds_2$. Thus, for static equilibrium

$$p_2 ds_1 ds_2 - p_1 ds_1 ds_2 + 2\sigma\sin\left(\frac{d\theta_1}{2}\right)ds_2 + 2\sigma\sin\left(\frac{d\theta_2}{2}\right)ds_1 = 0 \quad (3.56)$$

Here it is useful to observe that, because the angles $d\theta_1$ and $d\theta_2$ are assumed to be very small, the sines of these angles are very well approximated by the angles themselves; thus $\sin(d\theta_1/2) \approx d\theta_1/2$ and $\sin(d\theta_2/2) \approx d\theta_2/2$. Moreover note that $d\theta_1 = ds_1/R_1$ and $d\theta_2 = ds_2/R_2$. Substituting these expressions into (3.56), dividing all terms by the area $ds_1 ds_2$, and solving for the pressure difference,

$$\Delta p = p_1 - p_2 = \sigma\left(\frac{1}{R_1} + \frac{1}{R_2}\right) \quad (3.57)$$

Thus, the pressure loss across a curved interface between two fluids is proportional to the surface tension and the sum of the interfacial curvatures $1/R_1$ and $1/R_2$. For the special case of a spherical interface, $R_1 = R_2 = R$, and (3.57) reduces to

$$\Delta p = p_1 - p_2 = \frac{2\sigma}{R} \quad (3.58)$$

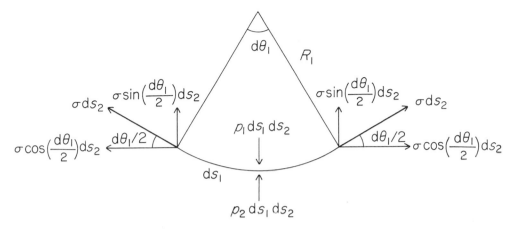

**Figure 3.14** Forces acting on curved interfacial surface segment; view is normal to arc $ds_1$ in Figure 3.13.

Consider now a spherical, gas bubble of radius $R$ that momentarily is in equilibrium at a depth $h$ beneath the surface of a liquid of density $\rho$. We will see in Chapter 6 that the liquid pressure $p_2$ at depth $h$ equals $\rho g h + p_0$, where $p_0$ is atmospheric pressure. The pressure $p_1$ within the bubble therefore is equal to $2\sigma/R + \rho g h + p_0$. As the bubble ascends to a shallower depth $h$, it encounters a lower liquid pressure. Assuming for the moment that no exchange of mass between the liquid and bubble occurs, evidently the pressure $p_1$ within the bubble must decrease, or its radius $R$ must increase, or both, for the bubble to remain in equilibrium. Note, however, that we do not have sufficient information here to fully analyze this problem; although we can calculate the liquid pressure $p_2$, we must specify either the bubble pressure $p_1$ or its radius $R$ to obtain the other.

Suppose that the bubble breaks the liquid surface and, in doing so, drags liquid with it to remain intact as a thin, spherical film bubble within the air, with film thickness $b$. The bubble has two interfaces: one between the air and liquid film, and one between the film and trapped gas. According to (3.58), evidently a pressure difference exists across each interface, which implies that the total difference in pressure between the air and the trapped gas is the sum of the differences across the two interfaces. Let $R$ denote the radius of the inner film; the radius of the outer film therefore equals $R + b$. The pressure loss $\Delta p_g$ across the inner liquid–gas interface is $2\sigma_g/R$, where $\sigma_g$ is the associated surface tension. The pressure loss $\Delta p_a$ across the outer liquid–air interface is $2\sigma_a/(R + b)$, where $\sigma_a$ is the associated surface tension. The total pressure loss $\Delta p$ across the bubble film therefore is

$$\Delta p = p_1 - p_0 = 2\left(\frac{\sigma_g}{R} + \frac{\sigma_a}{R + b}\right) \qquad (3.59)$$

In the case that $R \gg b$, $R + b \approx R$, and this reduces to $\Delta p = 2(\sigma_g + \sigma_a)/R$. Finally, in the case that the trapped gas is air, $\sigma_g = \sigma_a$, and

$$\Delta p = p_1 - p_0 = \frac{4\sigma_a}{R} \qquad (3.60)$$

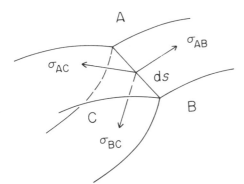

**Figure 3.15**   Forces due to surface tension acting on junction d$s$ among three immiscible fluids A, B, and C.

This demonstrates that the pressure $p_1$ within a bubble, say a soap-film bubble, is greater than atmospheric pressure $p_0$, and that $p_1$ increases with decreasing bubble size.

A related problem concerns the forces due to surface tension acting on a short segment d$s$ of the junction of interfaces between three immiscible fluids A, B, and C (Figure 3.15). Neglecting effects of body forces acting on the three fluids, we may observe that the segment d$s$ is in static equilibrium when the vector sum of the three forces $\sigma_{AB}$d$s$, $\sigma_{AC}$d$s$, and $\sigma_{BC}$d$s$ equals zero. Here $\sigma_{AB}$ denotes the coefficient of surface tension between substances A and B, and so forth for $\sigma_{AC}$ and $\sigma_{BC}$. One can envision that, for given values of surface tension, a static condition requires a specific angular relation among the three interfaces. In the simplest case where $\sigma_{AB} = \sigma_{AC} = \sigma_{BC}$, for example, the angle between any two interfaces must equal $2\pi/3$ radians. If the angular relation among the interfaces is influenced by other factors, for example, when body forces act on the fluids, such an equilibrium may not develop. This occurs when a light oil is placed on a water surface; the junction between the oil, water, and air is unstable, as manifest by the oil spreading over the water surface.

Consider now the capillary meniscus associated with a liquid that has risen to a height $h$ within a small, circular capillary tube of radius $R_c$, and which wets the tube wall with contact angle $\psi$ (Figure 3.16). This wetting, manifest as a curved air–liquid interface, occurs due to adhesion between the liquid and solid surface. The pressure of the liquid at its *free surface,* well outside the capillary tube (point $P_1$), is equal to the atmospheric pressure $p_0$. Likewise the pressure above the meniscus equals $p_0$. Considering point $P_2$, one may be tempted to infer that the pressure here must equal the sum of atmospheric pressure $p_0$ and the pressure due to the weight of the overlying fluid column. Observe, however, that such a state would tend to induce liquid motion from $P_2$ downward out of the tube toward the free surface. So, in fact, the pressure at $P_2$ also must equal the atmospheric pressure $p_0$ for a static condition. (This also implies that the mensicus is not a free surface; this idea is considered further when we treat Hubbert's potential in Chapter 9.)

It follows that a force must exist that is directed upward, and that is equal in magnitude to the weight of the fluid column within the tube. The contact length of the meniscus with the tube is $2\pi R_c$. Consider now a circle on the air–liquid interface that is infinitesimally close to this contact and concentric with it. The forces

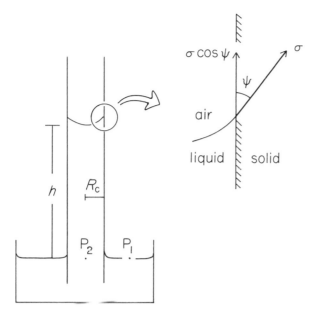

**Figure 3.16** Forces acting on meniscus associated with liquid that has risen to height $h$ within circular capillary tube of radius $R_c$.

associated with surface tension acting on this concentric circle toward and away from the wall of the tube are in equilibrium. In the limit as this circle goes to the contact, the liquid must adhere to the wall with a force per unit contact length equal to $\sigma$ in magnitude, and in the direction of the contact angle $\psi$. The upward component of this force per unit length is $\sigma \cos \psi$ (Figure 3.16), and the total upward force acting over the contact is $2\pi R_c \sigma \cos \psi$. For a static meniscus, then, this force must balance the weight of the fluid column, equal to $\pi R_c^2 h \rho g$ in magnitude. Thus

$$2\pi R_c \sigma \cos \psi - \pi R_c^2 h \rho g = 0 \tag{3.61}$$

Solving for $h$,

$$h = \frac{2\sigma \cos \psi}{\rho R_c g} \tag{3.62}$$

which indicates that the *capillary rise h* is proportional to the surface tension and the cosine of the *wetting angle* $\psi$, and is inversely proportional to the fluid density and the radius of the capillary tube.

Like surface tension, the wetting angle is characteristic of a particular liquid–solid combination, and must be determined experimentally. This angle, for example, is essentially zero for pure water in contact with clean quartz glass and most silicate minerals. A value of $\psi$ greater than $\pi/2$ radians implies that adhesive forces between a liquid and solid surface are weaker than cohesive forces within the liquid. This is referred to as a *nonwetting* or *hydrophobic* condition, in which case a capillary drop occurs. Water in contact with surfaces coated with certain organic compounds (humic substances, for example) can exhibit a nonwetting behavior.

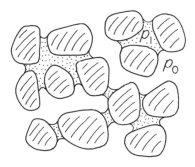

**Figure 3.17**   Capillary menisci within pores and pore throats of unsaturated porous medium.

The curvature of a meniscus within a capillary tube implies that a pressure loss exists across it. The magnitude of this loss is in fact equal to the weight of the fluid column divided by the cross-sectional area of the tube; thus $\Delta p = p_0 - p_1 = \rho g h$. Fluid beneath a meniscus is therefore at a pressure $p_1$ that is less than atmospheric pressure $p_0$, and is thus under tension relative to $p_0$. An important related problem involves unsaturated porous media, where moisture is held by capillarity within pores and pore throats (Figure 3.17). Because the menisci possess curvature, the liquid between them, based on (3.58), is at a pressure that is less than the surrounding (pore) atmospheric pressure. The value of this pressure loss between the nonwetting phase (air) and the wetting phase (water) is referred to as *capillary pressure* $p_c$ or *matric suction* in soil physics; thus $p_c = p_0 - p_1$.

## 3.7 EXAMPLE PROBLEMS

### 3.7.1 Velocity distributions in viscous fluids: Couette, conduit, and free-surface flows

We now have sufficient information to infer some details about flow velocities within fluids. Let us begin with a simple example involving a Newtonian fluid. Consider the experiment described above involving parallel plates separated by the distance $b$, where one is stationary and the other has a steady velocity $U$ due to a tangential force $F$ applied to it (Figure 3.3). Recall that this essentially is a small portion of the Couette experiment in the case where $b$ is much smaller than the cylinder radii. As before, assume that a no-slip condition exists at the fluid–plate boundaries.

The expression defining a Newtonian fluid, (3.6), is a condition that must be satisfied at each position $y$ within the fluid. Rearranging and integrating,

$$\int du = \frac{\tau_{yx}}{\mu} \int dy \qquad (3.63)$$

$$u = \frac{\tau_{yx}}{\mu} y + C \qquad (3.64)$$

The constant of integration $C$ may be determined from the no-slip condition, that is, $u = 0$ at $y = 0$. Thus

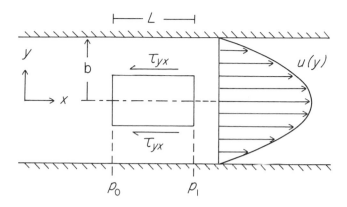

**Figure 3.18**   Steady viscous flow between two parallel, planar boundaries with aperture $2b$.

$$u = \frac{\tau_{yx}}{\mu} y \tag{3.65}$$

which indicates that the velocity is a linear function of $y$, so long as $\tau_{yx}$ is a constant, independent of $y$. Now, we have in (3.3) an experimental expression for $\tau_{yx}$; substituting the right side of this equation into (3.65) then retrieves the anticipated distribution of velocity, (3.4), across the fluid. But what occurs in situations in which $\tau_{yx}$ is not constant over $y$? Let us consider two related problems that bear directly on numerous geological flows, where we can deduce more about the stress $\tau_{yx}$.

Consider two stationary, parallel boundaries between which a Newtonian fluid is steadily flowing in the direction of positive $x$ (Figure 3.18). The no-slip condition exists at each boundary, and from this we may anticipate that the highest streamwise velocity will occur at the center of the conduit. Noting the symmetry of the problem, let us position the origin of the $y$-axis at the center of the conduit, and set its aperture equal to $2b$. Further consider a short length $L$ of this conduit with width $dz$ normal to the plane of the page, and select an arbitrary stream tube, centered within the conduit, having surfaces normal to the $y$-axis at positions $y$ and $-y$. The normal pressure on the inflow face of this stream tube is $p_0$ and that on the outflow face is $p_1$. Therefore the force on the inflow face is $2p_0 y dz$ and that on the outflow face is $2p_1 y dz$; the net force $F$ acting on the fluid in the stream tube due to this pressure difference is $2p_0 y dz - 2p_1 y dz$. Opposing this is a fluid shear force $T$ acting on the stream tube faces normal to the $y$-axis, given by $-2\tau_{yx} L dz$. The negative sign indicates that this force is acting in the direction of negative $x$. Because the flow is steady, the sum $F + T = 0$, whence

$$2\tau_{yx} L \, dz = 2(p_0 - p_1)y \, dz \tag{3.66}$$

$$\tau_{yx} = \frac{p_0 - p_1}{L} y = \frac{\Delta p}{L} y \tag{3.67}$$

Thus $\tau_{yx}$ varies linearly with $y$. By the definition (3.8) of a Newtonian fluid,

$$\frac{du}{dy} = -\frac{1}{\mu}\frac{\Delta p}{L}y \tag{3.68}$$

The negative sign indicates that the velocity $u$ decreases in both the positive and negative $y$-coordinate directions away from $y = 0$. Separating variables and integrating,

$$\int du = -\frac{1}{\mu}\frac{\Delta p}{L}\int y\,dy \tag{3.69}$$

$$u = -\frac{1}{2\mu}\frac{\Delta p}{L}y^2 + C \tag{3.70}$$

The constant of integration $C$ may be determined from the no-slip condition: $u = 0$ at $y = \pm b$, so

$$u = u(y) = \frac{1}{2\mu}\frac{\Delta p}{L}(b^2 - y^2) \tag{3.71}$$

which is parabolic in form. We may immediately compute the maximum velocity $u_{max}$ at the center of the conduit by setting $y = 0$:

$$u_{max} = \frac{b^2}{2\mu}\frac{\Delta p}{L} \tag{3.72}$$

Although intuition tells us that this is correct, it is left as an exercise to formally deduce that $u_{max}$ occurs at the center, as asserted above.

Of special interest in later problems is the volumetric discharge $Q(L^3 t^{-1})$ of fluid for width $dz$. Note first that, due to the symmetry of the velocity profile $u(y)$, we need only to determine the discharge over $0 \leq y < b$, then double this quantity. The discharge through any small flow tube of cross-sectional area $dy\,dz$ is $u(y)\,dy\,dz$. The total discharge over $-b < y < b$ therefore is

$$Q = 2\,dz\int_0^b u(y)\,dy \tag{3.73}$$

$$= \frac{2b^3}{3\mu}\frac{\Delta p}{L}\,dz \tag{3.74}$$

This is the "cubic law" for volumetric discharge through a parallel-wall conduit of uniform aperture $2b$. Finally, we can compute the average velocity $u_{ave}$ by dividing the volumetric discharge $Q$ by the cross-sectional area $2b\,dz$ of the aperture, whence

$$u_{ave} = \frac{b^2}{3\mu}\frac{\Delta p}{L} \tag{3.75}$$

These results can be applied to certain problems of water flow in rock fractures, and magma flow in sills or dikes. Be aware, however, that our development, although giving correct results, is not a general one. We later will consider a more exacting specification of how pressure may vary across the inflow and outflow faces, and the possibility that flow may be driven by forces other than those arising from pressure alone.

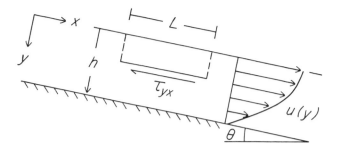

**Figure 3.19** Steady, uniform viscous flow down an inclined plane.

Consider a similar problem involving steady, uniform flow of liquid down a plane inclined at an angle $\theta$ (Figure 3.19). The no-slip condition exists at the lower boundary, and from this we may anticipate that the highest streamwise velocity will occur at or near the liquid surface. It is convenient (although not essential) to position the origin of the $y$-axis at the liquid surface, and let $y$ increase positively downward to a maximum of $h$, the flow depth. Consider a short length $L$ of flow with width $dz$ normal to the plane of the page, and consider an arbitrary plane within the liquid normal to the $y$-axis at position $y$.

Neglecting the weight of the atmosphere above the liquid, the weight $W$ of the liquid above the plane at $y$ is equal to the product of its mass $m$ and acceleration due to gravity $g$. The downslope component $W_x$ of this weight, which induces flow in the direction of $x$, is therefore

$$W_x = mg \sin \theta \tag{3.76}$$

$$= \rho V g \sin \theta \tag{3.77}$$

$$= \rho g \sin \theta \, L y \, dz \tag{3.78}$$

where $V$ is the volume of the liquid above the plane at $y$. Opposing this downslope component of weight $W_x$ is a fluid shear force $T$ acting on the plane, given by $-\tau_{yx} L dz$. The negative sign indicates that this force is acting in the direction of negative $x$. Because the flow is steady, the sum $W_x + T = 0$, whence

$$\tau_{yx} = \rho g \sin \theta \, y \tag{3.79}$$

Thus $\tau_{yx}$ varies linearly with $y$. By the definition (3.8) of a Newtonian fluid,

$$\frac{du}{dy} = -\frac{\rho}{\mu} g \sin \theta \, y \tag{3.80}$$

The negative sign indicates that the velocity $u$ decreases in the positive $y$-coordinate direction. Separating variables and integrating,

$$\int du = -\frac{\rho}{\mu} g \sin \theta \int y \, dy \tag{3.81}$$

$$u = -\frac{\rho}{2\mu} g \sin \theta \, y^2 + C \tag{3.82}$$

The constant of integration $C$ is again determined from the no-slip condition: $u = 0$ at $y = h$, whence

$$u = u(y) = \frac{\rho}{2\mu} g \sin \theta (h^2 - y^2) \tag{3.83}$$

which is parabolic in form. Note that in this development we have assumed the atmosphere exerts zero shear stress on the liquid surface. In addition, one may easily follow the example above for conduit flow to compute the maximum velocity, the discharge, and the average velocity.

Comparing (3.83) with the velocity distribution for conduit flow, we see that the velocity distribution of the free-surface flow essentially consists of half the former. The notable difference is that the quantity responsible for flow in the conduit example, $\Delta p/L$, is replaced by the quantity $\rho g \sin \theta$ in the free-surface flow. The significance of this will become clear when we reexamine these problems in a more formal way. Meanwhile, let us pursue the present line of reasoning and consider a similar problem involving flow of a Bingham plastic.

### 3.7.2 Free-surface flow of a Bingham plastic: nonturbulent lava streams and debris flows

There is evidence that a rock melt can behave like a Bingham plastic when it contains a few percent crystals by volume, which can reflect presence of significant chemical bonding within the melt as well as that manifest in crystalline form. Evidently this bonding together with interactions among crystals, and between crystals and melt, contribute to a finite yield strength below which the melt will not undergo shear strain. For this reason, the velocity profile of a flowing Bingham plastic may depart from the previous results for a purely viscous fluid. Reconsider the example of steady, uniform flow down an inclined plane, where flow involves a Bingham lava. This physically corresponds to flow near the center of a channel whose width is much larger than its depth, such that the channel banks are sufficiently far away that they do not affect velocities near the center.

Following the results of the previous problem, the shear stress $\tau_{yx}$ at position $y$ is $\rho g \sin \theta y$. In addition, the yield stress $\tau_0$ may be expressed as $\rho g \sin \theta y_0$, where $y_0$ denotes a critical depth above which no strain occurs (Figure 3.20). We imme-

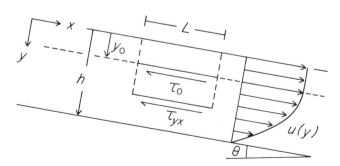

**Figure 3.20**   Steady, uniform flow of Bingham plastic down an inclined plane.

diately may anticipate that this critical depth coincides with the lower boundary of a rigid (yet molten) raft of lava that rides on the underlying deforming fluid. Substituting these quantities into the definition (3.28) of a Bingham plastic and rearranging,

$$\frac{du}{dy} = -\frac{1}{\mu}(\tau_{yx} - \tau_0) \tag{3.84}$$

$$= -\frac{1}{\mu}\rho g \sin \theta (y - y_0) \tag{3.85}$$

As before, the negative sign indicates that the velocity $u$ decreases in the positive $y$-coordinate direction. Separating variables and integrating,

$$\int du = -\frac{\rho}{\mu} g \sin \theta \int y \, dy + \frac{\rho}{\mu} g \sin \theta y_0 \int dy \tag{3.86}$$

$$u = -\frac{\rho}{2\mu} g \sin \theta y^2 + \frac{\rho}{\mu} g \sin \theta y_0 y + C \tag{3.87}$$

The constant of integration $C$ can be determined from the no-slip condition: $u = 0$ at $y = h$, whence

$$u = u(y) = \frac{\rho}{2\mu} g \sin \theta (h^2 - y^2) - \frac{\rho}{\mu} g \sin \theta y_0 (h - y) \tag{3.88}$$

Thus, the velocity profile is parabolic in form for $y > y_0$, where the local velocity $u(y)$ is reduced below that expected for a Newtonian flow by an amount equal to the last term in (3.88). Note that these results can be generalized to simple channel shapes that are not wide relative to their depth, where friction associated with banks becomes important. In addition, similar results have been applied to debris flows, for which there is laboratory and field evidence of Bingham behavior. In this case, the phenomenon of a rigid surface layer riding on the underlying fluid is referred to as "plug flow."

### 3.7.3 Surface tension in unsaturated flow: bubbling pressure

Consider an unconsolidated sand that it is saturated with water. Using what is known as a *pressure plate apparatus,* one can apply the equivalent of a suction to a sample of this type of sand medium. Specifically this means that the pressure within the liquid (water) phase is lowered to a value below the atmospheric pressure within the air-filled pores. (Actually, pressures greater than atmospheric are used in the apparatus.) Suction in this context then refers to the difference between the atmospheric pressure and the pressure within the water, which, as noted in Section 3.6.2, is referred to as capillary pressure or matric suction $p_c$. (Note that the suction produced by this apparatus is entirely analogous to the suction that roots of a plant apply to a soil to extract moisture from it.) Initially the sample retains all its moisture, despite the applied suction. As suction is increased, however, a value is reached where water is extracted from the sample; the sample begins to *dewater.* The water that is extracted is replaced by a second fluid phase, air in this case. As a consequence of the presence of two phases, water–air menisci develop, and the curvatures of these initial menisci reflect the pressure difference—the capillary pressure—between water and air phases. Moreover, due to the fact that a given capillary pressure can be sustained by pores with small radii more easily than pores with large radii, the largest pores begin to dewater first. The capillary pressure at which this initial dewatering of the largest pores occurs is the *bubbling pressure* or *air entry pressure* $p_b$, so called because air bubbles first enter the otherwise saturated medium.

The bubbling pressure is an important parameter in treating flow within unsaturated media. The reason is this: At capillary pressures less than $p_b$, the medium remains saturated, so all pores are potentially capable of conducting water in response to a hydraulic gradient. Once some of the pores are dewatered, however, they cannot conduct water. A consequence of this, as we shall see in Chapter 13, is that the permeability of a porous medium decreases dramatically below its saturated permeability with decreasing moisture content. The bubbling pressure thus characterizes the condition at which this decline in permeability first begins.

Now consider two mixtures of sand, one coarse and one fine, whose grain sizes and shapes are geometrically similar. A condition of similarity implies that one mixture is a scaled version of the other; for each size fraction in one mixture there exists a corresponding fraction in the other mixture representing an equivalent proportion of the whole (Figure 3.21). (This idea of geometrical similarity will be examined in detail in Chapter 5.) A mean grain diameter can be defined for each mixture, say $\bar{d}_1$ and $\bar{d}_2$. Let us now assume that the two mixtures are similarly packed; specifically that they possess the same porosity and bulk density. A consequence of this is that pore shapes and sizes of one mixture are a scaled version of those in the other mixture (Figure 3.21). Moreover, the effective radius $R_m$ of the largest pores in each mixture can be expressed as a constant proportion $k$ of the mean grain diameter. That is, the largest pore radius in mixture 1 is $k\bar{d}_1$, and the largest pore radius in mixture 2 is $k\bar{d}_2$. In general we may write $R_m = k\bar{d}$ without using a subscript to distinguish between mixtures.

With wetting angle $\psi \approx 0$, the pressure loss $p_c$ across the meniscus in a capillary tube of radius $R$ is given by $2\sigma/R$, where $\sigma$ is a coefficient of surface tension. The pores within a sand mixture are essentially small, irregular capillary tubes. We may therefore expect from this simple analogy that, as dewatering begins, the capillary pressure—specifically the bubbling pressure—in the largest pores will be given by $p_b = 2\sigma/R_m$. Substituting $k\bar{d}$ for $R_m$,

$$b_p = \frac{2\sigma}{k\bar{d}} \tag{3.89}$$

Thus, bubbling pressure should be inversely proportional to mean grain diameter if we restrict our attention to geometrically similar grain mixtures.

 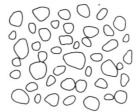

**Figure 3.21**  Schematic of geometrically similar mixtures of sand.

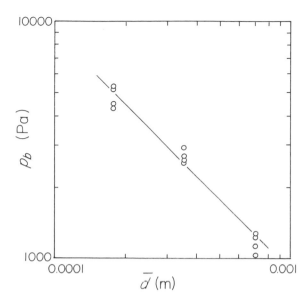

**Figure 3.22** Experimental (circles) and fitted log-linear relation (line) between bubbling pressure $p_b$ and mean grain diameter $\bar{d}$ of geometrically similar mixtures of sand; adapted from Benoit (1992).

Taking logarithms of (3.89),

$$\ln b_p = \ln \frac{2\sigma}{k} - \ln \bar{d} \tag{3.90}$$

$$= C - \ln \bar{d} \tag{3.91}$$

where the constant $C = \ln(2\sigma/k)$. A plot of $\ln p_b$ versus $\ln \bar{d}$ therefore should have a slope of minus one. Data for geometrically similar sand mixtures provided by Toby Benoit illustrate this (Figure 3.22), where the value of the slope estimated by statistical methods is $-1.02$. The vertical variation in $p_b$ at a given value of $\bar{d}$ reflects different variances of grain–size mixtures; each of four values of log-variance is represented by three mixtures, each with a different $\bar{d}$.

## 3.8 READING

Bear, J. 1988. *Dynamics of fluids in porous media*. New York: Dover. This text provides a good summary (pp. 27–38) of mechanical properties of fluids within the context of porous media flow. It also includes a thorough discussion of the concept of effective stress, and examines several approaches to the problem of defining the compressibility of porous media.

Benoit, A. T. 1992. Predicting unsaturated hydraulic conductivity of coarse unconsolidated sand. M.S. thesis, Department of Geology, Florida State University, Tallahassee. This work uses ideas of geometrical similarity of grain mixtures to systematically relate unsaturated conductivity to grain-size distributions.

DeFay, R., Prigogine, I., Bellemans, A., and Everett, D. H. 1966. *Surface tension and adsorption*. New York: Wiley, 432 pp. This text provides a systematic treatment of the effects of surface tension on mechanical and thermodynamical properties of fluids, including drops and bubbles. Chapter 1 provides a very clear explanation of various models for surface tension, adhesion, and contact angle, and of the thermodynamical work involved when a drop or bubble undergoes a change in phase at its interface with a surrounding gas.

Hess, H. H. and Poldervaart, A., eds. 1968. *Basalts: the Poldervaart treatise on rocks of basaltic composition, volumes 1 and 2.* New York: Wiley, 862 pp. Volume 2 contains a chapter by J. E. Nafe and C. L. Drake entitled "Physical properties of rocks of basaltic composition," which is a useful source for such data.

Hsieh, P. A., Bredehoeft, J. D., and Farr, J. M. 1987. Determination of aquifer transmissivity from Earth tide analysis. *Water Resources Research* 23:1824–32. The authors of this paper examine how the water level in an open well tapping an artesian aquifer fluctuates in response to compression and decompression of the aquifer due to Earth tides. They use this response to estimate the hydraulic conductivity of the aquifer.

Huppert, H. E. 1986. The intrusion of fluid mechanics into geology. *Journal of Fluid Mechanics* 173:557–94. This paper provides a general discussion of magma and lava rheology, including the tendency for viscosity to decrease with temperature and increase with $SiO_2$ content. It provides a figure of shear rate versus shear stress based on tests of magma in Makapouhi Lava Lake, Hawaii, and a lava flow on Mount Etna; both possess a finite yield strength. It also discusses, with reference to Marsh (1981), an empirical formula that characterizes how viscosity varies with crystal content: $\mu_p = \mu_m(1 - 1.67x)^{-2.5}$, where $\mu_p$ is the shear viscosity of the melt plus crystals, $\mu_m$ is the shear viscosity of the pure melt, and $x$ is the volume fraction of crystals.

Jaeger, J. C. and Cook, N. G. W. 1976. *Fundamentals of rock mechanics,* 2nd ed. New York: Halsted Press, 585 pp. This text provides a very thorough coverage of the basics of continuum mechanics for rocks.

Johnson, A. M. 1970. *Physical processes in geology.* San Francisco: Freeman, Cooper & Company, 577 pp. This text has a clear description, involving appealing diagrams (pp. 11–22), of various fluid rheologies. It also includes a basic treatment of flow of a Bingham substance in channels, and summarizes initial attempts to measure viscosity of pooled lava.

Kluitenberg, G. A. 1966. Application of the thermodynamics of irreversible processes to continuum mechanics. In Donnelly, R. J., Herman, R., and Prigogine, I., eds. *Non-equilibrium thermodynamics, variational techniques and stability.* Chicago: University of Chicago Press, pp. 91–99. This paper is a summary article of the author's previous work. It demonstrates how classical rheological formula can be derived from principles of irreversible thermodynamics. The material is presented at an advanced level.

Klute, A., ed. 1986. *Methods of soil analysis, part 1: physical and mineralogical methods,* 2nd ed. American Society of Agronomy and Soil Science Society of America, Madison, Wis., No. 9, Part 1. This very thorough handbook includes a description of the pressure cell apparatus used to apply the equivalent of a suction to soil samples, as well as a description of other standard methods of soil physics.

Meyer, R. E. 1971. *Introduction to mathematical fluid dynamics.* New York: Dover, 185 pp. This text outlines (pp. 78–84) the kinetic theory for the viscosity of a gas.

Middleton, G. V. and Wilcock, P. R. 1994. *Mechanics in the earth and environmental sciences.* Cambridge: Cambridge University Press, 459 pp. This text provides a nice introductory treatment of how the mechanical behavior of Earth materials, solid and fluid, bear on a wide breadth of geological processes.

Miller, F., Jr. 1977. *College physics,* 4th ed. New York: Harcourt Brace Jovanovich, 836 pp. This introductory text provides (pp. 281–84) a clear explanation of surface tension and capillary rise.

Neuzil, C. E. 1993. Low fluid pressure within the Pierre Shale: a transient response to erosion. *Water Resources Research* 29:2007–20. This article examines how flow in the low-permeability Pierre Shale is responding to unloading associated with long-term geomorphic erosion. Neuzil points out how the Pierre Shale behaves as a Kelvin substance in tests, but argues on theoretical grounds that it has likely responded elastically during erosion.

Paterson, W. S. B. 1981. *The physics of glaciers,* 2nd ed. Oxford: Pergamon, 380 pp. This text provides a clear and thorough summary of ice rheology and Glen's law.

Rojstaczer, S. 1988. Determination of fluid flow properties from the response of water levels in wells to atmospheric loading. *Water Resources Research* 24:1927–38. Rojstaczer provides in this paper a general theory for how the water level in a well tapping a semiconfined aquifer should fluctuate in response to compression and decompression of the aquifer due to changes in atmospheric pressure.

Terzaghi, K. 1925. *Erdbaumechanik auf bodenphysikalischer grundlage.* Vienna: Franz Deuticke. The concept of effective stress was first introduced by Terzaghi in this book, which translates to: *Mechanics of earthworks based on rudiments of soil physics.*

Turner, F. J. and Weiss, L. E. 1963. *Structural analysis of metamorphic tectonites.* New York: McGraw-Hill, 545 pp. This text outlines (pp. 275–83) the basic rheological behaviors of rocks, including the concept of a rheid.

Whipple, K. X. and Dunne, T. 1992. The influence of debris-flow rheology on fan morphology, Owens Valley, California. *Geological Society of America Bulletin* 105:887–900. The authors use the Bingham model for debris flows to examine their ability to transport sediment on low-angle fans.

Williams, J. and Elder, S. A. 1989. *Fluid physics for oceanographers and physicists: an introduction to incompressible flow.* Oxford: Pergamon, 300 pp. This text includes a clear introductory chapter (pp. 8–20) regarding mechanical properties of fluids.

# CHAPTER 4

# Thermodynamic Properties of Fluids

Fluid behavior in many geological problems is strongly influenced by extant thermal conditions and flow of heat. Recall, for example, that the coefficient $A$ in Glen's law for ice (3.40) varies over three orders of magnitude with a change in temperature of 30 °C. The effect of this is to strongly modulate the rate of ice deformation for a given level of stress. Recall further that we introduced several fluid properties—fluid compressibility, for example—where we asserted that our purely mechanical developments were incomplete inasmuch as they did not treat effects of varying temperature. The reasons for this will become clear in this chapter, including why it is difficult to maintain isothermal conditions when the pressure of a fluid is changing. In addition, many geological problems involve fluid flows that are induced by effects of variations in thermal conditions over time and space. These include buoyancy-driven convective motions that arise from variations in fluid density associated with variations in temperature (Chapter 16). Specific examples include convective overturning in a magma chamber, which can significantly influence how crystallizing minerals are distributed; convective circulations of water and chemical solutions in a sedimentary basin, which can influence where rock materials are dissolved and where they are precipitated as cements within pores; and convective circulation of water within the active layer above seasonally frozen ground, which may influence where patterned ground develops in periglacial environments.

These processes, and viscous flows in general, invariably involve conversions of mechanical energy to heat, or vice versa. So in considering problems involving heat energy, we should recall from introductory chemistry and physics that such conversions can involve work performed on the fluid or its surroundings, and anticipate that the effects of this ought be manifest in fluid behavior. This chapter, then, is concerned with fluid pressure, temperature, and density, and how these variables are related to heat, mechanical energy, and work. We will note in digressions how these macroscopic concepts, like fluid viscosity, often have clear interpretations at a molecular scale based on kinetic theory of matter. Although the emphasis is on fluids, this chapter also examines certain thermodynamic properties of rock, a needed part of understanding interactions between fluids and rocks in certain geological problems. Some of the material covered should be familiar to students from their previous work in chemistry, physics, petrology, and geochemistry. The coverage here of fluid thermodynamics is therefore intended to provide essential material complementary to understanding physical aspects of fluid flows, rather than to provide an exhaustive coverage of this topic.

## 4.1 SPECIFIC HEAT

Consider the quantity of heat $\Delta H$ that must be added to (or extracted from) a substance to produce a given change in its temperature $\Delta T$. Experience along with ideas from the kinetic theory of matter suggest that this quantity $\Delta H$ per unit change in temperature is proportional to the mass $m$ of the substance. That is

$$\frac{\Delta H}{\Delta T} \propto m \tag{4.1}$$

This is intuitively appealing: The thermal (heat) energy of a substance consists of mechanical energy associated with molecular motions. The quantity of heat required to produce a given change in the total energy of these motions, and thus a change in temperature, therefore ought to correlate with the number (mass) of thermally agitated molecules. In a limiting sense as $\Delta T$ approaches zero, we may rewrite (4.1) as

$$\frac{\mathrm{d}H}{\mathrm{d}T} = c(T)m \tag{4.2}$$

Here $c$ is a coefficient of proportionality, which varies with the substance, and in general is a function $c(T)$ of the temperature $T$ of the substance. Note that $c$ also may vary with pressure. The product $cm$ is the *thermal capacity* of a substance of mass $m$, and $c$ is the *specific heat*. SI units of $c$ are J kg$^{-1}$ °C$^{-1}$. Separating (4.2) and integrating,

$$H = \int \mathrm{d}H = m \int_{T_1}^{T_2} c(T)\,\mathrm{d}T \tag{4.3}$$

This describes the total quantity of heat $H$ required to produce a change in temperature from $T_1$ to $T_2$.

Consider now the function $c(T)$. From (4.2) we may write

$$c(T) \approx \frac{1}{m}\frac{\Delta H}{\Delta T} \tag{4.4}$$

As a point of reference, $\Delta H$ equals about 2.035 J for a one-degree change in temperature $\Delta T$ of 1 gram of ice at $-11$ °C, and $\Delta H$ equals about 1.641 J for a one-degree change in temperature at $-60$ °C. Such variations in $\Delta H$ with temperature are reflected by the function $c(T)$. For ice at atmospheric pressure, $c(T)$ varies from a minimum value of about 1,378 J kg$^{-1}$ °C$^{-1}$ at $-100$ °C to a maximum value of about 2,119 J kg$^{-1}$ °C$^{-1}$ at 0 °C (Figure 4.1a). On purely empirical grounds, $c(T)$ is very well approximated by the second-order polynomial

$$c(T)_{\mathrm{ice}} = 2119 + 8.243T + 0.007943\,T^2 \tag{4.5}$$

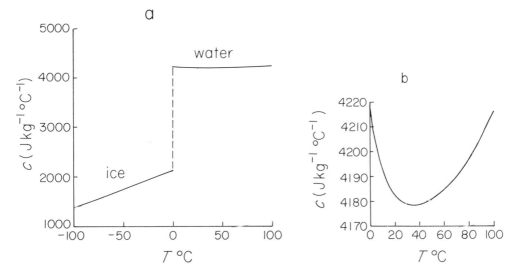

**Figure 4.1** Graph of specific heat $c(T)$ of ice and liquid water as a function of temperature for $-100\,°C \le T \le 100\,°C$ ($a$), and for $0\,°C \le T \le 100\,°C$ ($b$).

over the domain $-100\,°C \le T \le 0\,°C$, and by the simpler linear function

$$c(T)_{ice} = 2119 + 7.883T \tag{4.6}$$

over the domain $-40\,°C \le T \le 0\,°C$.

As a further example, $\Delta H$ in (4.4) equals about 4.187 J for a one-degree change in temperature $\Delta T$ of 1 gram of pure water at 15 °C; $\Delta H$ equals about 4.205 J for a one-degree change in temperature at 90 °C. For liquid water at atmospheric pressure, $c(T)$ exhibits a minimum value of 4,178 J kg$^{-1}$ °C$^{-1}$ at about 35 °C; its maximum value is about 4,218 at 0 °C (Figure 4.1b). The difference between these values is only about 1 percent of the maximum at 0 °C. Thus, $c$ for liquid water over the range 0 °C to 100 °C essentially is constant (Figure 4.1a). In cases such as this (4.3) reduces to

$$H = cm \int_{T_1}^{T_2} dT = cm(T_2 - T_1) = cm\Delta T \tag{4.7}$$

Before considering these formulae further, it is instructive to briefly examine how specific heats of fluids vary with their molecular makeup. Let us begin with gases.

The kinetic theory of gases suggests that the average, molecular kinetic energy of a gas is proportional to its absolute temperature $T$. Adding heat to a gas thus increases the kinetic energy of its molecules. For now, consider the case in which the volume of gas is constant; this obviates the need to take into account mechanical work performed in relation to a change in the volume of the gas. The average translational kinetic energy is $3kT/2$ per molecule, or $3R_0T/2$ per mole, where $k$ is Boltzmann's constant ($k = 1.38 \times 10^{-23}$ J K$^{-1}$) and $R_0$ is the universal gas constant ($R_0 = 8.314$ J mole$^{-1}$ K$^{-1}$). (See Young [1964] for a development of this.) A mole contains a mass equal to the molecular mass $M$. The average, translational kinetic energy per unit mass therefore is $3R_0T/2M$. Considering a unit mass of gas, we therefore must add a quantity of (heat) energy $\Delta H$ equal to $(3R_0/2M)\Delta T$ to raise the temperature of the gas by an amount $\Delta T$, assuming that this added heat takes only the form of an increase in the translational kinetic energy of the gas molecules. It follows from (4.4) that the specific heat at constant volume $c_V$ is

$$c_V = \frac{3}{2}\frac{R_0}{M} \tag{4.8}$$

This result can be simplified if we consider the *molar specific heat* $C_V$ (the specific heat per mole) rather than $c_V$ (the specific heat per unit mass). Then

$$C_V = \frac{3}{2}R_0 \tag{4.9}$$

which suggests the simple result that molar specific heats of gases generally ought to have the same value, equal to 12.47 (J mole$^{-1}$ K$^{-1}$), independent of temperature. Note for later use that $c_V = C_V/M$ from (4.8) and (4.9).

A comparison of this theoretical value with experimental values of $C_V$ for the inert gases, whose molecules are monatomic, is very good (Table 4.1). The theoretical value of 12.471, however, is significantly less than experimental values of $C_V$ for diatomic and more complicated molecules. Evidently one or more additional forms of energy must exist, in addition to translational energy, which consume the added heat in the case of diatomic, and more complex, molecules. That is, a larger molar specific heat indicates that more energy is required to raise the temperature by a given amount; and it must go into some other form, if not into translational kinetic energy. These additional forms of molecular energy include rotational and vibrational energy. Referring to the diatomic molecule in Figure 2.1, rotational kinetic energy is associated with the angular velocity of the molecule about the axis of rotation passing through its center of mass. In this regard, kinetic theory has been refined to predict partitioning of energy between translational and rotational forms, at least for diatomic and some other simple molecules. Historically this involved appealing to the *principle of equipartition of energy,* in which it is assumed that each distinct molecular motion (three translational components and two rotational components) possesses, on average, an equal proportion of the total energy. Further, if one envisions the two atoms in Figure 2.1 as being connected by a small spring representing interatomic forces, additional kinetic and potential energies would be associated with the stretching and compressing of this spring in conjunction with

**TABLE 4.1** Molar Specific Heats of Selected Gases at 25 °C (J mol$^{-1}$ K$^{-1}$)

| Gas | Formula | $C_V^a$ | $C_p^b$ |
|-----|---------|---------|---------|
| Helium | He | 12.45 | 20.77 |
| Neon | Ne | 12.46 | 20.77 |
| Argon | Ar | 12.41 | 20.73 |
| Krypton | Kr | 12.37 | 20.69 |
| Xenon | Xe | 12.45 | 20.77 |

$^a$ Molar specific heat at constant volume.
$^b$ Molar specific heat at constant pressure.

vibration of the two atoms. Unfortunately the principle of equipartition of energy is not universally valid; energies associated with rotation and vibration vary in discrete steps with temperature, for which a complete description requires principles of quantum mechanics.

Several points are noteworthy. Whereas the kinetic theory presented here offers insight to simple molecular systems, a complete understanding of the specific heat of real gases is beyond a kinetic theory based on Newtonian mechanics. Variations in the specific heat (and other thermodynamic properties) of gases evidently depend on transient arrangements of molecules involving intermolecular forces that arise during molecular collisions, in addition to simple molecular kinetics. The same conclusion emerges from kinetic treatments of solids and liquids. Herein arises a very practical reason why relations between specific heat and temperature (and possibly other variables) for real fluids are often described empirically, as in the examples of ice and water above. Nonetheless, the simple ideas from molecular kinetics presented above will provide insight regarding other aspects of the thermodynamic behavior of fluids.

## 4.2 HEAT CONDUCTION

Noticeably absent in the development above is any mention of the *rate* at which heat is transferred to a substance. Recall the example from Chapter 1 involving one-dimensional heat flow through a substance of uniform cross-sectional area $A$ (Figure 1.1; Example Problem 1.4.1). Experience suggests that the rate of heat flow $Q = dH/dt$ is proportional to the area $A$ normal to flow and to the temperature difference $\Delta T$ between the inflow and outflow faces, and inversely proportional to the length $\Delta s$ between these faces. In a limiting sense as $\Delta s$ approaches zero, we may write

$$Q = -K_T(p, T)A\frac{dT}{ds} \tag{4.10}$$

Here $K_T$ is a coefficient of proportionality, the *thermal conductivity*, which varies with the substance, and in general is a function $K_T(p, T)$ of the pressure $p$ and the temperature $T$ of the substance. SI units of $K_T$ are W m$^{-1}$ °C$^{-1}$. The ratio $dT/ds$ denotes a temperature gradient, the local rate of change in temperature $T$ with respect to the coordinate distance $s$. Equation (4.10) is *Fourier's law* for heat conduction in one dimension. Dividing (4.10) by the area $A$ we may write

$$q_T = \frac{Q}{A} = -K_T(p, T)\frac{dT}{ds} \tag{4.11}$$

Here $q_T$ denotes a rate of heat flow per unit area, or *heat flux density*. Note that the negative sign, a conscious choice, indicates flow of heat in the direction of decreasing temperature $T$. This will be examined further below, where (4.11) is generalized to the case of heat flow in more than one dimension.

As a point of reference, the thermal conductivity of water systematically increases with both temperature and pressure, at least over the domains 0 °C to 130 °C and about $1 \times 10^5$ Pa to $8 \times 10^8$ Pa (Figure 4.2). Thermal conductivities

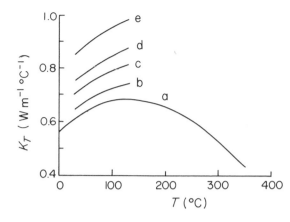

**Figure 4.2** Graph of thermal conductivity $K_T(p, T)$ of liquid water at pressure $p = 1 \times 10^5$ Pa for $0\,°C \leq T \leq 100\,°C$ and $p =$ saturation vapor pressure for $T \geq 100\,°C$ (a), $p = 9.81 \times 10^7$ Pa (b), $p = 2.45 \times 10^8$ Pa (c), $p = 3.92 \times 10^8$ Pa (d), and $p = 7.84 \times 10^8$ Pa (e); data from Clark (1966), Weast (1988).

of rocks and rock melts exhibit similar variations in thermal conductivities. Values of $K_T$, however, do not universally increase or decrease with $T$. For example, experimental measurements indicate that thermal conductivities of granites decrease with increasing temperature, whereas the conductivity of at least one diabase very slightly increases with temperature (Figure 4.3). Effective values of $K_T$ further may vary with porosity and interstitial fluid (Figure 4.4), a topic to be covered later. An important caveat is worth noting here: Discontinuities such as fractures can locally alter the rate of heat flow through rock from that which would occur in absence of the discontinuities. Laboratory tests, however, typically involve intact, homogeneous rock samples that do not contain fractures or other discontinuities. The *effective* thermal conductivity of a large mass of rock in its natural geological setting therefore can differ from values of conductivity estimated in laboratory tests.

The result embodied in Fourier's law, that flow of heat by conduction arises from a temperature gradient, has an appealing, qualitative explanation in kinetic theory. Recall that the thermal (heat) energy of a substance consists of the mechanical energy of its agitated molecules and atoms. Regions whose molecules have, on average, more kinetic energy than molecules in neighboring regions will tend

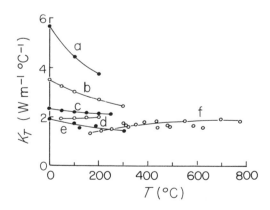

**Figure 4.3** Graph of thermal conductivity $K_T(T)$ of various rock materials: quartz sandstone (a), granite (b), diabase (c), diabase (d), shale (e), and basalt (f); data from Clark (1966), Nafe and Drake (1968), Weast (1988).

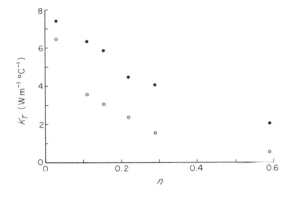

**Figure 4.4** Graph of effective thermal conductivity $K_T$ for six quartz sandstones of varying porosity $n$ saturated with water (closed circles) and air (open circles); data from Clark (1966).

to transfer energy to these neighboring regions with lower molecular energy via molecular collisions. Thus, energy tends to diffuse from regions of higher temperature (and greater molecular agitation) to regions of lower temperature (with less molecular agitation). Further, differences in thermal conductivities of substances reflect varying propensities to diffuse this energy due to differences in atomic structure and molecular arrangement.

The one-dimensional form of Fourier's law (4.11) is sufficient if temperature varies only in one specified coordinate direction $s$. We should, however, allow for the general possibility that temperature varies in any direction within a given coordinate system. Further, we should allow for the possibility that the thermal conductivity of a substance varies with position and orientation. Let us examine these possibilities first in two dimensions, then generalize our results to three, beginning with the idea of a temperature gradient.

Consider a conducting substance within a Cartesian coordinate system, and suppose that temperature varies as a function $T(x, y)$ over this substance. We may consider $x$ and $y$ to be map coordinates, then plot $T$ as a vertical coordinate (Figure 4.5). The function $T(x, y)$ appears as a surface over $x$ and $y$, where we note that contours of equal temperature (isotherms) on this surface project to vertical surfaces of equal temperature within the conducting substance. Now consider an arbitrary position $(x, y)$, and envision determining the direction in which the rate of decrease in $T$ is greatest. This defines the direction of the local gradient in $T$. Further, the rate of change of $T$ in this direction equals the magnitude of this gradient. This direction and magnitude are, borrowing from terminology of structural geology, analogous to the bearing and plunge of the true dip of the surface $T(x, y)$ at position $(x, y)$.

As both direction and magnitude of this gradient in $T$ are important, it may be considered a vector quantity. Associated with this vector is a component describing the rate of change in $T$ with respect to $x$, denoted by $\partial T/\partial x$, and a component describing the rate of change in $T$ with respect to $y$, denoted by $\partial T/\partial y$. The partial notation is used here to denote these as components in the $x$-coordinate and $y$-coordinate directions. The temperature gradient, denoted by $\nabla T$, then may be written in the vector form

$$\nabla T = \left\langle \frac{\partial T}{\partial x}, \frac{\partial T}{\partial y} \right\rangle \tag{4.12}$$

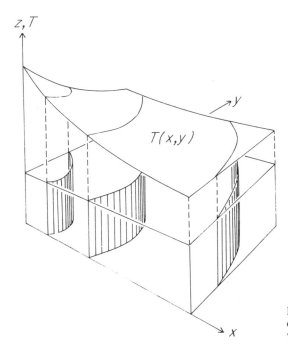

**Figure 4.5** Map projection of two-dimensional temperature field $T(x, y)$ within thermally conducting substance.

It is conventional to write this as

$$\nabla T = \frac{\partial T}{\partial x}\mathbf{i} + \frac{\partial T}{\partial y}\mathbf{j} \tag{4.13}$$

where $\mathbf{i}$ and $\mathbf{j}$ denote unit vectors: $\mathbf{i} = \langle 1, 0 \rangle$ and $\mathbf{j} = \langle 0, 1 \rangle$. To see that these statements are indeed equivalent, we need only to multiply the unit vectors by component gradients in (4.13), then sum:

$$\frac{\partial T}{\partial x}\langle 1, 0 \rangle + \frac{\partial T}{\partial y}\langle 0, 1 \rangle = \left\langle \frac{\partial T}{\partial x}, 0 \right\rangle + \left\langle 0, \frac{\partial T}{\partial y} \right\rangle = \left\langle \frac{\partial T}{\partial x}, \frac{\partial T}{\partial y} \right\rangle = \nabla T \tag{4.14}$$

The unit vectors $\mathbf{i}$ and $\mathbf{j}$ formally indicate the directions of the $x$ and $y$ components of $\nabla T$, and the quantities $\partial T/\partial x$ and $\partial T/\partial y$ determine the magnitude of $\nabla T$. The magnitude of the vector $\nabla T$, denoted by $|\nabla T|$, is given by $|\nabla T| = [(\partial T/\partial x)^2 + (\partial T/\partial y)^2]^{1/2}$.

Returning to (4.13), it is convenient to use the *vector differential operator* $\nabla$ to represent the procedure of taking the derivative of a vector, in the same sense that we use the operator $d/dx$ to represent the procedure of taking the derivative of an ordinary function with respect to $x$. Then in two dimensions,

$$\nabla = \mathbf{i}\frac{\partial}{\partial x} + \mathbf{j}\frac{\partial}{\partial y} \tag{4.15}$$

such that $\nabla T$ gives (4.13). By itself, the operator $\nabla$ has no physical meaning, just as the operator $d/dx$ has no physical meaning. Only when it operates on a variable quantity, for example, as denoted by $\nabla T$, does it take on physical significance. Further, it now should be apparent that, whereas $T$ is a scalar, $\nabla T$ is a vector.

Suppose now that temperature varies as a function $T(x, y, z)$ over three dimensions. A geometrical interpretation similar to that in Figure 4.5 can be envisioned; but now surfaces of equal temperature within the conducting medium, in general, may curve in any direction. Then the vector differential operator $\nabla$ becomes

$$\nabla = \mathbf{i}\frac{\partial}{\partial x} + \mathbf{j}\frac{\partial}{\partial y} + \mathbf{k}\frac{\partial}{\partial z} \tag{4.16}$$

such that

$$\nabla T = \frac{\partial T}{\partial x}\mathbf{i} + \frac{\partial T}{\partial y}\mathbf{j} + \frac{\partial T}{\partial z}\mathbf{k} \tag{4.17}$$

where unit vectors $\mathbf{i} = \langle 1, 0, 0 \rangle$, $\mathbf{j} = \langle 0, 1, 0 \rangle$ and $\mathbf{k} = \langle 0, 0, 1 \rangle$.

Let us now momentarily turn our attention to the thermal conductivity $K_T$, and consider, in turn, fluids and solids. As noted above, thermal conductivities of fluids in general vary with pressure and temperature. It therefore would be convenient if pressure and temperature could be specified, for a given problem, as explicit functions of coordinate position, independently of fluid flow. Then the conductivity of the fluid also could be specified as a function of position. Unfortunately, this generally is not possible, because pressure and temperature are not independent of fluid motion. Rather, pressure and temperature normally are determined, as we shall see later, by simultaneously solving a set of equations that describe, together, the motion and thermal state of a fluid, including an equation of the form $K_T = K_T(p, T)$. Thus $K_T$ is to be associated with the fluid rather than specified as a function of spatial coordinates. Fourier's law then becomes

$$\mathbf{q}_T = \langle q_{Tx}, q_{Ty}, q_{Tz} \rangle = -K_T(p, T)\left\langle \frac{\partial T}{\partial x}, \frac{\partial T}{\partial y}, \frac{\partial T}{\partial z} \right\rangle \tag{4.18}$$

$$\mathbf{q}_T = -K_T(p, T)\nabla T \tag{4.19}$$

where components of the heat flux $\mathbf{q}_T$ (a vector) in coordinate directions $x$, $y$, and $z$ are

$$q_{Tx} = -K_T(p, T)\frac{\partial T}{\partial x}; \qquad q_{Ty} = -K_T(p, T)\frac{\partial T}{\partial y}; \qquad q_{Tz} = -K_T(p, T)\frac{\partial T}{\partial z} \tag{4.20}$$

In general, thermal conductivities of fluids vary more with $T$ than with $p$; and in many situations where $T$ does not vary significantly, $K_T$ fortunately can be assumed to be constant, independent of $p$ and $T$. Then the notation of (4.19) may be simplified to

$$\mathbf{q}_T = -K_T\nabla T \tag{4.21}$$

and the functional notation similarly is removed from the equations (4.20) of component fluxes.

Thermal conductivities of solids also vary with pressure and temperature. But in addition, the conductivity of a solid may vary as an explicit function $K_T(x, y, z)$ of spatial coordinates $x$, $y$, and $z$ due to compositional variations with position—for example mineralogical variations over the extent of a rock unit. Assuming that $K_T$ is negligibly dependent on $p$ and $T$ relative to its dependence on coordinate position associated with composition, Fourier's law becomes

$$\mathbf{q}_T = -K_T(x, y, z)\nabla T \qquad (4.22)$$

When the medium is homogeneous with respect to thermal conductivity, $K_T(x, y, z) = K_T(x + \Delta x, y + \Delta y, z + \Delta z)$, where $\Delta x$, $\Delta y$, and $\Delta z$ are arbitrary distances, and (4.22) takes the form of (4.21).

Conductivities also may vary with orientation at any given position. For example, the thermal conductivity $K_{Tx}$ measured parallel to $x$ may be greater than the conductivity $K_{Tz}$ measured parallel to $z$. This describes a condition of *anisotropy* with respect to thermal conductivity, a topic to which we will return in Chapter 13 when we generalize Fourier's law for arbitrary orientation of the working coordinate axes.

The two forms of Fourier's law, (4.21) and (4.22), will suffice for our purposes. Students interested in further treatments of heat flow, including a full generalization of Fourier's law for anisotropic and heterogeneous conditions with arbitrary orientation of axes, are urged to examine the text on heat conduction by Carslaw and Jaeger. The case of heat flow within a fluid-filled porous medium, where heat conduction occurs in both solid and fluid phases, will be examined in Chapters 9 and 13.

## 4.3 FLUID PHASES

Many substances can exist in several forms, which, despite possessing the same chemical composition, have very different mechanical and thermal properties. Perhaps the most familiar example is water, $H_2O$, which exists as ice, liquid water, and water vapor (steam). In addition, certain substances can possess different states in any one form. Water ice, for example, can exist in six known, different crystalline states; five are denser than liquid water, and the highest pressure form is stable at 190 °C. Helium has two liquid states. (No known substance has more than one gaseous state.) These various states are referred to as *phases*, and the particular phase in which a substance exists depends on extant pressure and temperature conditions, as is typically summarized graphically by a pressure–temperature ($p$–$T$) *phase diagram* (Figure 4.6).

A transition of a substance from one of its phases to another is referred to as a *phase transition* or *change of phase*. A change of phase is accompanied by release or absorption of heat, although no change in temperature occurs. This heat associated with a phase change is *latent heat*, so named because early experimentalists perceived it as a "hidden" quantity of energy inasmuch as they observed no significant change in the temperature of a substance as it changed phases, despite its being continuously heated (or cooled) during the period of change in phase. The transition between solid and liquid phases involves *latent heat of fusion*, also referred to as *latent heat of crystallization* in literature concerning magmatic processes. The transition between liquid and vapor phases involves *latent heat of vaporization*, and the transition between solid and vapor phases involves *latent heat of sublimation*. As a point of reference, the latent heat of fusion of water at atmospheric pressure (0 °C) is about $3.35 \times 10^5$ J kg$^{-1}$, and the latent heat of vaporization (100 °C) is about $2.26 \times 10^6$ J kg$^{-1}$. Note that the term "latent" often is dropped, and that values for latent heats typically are expressed in chemistry texts as joules per mole.

Kinetic theory again provides useful, qualitative insight regarding latent heat and changes in phase. Consider for illustration the transition between liquid and

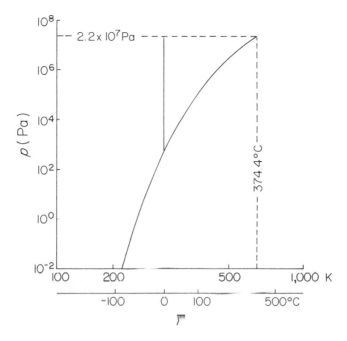

**Figure 4.6** Pressure-temperature diagram of water.

gas phases. In gases, molecular interactions largely consist of chance collisions, and the mean free path (Chapter 2) is sufficiently large that intermolecular forces are unimportant. In contrast, the mean free path in the liquid phase is sufficiently small that intermolecular (attractive) forces are important. To increase the mean free path during a change of phase from liquid to gas, work must be performed on the molecules to overcome these intermolecular forces—which is equivalent to saying that potential energy must be imparted to the molecules in "separating" them. The energy to do this must be supplied from the exterior. Further, as this added energy goes into a potential form rather than into kinetic forms of translation and rotation, no change in the temperature of the gas is observed. (Thus this energy is latent.) Once the change in phase is complete, further added energy goes into kinetic form and the temperature of the gas again increases. Conversely, energy is released during transition from a gaseous to liquid phase. An interesting example illustrating the importance of this release of latent heat involves warming of cold snowpacks in temperate, alpine regions of the Northern Hemisphere during spring and early summer. Snow that is initially below freezing becomes isothermal (0 °C) in part through release of latent heat as percolating meltwater, originating at the surface, refreezes in the deeper, colder part of a snowpack.

The lines in a phase diagram separating the various phases are referred to as *sublimation, fusion,* and *vaporization curves* (Figure 4.6). A substance existing at pressure and temperature conditions that fall on one of these curves is in a state of *phase equilibrium* such that molecules are continuously moving from one phase to the other. Given the distribution of molecular speeds that exists at any instant in a liquid phase, for example, some proportion of molecules has sufficient speed to overcome the attractive forces acting among molecules, and if these molecules are sufficiently close to a liquid–gas interface, they can move into a gaseous state. Likewise, thermally agitated molecules in the gaseous phase are continually colliding with this interface and can move into the liquid state. A phase equilibrium thus implies that the *net* rate of exchange of molecules back and forth between phases is zero. Similar remarks apply to equilibria between solid and liquid phases, and between solid and gaseous phases. The *triple point* is a special case of equilibrium between the solid, liquid, and gaseous phases. The triple point of pure water, for example, occurs at a temperature of 0.0098 °C and a pressure of 610.6 Pa. The *critical point* or *critical state* is where liquid has sufficient energy that physical properties of the liquid and gaseous phases are indistinguishable. No liquid phase can exist at a temperature greater than that associated with the critical point, regardless of pressure. The *critical temperature* of water is 374.4 °C; the associated *critical pressure* is about $2.2 \times 10^7$ Pa.

For given temperature, the pressure of a gas in phase equilibrium with its liquid state is the *vapor pressure* of the liquid, and is characteristic of the substance. Similarly the pressure of a gas in phase equilibrium with its solid state is, for given temperature, the *sublimation pressure*. The sublimation pressure increases with temperature up to the triple point, and the vapor pressure increases with temperature up to the critical point, as is evident by the sublimation and vaporization curves (Figure 4.6). The general form of these curves is examined below. Turning finally to the fusion curve, it is noteworthy that the slope of this curve for most substances is positive. A state of phase equilibrium therefore requires increasing pressure with increasing temperature. The slope of the fusion curve for water, however, is negative (albeit very steep) over ranges of temperature and pressure that coincide with conditions near Earth's surface. This property of water leads to the phenomenon of *pressure melting,* where a transition from the solid to liquid phase, with associated absorption of latent heat, can be induced by an increase in pressure at constant temperature (Example Problem 4.10.4). We will return to the topic of phase transitions in Section 4.6.2.

## 4.4 EQUATIONS OF STATE

An essential part of understanding the behavior of a fluid is knowing how its density varies with other physical and chemical quantities. These quantities usually include the temperature and pressure of the fluid; but it is sometimes of value to also know how the density of a fluid varies with its chemical composition, or an indirect measure of composition, for example, the salinity of a brine. Such a relation is an *equation of state*. Analytical equations of state for most liquids cannot be deduced, and typically are semiempirical or wholly empirical. Gases usually are much simpler, and in many cases their equation of state is very well approximated by the *ideal gas law*.

### 4.4.1 Ideal gases

The ideal gas law, which embodies the relation between the pressure $p$, temperature $T$, and density $\rho$ (or volume $V$) of an ideal gas, normally is expressed as

$$pV = nR_0T \qquad (4.23)$$

where $R_0$ is the universal gas constant equal to 8.314 joules per mole per degree kelvin, and $T$ is expressed as an absolute temperature (K). Noting that the number of moles $n$ may be expressed as $m/M$, where $m$ is the mass of the gas and $M$ is its molecular mass, then

$$pV = mRT \qquad (4.24)$$

and

$$\rho = \frac{m}{V} = \frac{p}{RT} \qquad (4.25)$$

where $R = R_0/M$ is the *specific gas constant*. Two underlying assumptions are that the temperature and pressure are constant throughout the gas volume, and the weight of the gas is sufficiently small that vertical variations in pressure or density associated with gravity are negligible. The ideal gas law well approximates the behavior of many gases at sufficiently high temperatures and low pressures. Although originally empirically based, the ideal gas law has a clear justification in kinetic theory; a good overview is presented by Young (1964). At low temperatures or high pressures, near the liquid-phase boundary, molecular interactions become more complex than the simple elastic collisions that dominate in ideal gases, and deviations from the ideal law can be significant.

### 4.4.2 Real gases

Equations of state for real gases take on several forms, and are normally semiempirical or wholly empirical. Note that the ideal gas law is independent of any specific properties of gases such as molecular size or attractive (cohesive) forces acting among molecules. Yet, we have suggested above that cohesive forces are important in the behavior of real gases, and as we will see now, molecular size is also. An early attempt to take these factors into account, using kinetic theory, is the *van der Waals equation of state* for real gases:

$$\left(p + a\frac{n^2}{V^2}\right)(V - bn) = nRT \qquad (4.26)$$

Here $a$ and $b$ are coefficients that depend on the particular gas (What are their units?), and other symbols have the same meaning as in the ideal gas law. The term involving $a$ accounts for additional pressure related to cohesive forces between molecules; thus $p$ in the ideal gas law is replaced by an "effective pressure" equal to the parenthetical part in (4.26) involving $a$. The term involving $b$ accounts for the idea that individual molecules at any instant between collisions are not free to move anywhere within the total volume. Rather, due to their finite size, molecules can move only into the volume not occupied by other molecules; thus $V$ in the ideal law is replaced by a "reduced volume" equal to the parenthetical part in (4.26) involving $b$. The coefficients $a$ and $b$ in principle are determined theoretically, but in practice experimentally. Their values for a given gas can be obtained from tables of physical and chemical constants. Further, in principle $a$ and $b$ are constants; in actuality, they vary with pressure and temperature. Although the van der Waals equation performs better than the ideal gas law with real gases far from phase boundaries, it performs poorly with gases near phase boundaries.

Other semiempirical formulae are available, some of which are quite precise, at least over restricted ranges of pressure and temperature. As a rule, increased precision translates to an increasing number of coefficients whose values must either be theoretically deduced or experimentally measured. Students interested in examining such formulae should consult standard, physical chemistry texts.

### 4.4.3 Liquids

The complexity of molecular interactions in liquids precludes development of a general analytical equation of state for liquids analogous to the ideal gas law. For this reason "equations" of state are normally *ad hoc* formulae based on curves fitted to experimentally measured values of density as functions of temperature and pressure, and possibly chemical factors. An important guide is to tailor the procedure to the particular fluid and problem of interest. For example, it is often possible to limit the analysis to restricted ranges of $T$ and $p$ that are likely to occur in a given problem, which can lead to a formula that is simpler, and more precise, than one based on wide ranges of $T$ and $p$. Further, for a sufficiently restricted range of pressure, the densities of most liquids vary more with temperature than with pressure such that pressure can be omitted from the analysis. The density of water, for example, varies insignificantly with pressure over the range of pressures normally encountered near Earth's surface. For this reason the density of water often is assumed to be constant, for example, in groundwater flow where temperature does not vary significantly, a point we will examine in Chapter 6.

Let us now develop a particularly useful empirical equation of state that can be used to approximate the behavior of many liquids (as well as gases and solids). Assume that the volume $V$ of a liquid at temperature $T$ and pressure $p$ can be approximated by the sum

$$V = V_0 + \mathrm{d}V_T + \mathrm{d}V_p \tag{4.27}$$

Here $V_0$ is a reference volume whose importance will become clear below; $\mathrm{d}V_T$ is a small change in volume associated with thermal expansion of the liquid at constant pressure; and $\mathrm{d}V_p$ is a small change in volume due to a change in pressure at constant temperature. As terms in temperature and pressure are distinct, we are assuming that thermal and pressure effects are decoupled. Thus a change in temperature, for example, does not lead to a significant change in pressure. The small changes in volume $\mathrm{d}V_T$ and $\mathrm{d}V_p$ associated with small changes in temperature $\mathrm{d}T$ and pressure $\mathrm{d}p$ can be approximated using a Taylor expansion with reference to the volume $V_0$. Without showing the necessary steps:

$$\mathrm{d}V_T = \frac{\partial V}{\partial T}\mathrm{d}T; \qquad \mathrm{d}V_p = \frac{\partial V}{\partial p}\mathrm{d}p \tag{4.28}$$

Qualitatively, the ratio $\partial V/\partial T$ denotes the rate of change in volume with temperature. Multiplying this by the total (small) change in temperature $\mathrm{d}T$ thus gives the total (small) change in volume $\mathrm{d}V_T$. Similar remarks pertain to the ratio $\partial V/\partial p$ and $\mathrm{d}p$. (Students are urged to reexamine this once the idea of a Taylor expansion is introduced in Chapter 6.) Substituting these expressions into (4.28),

$$V = V_0 + \frac{\partial V}{\partial T}\mathrm{d}T + \frac{\partial V}{\partial p}\mathrm{d}p \tag{4.29}$$

Now, let $\alpha$ denote a local isobaric *coefficient of thermal expansion,* defined by

$$\alpha = \frac{1}{V_0}\frac{\partial V}{\partial T} = -\frac{1}{\rho_0}\frac{\partial \rho}{\partial T} \tag{4.30}$$

and let $\beta$ denote a local *isothermal compressibility,* defined by

$$\beta = -\frac{1}{V_0}\frac{\partial V}{\partial p} = \frac{1}{\rho_0}\frac{\partial \rho}{\partial p} \tag{4.31}$$

Note that this latter expression is consistent with our earlier definition of compressibility (3.42). It is, however, a "local" compressibility inasmuch as it is defined in terms of the reference volume $V_0$ and associated density $\rho_0$. Implications of the difference in these definitions of compressibility are examined below. Meanwhile, it is useful to observe from (4.30) and (4.32) that $\partial V/\partial T = \alpha V_0$ and $\partial V/\partial p = -\beta V_0$; substituting into (4.29),

$$V = V_0 + \alpha V_0 dT - \beta V_0 dp \tag{4.32}$$

Assume that $\alpha$ and $\beta$ are constants, independent of $T$ and $p$, at least over sufficiently restricted ranges of $T$ and $p$. Then the small changes $dT$ and $dp$ may be replaced with $dT = T - T_0$ and $dp = p - p_0$, where $T_0$ and $p_0$ denote the temperature and pressure associated with the reference volume $V_0$, and $T$ and $p$ do not fall outside the ranges in temperature and pressure for which $\alpha$ and $\beta$ are constants. Then

$$V = V_0[1 + \alpha(T - T_0) - \beta(p - p_0)] \tag{4.33}$$

This is a linear equation of state for liquids. When $\beta$ is zero (an incompressible fluid), the last term is dropped. Note that $\alpha$ usually is positive; but for water, for example, it is negative below 4 °C and positive above 4 °C. Both coefficients $\alpha$ and $\beta$ are experimentally determined.

Usually it is more useful to have an expression of the form $\rho = \rho(T, p)$, analogous to the version of the ideal gas law involving density (4.25). It is left as an exercise to follow the development above and show that

$$\rho = \rho_0[1 - \alpha(T - T_0) + \beta(p - p_0)] \tag{4.34}$$

Note that the reference density $\rho_0$ is chosen so as to be in the midst of (or close to) the range of densities likely to occur in a problem of interest. This is to ensure that the coefficients $\alpha$ and $\beta$, defined with respect to $\rho_0$, are essentially constants over the associated ranges of $T$ and $p$. (We will examine a rigorous derivation of [4.34] in Chapter 16.) An empirical equation of state for water for large variations in pressure and temperature is examined in Example Problem 4.7.2.

## 4.5 THERMODYNAMIC STATE
## AND THE FIRST LAW OF THERMODYNAMICS

When speaking of a *system,* we normally are referring to a well-defined aggregate of matter. In simple mechanics, for example, we usually think of a system as being composed of a collection of particles or a rigid body. A *thermodynamic system* similarly is a well-defined aggregate of matter. Its boundaries, however, may be real, as in a fluid-filled container, or imaginary, as in a specified control volume embedded within a fluid. A thermodynamic system further may interact with matter outside its boundaries, its surroundings. This interaction takes two forms: exchange of heat and performance of mechanical work. Both should be viewed as two-way processes in the sense that heat may be added or extracted from the system, and work may be performed by the system on its surroundings, or performed by the surroundings on the system.

Just as we specify the state of a mechanical system using a set of coordinates— for example, the Cartesian coordinates and component velocities of a ballistic particle—we specify the state of a thermodynamic system using *thermodynamic coordinates*. These thermodynamic coordinates are macroscopic quantities in that they neglect the details of molecular behavior, yet specify molecular behavior in an average sense. We have seen above, for example, that the average, molecular kinetic energy of a gas is proportional to its temperature, a thermodynamic co-ordinate. In fact, we have thus far used pressure, temperature, volume, density, and mass as thermodynamic coordinates. Except in developing Fourier's law, we further have assumed existence of thermodynamic equilibrium: that the temper-ature is the same throughout the system. (This is referred to as the *zeroth law* of thermodynamics.) If this were not the case, we could not use a single coordinate value of temperature to specify the state of a system.

It also is useful to distinguish between *conservative* and *nonconservative* or *dissipative* systems. A conservative system is one in which energy in one form may take on another form without loss. A mechanical transformation of poten-tial energy to kinetic energy, or vice versa, such that the total mechanical en-ergy remains constant, is a conservative process. A dissipative system is one in which conversion between forms of energy involves an irreversible loss of me-chanical energy into the form of heat. A system in which kinetic energy of motion is partly consumed (heat is generated) through work performed against frictional forces is dissipative. The generated heat cannot be converted by the system back to mechanical energy.

Here we are concerned with the transformation between thermal energy and me-chanical work performed through interaction of a thermodynamic system with its surroundings. We will start by considering a *reversible, quasi-static process*. Such a process is one in which the force involved in performing mechanical work is at all instances infinitesimally greater than the force opposing the performance of this work, such that an infinitesimal increase in the opposing force leads to reversal of the process in the sense that it retraces its previous course in reverse. Further, the system is envisioned as always being infinitesimally close to a state of ther-mal equilibrium, or in a quasi-static state. A reversible, quasi-static process is a conservative one; no real physical systems truly exhibit this behavior.

Because of the conservative property of this reversible, quasi-static process, a general statement of conservation of energy that involves work performed is needed, such that the total energy of the system and its surroundings is conserved. To this end it is operationally convenient to define the internal energy $U$ in the fol-lowing sense: Suppose that a small quantity of heat $dH$ is added to the system in absence of work performed; then the internal energy must increase by a correspond-ing amount, $dU = dH$. Conversely, suppose that the system performs an amount of work $dW$ on its surroundings in absence of any heat exchange; then the internal energy must decrease by a corresponding amount, $dU = -dW$. In the case where both heat is exchanged and work is performed,

$$dU = dH - dW \qquad (4.35)$$

This alternatively may be written $dH = dU + dW$, which has the interpretation that heat added to the system is partitioned between a change in the internal energy and work performed by the system on its surroundings. Conversely, heat extracted from the system must be compensated by a loss in the internal energy and work performed by the surroundings on the system. Note that (4.35) provides a definition of the change in internal energy; it does not provide a way to compute the actual internal energy $U$ of the system. In addition, (4.35) reveals nothing about the details of the processes of heat exchange and work performed; thus the internal energy of this thermodynamic system depends only on its extant state, not on the history of how it derived that state.

## 4.6 ISOBARIC AND ISOTHERMAL PROCESSES

Let us now establish several important relations to be used with (4.35) below by examining a simple thermodynamic system: a gas within a cylinder of cross-sectional area $A$ that is closed at one end and confined by a piston at the other. First, suppose that the gas expands and displaces the piston a small distance $dx$ (Figure 4.7). The amount of work $dW$ performed during this displacement is $dW = Fdx = pAdx$ where $F = pA$ is the force exerted on the piston by the gas with pressure $p$. Noting that $Adx = dV$ is the change in volume of the gas,

$$dW = p\,dV \qquad (4.36)$$

Now, despite the simple geometry of the system, this result actually is a very general result that is independent of the shape of the container. (Why?) Integrating (4.36),

$$W = \int dW = \int_{V_1}^{V_2} p dV \qquad (4.37)$$

which is the total work performed by a gas on its surroundings (in this case the piston) in expanding from volume $V_1$ to $V_2$.

Implicit in (4.37) is the possibility that $p$ is not constant during expansion of the gas. Before considering this, however, suppose that the experiment is performed in such a way that $p$ remains constant, an *isobaric expansion*. Then evaluating the integral in (4.37) gives

$$W = p(V_2 - V_1) = p\Delta V \qquad (4.38)$$

If $p$ is not constant, we must have an expression that describes how it varies with $V$. In the special case of an ideal gas, such an expression is provided by the ideal gas

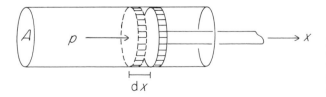

**Figure 4.7** Schematic of closed cylinder-gas system in which gas pressure $p$ performs small amount of work equal to $dW = pAdx = pdV$.

law (4.24). Now, suppose that instead of maintaining constant $p$, the temperature $T$ is held constant, an *isothermal expansion*. Substituting $mRT/V$ for $p$ in (4.37),

$$W = mRT \int_{V_1}^{V_2} \frac{1}{V} dV = mRT \ln \frac{V_2}{V_1} \tag{4.39}$$

To maintain constant $T$, however, requires adding heat to the system during expansion (as we will see below). Experiments indicate that the quantity of heat $dH$ that must be added to maintain isothermal conditions is equal to the work $dW$ performed during expansion. Based on (4.35), evidently $dU = 0$, or $U = $ constant. Thus, the internal energy of the system remains constant during isothermal expansion, which implies that the internal energy of an ideal gas depends only on its temperature and not on other factors such as pressure.

This simple result is not generally valid for real fluids. Nonetheless, certain real gases are sufficiently ideal in their behavior that this result can be applied to them without significant error. In any case, the more general expression (4.37) for calculating work is valid for any thermodynamic system.

### 4.6.1 Specific heat of an ideal gas

Let us now reconsider the specific heat of an ideal gas. Recall that the molar specific heat $C_V$ is defined in terms of isovolumetric conditions. Another useful measure of specific heat is the molar specific heat $C_p$ defined in terms of isobaric conditions, which takes into account the fact that as heat is added to a substance, its expansion involves work performed on the surroundings. For a unit change in temperature $\Delta T$, additional heat therefore must be added to compensate this work. We thus write

$$C_p = C_V + \frac{\Delta W}{\Delta T} \tag{4.40}$$

This indicates that the total heat $\Delta H$ required per mole to produce a unit change in temperature ($\Delta T = 1$) at constant pressure equals the heat required under isovolumetric conditions in absence of work, equal to $C_V$, plus heat required to compensate the quantity of work per mole $\Delta W$. From (4.38), the work performed is $p\Delta V$; substituting for $\Delta W$ in (4.40),

$$C_p = C_V + \frac{p\Delta V}{\Delta T} \tag{4.41}$$

Now, differentiating the ideal gas law (for $n = 1$ mole) with respect to $T$ at constant $p$ gives $pdV = R_0 dT$. Integrating between the limits $V_1$ and $V_2$, and $T_1$ and $T_2$, and comparing with (4.38),

$$W = \int_{V_1}^{V_2} pdV = \int_{T_1}^{T_2} R_0 dT \tag{4.42}$$

$$= p\Delta V = R_0 \Delta T \tag{4.43}$$

Substituting $R_0 \Delta T$ for $p\Delta V$ in (4.41),

$$C_p = C_V + R_0 \tag{4.44}$$

Comparing this with (4.9), evidently the molar specific heat $C_p$ for an ideal gas whose energy involves only translational, molecular kinetic energy is $5R_0/2 = 20.785$, which compares very well with experimental results for inert gases (Table 4.1). Note that $C_p$ is the molar specific heat usually reported in tables of physical constants. Also note for later use that, since $R = R_0/M$, $c_V = C_V/M$ and $c_p = C_p/M$, (4.44) may be expressed in terms of the specific gas constant $R$ and the specific heats $c_V$ and $c_p$ (per unit mass):

$$c_p = c_V + R \tag{4.45}$$

For an incompressible fluid, $c_V = c_p = c$.

## 4.6.2 Phase transitions

Consider now the three transitions of phase: solid to liquid, liquid to gas, and solid to gas. For illustration we will start with the liquid-to-gas transition, then generalize our results to the other two. Envision a cylinder that is closed at one end and confined by a piston at the other. Suppose that the cylinder, having volume $V_l$, is initially filled with exactly one unit mass of liquid at temperature $T$ and confined under a pressure $p$ equal to the vapor pressure of the liquid. Further suppose that a finite quantity of heat $\Delta H$ equal to the latent heat of vaporization is added to the system. As the latent heat of vaporization equals a quantity of heat per unit mass, the unit mass of liquid fully evaporates, and the cylinder volume expands to a volume $V_g$. From (4.38) the amount of work $W$ performed during expansion against the constant confining pressure $p$ is

$$W = p(V_g - V_l) \tag{4.46}$$

Now, using the chain rule to differentiate (4.46) with respect to $p$,

$$\frac{dW}{dp} = (V_g - V_l) + p\left(\frac{dV_g}{dp} - \frac{dV_l}{dp}\right) \tag{4.47}$$

The last term involving derivatives of volumes may be neglected since their magnitudes will in general be much less than the term $(V_g - V_l)$; thus

$$dW = (V_g - V_l)dp \tag{4.48}$$

Based on considerations of the Carnot cycle (which we will not develop here), the amount of work performed $dW$ in association with a reversible addition of heat $\Delta H$ and a change in temperature $dT$ can be expressed as

$$\Delta H = T\frac{dW}{dT} \tag{4.49}$$

Substituting (4.48) for $dW$ leads to

$$\frac{dp}{dT} = \frac{\Delta H}{T(V_g - V_l)} \tag{4.50}$$

This is the *Clapeyron equation*; in this form, $\Delta H$ is the *heat of transition* or latent heat $L$, $T$ is the temperature of transition (K), $V_l$ is the volume of a finite mass in its initial (liquid) phase, $V_g$ is the volume of the same mass in its final (vapor) phase, and $dp/dT$ is the variation in confining pressure with temperature.

More generally, let $V_1$ and $V_2$, respectively, denote specific volumes of the first and second phases of each transition pair (solid to liquid, liquid to gas, solid to gas). Then (4.50) may be written

$$\frac{dp}{dT} = \frac{L}{T(V_2 - V_1)} \tag{4.51}$$

where $L$ is the appropriate latent heat (fusion, vaporization, sublimation). Note, that that the Clapeyron equation gives the local slope of the fusion, vaporization, and sublimation curves.

Consider the vaporization curve further. At temperatures well below the critical temperature $V_l \ll V_g$ and $V_l$ can be neglected. Assuming that the vapor behaves like an ideal gas, the specific volume $V_g$ (per unit mass) equals $RT/p$; substituting this into (4.50) and neglecting $V_l$,

$$\frac{dp}{dT} = \frac{Lp}{RT^2} \tag{4.52}$$

In general, the heat of vaporization $L$ varies with temperature. Let us momentarily assume, however, that it is constant, at least over a small range of $T$. Separating variables then integrating,

$$\ln p = -\frac{L}{RT} + C \tag{4.53}$$

This form of the Clausius–Clapeyron equation predicts that the logarithm of the vapor pressure $p$ varies as a linear function of the reciprocal of the absolute temperature $T$, with slope equal to $-L/R$. This expression also is valid with regard to the sublimation curve, where $p$ then denotes the sublimation pressure. If $L$ is a known function of $T$, this can be incorporated into the integration above.

## 4.7 ADIABATIC PROCESSES

Consider a column of fluid that is heated at its base. Examples include the atmosphere just above a lake or land surface, and heating of water or of a gas. If heating is uneven, it is possible for a parcel of fluid that is ... its surroundings to begin rising due to buoyancy. This describes auto... a process we will examine later (Chapter 6). Once the parcel leaves the ... heat— the ground surface in the case of a heated air parcel—it becomes ... from this source and absorbs no more heat. Heat diffuses from the ... cooler surroundings. However, if this transfer of heat is dominated ... such that the heat flux is slow relative to the rate at which the parcel ... buoyancy, the parcel can be treated to a good approximation as an ... at least for a brief period. As the parcel ascends, it encounters ... pressure. As a consequence it expands and displaces this surroun... is the same as performing work on the surrounding fluid. Let us ... law of thermodynamics to this problem.

Writing (4.35) in the form

$$dH = dU + dW \qquad (4.54)$$

this law indicates that a small quantity of heat $dH$ added to a fluid parcel is partitioned between a change in its internal energy $dU$ and work performed by expansion $dW$. For a system in which heat is neither added nor extracted, $dH = 0$ in (4.54), and thus $dU = -dW$. An isolated system that performs a small quantity of work $dW$ on its surroundings therefore loses internal energy manifest as a decrease in its temperature $dT$. For a parcel of mass $m$ the change in internal energy therefore is equal to $dU = c_V m dT$ according to (4.2). The small amount of work $dW$ is given by $dW = pdV$ according to (4.36). Noting from the ideal gas law that $p = mRT/V$ or $dW = pdV = (mRT/V)dV$, substitution into (4.54) with $dH = 0$ gives

$$c_V m\, dT = -\frac{mRT}{V} dV \qquad (4.55)$$

$$\frac{1}{T} dT = -\frac{R}{c_V} \frac{1}{V} dV \qquad (4.56)$$

From (4.45), $R = c_p - c_V$, so $R/c_V = \gamma - 1$, where $\gamma = c_p/c_V$. Substituting into (4.56),

$$\frac{1}{T} dT = -(\gamma - 1)\frac{1}{V} dV \qquad (4.57)$$

Integrating this expression, then rearranging using properties of logarithms,

$$\ln T = -(\gamma - 1)\ln V + C \qquad (4.58)$$

$$TV^{\gamma-1} = \text{constant} \qquad (4.59)$$

This indicates that the product $TV^{\gamma-1}$ is constant at all instants of an *adiabatic process*; as the volume of a parcel of gas of finite mass increases with expansion, its temperature correspondingly decreases. Alternatively, integrating (4.57) between initial values $T_0$ and $V_0$ and arbitrary values $T_1$ and $V_1$ obtains

$$T_1 V_1^{\gamma-1} = T_0 V_0^{\gamma-1} \qquad (4.60)$$

Applying the chain rule to the ideal gas law, $d(pV) = pdV + Vdp = mRdT$. It is left as an exercise to use this expression to obtain:

$$p\rho^{-\gamma} = \text{constant} \qquad (4.61)$$

$$p_1 \rho_1^{-\gamma} = p_0 \rho_0^{-\gamma} \qquad (4.62)$$

which express the relation between density and pressure during adiabatic expansion or compression of an ideal gas.

Let $S$ denote the entropy of a thermodynamic system, then recall from introductory physics that an infinitesimal change in the entropy $dS$ of a system at temperature $T$ is given by $dS = dH/T$. Since $dH = 0$ for an adiabatic process, $dS = 0$; thus an adiabatic process is also an *isentropic process* (zero change in entropy).

## 4.8 COMPRESSIBILITY AND THERMAL EXPANSION

Recall the definition (3.42) for the isothermal compressibility $\beta$ of a fluid:

$$\beta = \frac{1}{\rho} \frac{d\rho}{dp} \tag{4.63}$$

For constant temperature, differentiating the ideal gas law (4.25) with respect to pressure gives $d\rho/dp = 1/RT$. Substituting this and the ideal law for $\rho$ into (4.63) then gives

$$\beta = \frac{1}{p} \tag{4.64}$$

Thus the isothermal compressibility of an ideal gas is inversely proportional to its pressure. An ideal gas under isothermal conditions therefore becomes more difficult to compress with increasing pressure in the sense that, with increasing pressure, a greater change in pressure is required to produce a given proportional change in the density.

Assuming an isentropic process, (4.61) may be rearranged to $\rho = kp^{1/\gamma}$, where $k$ is a constant. Differentiating this with respect to pressure gives $d\rho/dp = (k/p)p^{1/\gamma-1}$. Substituting these expressions into (4.63) then gives

$$\beta = \frac{1}{\gamma p} \tag{4.65}$$

Thus the compressibility of an ideal gas undergoing isentropic compression or expansion also is inversely proportional to its pressure.

Consider a general definition for the isobaric coefficient of thermal expansion $\alpha$:

$$\alpha = \frac{1}{V} \frac{\partial V}{\partial T} = -\frac{1}{\rho} \frac{\partial \rho}{\partial T} \tag{4.66}$$

For constant pressure, differentiating the ideal gas law (4.25) with respect to temperature gives $d\rho/dT = -p/RT^2$. Substituting this and the ideal law for $\rho$ into (4.66) then gives

$$\alpha = \frac{1}{T} \tag{4.67}$$

Thus the coefficient of thermal expansion of an ideal gas defined for constant pressure is inversely proportional to its temperature.

Consider, now, the two definitions, (3.42) and (4.31), for the compressibility of a liquid. For constant compressibility $\beta$, expression (3.42) suggests that density varies exponentially with pressure, as indicated by (3.48), since the proportional change in density for a given change in pressure is constant. The expression (4.31) for local compressibility, in contrast, suggests that density varies linearly with pressure. The latter definition (3.42) for local compressibility is therefore consistent with the intuitively appealing idea that a liquid becomes increasingly difficult to compress (in a proportional sense) with increasing density (and pressure).

## 4.9 BULK VISCOSITY

Consider a fluid element in the shape of a sphere. Suppose that the element is uniformly subjected to a normal stress. In absence of motion, this stress is equal in magnitude to the thermodynamic pressure $p$ of fluid within the element. Suppose that the stress is slowly increased in a quasi-static, reversible manner. Work is performed on the fluid element and, according to the first law of thermodynamics (4.35), this work must involve an increase in the internal energy of the fluid. Recalling the kinetic interpretation of thermal and internal energy (Sections 4.1 and 4.3), the internal energy consists of several modes of molecular motion: translational, rotational, and vibrational. The mechanical work performed on the fluid element during compression therefore is transferred to these modes of molecular motion. With slow compression, this energy is readily partitioned among all possible modes compatible with the equilibrium (thermodynamic) pressure state. From (4.36) the rate at which work is performed is given by

$$\frac{dW}{dt} = p\frac{dV}{dt} \tag{4.68}$$

Suppose, now, that the fluid element is rapidly compressed. The mechanical work performed is instantaneously transferred to the translational mode of molecular motion. But if the rate of compression is sufficiently rapid, a lag occurs as this increasing translational energy is repartitioned among other possible modes compatible with the equilibrium pressure state that otherwise would exist for the same quantity of energy added under quasi-static conditions. (This process of repartitioning is referred to as *relaxation.*) As the translational mode is responsible for exerting a pressure that opposes the normal stress, this pressure is momentarily higher (until energy is repartitioned among nontranslational modes) than would occur in quasi-static compression. The effect is that more energy is consumed— that is, more work is performed—due to the additional resistance to compression supplied by this momentarily elevated pressure, beyond the resistance that otherwise would be provided by the thermodynamic pressure during quasi-static compression. Moreover, this effect, in principle, occurs with isothermal compression and decompression. It therefore may be treated as an apparent viscous effect and characterized in terms of a *bulk viscosity* $\mu_b$. Further note that this effect vanishes with a monatomic gas, because, in this case, the only mode of molecular energy is translational.

The bulk viscosity, then, is a property that characterizes energy dissipation within a fluid undergoing isothermal volumetric strain at a constant rate, analogous to the viscosity $\mu$ for the case of shear strain. The question arises as to whether the bulk viscosity of a given fluid is sufficiently large that it must be included within mathematical treatments of flow. The bulk viscosity, based on kinetic theory, is essentially zero for low-density gases; and it evidently is nonzero, but very small, for dense gases and liquids, although the theory necessary to fully clarify this is not available. Moreover, because isothermal conditions are virtually impossible to maintain during compression (or expansion) of fluids, experiments to measure bulk viscosity are of limited help. In any case, the idea that bulk viscosity normally can be neglected in mathematical analyses of flow, at least for Newtonian fluids, is borne out by a long history of experimental results. We will return to this topic in Chapter 12.

## 4.10 EXAMPLE PROBLEMS

### 4.10.1 Change in ice temperature with steady heat inflow

Recall that the coefficient $A$ in Glen's law (3.40) depends on ice temperature $T$. An important part of treating ice flow in glaciological studies, therefore, is to consider how the temperature of glacier ice may vary with time due to solar and geothermal heating. This problem actually is a complex one; it involves describing variations in heat flow at the glacier surface and base, and characterizing heat flow within the glacier (Example Problem 9.3.2). In addition, it is possible that heat is generated within the glacier due to ice deformation. (We will examine viscous generation of heat in Chapter 12.) Let us learn something of this complexity by selecting a simplified part of the problem.

Consider the change in temperature produced by a constant heat flux $Q$ into a small mass of ice that is otherwise closed to exchange of heat with its surroundings. Assume that the mass $m$ of the ice is sufficiently small that heating over its entirety is essentially instantaneous; no significant time delay is involved due to heat flow within $m$. Using (4.2),

$$Q = \frac{dH}{dt} = mc(T)\frac{dT}{dt} \tag{4.69}$$

Suppose that the initial temperature $T_0$ of the ice is $-40\,°C$; this value is certainly possible for ice in polar glaciers. Substituting (4.6) for $c(T)$ in (4.69) and rearranging,

$$(2119 + 7.883\ T)\frac{dT}{dt} = \frac{Q}{m} \tag{4.70}$$

Separating variables and integrating,

$$\int (2119 + 7.883\ T)dT = \frac{Q}{m}\int dt \tag{4.71}$$

$$2119\ T + \frac{7.883}{2}T^2 = \frac{Q}{m}t + C \tag{4.72}$$

The constant of integration $C$ is evaluated using the initial condition that $T_0 = -40\,°C$ at $t = 0$. Thus, to four significant digits,

$$2119\ T + 3.942\ T^2 = \frac{Q}{m}t - 78450 \tag{4.73}$$

which is an *implicit solution* for $T(t)$. This can be rearranged in the form

$$aT^2 + bT + c = 0 \tag{4.74}$$

where $a = 3.942$, $b = 2119$, and $c = 78450 - Qt/m$. Applying the quadratic formula,

$$T = \frac{-b \pm \sqrt{b^2 - 4ac}}{2a} \tag{4.75}$$

Choosing the positive square root term (why?), and simplifying,

$$T = -268.8 + 0.1269\sqrt{3253000 + 15.77\frac{Q}{m}t} \qquad (4.76)$$

Thus, temperature $T$ increases with the square root of time $t$ for steady heat flow $Q$. This is intuitively appealing: since the specific heat $c$ of ice increases with temperature, a change in temperature $\Delta T$ for a given quantity of heat $\Delta H$ input to the ice decreases with increasing temperature. The temperature of the ice increases with time, so the rate at which it changes must therefore decrease. Note that the solution (4.76) is relevant only up to a temperature $T_m$ equal to the *pressure melting point;* for example $T_m = 0\,°C$ at atmospheric pressure. Suppose now that $Q = 1 \times 10^{-5}$ W. It is left as an exercise to determine how long it would take to heat 0.001 kg of ice at atmospheric pressure from $-40\,°C$ to $0\,°C$.

A similar problem arises in hydrological studies involving solar heating of a water body, such as a lake. This is important because the thermal energy of the water governs, in part, the rate of evaporation. It is left as an exercise to determine a formula for the change in temperature produced by a constant rate of heat flow $Q$ into a small mass of water that is otherwise closed to exchange of heat with its surroundings. Assume, as above, that the mass $m$ of the water is sufficiently small that heating over its entirety is essentially instantaneous. Further assume that the specific heat $c$ is constant.

### 4.10.2 An empirical equation of state for water

Sometimes it is possible to obtain approximate equations of state for compressible liquids for use in problems where effects of temperature and pressure on density must be considered. Such formulae typically are semiempirical or wholly empirical, and there is no universally accepted way to derive them. A strong guide is to devise a formula that is particularly suited to its eventual use; sometimes a purely empirical formulation may suffice.

Consider the case of water. Experimental measurements of the density of water at different temperatures and pressures are available from a variety of sources. Such data indicate that density varies significantly over the temperature range of $0\,°C$ to $350\,°C$, and slightly over the pressure range of $1 \times 10^5$ Pa to $2 \times 10^7$ Pa; the variation in density with pressure increases with temperature (Figure 4.8). These temperatures and pressures cover a wide range of conditions at and near Earth's surface, and, in particular, they cover conditions associated with thermal geysers, to be examined in the next Example Problem.

Now, we desire a formula that gives density $\rho$ as a function $\rho(p, T)$ of pressure $p$ and temperature $T$. Perhaps the most straightforward approach is to use multiple-regression techniques to empirically fit a function to the data, where the function $\rho(p, T)$ is envisioned as a surface over the $T$ and $p$ coordinates. However, we will adopt a slightly different approach to this problem.

Note that the data for $p = 2 \times 10^7$ Pa cover the full range in $T$; at lower pressures the data do not extend beyond values of $T$ at which water boils (Figure 4.8). Further note that for given $T$, $\rho$ increases linearly with $p$, at least for data between $20\,°C$ and $100\,°C$ (Figure 4.9). (Is this consistent with the equation of state [4.34]?) Assume that this linearity holds for higher $T(1 \times 10^7$ Pa $< p < 2 \times 10^7$ Pa); then

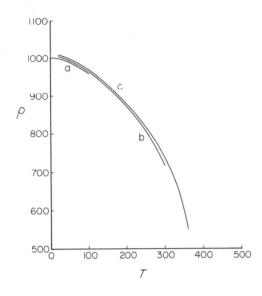

**Figure 4.8** Graph of density $\rho$ of liquid water as a function $\rho(T)$ of temperature at pressures $p = 1.013 \times 10^5$ Pa (a), $p = 1 \times 10^7$ Pa (b), and $p = 2 \times 10^7$ Pa (c).

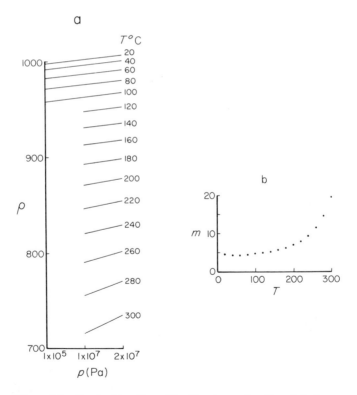

**Figure 4.9** Graph of density $\rho$ of liquid water as a function $\rho(p)$ of pressure for temperatures 20 °C to 300 °C (*a*), and plot of slopes *m* of curves in (*a*) as a function $m(T)$ of temperature (*b*).

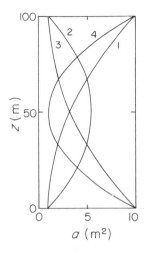

**Figure 4.10** Schematic diagram of hypsometric functions $a(z)$; numbered cases are described in text.

consider a plot of the slopes $m$ of the lines relating $\rho$ to $p$ as a function $m(T)$ of $T$ (Figure 4.10). Those familiar with multiple regression will know that one can fit a curve to such data with a polynomial, and improve the fit by increasing the order of the polynomial. In fact, a smooth curve through the data of $m(T)$ is very well approximated by a fifth-order polynomial:

$$m = m(T) = a_0 + a_1 T + a_2 T^2 + a_3 T^3 + a_4 T^4 + a_5 T^5 \qquad (4.77)$$

where $a_0 = 4.8863 \times 10^{-7}$, $a_1 = -1.6528 \times 10^{-9}$, $a_2 = 1.8621 \times 10^{-12}$, $a_3 = 2.4266 \times 10^{-13}$, $a_4 = -1.5996 \times 10^{-15}$, and $a_5 = 3.3703 \times 10^{-18}$. Similarly, a curve can be fit to the data relating $\rho$ to $T$ for $p = 2 \times 10^7$ Pa (Figure 4.8). This curve also is very well approximated by a fifth-order polynomial:

$$\rho(p,T) = \rho(2 \times 10^7, T) = b_0 + b_1 T + b_2 T^2 + b_3 T^3 + b_4 T^4 + b_5 T^5 \qquad (4.78)$$

where $b_0 = 1.0213 \times 10^3$, $b_1 = -7.7377 \times 10^{-1}$, $b_2 = 8.7696 \times 10^{-3}$, $b_3 = -9.2118 \times 10^{-5}$, $b_4 = 3.3534 \times 10^{-7}$, and $b_5 = -4.4034 \times 10^{-10}$. The density $\rho$ then is given by combining (4.77) and (4.78):

$$\rho(p,T; p = 2 \times 10^7) = (a_0 + a_1 T + a_2 T^2 + a_3 T^3 + a_4 T^4 + a_5 T^5)(p - 2 \times 10^7)$$
$$+ (b_0 + b_1 T + b_2 T^2 + b_3 T^3 + b_4 T^4 + b_5 T^5) \qquad (4.79)$$

where the notation $\rho(p,T; p = 2 \times 10^7)$ indicates that $p = 2 \times 10^7$ Pa is the reference pressure used in fitting (4.78). As a point of reference, note that when $p$ in (4.79) equals $2 \times 10^7$ Pa, this formula reduces to (4.78); when $p < 2 \times 10^7$ Pa, a value equal to the first term on the right of (4.79) is subtracted from the density $\rho(2 \times 10^7, T)$ at $p = 2 \times 10^7$ Pa. Expanding the term in (4.79) involving $p$ and rearranging gives

$$\rho(p,T; p = 2 \times 10^7) = (a_0 + a_1 T + a_2 T^2 + a_3 T^3 + a_4 T^4 + a_5 T^5)p$$
$$+ (A_0 + A_1 T + A_2 T^2 + A_3 T^3 + A_4 T^4 + A_5 T^5) \qquad (4.80)$$

where $A_i = b_i - 2 \times 10^7 a_i$. This formula can be abbreviated by

$$\rho = a_i(T)p + A_i(T) \qquad (4.81)$$

Thus (4.79) actually is the formula for a line, where the slope $m = a_i(T)$ and the intercept $A_i(T)$ happen to be functions of $T$.

How good is formula (4.80)? Note first that it is relevant only for the liquid phase of water. For example, computation of the density of water at 150 °C under one atmosphere of pressure (101,325 Pa) is inappropriate. At 20 °C, (4.80) predicts a density of 999.53 kg m$^{-3}$ for $p = 101{,}325$ Pa, and 1,004.07 kg m$^{-3}$ for $p = 1 \times 10^7$ Pa; measured values at these pressures are 999.20 kg m$^{-3}$ and 1,002.9 kg m$^{-3}$. At 100 °C, (4.80) predicts a density of 959.51 kg m$^{-3}$ for $p = 101{,}325$ Pa; the measured value is 958.36 kg m$^{-3}$. At 200 °C, (4.80) predicts a density of 869.09 kg m$^{-3}$ for $p = 1 \times 10^7$ Pa; the measured value is 870.85 kg m$^{-3}$. Errors in these predictions are on the order of one tenth of one percent of measured densities.

Why use polynomials, and why fifth order? Polynomials have three nice properties: they are robust in terms of fitting irregular curves to data; they are easy to use in line-fitting (regression) techniques; and they are easy to integrate. Certainly, polynomials are not the only class of functions that could be used. The choice of order is based on the observation that a sixth-order polynomial does not visually improve the fit, and a fourth- or lower-order polynomial does not faithfully match the data as well as a fifth-order polynomial. That both functions (4.77) and (4.78) are fifth order is a coincidence. In this regard, if the ranges of pressure and temperature of interest are more restricted than those presented here, one could fit curves only to those data falling within the restricted ranges, possibly reducing the order of the polynomials necessary to provide a close fit. Finally, no claim can be made regarding the physical significance of the polynomial fit; this is purely a curve-fitting technique. Indeed, the physical units of the polynomial terms do not obviously satisfy dimensional homogeneity (Chapter 5).

### 4.10.3 Water loss from a geyser: thermal effects

Explanations of how thermal geysers erupt sometimes appeal to a mechanism whereby heating and expansion of the fluid column initially lead to an increase in the water level of a geyser. Water at the surface of the geyser is decanted as it crests the geyser rim, so the weight, and therefore pressure, on the liquid lower in the fluid column is reduced. With this reduction in pressure, the possibility exists that somewhere in the water column the saturated vapor pressure suddenly exceeds the local water pressure and the water boils. Ascending bubbles grow and expand, thereby displacing more water; additional water is decanted at the surface, and the pressure is further lowered in the fluid column. This, then, marks the onset of catastrophic eruption. Let us examine the first part of this problem: how water may be decanted due to thermal expansion of the fluid column.

Let $a(z)$ denote a hypsometric curve that describes how the horizontal cross-sectional area $a$ of the plumbing of a geyser varies with elevation $z$. The shape of this curve embodies descriptions of geysers as having "pipe," "potbelly," and "wineglass" shapes (Figure 4.10). Because $a(z)$ is independent of how the area $a$ at any level $z$ is precisely distributed horizontally, this idea of a hypsometric function can be extended to a system whose plumbing consists largely of a fracture network. Also note that a given function $a(z)$ is not unique; the same function may describe the vertical distributions of local area $a$ for systems that are otherwise geometrically dissimilar.

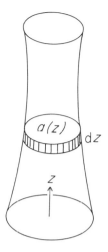

**Figure 4.11** Definition diagram for computing incremental mass $dm = \rho a(z)dz$ within geyser.

Further assume that the density of water depends only on temperature, and is negligibly influenced by pressure. (We will examine the more general case in which the density of water varies with both temperature and pressure in the next chapter.) In a geyser whose plumbing reaches a depth of 100 m, the highest pressures encountered are on the order of $1 \times 10^6$ Pa. Setting $p$ in (4.79) to this value, errors in computations of density are less than 1 percent near the surface; this error decreases with depth. With this approximation, (4.79) becomes

$$\rho(T) = c_0 + c_1 T + c_2 T^2 + c_3 T^3 + c_4 T^4 + c_5 T^5 \tag{4.82}$$

where $c_i = b_i - 1.9 \times 10^7 a_i$.

Consider a geyser such that $z = 0$ at its bottom and $z = h$ at its surface. The mass of water $dm$ within the small vertical interval $dz$ is $\rho[T(z)]a(z)dz$ (Figure 4.11), and the mass of water $m$ within the entire fluid column of height $h$ is

$$m(h) = \int_0^h \rho[T(z)]a(z)dz \tag{4.83}$$

Here $\rho[T(z)]$ indicates that, whereas $\rho$ varies with $T$ as in (4.82), $T$ is to be expressed as a function of $z$; so $\rho$ will then be a function of $z$ also.

Suppose for illustration that $T$ is a linear function of $z$, $T(z) = T_0 + \gamma z$, where $T_0$ is the temperature at the bottom of the geyser and $\gamma$ is the rate of change in $T$ with $z$. For our purpose here, we will neglect the details of how $T(z)$ might at any instant possess this linear form. In general $T(z)$ varies with time and depends on the vertical distribution of heat flow from surrounding rock into the water column, the vertical distribution of the heat capacity of water (and therefore the plumbing geometry), and any vertical redistribution of heat within the water column due to conduction and convection. Further suppose that the hypsometry of the geyser has the form $a(z) = \alpha_0 + \alpha_1 z + \alpha_2 z^2$ (Figure 4.10). The actual form of the curve $a(z)$ for "typical" geysers is largely unknown; the choice here of a polynomial is therefore heuristic. Substituting these expressions for $T(z)$ and $a(z)$ in (4.83),

$$m(h) = \int_0^h [c_0 + c_1(T_0 + \gamma z) + c_2(T_0 + \gamma z)^2 + c_3(T_0 + \gamma z)^3 + c_4(T_0 + \gamma z)^4$$

$$+ c_5(T_0 + \gamma z)^5](\alpha_0 + \alpha_1 z + \alpha_2 z^2)dz \qquad (4.84)$$

As complex as this formula might appear, it can be integrated analytically. It is left to persistent students to show that the solution is

$$m(h) = c_0 \sum_{i=1}^{3} \frac{\alpha_{i-1}}{i} h^i + \frac{\alpha_0}{\gamma} \sum_{i=1}^{5} \frac{c_i}{i+1}[(T_0 + \gamma h)^{i+1} - T_0^{i+1}]$$

$$+ \frac{\alpha_1}{\gamma^2} \sum_{i=1}^{5} c_i \left[ \frac{1}{i+2}(T_0 + \gamma h)^{i+2} - \frac{T_0}{i+1}(T_0 + \gamma h)^{i+1} + \left(\frac{1}{i+1} - \frac{1}{i+2}\right) T_0^{i+2} \right]$$

$$+ \frac{\alpha_2}{\gamma^2} \sum_{i=1}^{5} c_i \left[ \frac{1}{i+3}(T_0 + \gamma h)^{i+3} - \frac{2T_0}{i+2}(T_0 + \gamma h)^{i+2} + \frac{T_0^2}{i+1}(T_0 + \gamma h)^{i+1} \right]$$

$$+ \frac{\alpha_2}{\gamma^3} \sum_{i=1}^{5} c_i \left( -\frac{1}{i+1} + \frac{2}{i+2} - \frac{1}{i+3} \right) T_0^{i+3} \qquad (4.85)$$

Consider a geyser whose plumbing reaches a depth of 100 m and has a total volume of 400 m$^3$, and let us choose four cases for the hypsometric function $a(z)$: 1) $\alpha_0 = 1, \alpha_1 = 0$, and $\alpha_2 = 0.0009$; 2) $\alpha_0 = 1, \alpha_1 = 0.18$, and $\alpha_2 = -0.0018$; 3) $\alpha_0 = 10, \alpha_1 = -0.18$, and $\alpha_2 = 0.0009$; and 4) $\alpha_0 = 10, \alpha_1 = -0.36$, and $\alpha_2 = 0.0036$. These, respectively, coincide with geysers where $a(z)$ increases upward, represents a "potbelly" geometry, decreases upward, and represents a "wineglass" geometry (Figure 4.10). Note also that these values of $\alpha_i$ could equally represent more complex geyser geometries composed of fracture networks. Further suppose that initially $T_0 = 150\,°C$ and $\gamma = -0.7$; this gives a surface temperature ($h = 100$ m) of 80 °C. Substituting these values together with the four cases of $\alpha_i$ into (4.85), the water masses $m(h)$ respectively equal about 383,100 kg, 379,200 kg, 374,800 kg, and 378,700 kg. Assume that further heat flow into the water column after a finite period changes the temperature distribution such that $T_0 = 170\,°C$ and $\gamma = -0.8$; this gives a surface temperature of 90 °C. Substituting these values into (4.85), the new masses $m(h)$ equal about 379,100 kg, 374,100 kg, 368,600 kg, and 373,500 kg. Evidently, masses $\Delta m$ equal to 4,000 kg, 5,100 kg, 6,200 kg, and 5,200 kg were lost as overflow.

The fluid pressure at any instant in a geyser column is independent of $a(z)$. As we will see in Chapter 6, the pressure can be computed from the function $\rho[T(z)]$; the pressure at the base of the four geysers is about $p_0 + 926,000$ Pa with the first temperature distribution, and about $p_0 + 911,200$ Pa with the second distribution, where $p_0$ is atmospheric pressure. Thus the change in pressure at the base of the geysers is equal to about 14,800 Pa, or only about 0.15 atmospheres.

This simple analysis suggests that the loss of mass by overflow due to thermal expansion of the water column may be sensitive to the vertical distribution of the water mass, as embodied in $a(z)$, in relation to the temperature distribution $T(z)$. It should be evident that expansion is greatest when most of the fluid mass is located at a level where temperatures are highest, as in the third case above. Keep in mind, however, that we have neglected treating how $T(z)$ might at any instant arise; in general, this function should be written as $T(z,t)$ to denote its dependence on time.

For a small interval d$z$, the local thermal capacity of the fluid column is equal to $cpa(z)$d$z$ according to (4.2). The specific form of $a(z)$ therefore influences the rate at which the fluid column is locally heated, and how heat is redistributed within the geyser. With identical initial conditions $T(z, 0)$, different forms of $T(z, t)$ would likely occur with different hypsometries $a(z)$ at any instant following the onset of heating.

Interestingly, the associated loss of pressure within the fluid column is unaffected by differences in the loss of mass. Moreover, the slight loss of pressure in the examples above suggests that conditions have to be very close to the critical pressure–temperature state for boiling, if unloading associated with decanting of liquid water at the surface is to have any significant role in precipitating an eruption. (The boiling temperature of water at $1 \times 10^6$ Pa is about 180 °C. Is it likely that somewhere in the water column the saturated vapor pressure exceeds the local water pressure in the examples above?) In contrast, suppose that fluid in the upper part of the column reaches a critical state and flashes to steam. This may occur when superheated bubbles rise and condense and, in doing so, release latent heat to the surrounding fluid, thereby bringing it to a critical state; or when superheated water is convected upward, encounters lower pressure conditions, and boils. Ejection of mass from the vent during this initial stage of eruption very effectively unloads the fluid column, such that flashing propagates downward.

Finally, keep in mind that relevant complications in real geyser systems include effects of local constrictions in plumbing, dissolved solids and gases, and bubble dynamics. In fact, it is interesting that eruptions of "cold" carbon dioxide geysers chiefly depend on bubble dynamics. In addition, we have neglected heat transfer between the country rock and geyser, within the water column, or across the water surface. These factors will be strongly influenced by how the local area $a(z)$ is geometrically distributed—whether at any given level $a(z)$ represents cavernous space or consists of area within fractures. For example, a constricted plumbing geometry tends to suppress fluid convection, thereby maintaining thermally unstable conditions conducive to boiling at depth; cavernous space enhances the possibility of convective overturning that brings hot water up and cool water down, whereby heat may be dissipated at the surface without eruption (a hotspring). Plumbing geometry also determines the area of contact between water and rock, and therefore the efficiency with which the water mass is heated by conduction from the surrounding rock.

That the hypsometry $a(z)$ has an important effect on geyser dynamics seems clear; unfortunately, how $a(z)$ is actually distributed in real geysers is only qualitatively known. Some information is available for the Old Faithful geyser in Yellowstone National Park, Wyoming; the shape of the vent together with video-camera photography obtained by S. Kieffer and J. Westphal suggest that the upper part of the plumbing of Old Faithful consists of a tabular fissure. Also note that unloading of the geyser fluid in Old Faithful is associated with flashing of the water to steam near the surface, rather than decanting of liquid water.

### 4.10.4 Ice regelation and glacier movement

If you have walked over the floor of a glaciated valley, you probably are aware that areas where bedrock is extensively exposed can be very bumpy. This raises a very basic question that has been the focus of numerous glaciological studies over the past four decades: How does a glacier move over and around the bumps at its base?

A glacier moves over a bumpy bedrock surface by two mechanisms: ice deformation (Chapter 12), and basal sliding. Basal sliding, in turn, involves several mechanisms, one of which is *regelation*. This mechanism was first formally treated by Weertman in 1957; and although the details have been refined significantly since then (see Reading list), the essential ingredients are contained in the original work of Weertman. For historical as well as pedagogical reasons, we therefore will follow Weertman's work.

Let us assume that a glacier consisting of clean ice is flowing over an undeformable bedrock surface. Flow occurs with a steady velocity $u$ in the direction of positive $x$. The ice is at the pressure melting point (a *temperate glacier*), and a thin film of water everywhere separates the ice from the underlying bed. Hydrostatic conditions exist at the base of the ice. This idea will be clarified in Chapter 6; for now, envision that the hydrostatic pressure is equal to the weight of overlying ice per unit area of bed. The essential point is that the pressure conditions at the base of the ice coincide with the pressure melting point.

Now consider, as Weertman did, the simplistic case where the bedrock bumps are cubes with edges of length $\lambda$. One cubical bump occurs per area of bed equal to $\Lambda^2$, where $\Lambda$ is the linear spacing between neighboring bumps. The average shear stress at the bed is $\tau_b$; in the simplest case this stress arises from a downslope component of the weight of the overlying ice, acting tangentially to the bed. Due to the thin film of water separating the ice and bed, surfaces parallel to flow (between bumps, and on the tops and sides of bumps) do not experience this shear stress. This is because the water film, although possessing finite viscosity, offers negligible shear resistance; thus the water film acts in essence like a lubricant. The stress $\tau_b$, in contrast, leads to a compressive (normal) force on the upstream face of a bump and a tensile force on the downstream face. Thus $\tau_b$ is a nominal quantity that actually is distributed in the form of normal stresses acting over bump faces that are normal to flow.

The stress $\tau_b$ may be "partitioned" among bumps. The average force associated with $\tau_b$ on each cubical bump is $\tau_b \Lambda^2$. Assuming the compressive and tensile forces acting on a bump are of equal magnitude, the extant hydrostatic (normal) stress on the upstream face is enhanced by an amount equal to $\frac{1}{2}\tau_b \Lambda^2/\lambda^2$, and the hydrostatic stress on the downstream face is decreased by an amount equal to $-\frac{1}{2}\tau_b \Lambda^2/\lambda^2$. The drop in pressure $\Delta p$ across the cube is therefore on the order of $-\tau_b \Lambda^2/\lambda^2$.

Based on the negative slope of the fusion curve for water (Figure 4.6), the high pressure on the upstream face means that the melting temperature of ice is lower here than on the downstream face. The difference in melting temperature $\Delta T$ is obtained from the Clapeyron equation (4.51). Rearranging this equation,

$$dT = \frac{T(V_l - V_s)}{\Delta H}dp \qquad (4.86)$$

Before integrating this to obtain an expression relating temperature $T$ to pressure $p$, we must note that the heat of transition $\Delta H$ and the specific volumes, $V_l$ and $V_s$, are generally functions of both $p$ and $T$. If, however, we assume that the pressure difference $\Delta p$ and the temperature difference $\Delta T$ are sufficiently small, then $\Delta H$, $V_l$, and $V_s$ may be assumed to be approximately constants. Moreover, we may treat the temperature $T$ at which transition occurs, appearing in the numerator of (4.86), as a constant parameter. (How is this justified?) Then,

$$\int_{T_0}^{T} dT' = \frac{T(V_l - V_s)}{\Delta H}\int_{p_0}^{p} dp' \qquad (4.87)$$

$$T = T_0 + G(p - p_0) \qquad (4.88)$$

where $G = T(V_l - V_s)/\Delta H$, and $T_0$ is a reference temperature associated with the reference pressure $p_0$. (These are normally taken to be the temperature and pressure at the triple point; thus $T_0 = 273.15$ K and $p_0 = 610.6$ Pa for pure water. However, these values will cancel out in the next step below.)

Since the ice is everywhere at the temperature melting point, the temperature $T_u$ on the upstream face at pressure $p_u$ is

$$T_u = T_0 + G(p_u - p_0) \tag{4.89}$$

and the temperature $T_d$ on the downstream face at pressure $p_d$ is

$$T_d = T_0 + G(p_d - p_0) \tag{4.90}$$

The temperature difference $\Delta T = T_d - T_u$ associated with the pressure difference $\Delta p = p_d - p_u = -\tau_b \Lambda^2/\lambda^2$ thus is given by

$$\Delta T = -G\tau_b \frac{\Lambda^2}{\lambda_2} \tag{4.91}$$

where $G$ is negative for the case of ice and water ($V_l < V_s$), making $\Delta T$ positive.

The temperature difference $\Delta T$ provides a gradient across the bump with average magnitude $\Delta T/\lambda$; thus heat flows through the bump from the downstream side toward the upstream side (neglecting heat flow through the ice). Assuming steady, approximately one-dimensional heat flow, the quantity of heat $\Delta H_R$ delivered to the upstream face during the small interval $dt$ is obtained from Fourier's law (4.21).

$$\Delta H_R = -K_T \lambda^2 \frac{\Delta T}{\lambda} dt = -K_T \lambda \Delta T dt \tag{4.92}$$

This supply of heat tends to drive the temperature of the ice at the upstream face toward the liquid side of the fusion curve; thus melting occurs. Let $u$ denote the ice velocity associated with regelation. Then a volume of ice equal to $u\lambda^2 dt$ is melted on the upstream face during $dt$, producing a mass of water equal to $\rho u\lambda^2 dt$, where $\rho$ is the ice density. This water migrates as part of the film to the lower pressure site at the downstream face. Here, conduction of heat away from this site tends to drive the temperature of the water at the face toward the solid side of the fusion curve such that water in the film refreezes. Latent heat is released, which is the source of heat conducted upstream through the bump. Assuming the quantity of water melted on the upstream side equals the quantity that refreezes on the downstream side, the latent heat released during $dt$ is equal to $-\rho L u\lambda^2 dt$, where $L$ is the latent heat of fusion. This released heat equals that conducted upstream, $\Delta H_R$; thus

$$\rho L u\lambda^2 = K_T \lambda \Delta T \tag{4.93}$$

Substituting (4.91) for $\Delta T$ and solving for $u$,

$$u = -\frac{K_T G\tau_b}{\rho L\lambda} \frac{\Lambda^2}{\lambda^2} \tag{4.94}$$

Thus ice velocity associated with regelation is proportional to the thermal conductivity of the rock bumps and to the area $\Lambda^2$, and inversely to the area $\lambda^2$. A low areal density of bumps (large $\Lambda^2$) means that the stress $\tau_b$ is partitioned among fewer bumps, thereby enhancing the pressure drop $\Delta p$ and associated temperature drop $\Delta T$ across each bump. Likewise, for a given bump density, smaller bumps (small $\lambda^2$) must bear a greater stress, thereby enhancing $\Delta p$ and $\Delta T$ across each bump. (The ratio $\lambda/\Lambda$ is a roughness parameter.) In addition, $u$ is inversely proportional to $\lambda$; a small bump has a greater temperature gradient $\Delta T/\lambda$ than a large bump, for given $\Delta T$. This provides a faster delivery of heat from the downstream face to the upstream face, thereby enhancing both refreezing and melting. Thus regelation is increasingly more effective with decreasing bump size.

Based on tabled values of $\Delta H$, $V_l$, and $V_s$ available from a variety of sources, the quantity $G$ has a value of about $-7.4 \times 10^{-8}$ K Pa$^{-1}$ for pure ice and water. In actuality, glacier ice can be air saturated; then $G$ is equal to about $-9.8 \times 10^{-8}$ K Pa$^{-1}$. In addition, solutes can lower the melting temperature at the downstream face of a bump, thereby decreasing the temperature gradient and the rate of heat conduction to the upstream face.

Other factors also bear on the regelation mechanism. These include: nonhydrostatic conditions can exist at the base of a glacier, which influence the pressure conditions at the upstream and downstream faces of a bump; the ice and bed may not be separated everywhere by a film of water, so basal shear stresses are not distributed as in the idealized case above; basal ice can contain rock debris, which alters its rheological behavior; air-filled cavities can occur on the downstream sides of bumps (cavitation), so that the ice is not in contact with the bed; heat conduction is not merely one-dimensional; and real bumps are more complicated than the simple cube model above.

### 4.10.5  Growth of carbon dioxide bubbles in a liquid

Recall that in Example Problem 1.4.2 we neglected any exchange of mass that might occur between a bubble and liquid surrounding it to estimate the rate at which the bubble ascends. Let us now examine this exchange by extending the simple results reported by Shafer and Zare (1991) with reference to growth of $CO_2$ bubbles ascending within a glass of beer. This will be the first of three treatments of this problem; we will subsequently improve our description of bubble growth and ascent in Example Problems 6.7.3 and 8.4.4.

Growth of bubbles in a liquid requires nucleation centers for gas molecules to accumulate and coalesce. These centers may be small particles suspended in the liquid, small irregularities on solid surfaces, or the bubbles themselves. Once suspended within the liquid, a bubble can continue to grow or it can collapse; this depends on thermal conditions of the liquid, the concentration of dissolved gases within the liquid, the thermodynamic state of gases within the bubble, and the rate at which the bubble ascends. For simplicity, consider bubbles growing and rising within an isothermal liquid with a free surface at atmospheric pressure $p_0$. Under these conditions, a bubble grows by expansion as it ascends and encounters decreasing hydrostatic pressure conditions, and due to a high partial pressure of dissolved gas within the liquid relative to that within the bubble such that gas diffuses from the liquid into the bubble. We also will assume that the temperature of the bubble remains essentially constant as it ascends through the isothermal liquid, so thermal expansion or contraction are negligible.

Gaseous $CO_2$ dissolved in water is partitioned among several aqueous species. Henry's law relating the partial pressure of $CO_2(g)$ to the concentration of dissolved $CO_2(aq)$ therefore is not exact, but holds reasonably well since very little of the $CO_2(aq)$ actually is involved in reactions. There are several mechanisms by which natural waters are charged with $CO_2(aq)$. Perhaps the most important in near-surface hydrological systems involve plant respiration and oxidation of organic matter in soils; these processes can lead to $CO_2(g)$ partial pressures in a soil atmosphere that are two to three times greater than pressures in Earth's atmosphere. These high partial pressures then lead to dissolving of $CO_2$ in waters percolating downward. Dissolved $CO_2$ also is important in other geological systems; for example, $CO_2$ can be an important ingredient of magmas, for which the basic ideas described here are relevant.

The rate at which a bubble grows is proportional to the rate at which mass is added to it, and it also depends on the rate at which it ascends and expands in response to decreasing hydrostatic pressure conditions. Let us assume, then, that the rate at which mass ($CO_2$) is added to a bubble is proportional to its surface area, and to the difference between the partial pressure of $CO_2$ in the liquid and its partial pressure within the bubble:

$$\frac{dm}{dt} = \alpha 4\pi r^2 (p_{CO_2} - p_b) \tag{4.95}$$

Here $r$ is the bubble radius, $p_{CO_2}$ is the partial pressure of $CO_2(aq)$, $p_b$ is the partial pressure of $CO_2(g)$ in the bubble, and $\alpha$ is a proportionality constant having properties of a diffusion coefficient. (We will examine $\alpha$ below. Meanwhile, what are the dimensions of $\alpha$?) Let us further assume that the gas in the bubble is entirely $CO_2$; then $p_b$ is equal to the total pressure in the bubble. (How does this formulation change when more than one gas is involved?) From (3.58), $p_b = p_l + 2\sigma/r$, where $p_l$ is the hydrostatic pressure of the liquid and $\sigma$ is a coefficient of surface tension for the interface between $CO_2(g)$ in the bubble and the liquid. The coefficient $\sigma$ can vary with $r$; in the case of pure water vapor in contact with pure water liquid, $\sigma$ is virtually constant for $r$ greater than about $10^{-7}$ m, but decreases slightly when $r < 10^{-7}$ m. We therefore may reasonably assume that $\sigma$ is constant by restricting our attention to "large" bubbles. The hydrostatic pressure $p_l$ is equal to $p_0 + \rho g(h - z)$, where $z$ is a vertical coordinate, positive upward, such that $z = 0$ is positioned at the bottom of the liquid column with total depth $h$. Substituting these expressions into (4.95) and expanding,

$$\frac{dm}{dt} = \alpha 4\pi (p_{CO_2} - p_0 - \rho gh + \rho gz)r^2 - \alpha 8\pi\sigma r \tag{4.96}$$

which we will return to momentarily.

Now assume that the $CO_2$ gas within the bubble obeys the ideal gas law; from (4.24),

$$\frac{dm}{dt} = \frac{p_b}{RT}\frac{dV}{dt} + \frac{V}{RT}\frac{dp_b}{dt} \tag{4.97}$$

Since $V = \frac{4}{3}\pi r^3$, $dV/dr = 4\pi r^2$ and $dV/dt = 4\pi r^2 dr/dt$. Also, since both $r$ and $z$ may vary with time, $dp_b/dt = -2\sigma/r^2 dr/dt - \rho g dz/dt$, where the term involving $r$ has been differentiated implicitly. Substituting these expressions for $V$, $p_b$, $dV/dt$, and $dp_b/dt$ into (4.97), then equating (4.96) with (4.97) and rearranging, we obtain

$$\frac{dr}{dt} = \frac{\alpha(p_{CO_2} - p_0 - \rho gh + \rho gz)r - \alpha 2\sigma + \dfrac{\rho g}{3RT}r^2\dfrac{dz}{dt}}{\dfrac{4\sigma}{3RT} + \dfrac{1}{RT}(p_0 + \rho gh - \rho gz)r} \tag{4.98}$$

The denominator and the first two terms in the numerator describe the rate of bubble growth in response to diffusion of $CO_2$ gas into the bubble, in terms of instantaneous bubble size and pressure conditions. The last term in the numerator describes the influence of ascent speed $dz/dt$ as the bubble encounters decreasing pressure conditions.

Regarding the experiment of Shafer and Zare (1991), note that the total depth $h$ in a beer glass is small, that bubble radii are relatively large, and that $p_{CO_2}$ is about $2 \times 10^5$ Pa (about 2 atmospheres) for typical beer brewed in the United States, at drinking temperature (in the U.S.). The first of these points means that the decrease in hydrostatic pressure over $h$ has a negligible effect on bubble growth. You may have noticed, as Shafer and Zare point out, that beer bubbles grow more than twice their original size over the depth $h$. This would require about 10 m of depth, if explained by decreasing hydrostatic pressure alone. The second point implies that bubble pressure due to curvature of the gas–liquid interface is relatively small, so bubble pressure is approximately equal to $p_0$ (neglecting the small hydrostatic pressure). Thus, as a good approximation, (4.98) may be simplified to

$$\frac{dr}{dt} \approx \alpha RT \frac{p_{CO_2} - p_0}{p_0} \tag{4.99}$$

(It also is possible to demonstrate that neglected terms are very small for typical beer bubbles. It is left as an exercise to estimate the magnitude of $dm/dt$ using the ideal gas law, back calculate the magnitude of $\alpha$ from [4.95], then use typical values of the physical quantities in [4.98] to estimate the magnitude of each term in [4.98]).

Assuming that $p_{CO_2}$ is essentially constant over the duration of the ascent of a bubble, and that the temperature $T$ of a bubble also is constant due to the surrounding beer, (4.99) reduces to $dr/dt = \alpha_0$, where $\alpha_0$ equals the term on the right side. Integrating, then setting the constant of integration to the initial bubble radius $r_0$,

$$r = r_0 + \alpha_0 t \tag{4.100}$$

Data collected by Shafer and Zare for growth of $CO_2$ bubbles in a beer glass are consistent with this linear relation for $r_0 = 0.00018$ m and $\alpha_0 = 0.00004$ m s$^{-1}$ (Figure 4.12a). Also note that bubbles accelerate during their ascent (Figure 4.12b). This is due to the increasing buoyancy of a bubble as it grows. We will examine this part of the problem, including the hydrostatic effects neglected above, in Example Problem 6.7.3.

Addition of $CO_2$ mass to a bubble actually depends on the gradient of the $CO_2$ partial pressure in the liquid near the bubble–liquid interface, which induces diffusion of $CO_2(aq)$ toward the bubble. Formula (4.95) is therefore equivalent to assuming that this gradient is purely a function of the partial pressure difference $p_{CO_2} - p_b$. A more rigorous treatment of this problem involves describing the partial pressure gradient near the bubble, and makes use of a mass diffusion coefficient as embodied in Fick's law (Example Problem 8.4.4). This gradient in general is a function of bubble radius, and can be further influenced by bubble growth and ascent, inasmuch as these processes influence the local fluid velocity field around the bubble. For example, the magnitude of the partial-pressure gradient above a rapidly ascending bubble increases relative to that above an otherwise static bubble; the effect of an ascending bubble is to "compress" the pressure difference $p_{CO_2} - p_b$ over a shorter radial distance above the bubble. The coefficient of proportionality $\alpha$ in (4.95) therefore loosely embodies these factors, and includes

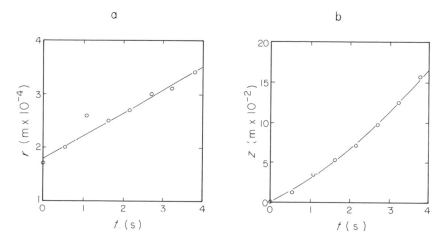

**Figure 4.12**  Graph of measured (*open circles*) and predicted (*lines*) relations between carbon dioxide bubble radius $r$ and time $t$ (*a*), and ascent height $z$ and time $t$ (*b*); adapted from Shafer and Zare (1991), by permission of the American Institute of Physics.

a mass diffusion coefficient as well as a factor that expresses the relation between $CO_2$ partial pressure and concentration.

The dynamics of bubbles where thermal effects are important, including boiling, are significantly more complicated. Students interested in this topic are urged to consider the relevant references listed below. An important application of this topic involves the growth and collapse of bubbles in thermally driven geysers. For example, suppose that hot, water-vapor bubbles form at some depth and ascend into cooler liquid above. Such bubbles may then be in a condition of thermodynamic disequilibrium, so that the hot vapor condenses, driving mass from the bubbles into the liquid. Heat is advected upward in the liquid column by this mechanism. In addition, the bubbles may collapse catastrophically, giving off shock waves that can be monitored before and during geyser eruption.

## 4.11  READING

Anderson, D. A., Tannehill, J. C., and Pletcher, R. H. 1984. *Computational fluid mechanics and heat transfer.* New York: McGraw-Hill, 599 pp. This advanced text covers a counterpart to the analytical treatments presented herein. It provides a thorough development of the principles and methods by which flows of heat and fluids are treated numerically. An appealing aspect of the text is its use of structural homologies to examine different problems.

Clark, S. P., Jr., ed. 1966. Handbook of Physical Constants. New York: Geological Society of America, Memoir 97, 587 pp. This memoir provides a thorough compilation of various physical and thermodynamic properties of water, minerals, rocks, and mineral and rock melts.

DeFay, R., Prigogine, I., Bellemans, A., and Everett, D. H. 1966. *Surface tension and adsorption.* New York: Wiley, 432 pp. Chapter 15 of this book provides a systematic treatment of the thermodynamics of bubbles and drops, including how fluid–interface curvature influences vapor pressure and surface tension. The chapter also examines how the curvatures of menisci in small pores affect the thermodynamic properties of the wetting fluid.

Guggenheim, E. A. 1959. *Thermodynamics,* 4th ed. Amsterdam: North-Holland, 476 pp. This advanced text (relative to Young's text) provides an alternative derivation of the Clapeyron and Clausius–Clapeyron equations, and has a clear, general discussion of fluid phases, phase relations, and phase transitions.

Hallet, B. 1976. The effect of subglacial chemical processes on glacier sliding. *Journal of Glaciology* 17:209–21. Hallet uses deposits of calcite and silica on the downstream sides of bumps as evidence that these compounds are expelled with refreezing during regelation, and argues that these solutes lower the melting temperature of the ice at these locations, thereby suppressing the temperature gradient and conduction of heat to upstream faces.

Ingebritsen, S. E. and Rojstaczer, S. A. 1993. Controls on geyser periodicity. *Science* 262:889–92. The authors use equations governing heat and mass transfer in porous media (see Chapters 9 and 13) to examine how the periodicity of thermal-geyser eruptions responds to seismic-induced strains, atmospheric loading, and Earth tides. A geyser is conceptualized as a highly permeable zone of intensely fractured rock surrounded by less permeable rock. The paper provides a good list of references on the general topic of geyser dynamics.

Kieffer, S. W. 1984. Seismicity at Old Faithful geyser: an isolated source of geothermal noise and possible analogue of volcanic seismicity. *Journal of Volcanology and Geothermal Research* 22:59–95. This paper describes the complex set of conditions involved in eruptions of Old Faithful, including mechanisms of fluid heating, and provides a brief description of its plumbing geometry. It examines the importance of bubbles during eruptions, as inferred partly from monitoring of shock waves produced by bubble collapse.

Loper, D. E. 1992. A nonequilibrium theory of a slurry. *Continuum Mechanics and Thermodynamics* 4:213–45. This paper is the third of a series that examines a problem related to bubble growth, namely the growth and melting of crystals in a melt due to diffusion of mass and energy between the crystals and surrounding liquid. The material is presented at an advanced level.

Paterson, W. S. B. 1981. *The physics of glaciers,* 2nd ed. Oxford: Pergamon, 380 pp. This text summarizes work on the regelation mechanism for glacier movement (as well as other mechanisms). It also provides a clear summary discussion of the assumptions and complications that are involved in treating the regelation mechanism, and a comparison of theoretical treatments with field observations. Students wishing to pursue this topic should especially examine referenced papers by Hallet, Kamb, LaChapelle, Lliboutry, Nye, and Weertman.

Shafer, N. E. and Zare, R. N. 1991. Through a beer glass darkly. *Physics Today* 44(10):48–52. This delightful article uses elementary physics to examine growth of carbon dioxide bubbles during their ascent in beer, and is the basis for Example Problem 4.10.5.

Weertman, J. 1957. On the sliding of glaciers. *Journal of Glaciology* 3:33–38. This paper provides the original treatment of the regelation mechanism applied to glacier movement, and is the basis for Example Problem 4.10.4.

Young, H. D. 1964. *Fundamentals of mechanics and heat.* New York: McGraw-Hill, 638 pp. This text presents the clearest and most thoroughly understandable introduction to elementary thermodynamics that I have read. A significant strength of the presentation is its successful interweaving of kinetic theory into the discussion of thermodynamic properties and fluid behavior, including a kinetic interpretation of the first law of thermodynamics for simple gases.

Zemansky, M. W. 1981. *Temperatures very low and very high.* New York: Dover, 127 pp. This text gives an historical perspective of efforts to experimentally attain extreme temperatures, and it provides a very readable discussion of temperature as a state variable, isothermal and adiabatic processes, and the three laws of thermodynamics, including the zeroth law.

# CHAPTER 5

# Dimensional Analysis and Similitude

Some fluid flow problems are sufficiently simple that they can be treated mathematically in a straightforward way, making use of definitions of physical quantities, and taking into account initial and boundary conditions. For example, our derivation of the average velocity in a conduit with parallel walls (Example Problem 3.7.1) was obtained in a straightforward way once we specified the geometry of the problem, then made use of the definition of a Newtonian fluid and the no-slip condition. Whereas this type of analysis may work for some problems, it would be misleading to think that such direct approaches to solving problems are, in principle, always possible, hinging only on one's mathematical skills and adeptness in specifying the geometry of a problem. Herein arise two noteworthy points.

First, when initially examining a problem, one can sometimes obtain a clear idea of the desired solution before attempting a formal mathematical analysis. The means to do this, as we shall see below, is supplied by *dimensional analysis,* and it is a strategy that ought to be adopted in many circumstances. In fact, it is worth noting that dimensional analysis underlies many of the problems presented in this text. The advantage of knowing the form of a desired solution, of course, is that one has a clear target to guide the subsequent mathematical analysis. Indeed, this is the vantage point from which many classic problems, for example Stokes's law for settling spheres, were initially examined.

Second, a complete mathematical formulation of a problem may not be possible, due to the complexity of the problem, or due to absence of information required to constrain the mathematics of the problem. As a simple example, suppose that we were unaware of the no-slip condition in our analysis of the conduit-flow problem (Example Problem 3.7.1). Our analysis in this case would have essentially ended with (3.70), with the constant of integration $C$ undetermined. Nevertheless, we could get close to our result (3.75) for the average velocity by another way. We could infer from dimensional analysis, as we shall see below, that the average velocity is proportional to the pressure loss $\Delta p$ and the square of the aperture, and inversely proportional to the viscosity $\mu$ and the length $L$. This includes all information contained in the result (3.75) except the constant 1/3. We may anticipate from this that dimensional analysis can be used to gain an understanding of the essential physical quantities involved in a problem, which then could be examined further by other means, perhaps experimentally.

An essential ingredient of dimensional analysis is to bring good physical insight to a problem. Indeed, dimensional analysis has limitations, and can lead to incorrect answers without this physical insight. A good way to introduce these ideas is with examples. Let us begin with a brief review of dimensional homogeneity, the cornerstone of dimensional analysis.

## 5.1 DIMENSIONAL HOMOGENEITY

The *principle of dimensional homogeneity* refers to a simple idea: A necessary, but insufficient, condition for a mathematical expression relating physical quantities to be physically sound is that each term in it must have the same dimensions. Consider, for example, Stokes's law for the velocity of a settling spherical particle (Example Problem 1.4.2):

$$w = -\frac{2}{9\mu}(\rho_s - \rho)R^2 g \tag{5.1}$$

Dimensional homogeneity refers to the fact that each term in this equation has the dimensions of velocity, $Lt^{-1}$. Furthermore, it should be apparent that regardless of how we organize this equation, say

$$\frac{w}{R^2} = -\frac{2}{9\mu}\rho_s g - \frac{2}{9\mu}\rho g \tag{5.2}$$

the dimensions of each term are the same, in this case $L^{-1} t^{-1}$. In this regard, you should recall from introductory physics that a simple, straightforward check of whether one's manipulation of such formulae is correct is to write out the dimensions of each term and test for homogeneity.

That dimensional homogeneity is an insufficient condition for a formula to be physically sound can be readily illustrated by an example such as

$$c = \left(\frac{\lambda^3}{A^2}g\right)^{\frac{1}{2}} \tag{5.3}$$

which might purport to express how the speed $c$ of a shallow water wave is related to its amplitude $A$, its wavelength $\lambda$, and acceleration due to gravity $g$. A check of dimensions reveals that this formula is homogeneous, but it is physically incorrect (Example Problem 10.3.3).

Consider now the definition of a Newtonian fluid from section 3.4.1:

$$\tau_{yx} = -\mu\frac{\partial u}{\partial y} \tag{5.4}$$

This also is dimensionally homogeneous. But recall that this definition originated with experiments involving Couette flow; the coefficient of viscosity $\mu$ was an empirical result rather than a given quantity. Thus, in contrast to Stokes's law, which was derived in part by making use of this definition of a Newtonian fluid, the viscosity $\mu$ was, from the original empirical point of view, only a coefficient of proportionality. The dimensions $ML^{-1}t^{-1}$ were assigned to it to make the formula dimensionally homogeneous.

## 5.2 DIMENSIONAL QUANTITIES

### 5.2.1 Viscosity of a gas

Suppose that we wished to infer which molecular-scale quantities are essential in defining the viscosity $\mu$ of a gas, to guide the mathematical analysis in Section 3.4.1 leading to the result (3.21). Our physical insight regarding this problem might suggest that $\mu$ depends on molecular mass $M$, molecular size $r$, molecular speed $c$, and temperature $T$. We thus may propose a function of the form

$$\mu = f(M, r, c, T) \tag{5.5}$$

Further assume that this function involves the product of $M$, $r$, $c$, and $T$, each raised to an exponent $A$, $B$, $C$, and $D$. Thus

$$\mu \propto M^A r^B c^C T^D \tag{5.6}$$

We then write a dimensional equation expressing the dimensions of each variable in terms of dimensions of the fundamental quantities: mass, length, time, and temperature,

$$]\mu[ = ML^{-1}t^{-1} = (M)^A(L)^B(Lt^{-1})^C(T)^D \tag{5.7}$$

$$= M^A L^B L^C t^{-C} T^D \tag{5.8}$$

Here $]\mu[$ denotes "the dimensions of $\mu$." We now have an equation with the four unknown exponents $A$, $B$, $C$, and $D$.

Dimensional homogeneity requires that the dimensions of the product on the right side of this proportionality equal those of viscosity on the left. This means that the exponent of each fundamental quantity resulting from the product on the right side must equal the exponent of that on the left. For example, since the exponent of mass on the left side equals one, it must also equal one on the right side, and noting that mass appears only with the exponent $A$ on the right side,

$$\text{mass:} \quad 1 = A \tag{5.9}$$

Considering length, we observe that the exponent of length on the left side equals $-1$, so using rules of exponents for quantities involving length on the right side,

$$\text{length:} \quad -1 = B + C \tag{5.10}$$

Considering time, we note that

$$\text{time:} \quad -1 = -C \tag{5.11}$$

Finally turning to temperature,

$$\text{temperature:} \quad 0 = D \tag{5.12}$$

which indicates that temperature is not an essential quantity in the problem. We thus note that $A = 1$ and $C = 1$; and substituting for $C$ in the equation involving length, we obtain $B = -2$. Evidently we may now rewrite the proportionality (5.6) as

$$\mu = k\frac{cM}{r^2} \tag{5.13}$$

where $k$ is a dimensionless coefficient. It cannot be determined by dimensional analysis, but rather must be determined by other means. Evidently $k = k_2/4\pi\sqrt{2}$ from (3.21).

What is essential to note here is this: using only dimensional analysis, we have learned how viscosity $\mu$ is related to $M$, $c$, and $r$. Equally important, we have learned that viscosity does not necessarily depend on temperature, which is consistent with (3.21). (For what reasons should this result have been anticipated in the first step above? How would one obtain from dimensional analysis the result [3.22], which does involve temperature?)

### 5.2.2 Average velocity in conduit flow

Suppose now that we wanted to determine how the average velocity $u_{ave}$ in a conduit with parallel walls depends on the physical quantities depicted in Figure 3.18. Physical insight might suggest a function of the form

$$u_{ave} = f(b, L, \mu, \Delta p) \tag{5.14}$$

where $b$ is the half-aperture, $L$ is the conduit length, $\mu$ is the fluid viscosity, and $\Delta p$ is the pressure loss over $L$. As before, assume a proportionality involving the product of $b$, $L$, $\mu$, and $\Delta p$, each raised to an exponent $A$, $B$, $C$, and $D$:

$$u_{ave} \propto b^A L^B \mu^C (\Delta p)^D \tag{5.15}$$

Writing a dimensional equation,

$$]u_{ave}[ = Lt^{-1} = L^A L^B (ML^{-1}t^{-1})^C (ML^{-1}t^{-2})^D \tag{5.16}$$

$$= L^A L^B M^C L^{-C} t^{-C} M^D L^{-D} t^{-2D} \tag{5.17}$$

Here we note that relevant fundamental quantities include mass, length, and time; temperature may be neglected since we are treating the problem entirely as a mechanical one. We then organize our work as follows:

$$\text{mass:} \quad 0 = C + D \tag{5.18}$$

$$\text{length:} \quad 1 = A + B - C - D \tag{5.19}$$

$$\text{time:} \quad -1 = -C - 2D \tag{5.20}$$

But now we have a dilemma: We have three equations and four unknowns. Nevertheless, let us proceed as far as possible. From the first equation we note that $C = -D$. Substituting this for $C$ in the third equation we obtain $D = 1$. Finally, substituting this back into the first equation we obtain $C = -1$. Other than using the first equation to simplify the second to $1 = A + B$, we can proceed no further. Evidently we may now rewrite the proportionality (5.15) as

$$u_{ave} = \frac{k}{\mu} \Delta p b^A L^B = \frac{k}{\mu} \Delta p f(b, L) \tag{5.21}$$

where $k$ is a dimensionless coefficient. Thus, although our solution is incomplete, we nonetheless have learned that the average velocity is proportional to $\Delta p$ and inversely to $\mu$, and probably depends further on some function of $b$ and $L$.

Can we somehow ensure that the dilemma above will not occur—that we will not obtain more unknowns than equations? The safest way is to consider problems that involve no more independent quantities than fundamental quantities, and therefore no more exponents than equations. But this "rule" seems unsatisfactory because it places a severe restriction on the complexity of problem that can be examined, and because it sometimes works out that the number of exponents is reduced, as we saw in the example above for viscosity. Herein one can sometimes make further use of physical insight. For example, experience suggests that with one-dimensional flow problems the pressure loss $\Delta p$ can be combined with the flow distance $L$ to form the pressure loss per unit flow distance, $\Delta p/L$ (the pressure gradient, which we will define later). Thus, reconsider the problem starting with a function of the form

$$u_{\text{ave}} = f\left(b, \mu, \frac{\Delta p}{L}\right) \tag{5.22}$$

Then the dimensional equation becomes

$$]u_{\text{ave}}[ = Lt^{-1} = L^A(ML^{-1}t^{-1})^B(ML^{-2}t^{-2})^C \tag{5.23}$$

where we have reduced the number of unknown exponents from four to three. It is left as an exercise to obtain the results $A = 2$, $B = -1$, and $C = 1$, whence

$$u_{\text{ave}} - k\frac{b^2}{\mu}\frac{\Delta p}{L} \tag{3.24}$$

where the coefficient $k$ evidently equals 1/3 from (3.75) in Example Problem 3.8.1.

### 5.2.3 Drag on a spherical particle: the Reynolds number

Consider now the drag force $F_D$ on a settling spherical particle. Physical insight might suggest that the drag is some function of the radius $R$ of the particle, the settling velocity $w$, the fluid density $\rho$, and the fluid viscosity $\mu$. Assuming a proportionality of the form

$$F_D \propto R^A w^B \rho^C \mu^D \tag{5.25}$$

the dimensional equation becomes

$$]F_D[ = MLt^{-2} = L^A(Lt^{-1})^B(ML^{-3})^C(ML^{-1}t^{-1})^D \tag{5.26}$$

Organizing exponents associated with fundamental quantities,

$$\text{mass:} \quad 1 = C + D \tag{5.27}$$

$$\text{length:} \quad 1 = A + B - 3C - D \tag{5.28}$$

$$\text{time:} \quad -2 = -B - D \tag{5.29}$$

Once again we have a dilemma: three equations and four unknowns. Observe, however, that each of the four exponents, except $A$, appears in at least two of the three equations, and $D$ appears in all three. Solving for $A$, $B$, and $C$, in terms of $D$,

$$A = 2 - D \tag{5.30}$$

$$B = 2 - D \tag{5.31}$$

$$C = 1 - D \tag{5.32}$$

Substituting into (5.25),

$$F_D \propto R^{2-D} w^{2-D} \rho^{1-D} \mu^D \tag{5.33}$$

$$\propto (\rho R^2 w^2) \left( \frac{\mu}{\rho R w} \right)^D \tag{5.34}$$

Evidently we may now rewrite the proportionality (5.25) as

$$F_D = (\rho R^2 w^2) f \left( \frac{\mu}{\rho R w} \right) \tag{5.35}$$

Thus, we have obtained for the drag $F_D$ on a settling spherical particle an expression that contains an unknown function of the dimensionless quantity $\mu / \rho R w$. The reciprocal of this dimensionless quantity, symbolized as $Re = \rho R w / \mu$, is known as the *Reynolds number* for a sphere. This definition leads to a brief, but important, digression.

Note that $Re$ here pertains specifically to a sphere—and only a sphere—since it has been defined in terms of the sphere radius $R$. Had we, for example, been concerned with another geometrical form, for example, a cube with sides of length $L$, we might have defined the Reynolds number as $\rho L w / \mu$. However, the two definitions are not directly comparable; flow around a cube is quite different from that around a sphere, and therefore the two definitions characterize very different particle-flow systems. This topic is examined further in Sections 5.4 and 5.5. Meanwhile, let us note that the Reynolds number may in general be symbolized as $Re = \rho L U / \mu$, where $L$ refers to a characteristic length and $U$ refers to a characteristic velocity of the specific flow system being examined. As we shall see below, the Reynolds number is one of several important dimensionless quantities that appear frequently in fluid physics problems.

Returning now to the expression (5.35) for drag on a spherical particle, consider one of its implications. This expression holds for any given spherical particle and Newtonian fluid. That is, a given value of drag can arise from an infinity of combinations of sphere size and velocity, and fluid density and viscosity. This suggests that, if the function $f(1/Re)$ in (5.35) is to be determined experimentally, it is not necessary to undertake a large number of experiments using different sizes of spheres and fluid properties. Rather, one has only to vary one or more of these, as experimental materials and apparatus permit, so as to cover a range of Reynolds number. Then, the results from such experiments should hold for any unmeasured combination of sphere size and fluid properties whose Reynolds number falls within the range of those examined.

In this regard, the ratio $C_D = F_D / \rho R^2 w^2$ is referred to as a "coefficient of drag" (Example Problem 5.7.1), and we may write

$$\frac{F_D}{\rho R^2 w^2} = C_D = f(Re) \tag{5.36}$$

Suppose that we can determine this function relating $C_D$ to $Re$ (experimentally or theoretically). Then knowing $Re$ and $\rho R^2 w^2$, say by measuring these quantities, we immediately can compute the force of drag on a sphere. This procedure, for example, is used in describing the forces acting on sediment grains in a current, and is implicit in Stokes's law (1.24) for settling particles. Let us now place some of these ideas of dimensional analysis on a more formal basis.

## 5.3 BUCKINGHAM PI THEOREM

Reconsider formula (5.24) for the average velocity $u_{ave}$ in a conduit with parallel walls. Let us rewrite this formula as

$$1 = k\frac{b^2}{\mu u_{ave}}\frac{\Delta p}{L} \tag{5.37}$$

and observe that now each side is dimensionless. We shall refer to the dimensionless grouping of physical quantities on the right side as a "Pi term," and denote it by $\Pi_1$. Subtracting one from both sides, we may then write

$$k\Pi_1 - 1 = 0 \tag{5.38}$$

$$f(\Pi_1) = 0 \tag{5.39}$$

which indicates that the relation between the physical quantities in this problem can be expressed in terms of one dimensionless product $\Pi_1$. This example leads to a general statement of the Buckingham Pi theorem, named after Lord Buckingham, who in 1915, introduced it as a strategy for reducing the number of unknown quantities in a physical problem.

> For $n$ physical quantities composed of $m$ fundamental quantities, the relation between the physical quantities can be expressed as a function of $n - m$ independent dimensionless products—$\Pi$ terms—composed of the $n$ quantities:
>
> $$f(\Pi_1, \Pi_2, \Pi_3, \ldots, \Pi_{n-m}) = 0 \tag{5.40}$$
>
> where any one of the $\Pi$ terms consists of no more than $m + 1$ physical quantities.

To use this theorem, two essential rules must be followed. First, one selects $m$ *repeating quantities,* none of which is dimensionless, and no two having the same dimensions. All $m$ fundamental quantities must be included collectively in these repeating quantities. Second, each $\Pi$ term is expressed as the product of the repeating quantities, each to an unknown exponent, and one other quantity to a known power, normally one.

Consider the problem of determining the volumetric discharge $Q$ of a Newtonian fluid within a horizontal circular tube of constant radius $R$—a problem that is closely related to that of flow within a conduit with parallel walls. Assume that flow is steady and arises from a uniform pressure loss $\Delta p$ per unit tube length $L$, that is, $\Delta p/L$. Further assume that, in addition to $Q$, essential quantities include $R$, $\mu$, and $\Delta p/L$, so $n = 4$. Because this is a purely mechanical problem, fundamental quantities include mass, length, and time, so $m = 3$. Thus, $n - m = 4 - 3 = 1$, so the problem involves one $\Pi$ term. Listing dimensions of the four quantities:

$$Q: \quad L^3 t^{-1}$$

$$R: \quad L$$

$$\mu: \quad ML^{-1}t^{-1}$$

$$\Delta p/L: \quad ML^{-2}t^{-2}$$

Now, according to the first rule of the Pi theorem above, one of the three repeating quantities must be either $\mu$ or $\Delta p/L$ because these are the only quantities that include the dimension of mass. Otherwise, the two other repeating quantities may include any of those remaining since either choice of the first repeating quantity includes all three fundamental quantities. Suppose we choose for repeating quantities $Q$, $R$, and $\mu$. Then, according to the second rule of the Pi theorem,

$$\Pi_1 = Q^A R^B \mu^C \frac{\Delta p}{L} \tag{5.41}$$

Writing a dimensional equation,

$$\Pi_1 = (L^3 t^{-1})^A L^B (ML^{-1} t^{-1})^C ML^{-2} t^{-2} \tag{5.42}$$

Solving for the unknown exponents associated with mass, length, and time, as in the preceding section:

$$\text{mass:} \quad 0 = C + 1 \tag{5.43}$$

$$\text{length:} \quad 0 = 3A + B - C - 2 \tag{5.44}$$

$$\text{time:} \quad 0 = -A - C - 2 \tag{5.45}$$

From these we obtain $A = -1$, $B = 4$, and $C = -1$. Substituting into expression (5.41) above,

$$\Pi_1 = \frac{R^4}{\mu Q} \frac{\Delta p}{L} \tag{5.46}$$

It follows, by reversing the steps leading to (5.39), that

$$Q = k \frac{R^4}{\mu} \frac{\Delta p}{L} \tag{5.47}$$

We thus may anticipate that the discharge $Q$ is proportional to the fourth power of the radius $R$. This is the "fourth-power law" for discharge within a circular tube, analogous to the "cubic law" (3.74) for discharge within a parallel-wall conduit. We shall see later that the dimensionless coefficient $k$ equals $\pi/8$, and that the fourth-power law has important implications regarding flow through porous media.

## 5.4 GEOMETRICAL SIMILITUDE

### 5.4.1 Exact

It should be evident that two spheres, although possibly having different radii, nonetheless are exactly geometrically similar. Likewise, for example, all cubes define a set of exactly, geometrically similar objects. This idea of exact geometrical similitude immediately can be extended to objects possessing shapes more complicated than spheres and cubes. In particular, two objects are geometrically similar if they can be arranged such that any position vectors $\mathbf{r}_1$ and $\mathbf{r}_2$ for corresponding points of the two objects are related by

$$\mathbf{r}_2 = l\mathbf{r}_1 \tag{5.48}$$

where $l$ is a scaling factor. This implies also that any geometrically corresponding linear dimensions of the two objects form a fixed ratio equal to $l$ (Figure 5.1). One object possesses all essential characteristics of the other, reproduced faithfully to

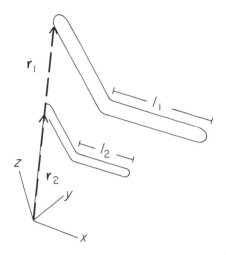

**Figure 5.1** Sketch of two boomerangs with examples of corresponding position vectors $\mathbf{r}_1$ and $\mathbf{r}_2$ and linear dimensions $L_1$ and $L_2$, related by scale factor $l$.

scale. Note, moreover, that the same idea can be applied to objects other than three-dimensional solids. For example, right-circular cylinders are geometrically similar objects, and conduits having rectangular cross sections, whose width-to-breadth ratio is a constant, are geometrically similar. In addition, one can refer to geometrical similarity of streamlines in a flowing fluid, a condition we will examine below with regard to dynamical similitude.

These ideas of geometrical similitude can be extended to *systems* of objects. For example, a system composed of two spheres having radii of 1 and 1.5, whose centers are separated by a distance of 4, is geometrically similar to a system composed of two spheres having radii of 2 and 3, whose centers are separated by a distance of 8. Note in this example that each of the two spheres is scaled in size by a factor of $l = 2$, and so too is the distance separating them. Similarly, a system composed of two juxtaposed right-circular cylinders having radii of 1 and 3, whose centers are separated by a distance of 4, is geometrically similar to a system composed of correspondingly arranged cylinders having radii of 0.5 and 1.5, whose centers are separated by a distance of 2. Extending this idea further, consider an idealized porous medium composed of regularly packed spheres with equal radii. One should recall from sedimentology that geometrical similitude is the essential feature defining the unit cell of such packing. For example, in cubic or "open" packing the unit cell is a cube whose corners coincide with the centers of the eight spheres involved; in "closest" packing the unit cell is a rhombohedron. One can easily envision that two similar unit cells involving two different sizes of spheres readily can be arranged such that condition (5.48) is satisfied; and it may be noted that all geometrically corresponding linear dimensions of the unit cells, distances between sphere centers, are scaled by the same factor $l$ as the sphere radii. A consequence is that the pore shapes within one cell are geometrically similar to those in the other.

A similar idea arises in considerations of fluid flow next to roughened surfaces, where the roughness is composed of regularly spaced elements. Examples you may have observed include the roughness produced by protruding rivets placed in a regular pattern on an airplane wing, or the roughness produced by baffles in a sluice. Like packed spheres, one can define unit cells for these patterns of roughness, but in an areal rather than volumetric sense. Characteristic dimensions of such unit

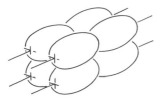

**Figure 5.2** Idealized porous medium composed of rectangularly packed, prolate spheroids.

cells can involve the size, shape, orientation, and spacing of roughness elements. Herein two observations are particularly noteworthy:

First, properties that indirectly indicate whether geometric similitude exists between two systems of elements sometimes can be identified. We may note, for example, that two systems of spheres with cubic packing possess the same porosity and dry bulk density (solid mass per unit volume). Similarly, two systems of roughness produced by rivets whose size and spacing of the rivets scale together possess the same areal density (number of elements per unit-cell area). Herein it is tempting to assume that two systems for which such properties are equal are geometrically similar. However, it does not necessarily follow, for example, that two sphere arrangements are geometrically similar if they have the same porosity and bulk density. In fact, equality of such properties is a necessary, but insufficient, condition for geometrical similitude.

The second noteworthy observation pertains to system dimensions. In the case of a system of regularly packed spheres, for example, the elements are themselves geometrically similar. As a consequence, the geometries of such systems are entirely specified by the sphere radii and unit cell dimensions in the sense that, by knowing the type of packing *a priori,* the only other information needed to distinguish between two systems is a single linear dimension. In cubic packing we require only the sphere radius, from which we can immediately extract any other characteristic dimension, for example, the unit cell size. Suppose that, instead of spheres, we consider an idealized porous medium composed of rectangularly packed, prolate ellipsoids whose long axes are parallel (Figure 5.2). If the eccentricity of the ellipsoids is specified, then again, knowing one characteristic linear dimension is sufficient to distinguish one geometrically similar system from another. If eccentricity is not specified, however, we require two characteristic linear dimensions—the lengths of the long and short ellipsoid axes—to determine whether two systems are similar. Moreover, we must increase the required number of characteristic dimensions with further complexity, for example, in the case of systems of ellipsoids where the lengths of all three axes vary. The lesson is this: With increasing complexity, the number of dimensions required to determine whether two systems are geometrically similar also increases. For systems of elements like the examples above, this number can increase rapidly with only a few additional variations in the shape, size, orientation, and spacing of the elements. This is one motivation for considering the idea of *statistical similarity* in the next section. Finally, reconsider the fractal model of a porous medium introduced in Example Problem 2.4.2; note that it possesses a *self-similar* geometry that is independent of scale.

## 5.4.2 Statistical

Few geological flow problems involve systems of simply shaped, regularly arranged elements. So, often it is necessary to replace the exacting idea of geometrical similitude above with a statistical one. Two examples serve to illustrate this.

Consider a system composed of many spheres with different radii. Suppose that the spheres are randomly packed, as opposed to the regular arrangement associated with cubic or rhombohedral packing. Further consider a finite sample volume that includes many spheres. Whereas the individual spheres are geometrically similar, any two such sample volumes may contain spheres that do not correspond, in the sense that for each sphere in one volume, there may not be a corresponding sphere with exactly the same radius in the other volume. In addition, due to the random packing, any specific arrangement—the relative positions and spacing—of two or more spheres in one volume is unlikely to have a corresponding match in the other volume. Thus, the idea of identical similarity is useless here.

With a sufficient number of spheres in each sample volume, however, one can envision that certain characteristic properties of the ensembles of spheres in both volumes may be the same in a statistical sense. For example, the distribution (histogram) of sphere radii in the two volumes can be the same. Then, each sphere radius in one volume possesses, in a statistical sense, a counterpart in the other. Similarly, equal porosities reflect, in a statistical sense, a similar average spacing between spheres in each volume.

Extending this idea, consider two systems of randomly packed spheres whose distributions of sphere radii have different means and values of dispersion (for example the standard deviation). A first necessary condition for geometric similarity between the systems is that each sphere radius in a sample volume from one system has a scaled counterpart in a sample volume from the other system. Without proof, these conditions are satisfied if the *form* of the distribution of radii in log-transformed space is the same for each system, and if the standard deviation in log-transformed space is equal. For example, two sphere mixtures whose radii are log-normally distributed are geometrically similar if their standard deviations (log coordinates) are equal. In addition, geometric similarity requires equal porosities (or bulk densities). Finally, one may add the requirement of geometrically similar grain shapes if they are not spheres. Then, similarity exists when, for any given $b$-axis value, the same proportion of grain eccentricities, as measured by ratios of the $a$ and $c$ axes to the $b$ axis, are present in both mixtures. These ideas can be formalized using the mathematics of probability; those familiar with probability density functions may wish to examine Example Problem 5.7.3.

Similar ideas can be applied to random roughness. Consider the experiment in which well-sorted, equidimensional sand grains are closely, but randomly, packed in a single-grain layer and glued to an underlying smooth surface. Because the grains are closly packed, the number of grains per unit area essentially is constant over the roughness. One then can envision that two such layers composed of different grain sizes ought to be visually, geometrically similar, despite the fact that the detailed textures produced by the two roughnesses are everywhere distinct at the scale of individual grains. Such roughness is referred to as *Nikuradse sand roughness,* named after J. Nikuradse, who in 1933, was the first to systematically examine the effects of roughened surfaces on adjacent fluid flows. A single linear dimension equal to one grain diameter, the roughness height $k_s$, is sufficient to distinguish one surface roughened in this manner from another. We shall see later that the roughness height is an important parameter used in describing the velocity profile of a turbulent flow next to a roughened surface. As with porous media composed of different grain sizes, the number of dimensions required to define geometrical similitude increases for roughness produced by poorly sorted grain mixtures, particularly if the grains are imbricated, as on a streambed.

## 5.5 DYNAMICAL SIMILITUDE

Let us examine the conditions under which two flows, involving different fluids with different velocities, exhibit geometrically similar streamlines in the vicinity of geometrically similar objects with different linear dimensions. Such fluid motions are *dynamically similar;* at all geometrically corresponding positions within them, the forces acting on the fluids exhibit a fixed ratio. For convenience we shall consider only viscous and inertial forces; elastic forces related to changes in fluid volume and gravitational forces are to be excluded. Thus we shall consider flow of an incompressible fluid in absence of a free surface. In addition, we shall assume steady flow; fluid properties therefore do not vary with time. Under these conditions, dynamical similitude is satisfied if at all corresponding positions within the flows the ratio of inertial and viscous forces is constant.

Consider flow in the vicinity of a sphere. For fluid motion parallel to the $x$-coordinate direction, the inertial force per unit volume at a given position in the flow is $\rho u \partial u/\partial x$. Here $\rho$ is the fluid density and $u$ is the fluid velocity. Thus the inertial force at each position is proportional to the product of the local velocity $u$ and the local change in $u$ with respect to position $x$. The quantity $u \partial u/\partial x$ denotes what we shall describe later as a convective acceleration. The viscous force per unit volume is $\mu \partial^2 u/\partial y^2$, where $\mu$ is the fluid viscosity. Thus the viscous force at each position is proportional to the second derivative of the velocity $u$ with respect to the transverse coordinate $y$. (Students are urged to reexamine this section once ideas of convective accelerations and viscous forces are covered in later chapters.) A condition of dynamical similarity for corresponding positions within two flows thus is defined by the ratio

$$\frac{\text{inertial force}}{\text{viscous force}} = \frac{\rho u \dfrac{\partial u}{\partial x}}{\mu \dfrac{\partial^2 u}{\partial y^2}} = \text{constant} \qquad (5.49)$$

Let us assume that the local velocity $u$, the velocity gradient $\partial u/\partial x$, and the second derivative of velocity $\partial^2 u/\partial y^2$ vary in proportion to the magnitudes of characteristic quantities of the flow. These include, in addition to the density $\rho$ and the viscosity $\mu$, a characteristic velocity, for example, the free stream velocity $U$, and a characteristic length, for example, the sphere radius $R$. Formally applying dimensional analysis to the quantities $u \partial u/\partial x$ and $\partial^2 u/\partial y^2$, one obtains the result that $u \partial u/\partial x$ is proportional to $U^2/R$ and $\partial^2 u/\partial y^2$ is proportional to $U/R^2$. Comparing with (5.49),

$$\frac{\rho \dfrac{U^2}{R}}{\mu \dfrac{U}{R^2}} = \frac{\rho U R}{\mu} = \text{constant} \qquad (5.50)$$

which will be recognized from Section 5.2.3 as the Reynolds number $Re$ for a sphere. Thus two flows are dynamically similar when their Reynolds numbers are equal.

Now consider flow in the vicinity of an object other than a sphere, for example, a cube. Streamlines around a cube oriented with two faces normal to the direction of the free stream velocity $U$ will appear different from streamlines around a cube oriented such that none of its faces are normal to the direction of $U$—regardless of cube size and fluid viscosity. This leads to an important point regarding the orientation of geometrically similar objects with respect to flow direction: Whereas two geometrically similar objects can be arranged to satisfy the condition (5.48), unless this condition *is* satisfied for a given flow problem, such that the two objects have the same orientation with respect to the characteristic direction of flow, they cannot be dynamically similar. Geometric similitude is therefore a necessary, but insufficient, condition for dynamical similitude. As a consequence, a Reynolds number defined for one orientation of a nonspherical object is not directly comparable with a Reynolds number defined for another orientation; each characterizes flow in the vicinity of a distinct system.

Considering systems composed of many randomly arranged elements, one cannot hope for geometrical similarity of streamlines, for the same reasons that one cannot expect exact similarity of the elements. Dynamical similarity in a strict sense therefore is not possible. Nonetheless, experience suggests that for systems composed of a sufficiently large number of elements, where geometrical similarity is satisfied in a statistical sense, a Reynolds number can usefully be defined based on characteristic (statistical) properties of the system. For example, a Reynolds number for flow through porous media can be defined using the specific discharge $q_h$ as a characteristic velocity, and the mean pore radius $\overline{R}$ as a characteristic length (Chapter 13). (Alternatively one may choose the average molecular velocity $q_h/n$, where $n$ is the porosity, and the average grain diameter $\overline{d}$.) Analogous to previous considerations of distinct shapes of objects, a Reynolds number defined for a medium composed of coarse, spherical sand grains is in a strict sense not directly comparable to a Reynolds number defined for a medium composed of fine angular sand and silt. Likewise, in analogy to the example of cubes with different orientations, streamlines associated with flow parallel to a preferential orientation of platy grains will be different from streamlines associated with flow normal to such a grain fabric; Reynolds numbers in each case are distinct.

## 5.6 CHARACTERISTIC DIMENSIONLESS QUANTITIES

It should be apparent from the discussion above that the Reynolds number frequently appears in fluid flow problems. Its importance and use in specific cases will be examined in subsequent sections. Meanwhile, let us briefly examine several other important dimensionless quantities that are used to characterize fluid systems and flows, and which will appear in numerous places throughout the text.

The *Froude number,* denoted by $Fr$, is defined as the ratio of inertial and gravitational forces. As in defining the Reynolds number, we shall assume that local quantities vary in proportion to the magnitudes of characteristic quantities of the flow. We then can infer that the inertial force, which must equal the product of mass and acceleration, may be expressed as a product of density, volume, and the ratio of velocity over time, or $\rho U^2 L^2$, where $U$ is a characteristic velocity and $L$ is a characteristic length. The gravitational force (weight), which must equal the product of mass and acceleration due to gravity $g$, may be expressed as $\rho L^3 g$. Thus the Froude number is

$$Fr = \frac{\rho U^2 L^2}{\rho L^3 g} = \frac{U^2}{Lg} \tag{5.51}$$

The Froude number is relevant to flow problems involving a free surface. In this context it alternatively may be expressed as

$$Fr = \frac{U}{\sqrt{Lg}} \tag{5.52}$$

The speed of shallow-water waves, where the depth of water is small relative to the wavelength, is proportional to the square root of the product of water depth and $g$ (Example Problem 10.3.3). The speed of deep-water waves, where the depth of water is large relative to the wavelength, is proportional to the square root of the product of wavelength and $g$. Letting $L$ denote water depth in the case of shallow-water waves, and wavelength in the case of deep-water waves, evidently the Froude number in the form of (5.52) is the ratio of the flow velocity and the wave speed. This has an intuitively appealing physical interpretation. Consider the waves produced by throwing a pebble into a flowing stream. When $Fr < 1$, the wave speed is greater than the flow velocity, so waves can move upstream; when $Fr \geq 1$ they cannot.

The *Mach number,* denoted by $M$, is defined as the ratio of the fluid velocity $U$ to the speed of sound $c_s$ in the fluid. The speed of sound equals $\sqrt{E/\rho}$, where $E$ is the modulus of elasticity. Thus the Mach number is

$$M = \frac{U}{c_s} = \frac{U}{\sqrt{E/\rho}} \tag{5.53}$$

This has a qualitative interpretation similar to that for the Froude number. When $M < 1$, pressure waves (sound) can propagate upstream; when $M \geq 1$ they cannot. Recall from Section 3.5.1 that we used the Mach number to define conditions under which it is permissible to treat the flow of a compressible fluid as incompressible.

The *Peclet number,* denoted by $Pe$, is important in coupled flows involving mass, heat, and solutes, and is defined as the ratio of advective and diffusive transport rates. In the case of heat, it is the ratio of the rate of heat transport by advection to the rate of transport by conduction. In the case of solutes (for example, in ground water), the Peclet number is the ratio of the rate of solute transport by advection to the rate of transport by molecular diffusion. In relation to such coupled flows, we also will examine four dimensionless quantities—the *Rayleigh, Grashof, Nusselt,* and *Prandtl numbers*—in our coverage of buoyancy-driven convection (Chapters 6 and 16).

## 5.7 EXAMPLE PROBLEMS

### 5.7.1 Drag on a spherical particle using Buckingham Pi theorem

Let us reconsider the problem of drag on a spherical particle, making use of the Buckingham Pi theorem. Recall that essential quantities include, in addition to the drag force $F_D$, the sphere radius $R$, the velocity $w$, fluid density $\rho$, and fluid viscosity $\mu$, so $n = 5$. Fundamental quantities include mass, length, and time, so $m = 3$. Thus $n - m = 2$, so the problem involves two $\Pi$ terms. Listing dimensions of the five quantities:

$$F_D: \quad MLt^{-2}$$

$$R: \quad L$$

$$w: \quad Lt^{-1}$$

$$\rho: \quad ML^{-3}$$

$$\mu: \quad ML^{-1}t^{-1}$$

Choosing repeating quantities $\mu$, $R$, and $w$, and the additional quantity $F_D$, a first $\Pi$ term is

$$\Pi_1 = \mu^A R^B w^C F_D \tag{5.54}$$

Writing a dimensional equation,

$$\Pi_1 = (ML^{-1}t^{-1})^A L^B (Lt^{-1})^C MLt^{-2} \tag{5.55}$$

Solving for unknown exponents associated with mass, length, and time, one obtains $A = -1$, $B = -1$, and $C = -1$. Substituting into (5.54),

$$\Pi_1 = \frac{F_D}{\mu R w} \tag{5.56}$$

A second $\Pi$ term similarly is determined from the three repeating quantities and the additional quantity $\rho$:

$$\Pi_2 = \mu^A R^B w^C \rho \tag{5.57}$$

Again writing a dimensional equation,

$$\Pi_2 = (ML^{-1}t^{-1})^A L^B (Lt^{-1})^C ML^{-3} \tag{5.58}$$

Solving for unknown exponents, one obtains $A = -1$, $B = 1$, and $C = 1$. Substituting into (5.57),

$$\Pi_2 = \frac{\rho R w}{\mu} \tag{5.59}$$

The relation between the two $\Pi$ terms then may be expressed as

$$f(\Pi_1, \Pi_2) = 0 \tag{5.60}$$

or alternatively as

$$\Pi_1 = g(\Pi_2) \tag{5.61}$$

Substituting for the $\Pi$ terms,

$$\frac{F_D}{\mu R w} = g\left(\frac{\rho R w}{\mu}\right) = g(Re) \tag{5.62}$$

The ratio $F_D/\mu R w$ evidently is a function of the Reynolds number $Re$ for a sphere. Comparing this result with (5.36), we observe that they do not agree aside from their functional relation to $Re$. Herein is an important lesson:

Dimensional analysis, as we have used it so far, offers little guidance as to which of the two expressions, (5.36) or (5.62), is the appropriate one to use. Further dimensional arguments, however, indicate that the drag $F_D$, involving both viscous and pressure forces, is proportional to $\rho R^2 w^2$, making (5.36) the convenient choice (consider, for example, the discussion by Tritton [1988], pp. 93–94). It is left as an exercise to derive (5.36) using the Buckingham Pi theorem. To do this, choose for repeating quantities $\rho$, $R$, and $w$. In addition, experimentally the clear choice is (5.36), as we shall now see.

It has become conventional to multiply the denominator of (5.36) by the factor $\frac{1}{2}\pi$, because the quantity $\frac{1}{2}\rho w^2$ equals what we will see later is an important quantity, the kinetic energy per unit volume, and because $\pi R^2$ is the silhouette area of the sphere. The drag coefficient $C_D$ then is

$$C_D = \frac{F_D}{\frac{1}{2}\pi\rho R^2 w^2} \tag{5.63}$$

This coefficient has been determined from settling experiments by several investigators. Evidently values of $C_D$ plot as a straight line in log–log coordinates for low Reynolds numbers (Figure 5.3). On purely empirical grounds, we may write a relation of the form

$$\log C_D = m \log Re + b \tag{5.64}$$

where $m$ is the slope of this line in base-ten log coordinates, and $b$ is the log-value of $C_D$ when $Re = 1$. A line fit by eye projected to $Re = 1$ gives a value for $C_D$ of about 12.5, so $b \approx \log 12.5$. The value of $m$ evidently is very close to $-1$ (Figure 5.3). Using these values and performing the following operation on (5.64),

$$10^{\log C_D} = 10^{m \log Re} 10^b \tag{5.65}$$

$$C_D = \frac{12.5}{Re} \tag{5.66}$$

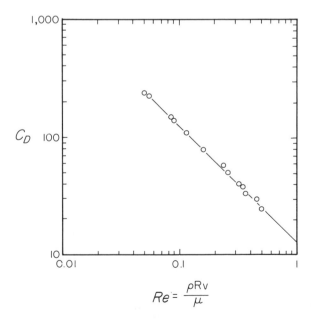

$$Re = \frac{\rho R v}{\mu}$$

**Figure 5.3** Experimental (*circles*) and fitted log-linear relation (*line*) between drag coefficient $C_D$ and low Reynolds number $Re$ for a sphere; data from Schlichting (1979).

From (5.63),

$$F_D = C_D \left( \frac{1}{2} \pi \rho R^2 w^2 \right) \tag{5.67}$$

$$F_D = \frac{12.5}{Re} \left( \frac{1}{2} \pi \rho R^2 w^2 \right) \tag{5.68}$$

With $Re = \rho R w / \mu$, this becomes $F_D = 6.25 \pi \mu R w$, which is very close to the theoretical solution due to Stokes: $F_D = 6 \pi \mu R w$. Perhaps a bit more care in fitting the line (Figure 5.3) would obtain

$$C_D = \frac{12}{Re} \tag{5.69}$$

which leads to the result $F_D = 6 \pi \mu R w$.

We possess in this example the luxury of having the theoretical solution of Stokes, with which we can compare our empirical solution for the drag on a sphere. In more difficult problems, however, such theory may not be available, so the value of an empirical approach of this sort, made available to us by dimensional analysis, is evident. Let us now turn to such a problem.

### 5.7.2 Drag on spinose foraminifera

Planktonic foraminifera need to stay near the ocean surface to capture energy from sunlight. To counter gravitational settling, these foraminifera adjust their buoyancies, in part, by manufacturing low-density lipids or gases. This requires expending biochemical energy. In absence of this expense, a foraminifer must settle (or ascend) in response to buoyant and viscous forces that are determined by its purely mechanical state, as does any particle. Thus the biochemical energy that a foraminifer expends to adjust its buoyancy is, in principle, related to the rate at which it would otherwise settle if it did not expend this energy. Since settling velocity, in turn, depends on the material properties and shape of a foraminifer, the possibility exists that these attributes are adapted to modulate settling, a hypothesis first suggested by Häcker in 1908.

Consider foraminifera that have spines, for example *Globigerinoides sacculifer* and *Orbulina universa* (Figure 5.4). Growing spines produces two counteractive effects. Spines increase the weight of a foraminifer, and therefore tend to increase its settling speed; but they also increase the fluid drag on the foraminifer, and therefore tend to decrease its settling speed. This suggests that growing spines may help impede settling, insofar as increasing drag by growing spines outweighs the disadvantage of increasing weight.

True spines, composed of $CaCO_3$, characteristically are acicular, with the crystallographic *c*-axis parallel to spines axes. Among true spines, cross-sectional shapes vary from circular to triangular to triradiate. In some species, circular spines become triangular near their tips, and triangular spines become triradiate. Adaptive explanations for spines include defense from predation, and structural support during test growth. Spines also support photosynthetic endosymbionts. But regardless of historical reasons for their existence, spines must influence the settling speed of a foraminifer due to their finite weight and extension into the surrounding fluid.

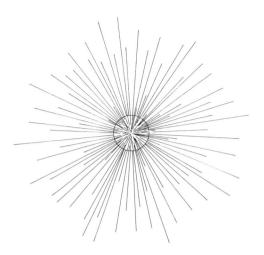

**Figure 5.4** Simplified sketch of spinose foraminifer *Orbulina universa*; after H. B. Brady in his report on the foraminifera dredged by H. M. S. *Challenger* during the years 1873–1876 (Barker, 1960).

For simplicity envision a foraminifer test as a sphere with radius $R_t$, and a spine as a right circular rod with length $l_s$ and radius $r$. Any protoplasm surrounding the test may be envisioned as being geometrically part of the test; then $R$ denotes a radius for the test and protoplasm combined, and $l$ denotes the length of a spine that extends beyond the protoplasm. Thus $l = l_s - (R - R_t)$; and in absence of protoplasm, $R = R_t$ and $l = l_s$. Let $\rho$ denote the density of the liquid in which a foraminifer is immersed, and let $\rho_e$ denote the density of the foraminifer. Since a test actually consists of a porous shell with protoplasm, test and spine densities may differ; $\rho_e$ therefore is an effective density computed from the total mass and total volume of spines, test, and surrounding protoplasm.

The first task is to define an appropriate Reynolds number $Re$ and coefficient of drag $C_D$ for a foraminifer, from which the fluid drag associated with spines can be computed. Here arises the temptation to compute the drag on a spine of given length, multiply by the number of spines, then add this quantity to the drag on a (spherical) test. These drag forces, however, are in general not additive because the flow field around an individual spine on a test is strongly influenced by neighboring spines and by its orientation with respect to the direction of settling. This flow field, and therefore drag on the spine, are very different from the flow field and drag that otherwise would occur if the spine was isolated within a flow. For similar reasons, drag arising from viscous forces in the immediate vicinity of a spinose test (near the base of the spines) will generally not equal the drag on an otherwise identical test isolated from its spines.

Consider geometrically similar foraminifera. This means that $R$, $l$, and $r$ possess a fixed ratio, $R : l : r$, for constant $n$; each foraminifer is a scaled version of another. For this situation the Reynolds number is

$$Re = \frac{\rho R w}{\mu} \qquad (5.70)$$

Note that the Reynolds number defined by (5.70) for a foraminifer with a specific ratio $R : l : r$ and value of $n$ does not represent the same thing as the Reynolds number defined by (5.70) for a foraminifer possessing a different ratio $R : l : r$ or different value $n$. That is, a particular value of $Re$ represents a different physical abstraction when comparing geometrically dissimilar systems. We may then define a coefficient of drag $C_D$ as in (5.63). Although it is not essential for the quantity $\pi R^2$ in the denominator of (5.63) to equal the silhouette area of a foraminifer, note that this condition is satisfied in the limiting case of a spherical test (and surrounding protoplasm) without spines.

For a specified Reynolds number, dynamical similitude is not necessarily satisfied with respect to spine arrangements on foraminifera. In the case of geometrically similar foraminifera, nonetheless, an enlightened guess suggests that their quasi-spherical symmetry ought to lead to a drag law like that for spheres. Indeed, a foraminifer has the shape of a sphere in the limit where $n$, $l$, or $r$ approaches zero. With one to three spines, however, associated planes of symmetry do not separate a foraminifera into hemispheres having spherical symmetry. This means, for example, that the drag on a foraminifera that is settling parallel to a plane containing three spines is in general different than the drag when it is settling normal to this plane. Dynamical similitude is not satisfied for arbitrary rotation of the foraminifer. But with an increasing number of spines, we might expect that the exact orientation of a foraminifer during settling becomes less significant in terms of its influence on the total drag. That is, geometrical similitude may exist in a statistical sense with moderate to large $n$, which provides a condition of (statistical) dynamical similitude for arbitrary rotation.

In general, the curve relating $C_D$ to $Re$ ought to be displaced vertically above that for spheres. For geometrically similar foraminifera, this relation is

$$C_D = \frac{C}{Re} \tag{5.71}$$

where the constant $C$ pertains to a specific ratio $R : l : r$ and number $n$, and replaces the factor 12 in (5.69) for spheres. Thus, knowing how $C$ varies with the ratio $R : l : r$ and the number $n$ completely specifies the influence of $l$, $r$, and $n$ on drag and settling velocity for given $Re$.

By Archimedes's principle the buoyant weight $W$ of a spherical test with $n$ spines is

$$W = \pi g (\rho_e - \rho)\left(\frac{4}{3}R^3 + nlr^2\right) \tag{5.72}$$

where $g$ is acceleration due to gravity. Assuming zero acceleration, $W = F_D$; substituting (5.72) into (5.63) and (5.70) into (5.71), then combining these and solving for $w$,

$$w = \frac{8g(\rho_s - \rho)R^2\left(1 + \dfrac{3nlr^2}{4R^3}\right)}{3\mu C} \tag{5.73}$$

When $n$, $l$, or $r$ equals zero, by definition $C = 12$, and Stokes's law (1.24) for the settling velocity $w$ of a sphere is retrieved.

The next task involves inferring how $C$ depends on $n$, $l$, and $r$ as these vary independently, a situation in which foraminifera are not geometrically similar. Dimensional analysis suggests for this situation that the coefficient of drag $C_D$ varies as a function of the Reynolds number $Re$ and a set of dimensionless groupings involving variable quantities, including ratios formed from variable linear dimensions and the characteristic linear dimension used in defining the Reynolds number (for example, see Tritton [1988], pp. 90–95). Thus,

$$C_D = C_D\left(Re, n, \frac{l}{R}, \frac{r}{R}\right) \tag{5.74}$$

For geometrically similar foraminifera, $n$, $l/R$, and $r/R$ are constants, and (5.74) reduces to (5.71).

Dimensional analysis cannot provide the form of the function (5.74). In absence of a theory based on the equations of motion, one must therefore obtain the function (5.74) empirically, based on experimental data relating $C_D$ to the variable quantities in (5.74). This is purely a curve-fitting exercise, perhaps guided only by parsimony. Let us therefore propose a simple power function of the form

$$C_D = \frac{12}{Re}\left[1 + C_0(n - n_1)^a \Lambda^b \mathrm{P}^c\right] \tag{5.75}$$

where

$$\Lambda = \frac{l}{R}; \quad \mathrm{P} = \frac{r}{R} \tag{5.76}$$

The coefficients $C_0$ and $n_1$, and the exponents $a$, $b$, and $c$, are to be determined experimentally. Notice that when $\Lambda$ or $\mathrm{P}$ equals zero, or when $n$ equals $n_1$, (5.75) reduces to Stokes's law (5.69). The coefficient $n_1$ is an artifice involved in the curve-fitting procedure; its significance is examined below. Also notice that the exponents $a$, $b$, and $c$ may, in general, possess noninteger values.

Now, equating (5.71) and (5.75) and solving for $C$, then substituting this into (5.73),

$$w = \frac{2g(\rho_e - \rho)R^2\left(1 + \frac{3}{4}n\Lambda\mathrm{P}^2\right)}{9\mu\left[1 + C_0(n - n_1)^a \Lambda^b \mathrm{P}^c\right]} \tag{5.77}$$

This formula describes the settling speed $w$ of a spinose foraminifer in terms of its geometrical properties and fluid properties.

Under natural conditions, it is important to distinguish the components of the effective density $\rho_e$. It is left as an exercise to demonstrate that (5.77) becomes

$$w = \frac{2g(\rho_R - \rho)R^2\left[1 + \dfrac{3(\rho_s - \rho_p)(R - R_t)}{4(\rho_R - \rho)R}n\mathrm{P}^2 + \dfrac{3(\rho_s - \rho)}{4(\rho_R - \rho)}n\Lambda\mathrm{P}^2\right]}{9\mu[1 + C_0(n - n_1)^a \Lambda^b \mathrm{P}^c]} \tag{5.78}$$

where $\rho_R$ is the effective density of the test and surrounding protoplasm (with combined radius $R$) in absence of spines:

$$\rho_R = \rho_p + (\rho_t - \rho_p)\left(\frac{R_t}{R}\right)^3 \tag{5.79}$$

and $\rho_t$, $\rho_s$, and $\rho_p$ denote the densities of the test, spines, and protoplasm individually. When $\rho_s = \rho_R = \rho_e$, (5.78) reduces to (5.77); and when $\Lambda$ or $\mathrm{P}$ equals zero, (5.78) reduces to Stokes's law for a sphere with radius $R$.

Insofar as the form of (5.75) is correct, a comparison of (5.75) and (5.70) suggests that we may define a new Reynolds number $\mathbb{R}e$,

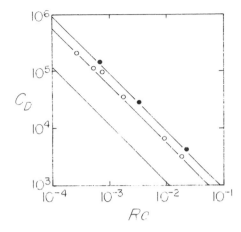

**Figure 5.5** Log-linear relations between drag coefficient $C_D$ and Reynolds number $Re$ for two geometrically similar models ($n = 6$) settled in three liquids (*open circles*) and a single model ($n = 26$) settled in three liquids (*black circles*); lower line is Stokes's law for spheres.

$$\mathbb{R}e = \frac{\rho \mathbb{R} w}{\mu} \tag{5.80}$$

where $\mathbb{R}$ is given by

$$\mathbb{R} = R \left[ 1 + C_0 (n - n_1)^a \Lambda^b \mathbb{P}^c \right]^{-1} \tag{5.81}$$

Data for geometrically dissimilar foraminifera, in principle, should collapse to the same line given by (5.69) in a plot of $C_D$ versus $\mathbb{R}e$.

Experiments to test the formulae above involved settling large models of spinose foraminifera constructed from beeswax and pins within viscous liquids, varying $R$, $l$, $r$, $n$, $\rho_e$, $\rho$, and $\mu$. Experimental liquids included corn syrup and mixtures of four brands of shampoo. For two geometrically similar models ($n = 6$) settled in three liquids, and a single model ($n = 26$) settled in three liquids, values of the logarithm of the drag coefficient $C_D$ are indeed linearly related to the logarithm of the Reynolds number $Re$, with a slope of $-1$ (Figure 5.5). This confirms that the quasi-spherical symmetry of geometrically similar models with moderate $n$ leads to a drag law like that for spheres.

It is straightforward to linearize (5.75) for the purpose of estimating the coefficients $C_0$ and $n_1$, and the exponents $a$, $b$, and $c$, using regression analysis. The settling experiments provided estimates of $C_0 = 0.58$, $n_1 = 4.5$, $a = 0.48$, $b = 2.0$, and $c = 0.60$. A view of the fit of the data is obtained by forming the quantities

$$C_n = \left( \frac{C}{12} - 1 \right) (C_0 \Lambda^b \mathbb{P}^c)^{-1} \tag{5.82}$$

$$C_\Lambda = \left( \frac{C}{12} - 1 \right) \left[ C_0 (n - n_1)^a \mathbb{P}^c \right]^{-1} \tag{5.83}$$

then plotting these quantities against $n$ and $\Lambda$ (Figure 5.6). The significance of the coefficient $n_1$ is now apparent; this is merely an artifice to provide a good fit to the data involving varying $n$. The actual curve must for physical reasons go to zero (broken line, Figure 5.6).

The range in $\Lambda$ used in constructing the models covers that observed in modern foraminifera; a value of $b = 2$ in (5.78) therefore should be valid for real

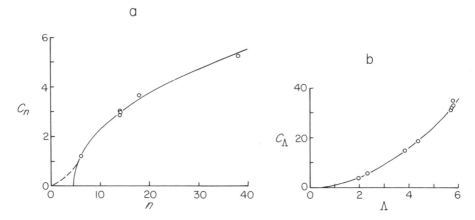

**Figure 5.6**   Plots of settling-speed model fit to experimental data for varying spine number $n$ ($a$) and dimensionless spine length $\Lambda$ ($b$).

foraminifera. The range in P is slightly larger than values for modern foraminifera; a value of $c = 0.60$ therefore is reasonable for real foraminifera assuming that the simple power relation involving P extends to values of P that are slightly smaller than those used in the experiments. The estimated value $a = 0.48$, in contrast, may be limited to moderate $n$ because the largest experimental value of $n$ is still much less than $n$ for some real foraminifera, for example, *Orbulina universa* (although it covers the range in $n$ of juvenile *Orbulina universa*). The actual form of the function (5.75) may be more complicated than a power relation, at least that part of it involving $n$.

A simple test of the form of (5.75), including the estimated values of coefficients and exponents, involves plotting all experimental values of $C_D$ against the modified Reynolds number $\mathbb{R}e$ given by (5.80). The fit about a line given by Stokes's law (5.69) is good, including the datum for a model whose beeswax test was shaped like *Globigerinoides trilobus* (Figure 5.7).

The form of $w(n, \Lambda, P)$ given by (5.78) possesses several interesting properties. Namely, this function may have one or more extrema. For example, taking the partial derivative $\partial w/\partial \Lambda$ of (5.78) and setting this result to zero, any local extremum associated with variations in $\Lambda$ for constant $n$ and P must satisfy the condition:

$$\Lambda^b + \left[ \frac{4b(\rho_r - \rho)}{3(b-1)(\rho_s - \rho)n\mathrm{P}^2} + \frac{b(\rho_s - \rho_p)}{(b-1)(\rho_s - \rho)}\left(1 - \frac{R_t}{R}\right) \right]\Lambda^{b-1}$$

$$- \frac{1}{(b-1)C_0(n - n_1)^a\mathrm{P}^c} = 0 \qquad (5.84)$$

With $b = 2$, the dimensionless spine length $\Lambda_0 = l_0/R$ associated with an extremum can be obtained from (5.84) using the quadratic formula, whence

$$\Lambda_0 = \left( \left[ \frac{4(\rho_r - \rho)}{3(\rho_s - \rho)n\mathrm{P}^2} + \frac{(\rho_s - \rho_p)}{(\rho_s - \rho)}\left(1 - \frac{R_t}{R}\right) \right]^2 + \frac{1}{(b-1)C_0(n - n_1)^a\mathrm{P}^c} \right)^{\frac{1}{2}}$$

$$- \left[ \frac{4(\rho_r - \rho)}{3(\rho_s - \rho)n\mathrm{P}^2} + \frac{(\rho_s - \rho_p)}{(\rho_s - \rho)}\left(1 - \frac{R_t}{R}\right) \right] \qquad (5.85)$$

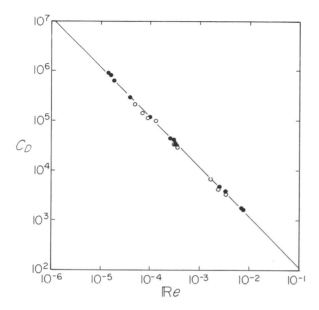

**Figure 5.7** Log-linear relation between coefficient of drag $C_D$ and modified Reynolds number $\mathbb{R}e$ for foraminifera models; data involve experiments used to develop settling-rate model (*black circles*), test cases (*open circles*), and model whose test was shaped like *Globigerinoides trilobus (triangle)*.

which is a positive real root. It is left as an exercise to demonstrate that the extremum associated with this root is a maximum settling speed. This suggests that for given $n$ and P, there is disadvantage in possessing spines whose dimensionless length $\Lambda_0 = l_0/R$ is associated with this maximum, inasmuch as settling rapidly is undesirable. Students are urged to read Takahashe and Bé (1984), who examine settling rates of foraminifera captured in plankton tows.

### 5.7.3 Geometrical similitude of randomly packed, unconsolidated sand grains

Consider the problem of determining whether two porous media composed of randomly packed grains are geometrically similar. For consolidated media, this task involves petrographic measurements of pore sizes and shapes in combination with use of techniques referred to as *porosimetry*. Porosimetry makes use of the relation (3.62) between capillary tube size and capillary pressure, wherein the pore-size distribution of a medium is inferred from measurements of capillary pressure with water desorption, or alternatively, by injecting into the medium known volumes of a fluid with high surface tension, usually mercury. Porosimetry also is used with unconsolidated media. However, these techniques involve uncertainty in applying the relation (3.62) for a cylindrical tube to irregular pores and pore throats, and it is sometimes more convenient to treat pore geometry indirectly by considering variability in grain shapes and sizes, and combining this with measures of packing, such as porosity and bulk density. For simplicity we shall here exclude considerations of grain shape by assuming that the media are composed of different mixtures of well-rounded, albeit unsorted, sand. Students unfamiliar with mathematical statistics may qualitatively substitute the familiar idea of a "histogram" for that of a "probability density function" in the development below (also see Example Problem 1.4.3).

Suppose that a sand mixture consists of three grain diameters—0.5 mm, 1 mm, and 1.5 mm—in the respective proportions 0.5, 0.3, and 0.2. Disregarding grain packing for the moment, it should be evident that a geometrically similar mixture of grains might consist of the three grain diameters 1 mm, 2 mm, and 3 mm, also represented in the respective proportions 0.5, 0.3, and 0.2. Here we observe that each grain in the first mixture has a counterpart, whose diameter is scaled by a factor $k = 2$, in the second mixture. But let us examine this another way.

The probability of randomly selecting a grain with diameter 0.5 mm from the first mixture is 0.5, the probability of selecting a grain with diameter 1 mm is 0.3, and the probability of selecting one with diameter 1.5 mm is 0.2. To create the second geometrically similar mixture, we therefore evidently require that the probability of randomly selecting a grain with diameter $0.5k$ mm $= 1$ mm must equal 0.5, the same as the probability of selecting a grain with diameter 0.5 from the first mixture. Similarly, the probability of selecting a grain with diameter $1k$ mm $= 2$ mm must equal 0.3, and the probability of selecting one with diameter $1.5k$ mm $= 3$ mm must equal 0.2. Let us now generalize this idea to mixtures composed of many different grain sizes.

Let $b$ denote the diameter of a grain, and let $B$ denote a random variable to be associated with $b$. Further let $f_B(b)$ denote a probability density function of grain diameters $B$ in a sand mixture. The probability of selecting from this mixture a grain having a diameter $B$ within the infinitesimal interval $b$ to $b + db$ is $f_B(b)db$. Envision now a second, geometrically similar mixture, created by multiplying the diameter of each grain in the first mixture by a constant $k$. The probability $f_B(b)db$ must be represented in this second mixture by the same proportion of grains having a diameter within the interval $kb$ to $kb + d(kb)$, that is $kb$ to $k(b + db)$. To determine the distribution of this new mixture, such that this condition is satisfied, one must determine the probability density function of the product $C = kB$. It follows from a theorem of probability that the probability density function $f_C(c)$ of the second mixture is given by

$$f_C(c) = f_B(c/k)\left|\frac{d}{dc}(c/k)\right| \tag{5.86}$$

$$= \frac{1}{k}f_B(c/k) \tag{5.87}$$

Now consider the case where grain diameter is expressed as a logarithmic value, that is, $\ln b$, and let $\ln B$ denote a random variable to be associated with $\ln b$. Then let $f_{\ln B}(\ln b)$ denote the probability density function of the logarithms of grain diameters. The probability of selecting from this mixture a grain having a diameter $\ln B$ within the infinitesimal interval $\ln b$ to $\ln b + d(\ln b)$ is

$$f_{\ln B}(\ln b)d(\ln b) \tag{5.88}$$

Consider a second geometrically similar distribution, again created by multiplying the diameter of each grain in the first mixture by a constant $k$, and denote its probability density function by $f_{\ln kB}(\ln kb)$. The probability (5.88) must be represented in this second mixture by the proportion

$$f_{\ln kB}(\ln kb)d(\ln kb) \tag{5.89}$$

Observe that the interval $d(\ln kb)$ equals $d(\ln k + \ln b) = d(\ln b)$, so

$$f_{\ln kB}(\ln kb)d(\ln kb) = f_{\ln kB}(\ln kb)d(\ln b) \tag{5.90}$$

Therefore if the values of $f_{\ln B}(\ln b)$ and $f_{\ln kB}(\ln kb)$ are the same, the probabilities represented by (5.88) and (5.89) are equal. It follows that mixtures whose

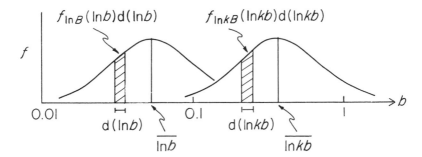

**Figure 5.8** Probability density functions $f_{\ln B}(\ln b)$ and $f_{\ln kB}(\ln kb)$ of logarithms of grain diameters $\ln b$ and $\ln kb$, where $k$ is an arbitrary constant; the two density functions possess the same variance in log-space, and the probability $f_{\ln B}(\ln b)\mathrm{d}(\ln b) = f_{\ln kB}(\ln kb)\mathrm{d}(\ln kb)$.

probability density functions in logarithmic coordinates have the same form, although possibly having different means, are geometrically similar (Figure 5.8). Note that this requires the same dispersion in logarithmic coordinates, for example, the same log-variance. As an example, if $f_{\ln B}$ is a normal probability density function—that is, $f_B$ is log-normally distributed—all mixtures that are log-normally distributed with the same log-variance are geometrically similar.

The condition that two mixtures have the same log forms of probability density functions, however, does not guarantee that two media composed of the mixtures are geometrically similar. This is due to the possibility that the mixtures are packed differently. So an additional requirement for two media to be geometrically similar, assuming a random arrangement of grains, is that their porosities (or bulk densities) must be equal (Figure 3.21). This is analogous to the situation where unit cells of cubically packed spheres, regardless of sphere size, possess the same porosity. Finally, note that these ideas can be generalized to include considerations of grain shape; however, the mathematics involve conditional, multidimensional probability density functions, which are beyond the introductory level intended here.

## 5.8 READING

Barker, R. W. 1960. Taxonomic Notes on the Species Figured by H. B. Brady in his Report on the Foraminifera Dredged by H. M. S. *Challenger* During the Years 1873–1876. Society of Economic Paleontologists and Mineralogists, Special Publication No. 9, Tulsa, Okla. This monograph contains reproduced plates of many beautifully illustrated foraminifera sampled on the *Challenger* voyage.

Furbish, D. J. 1987. Conditions for geometric similarity of coarse stream-bed roughness. *Mathematical Geology* 19:291–307. This paper describes the necessary conditions for roughnesses created by coarse sediment to be geometrically similar, and provides a qualitative description of the relevance of this to boundary drag in turbulent flow.

Häcker, V. 1908. Tiefseeradiolarien. Wiss. Ergebn Deutsch. Tiefsee Exp. *Valdivia* 417–706. Häcker evidently was the first to suggest that foraminiferal morphologies, including appendages, might be adapted to modulate settling speeds.

Nikuradse, J. 1933. Strömungsgesetze in rauhen Rohren. Forschung auf dem Gebiete des Ingenieur-Wesens, 361, Berlin. This paper reports Nikuradse's systematic experiments concerning the influence of roughened surfaces on flow within circular pipes.

May, M. 1991. Aerial defense tactics of flying insects. *American Scientist* 79:316–28. This paper describes an ingenious experiment to measure the drag on a live cricket suspended from a pendulum tether, an experimental procedure that is precluded in the case of spinose foraminifera due to their fragility.

Reynolds, O. 1983. An experimental investigation of the circumstances which determine whether the motion of water shall be direct or sinuous, and the law of resistance in parallel channels. *Philosophical Transactions of the Royal Society* 174:935–82. This is the paper in which Osborne Reynolds originally examined the ratio of inertial and viscous forces, which now bears his name, to clarify the conditions leading to the onset of turbulent flow in a conduit.

Schlichting, H. 1979. *Boundary-layer theory,* 7th ed. New York: McGraw-Hill, 817 pp. This text, translated from the German by J. Kestin, is a classic treatise on flows near boundaries. It provides a clear description of the dimensional arguments leading to computations of the drag on a sphere (pp. 12–18), and a discussion of theoretical solutions due to Stokes and Oseen (pp. 112–16). Data in Figure 5.3 are taken from Figure 1.5 (p. 17) in Schlichting.

Takahashi, K. and Bé, A. W. H. 1984. Planktonic foraminifera: factors controlling sinking speeds. *Deep-Sea Research* 31:1477–1500. This paper reports on settling experiments performed with foraminifera captured in plankton tows and recovered from deep sea sediments.

Tritton, D. J. 1988. *Physical fluid dynamics,* 2nd ed. Oxford: Oxford University Press. This text includes a clear discussion (pp. 93–94) of the dimensional arguments leading to the definition of the drag coefficient of a sphere.

Vogel, S. 1981. *Life in moving fluids: the physical biology of flow.* Boston: Willard Grant Press, 352 pp. This delightful book, written from a biologist's perspective, provides a systematic, mechanically based treatment of the fluid conditions with which aquatic creatures must cope during their lives.

Williams, J. and Elder, S. A. 1989. *Fluid physics for oceanographers and physicists: an introduction to incompressible flow.* Oxford: Pergamon, 300 pp. This text includes a clear discussion (pp. 118–20, 231–32) of dimensional considerations involved in describing shallow-water and deep-water waves.

# CHAPTER 6

# Fluid Statics and Buoyancy

Fluid statics concerns the behavior of fluids that possess no linear acceleration within a global (Earth) coordinate system. This includes fluids at rest as well as fluids possessing steady motion such that no net forces exist. Such motions may include steady linear motion within the global coordinate system as well as rotation with constant angular velocity about a fixed vertical axis. In this latter case, centrifugal forces must be balanced by centripetal forces (which arise, for example, from a pressure gradient acting toward the axis of rotation). Moreover, we assert that no relative motion between adjacent fluid elements exists. Fluid motion, if present, is therefore like that of a rigid body. In addition, we neglect molecular motions that lead to mass transport by diffusion. Thus, the idea of a static fluid is a macroscopic one.

The developments in this chapter clarify how pressure varies with coordinate position in a static fluid. Both compressible and incompressible fluids are treated. In the simplest case in which the density of a fluid is constant, we will see that pressure varies linearly with vertical position in the fluid according to the *hydrostatic equation*. In addition, we will consider the possibility that fluid density is not constant. Then, variations in density must be taken into account when computing the pressure at a given position in a fluid column; the pressure arising from the weight of the overlying fluid no longer varies linearly with depth. In the case of an *isothermal* fluid, whose temperature is constant throughout, any variation in density must arise purely from the compressible behavior of the fluid in response to variations in pressure. In the case where temperature varies with position, fluid density may vary with both pressure and temperature. We will in this regard consider the case of a *thermally stratified* fluid whose temperature varies only with the vertical coordinate direction.

Because fluid statics requires treating how fluid temperature, pressure, and density are related, the developments below make use of thermodynamical principles developed in Chapter 4. Further, the developments involving fluid pressure provide a basis for considering the buoyancy forces that act on solid objects or fluid elements partially or wholly immersed within a fluid. This, in turn, provides a basis for examining the mechanical stability of a thermally stratified fluid: whether stratification leads to convective overturning of the fluid.

## 6.1 STATIC PRESSURE

Consider a fluid at rest in a global $xyz$-coordinate system, and imagine a control volume in the shape of a wedge whose edges are of lengths $dx$, $dy$, $dz$, and $ds$, with an arbitrary acute angle $\theta$ between faces $dx\,dy$ and $ds\,dy$ (Figure 6.1). The forces acting on this control volume include both body and surface forces. The body force present is the weight of the control volume, equal in magnitude to the product of its mass $m$ and acceleration due to gravity $g$. Written in terms of the volume $V$ and density $\rho$ of the fluid,

$$-mg = -\rho V g = -\frac{1}{2}\rho g\,dx\,dy\,dz \qquad (6.1)$$

where the negative sign indicates that this force acts in the direction of negative $z$. Because the fluid is not undergoing deformation, tangential surface stresses on the control volume are everywhere zero. Normal stresses, however, may be nonzero. Let normal stresses be designated as $p_1$, $p_2$, $p_3$, $p_4$, and $p_5$, and further assume that the control volume is sufficiently small that each of these stresses is essentially constant over the face on which it acts. Stresses $p_1$ and $p_2$ act parallel to the $y$-axis in positive and negative directions, respectively, on the two faces $dx\,ds\,dz$. Stress $p_3$ acts parallel to the $x$-axis on face $dx\,dz$; $p_4$ acts vertically upward on face $dx\,dy$; and $p_5$ acts normally to the inclined face $ds\,dy$.

The static condition of the fluid implies that forces acting on the control volume in each of the coordinate directions must sum to zero. Observe that the force due to stress $p_1$ acting in the direction of positive $y$ equals $\frac{1}{2}p_1\,dx\,dz$, and the force due to stress $p_2$ acting in the direction of negative $y$ equals $-\frac{1}{2}p_2\,dx\,dz$. Thus

$$\frac{1}{2}p_1\,dx\,dz - \frac{1}{2}p_2\,dx\,dz = 0 \qquad (6.2)$$

$$p_1 = p_2 \qquad (6.3)$$

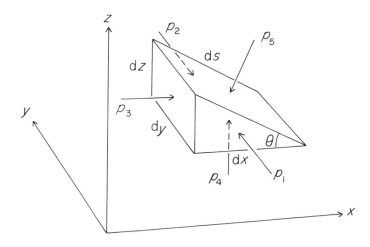

**Figure 6.1**  Elementary wedged-shape control volume within static fluid, acted on by normal stresses and gravity.

Now consider forces parallel to the $x$-axis. The force acting in the direction of positive $x$ equals $p_3 \, dy \, dz$. A force acting in the direction of negative $x$ arises from $p_5$, and equals $-p_5 \sin \theta \, ds \, dy$ (Figure 6.1). Observing that $ds = dz/\sin \theta$,

$$p_3 dy \, dz - p_5 \sin \theta ds \, dy = 0 \tag{6.4}$$

$$p_3 = p_5 \tag{6.5}$$

Finally consider forces parallel to the $z$-axis. The force acting in the direction of positive $z$ equals $p_4 \, dx \, dy$. A force acting in the direction of negative $z$ arises from $p_5$, and equals $-p_5 \cos \theta \, ds \, dy$. Observing that $ds = dx/\cos \theta$, and recalling that the weight of the control volume (6.1) acts in the direction of negative $z$,

$$p_4 dx \, dy - p_5 \cos \theta ds \, dy - \frac{1}{2}\rho g \, dx \, dy \, dz = 0 \tag{6.6}$$

$$p_4 - p_5 - \frac{1}{2}\rho g \, dz = 0 \tag{6.7}$$

In the limit as the control volume becomes infinitesimally small—that is, as the differential $dz$ approaches zero—the third term in (6.7) involving $dz$ becomes vanishingly small, and

$$p_4 = p_5 \tag{6.8}$$

Evidently from (6.5) and (6.8), $p_3 = p_5$. Moreover, by using a similar argument involving a slightly more complicated shape of control volume, we can demonstrate that

$$p_1 = p_2 = p_3 = p_4 = p_5 \tag{6.9}$$

Thus the normal stresses acting on any infinitesimal, static fluid element are equal, regardless of their orientations. We may therefore associate a scalar quantity, the static fluid pressure $p$, with the normal stress at each coordinate position in a static fluid.

---

## 6.2 EQUATION OF FLUID STATICS

Let us examine how pressure varies with respect to coordinate position within a static fluid. Consider a rectangular control volume with edges having lengths $dx$, $dy$, and $dz$ embedded within a Cartesian coordinate system (Figure 6.2). Again assume that the volume is sufficiently small that the pressure acting on each face is essentially constant over the face. The static condition of the fluid implies that all body and surface forces acting on the control volume in each of the coordinate directions must sum to zero.

Consider first those forces acting parallel to the $x$-axis. Let us center the control volume on the coordinate position $x = a$, and associate a fluid pressure $p$ specifically with this position $a$. We shall leave open the possibility that values of pressure at the left and right faces $dy \, dz$ are different from $p$. These pressures are then specified in terms of $p$ using a Taylor expansion.

Recall Taylor's formula, with which we can approximate the value of a function $f(x)$ near the coordinate position $a$, where $f(a)$ is known:

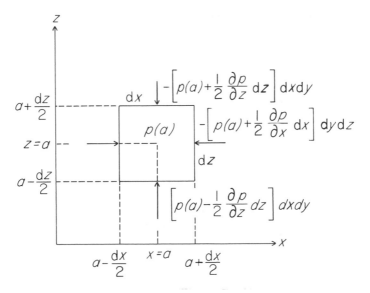

**Figure 6.2** Elementary rectangular control volume within static fluid, acted on by normal pressures and gravity.

$$f(x) = f(a) + \frac{f'(a)}{1!}(x - a) + \frac{f''(a)}{2!}(x - a)^2 + \cdots$$

$$+ \frac{f^{(n)}(a)}{n!}(x - a)^n + \frac{f^{(n+1)}(z)}{(n + 1)!}(x - a)^{n+1} \qquad a < z < x \qquad (6.10)$$

We wish to approximate the pressure at the coordinate position of the left face, $a - dx/2$, and at the coordinate position of the right face, $a + dx/2$. Starting with the left face, note that $f(a) = p(a)$ and the quantity $x - a$ in (6.10) equals $-dx/2$. Thus

$$p(a - dx/2) = p(a) - \frac{1}{2}\frac{\partial p}{\partial x}dx + \frac{1}{8}\frac{\partial^2 p}{\partial x^2}(dx)^2 + \cdots \qquad (6.11)$$

We may write this as

$$p(a - dx/2) = p(a) - \frac{1}{2}\frac{\partial p}{\partial x}dx + TE \qquad (6.12)$$

where TE refers to the *truncation error* involved in approximating $p(a - dx/2)$ with the first two terms to the right of the equal sign in (6.11), neglecting the quadratic and higher degree terms. The magnitude of this truncation error is described in terms of some power $m$ of the differential quantity $dx/2$; we state that it is *of the order* $O[(dx/2)^m]$, and we write

$$p(a - dx/2) = p(a) - \frac{1}{2}\frac{\partial p}{\partial x}dx + O\left[\left(\frac{dx}{2}\right)^m\right] \qquad (6.13)$$

In the case here $m$ equals 2. This means that for some positive constant $K$,

$$|TE| \leq K \left| \left( \frac{dx}{2} \right)^2 \right| \qquad (6.14)$$

for a "sufficiently small" $dx$.

Evaluating the magnitude of a truncation error is important in numerical analyses that involve approximating derivatives of a quantity using values of the quantity at positions separated by a finite distance. For our purposes, however, the value of $K$ in (6.14) does not actually need to be specified; it is sufficient to note that in the limit as $dx$ approaches zero, the quadratic and higher-degree terms in (6.11) approach zero faster than the term involving $dx$, and the truncation error becomes vanishingly small. Then, the force $p(a - dx/2) \, dy \, dz$ acting on the left face $dy \, dz$ in the direction of positive $x$ is *to first order*

$$p(a - dx/2) \, dy \, dz = \left[ p(a) - \frac{1}{2} \frac{\partial p}{\partial x} dx \right] dy \, dz \qquad (6.15)$$

For the right face, $x - a$ in (6.10) equals $dx/2$, so using a similar expansion the force $-p(a + dx/2) dy \, dz$ acting on the right face in the direction of negative $x$ is to first order

$$-p(a + dx/2) \, dy \, dz = - \left[ p(a) + \frac{1}{2} \frac{\partial p}{\partial x} dx \right] dy \, dz \qquad (6.16)$$

Since the static condition requires that these forces sum to zero,

$$\left[ p(a) - \frac{1}{2} \frac{\partial p}{\partial x} dx \right] dy \, dz - \left[ p(a) + \frac{1}{2} \frac{\partial p}{\partial x} dx \right] dy \, dz = 0 \qquad (6.17)$$

$$\frac{\partial p}{\partial x} = 0 \qquad (6.18)$$

This indicates the simple result that pressure does not vary in the horizontal $x$-coordinate direction. Making use of a similar expansion involving pressure parallel to the $y$-coordinate axis, we also can demonstrate that pressure does not vary—that is, $\partial p / \partial y = 0$—in this horizontal direction.

Consider now the forces acting on the control volume parallel to the $z$-axis. A Taylor expansion involving pressure leads to expressions that are similar to (6.15) and (6.16) for the force $p(a - dz/2) \, dx \, dy$ acting vertically upward on the lower face $dx \, dy$, and for the force $-p(a + dz/2) \, dx \, dy$ acting vertically downward on the top face $dx \, dy$:

$$p(a - dz/2) \, dx \, dy = \left[ p(a) - \frac{1}{2} \frac{\partial p}{\partial z} dz \right] dx \, dy \qquad (6.19)$$

$$-p(a + dz/2) \, dx \, dy = - \left[ p(a) + \frac{1}{2} \frac{\partial p}{\partial z} dz \right] dx \, dy \qquad (6.20)$$

In addition, we must consider the weight of the fluid in the control volume, which equals $-\rho g \, dx \, dy \, dz$. The static condition thus requires that

$$\left[ p(a) - \frac{1}{2}\frac{\partial p}{\partial z}dz \right]dx\,dy - \left[ p(a) + \frac{1}{2}\frac{\partial p}{\partial z}dz \right]dx\,dy - \rho g\,dx\,dy\,dz = 0 \tag{6.21}$$

$$\frac{\partial p}{\partial z} = -\rho g \tag{6.22}$$

Equation (6.22) is the *equation of fluid statics,* which, with (6.18) and the result that $\partial p/\partial y = 0$, indicates that pressure in a static fluid varies only in the vertical coordinate direction $z$. Because we thus are concerned with pressure variations over only the vertical coordinate $z$, (6.22) may be written as

$$\frac{dp}{dz} = -\rho g \tag{6.23}$$

We will use this important relationship many times throughout the text.

## 6.3 HYDROSTATIC EQUATION

Consider a static isothermal fluid with a free surface. The simplest case is a liquid at rest with a horizontal surface at atmospheric pressure; as a whole the liquid is not influenced by surface tension (unlike fluid that has risen in a capillary tube). We further will restrict our attention to vertical distances over which fluid density can be considered constant. The actual distance over which this assumption is reasonable depends on fluid compressibility and depth within the fluid; an idea of the physical dimension involved is developed in the next section. With constant density, the equation of fluid statics (6.23) may be rewritten and integrated:

$$\int dp = -\rho g \int dz \tag{6.24}$$

$$p = -\rho g z + C \tag{6.25}$$

This is the *hydrostatic equation,* where the constant of integration $C$ depends on the vertical distance within the fluid over which we are assuming constant density, and on the position of the origin of the coordinate system used.

Consider pressure conditions near the surface of a fluid column having depth $h$, with the origin $z = 0$ positioned at the base of the fluid column. Noting the boundary condition that $p$ equals atmospheric pressure $p_0$ at $z = h$, then $C = p_0 + \rho g h$ and

$$p = \rho g(h - z) + p_0 \tag{6.26}$$

Inasmuch as $h$ is less than the vertical distance over which $\rho$ is constant, (6.26) applies to the entire fluid column. We will see below that this indeed is a very good approximation for water to a considerable depth.

Often it is convenient to set the origin $z = 0$ at the fluid surface with the direction of positive $z$ downward. Then pressure is specified with respect to depth $z$ below the surface. The equation of fluid statics is then

$$\frac{dp}{dz} = \rho g \tag{6.27}$$

since pressure increases in the direction of positive $z$. The hydrostatic equation then has the form

$$p = \rho g z + C \tag{6.28}$$

The constant of integration $C$ is evaluated using the boundary condition that $p = p_0$ at $z = 0$. Thus $C = p_0$ and the pressure with respect to depth $z$ becomes

$$p = \rho g z + p_0 \tag{6.29}$$

This demonstrates the simple result that the pressure within a static fluid with constant density increases linearly with depth at a rate equal to $\rho g$.

## 6.4 HYPSOMETRIC EQUATIONS

Densities of compressible fluids, including liquids, are not necessarily constant, and as a consequence pressure does not vary linearly with depth. Because pressure generally increases with depth, fluid deep in the column is compressed, and therefore possesses a higher density than fluid above it. This means that at any given depth $z$, fluid beneath $z$ bears a weight of overlying fluid that is greater than if the fluid above $z$ was not compressed to a higher density. In turn, the pressure at $z$ is higher than what would otherwise occur within an incompressible fluid.

Consider, then, a sufficient vertical distance within a static fluid over which density is not necessarily constant. We will examine here both liquids and gases, starting with an isothermal liquid. Again imagine a liquid with a free surface at atmospheric pressure $p_0$, and set the origin $z = 0$ at the liquid surface with the direction of positive $z$ downward. Recall the equation of state (3.48) for density in relation to fluid compressibility $\beta$ and pressure. Letting $\rho_0$ denote a reference density to be associated with $p_0$ at $z = 0$, we may substitute (3.48) for $\rho$ in the equation of fluid statics (6.27):

$$\frac{dp}{dz} = \rho g = \rho_0 g e^{\beta(p-p_0)} \tag{6.30}$$

Separating variables and rearranging, integrating over the limits $p_0$ to $p$ and $z = 0$ to $z$, then finally solving for $p$,

$$e^{-\beta(p-p_0)} dp = \rho_0 g \, dz \tag{6.31}$$

$$\int_{p_0}^{p} e^{-\beta p'} dp' = \rho_0 g e^{-\beta p_0} \int_{0}^{z} dz' \tag{6.32}$$

$$p = p_0 - \frac{1}{\beta} \ln(1 - \beta \rho_0 g z) \tag{6.33}$$

where the primes denote variables of integration. Thus, pressure varies logarithmically with depth $z$ in an isothermal liquid with constant compressibility. To compare (6.33) with the hydrostatic equation (6.29), let $p_A$ denote the value of $p$ obtained by (6.33), and let $p_B$ denote the value obtained by (6.29). An idea of the percent "error" involved in using $p_B$ to describe how pressure varies with depth then can be obtained by forming the ratio $100(p_A - p_B)/p_A$. As a point of reference, the compressibility of water $\beta$, although temperature and pressure dependent (Chapter 4), is about $4.4 \times 10^{-10}$ Pa$^{-1}$ at 15 °C. With $\rho = 999.1$ kg m$^{-3}$ at atmospheric pressure $p_0$ = 101.3 kPa, our ratio equals about 0.00001 at $z$ = 10 m, 0.0002 at $z$ = 100 m, and 0.002 at $z$ = 1000 m. Thus, the difference in the two formulae varies from about one one-hundred-thousandth of a percent at 10 m to two thousandths of a percent at 1000 m. This illustrates that the hydrostatic equation is a very good approximation of the pressure within liquids, including water, with small compressibilities. In addition, this provides an idea of the vertical distance over which density can be considered constant for isothermal water. Keep in mind, however, that we have neglected variations in density associated with temperature.

Consider now the more general case in which fluid density $\rho$ varies as a function $\rho(p, T)$ of pressure $p$ and temperature $T$. The equation of fluid statics (6.23) then becomes

$$\frac{dp}{dz} = -\rho(p, T)g \qquad (6.34)$$

If the fluid behaves like an ideal gas, $\rho(p, T)$ is given by the ideal law (4.25). Substituting this into (6.34),

$$\frac{dp}{dz} = -\frac{g}{R}\frac{p}{T} \qquad (6.35)$$

Before manipulating this equation to determine how $p$ varies with $z$, we first must specify how $T$ varies with $z$. In general, $T$ may be an arbitrary function $T(z)$ of $z$. Here it is convenient to define a reference pressure $p_0$ and a reference elevation $z_0$, and note that $p = p_0$ at $z = z_0$. Substituting $T(z)$ into (6.35), rearranging and integrating between the limits $p_0$ and $p$, and $z_0$ and $z$,

$$\int_{p_0}^{p} \frac{1}{u}\,du = -\frac{g}{R}\int_{z_0}^{z}\frac{1}{T(v)}\,dv \qquad (6.36)$$

$$\ln\frac{p}{p_0} = -\frac{g}{R}\int_{z_0}^{z}\frac{1}{T(v)}\,dv \qquad (6.37)$$

where $u$ and $v$ denote variables of integration. This general result can be evaluated once the function $T(z)$ is specified. For illustration let $T$ vary linearly with $z$ such that

$$T = T_0 + \Gamma(z - z_0) \qquad (6.38)$$

from which it is apparent that $T = T_0$ when $z = z_0$. The quantity $\Gamma$, the rate of change in $T$ with respect to $z$, is referred to in meteorology as the *lapse rate* when the problem specifically involves the atmosphere. In this case the reference values $p_0$, $T_0$, and $z_0$ may refer to conditions at Earth's surface. Substituting (6.38) into (6.37) and evaluating the integral, then using the exponential function,

$$\frac{p}{p_0} = \left[1 + \frac{\Gamma(z - z_0)}{T_0}\right]^{-g/R\Gamma} \qquad (6.39)$$

which is a hypsometric equation expressed in terms of the reference values $p_0$, $T_0$, and $z_0$. Thus the pressure of an ideal gas varies as a power function of $z$ when $T$ varies linearly with $z$. Note that the linear form of $T(z)$, which has been independently specified, may be determined, for example, from experimental measurements.

When density varies with temperature, but negligibly with pressure (an incompressible fluid), $\rho(p, T)$ in (6.34) symbolically reduces to $\rho(T)$. Assume for this case that a suitable equation of state is given by

$$\rho = \rho_0[1 - \alpha(T - T_0)] \qquad (6.40)$$

which should be recognized as a reduced form of (4.34); $\alpha$ is a coefficient of thermal expansion, $\rho_0$ and $T_0$ are reference values of density and temperature used in defining (6.40). Substituting (6.40) into (6.34),

$$\frac{dp}{dz} = -\rho_0 g - \rho_0 g \alpha T_0 + \rho_0 g \alpha T \qquad (6.41)$$

Letting temperature be an arbitrary function $T(z)$ of $z$, then rearranging and integrating between the limits $p_0$ to $p$, and $z_0$ to $z$,

$$\int_{p_0}^{p} du = \int_{z_0}^{z} [-\rho_0 g - \rho_0 g \alpha T_0 + \rho_0 g \alpha T(v)] dv \qquad (6.42)$$

$$p - p_0 = \int_{z_0}^{z} [-\rho_0 g - \rho_0 g \alpha T_0 + \rho_0 g \alpha T(v)] dv \qquad (6.43)$$

where $u$ and $v$ denote variables of integration. This general result, like (6.37), can be evaluated once the function $T(z)$ is specified. Again for illustration let temperature $T$ vary linearly with $z$ according to (6.38), and for convenience let the reference temperature $T_0$ in (6.38) coincide with that in the equation of state (6.40). Substituting into (6.43) and evaluating the integral then leads to

$$p - p_0 = \rho_0 g z_0 + \frac{1}{2}\rho_0 g \alpha \Gamma z_0^2 - (\rho_0 g + \rho_0 g \alpha \Gamma z_0)z + \frac{1}{2}\rho_0 g \alpha \Gamma z^2 \qquad (6.44)$$

which is a hypsometric equation expressed in terms of the reference values $p_0$, $T_0$, and $z_0$. Thus the pressure of an incompressible fluid varies as a second degree polynomial in $z$ when $T$ varies linearly with $z$. Keep in mind that $\alpha$ is assumed to be independent of $T$ in this development, such that the equation of state (6.40) is valid inasmuch as temperatures do not deviate markedly from the reference temperature $T_0$.

## 6.5 BUOYANCY

The idea that a solid object wholly or partially immersed in a fluid possesses buoyancy ought to be familiar. In particular, the apparent weight loss of such an object, from Archimedes's principle, is equal to the weight of fluid it displaces. Let us formalize this principle and, in doing so, obtain a general treatment that is applicable to immersed fluid elements as well as solid objects. In addition, we will learn that buoyancy derives from the presence of a hydrostatic pressure gradient.

Consider a wedge-shaped object immersed within a fluid under hydrostatic conditions (Figure 6.3). Suppose that the object, with edges of lengths $X$, $Y$, $Z$, and $S$, is sufficiently large that fluid pressure may vary over each of its faces. One corner of the wedge is positioned at the origin of a Cartesian coordinate system, such that the $z$-coordinate of the surface $SY$ is locally given by

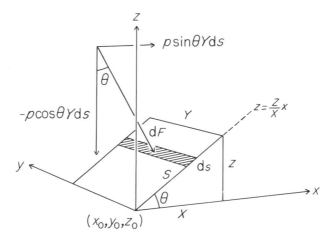

**Figure 6.3**    Finite wedge-shaped control volume within static fluid, acted on by normal pressures and gravity.

$$z = \frac{Z}{X}x \tag{6.45}$$

The task now consists of determining whether a net force arising from fluid pressure acts on the wedge and, if so, the coordinates of the point through which this force acts. To do this it is convenient to consider, in turn, the forces acting parallel to each of the coordinate directions. This amounts to integrating the pressure over each of the wedge faces.

Observe first that the normal stress (pressure) at a given vertical position $z$ acting in the direction of positive $y$ on face $XZS$ is opposed by a normal stress (pressure) of equal magnitude acting in the direction of negative $y$. It should be intuitively clear that these stresses, when integrated over each of the two faces, give forces acting on the wedge parallel to $y$ that sum to zero.

Now consider forces parallel to the $x$-coordinate. A pressure $p$ acts on the small area $Y\,dz$ on face $YZ$ such that the normal force $dF_{x-}$ acting on this area $Y\,dz$ is $-pY\,dz$; the negative sign indicates that this force acts in the direction of negative $x$. Recalling the hydrostatic equation (6.25) with $p = p_0$ at $z = z_0$, the total force $F_{x-}$ acting on face $YZ$ is then given by

$$F_{x-} = -Y\int_0^Z p\,dz \tag{6.46}$$

$$F_{x-} = -Y\int_0^Z (-\rho g z + p_0)\,dz \tag{6.47}$$

$$F_{x-} = \frac{1}{2}\rho g Y Z^2 - p_0 Y Z \tag{6.48}$$

Likewise a pressure $p$ acts on the small area $Y\,ds$ on face $YS$ such that a force $dF$ of magnitude $pY\,ds$ acts on this area $Y\,ds$. The force $dF$ can be resolved into a component acting parallel to $x$, $dF_{x+} = p\sin\theta Y\,ds$, and a component acting parallel to $z$, $dF_{z-} = -p\cos\theta Y\,ds$ (Figure 6.3). Considering the $x$-component, observe that $\sin\theta\,ds = dz$, so $dF_{x+} = pY\,dz$. Following the same steps as in (6.46), (6.47), and

(6.48) above, $F_{x+} = -\frac{1}{2}\rho g Y Z^2 + p_0 Y Z$. Summing the expressions for $F_{x+}$ and $F_{x-}$ then indicates that no net force parallel to the $x$-coordinate acts on the wedge.

A pressure $p_0$ acts on the face $XY$, so the total force $F_{z+}$ on $XY$ acting in the direction of positive $z$ is $p_0 XY$. Considering the $z$-component $dF_{z-}$ of $dF$ acting on the small area $Yds$, observe that $\cos\theta ds = dx$, so $dF_{z-} = -pY dx$. From (6.45), $dz/dx = Z/X$, so

$$dx = \frac{X}{Z}dz \qquad (6.49)$$

Making use of this, the total force $F_{z-}$ acting on $YS$ is given by

$$F_{z-} = -\frac{XY}{Z}\int_0^Z (-\rho g z + p_0)dz \qquad (6.50)$$

$$= -p_0 XY + \frac{1}{2}\rho g XYZ \qquad (6.51)$$

The sum of the expressions $F_z = F_{z+} + F_{z-}$ thus gives

$$F_z = \frac{1}{2}\rho g XYZ = \rho g V \qquad (6.52)$$

$F_z$ is the buoyancy force acting on the wedge and, as anticipated from Archimedes's principle, is numerically equal to the weight of the displaced fluid volume $V$. This example illustrates the point that the buoyancy force acting on an immersed body arises as the resultant of surface forces. Moreover, this resultant derives from the presence of a pressure gradient. To emphasize this latter point for the case of a hydrostatic fluid, (6.52) may be rewritten using the equation of fluid statics (6.23):

$$F_z = -V\frac{dp}{dz} \qquad (6.53)$$

Note that the general results (6.52) and (6.53), in principle, could be derived using arbitrarily shaped objects.

Equation (6.52) is independent of coordinate position, and contains no specific dimensions of the object. This result suggests that the buoyancy force $F_z$ may be considered as acting through a point whose coordinates coincide with the centroid (center of gravity) of the object. This is indeed the case, a conclusion that bears on, among other things, computing the forces that act on sediment particles in moving fluids (Example Problem 6.7.1). In this regard, recall that the "dry weight" of an object is $-\rho_s g V$, where $\rho_s$ is the density of the object and the negative sign indicates that this force is directed vertically downward. The buoyant weight $W$ of the object when submerged, regardless of its orientation, is then given by

$$W = -\rho_s g V + \rho g V = -(\rho_s - \rho)g V \qquad (6.54)$$

which should be recognized as a general form of (1.20), which pertains to the specific case of a sphere.

Consider now an immersed object of density $\rho_s$ that consists of a fluid control volume of arbitrary shape. The existence of a pressure gradient suggests that a buoyancy force equal to (6.53) acts on the control volume. As the fluid within the control volume is the same as that surrounding it, $\rho_s = \rho$ in (6.54), giving $W = 0$. That is, the weight of the fluid control volume is exactly equal in magnitude to the buoyancy force acting on it. Thus gravitational and buoyancy forces acting on fluid elements under hydrostatic conditions are balanced.

## 6.6  STABILITY OF A THERMALLY STRATIFIED FLUID

The introductory comments in Chapter 4 outline several geological processes that
are strongly influenced by convective circulation of fluids: crystal dispersion in
magmas, diagenesis within sedimentary basins, and development of patterned
ground in periglacial environments. As fluid convection is a dynamic process,
treatment of such motions is deferred to Chapter 16. Nonetheless, describing the
conditions that determine whether a fluid is susceptible to convection is, in part,
a problem of fluid statics. In general, if pressure and density distributions within
a fluid are such that the buoyancy—the vector sum of the weight and buoyancy
force acting on a fluid volume—is positive and larger for fluid lower in the fluid
column than for fluid above, a potentially unstable configuration exists since the
lower, lighter fluid will tend to rise, and the higher, denser fluid will tend to settle.
Thus, convective overturning arises due to differing gravitational forces per unit
volume acting on varying densities of fluid.

### 6.6.1  Liquids

Fluid convection is a complex phenomenon, for which significant insight has been
provided by laboratory experiments. Particularly relevant to many geological problems are experiments involving two parallel solid boundaries maintained at temperatures $T_1$ and $T_2$ and separated by a vertical distance $Z$ (Figure 6.4); this is referred
to as a Rayleigh–Bénard configuration. These conditions mimic, for example, the
case in which a steady geothermal heat flux maintains an elevated temperature at
the lower boundary of a stratified sedimentary layer, with a lower temperature at
the upper boundary of the layer.

Consider first the case in which fluid pressure does not vary significantly over the
distance $Z$ (as in small laboratory experiments), or where density varies negligibly
with pressure (incompressible liquids). Then a necessary condition for instability
is

$$\frac{\mathrm{d}T}{\mathrm{d}z} < 0 \qquad (6.55)$$

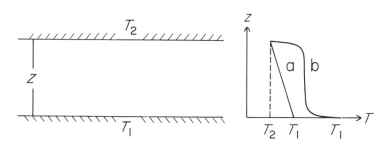

**Figure 6.4**  Rayleigh–Bénard configuration with thickness $Z$, constant lower
boundary temperature $T_1$, and constant upper boundary temperature $T_2$; temperature distribution for static fluid (a) and mean temperature distribution for high
Rayleigh number (b), where averaging is over time or along horizontal planes.

when density $\rho(T)$ decreases with $T$ over the range of $T$ considered, and

$$\frac{dT}{dz} > 0 \tag{6.56}$$

when $\rho(T)$ increases with $T$, as with water between 0 and 4 °C. As density is a function $\rho(T)$ only of temperature, condition (6.55) or (6.56) ensures existence of a potentially unstable density gradient such that dense fluid overlies less dense fluid.

Whereas (6.55) or (6.56) defines a necessary condition for the onset of convective motion, neither is a sufficient condition. To see why this is the case, it is useful to consider forces operating within a convecting fluid, despite the fact that we wish to describe conditions at the onset of fluid motion. Resisting this motion are frictional effects of fluid viscosity, which are manifest once shearing of fluid actually begins. In addition, the thermal conductivity of the fluid influences the motion. For the case in which (6.55) is relevant, the effect of a large thermal conductivity is to efficiently dissipate heat upward, such that momentarily elevated gradients in $T$, and thus unstable density gradients, are suppressed. In addition, a high thermal conductivity tends to suppress differences in temperature between rising (hot) fluid and descending (cool) fluid. In fact, with sufficiently high viscosity and thermal conductivity, it is possible to develop a nonzero thermal gradient without onset of convection. Then, the steady distribution of temperature $T(z)$ is linear, such that the fluid conducts heat like a solid.

This raises an important point: The onset of convective motion from a static condition with nonzero temperature gradient is not due to absence of an equilibrium (static) configuration, but rather due to the fact that the equilibrium configuration is an unstable one. Recall, for example, the parabolic distribution of pressure in a liquid (6.44) associated with a linear increase in temperature with depth. This was derived from the equation of fluid statics (6.23), which is an explicit expression of the balance of vertical forces in a static fluid. As density decreases with depth in (6.44), the weight per unit volume also decreases. But as a balance of forces is required, the pressure gradient, and therefore the buoyancy force per unit volume, decreases in compensation. Thus, whereas (6.55) or (6.56) is a necessary condition for the onset of convection, the destabilizing influence of the temperature gradient must be sufficient to overcome the stabilizing influences of viscosity and thermal conductivity.

In view of these points, let us introduce an important quantity used in characterizing convective flows, the dimensionless *Rayleigh number Ra*:

$$Ra = \frac{\rho^2 g c_p \alpha |T_1 - T_2| Z^3}{\mu K_T} \tag{6.57}$$

where, from Chapter 4, $c_p$ is the specific heat at constant pressure, $\alpha$ is the coefficient of thermal expansion, and $K_T$ is the thermal conductivity of the fluid. (Recall that $c_p = c$ for an incompressible fluid.) This has a slightly different appearance when the fluid is within a porous medium of thickness $Z$:

$$Ra = \frac{\rho^2 g c_e \alpha |T_1 - T_2| k_h Z}{\mu K_{Te}} \tag{6.58}$$

where $c_e$ is the effective specific heat and $K_{Te}$ is the effective thermal conductivity, which characterize how the fluid and solid matrix act together thermally (see

Chapter 9). The quantity $k_h$ is the *intrinsic permeability* of the porous medium. Dimensions of $k_h$ are $L^2$. In principle, $k_h$ is a property of the porous medium, independent of fluid properties; its relation to hydraulic conductivity $K_h$ is given by $K_h = k_h \rho g / \mu$ (see Chapter 13). The numerator in each of these dimensionless numbers characterizes factors that enhance buoyancy variations and hence instability, whereas the denominator characterizes factors that suppress them. As a point of reference, free convection in a Rayleigh–Bénard configuration begins when $Ra$ exceeds a critical value of about 1700. In porous media, $Ra$ must exceed a critical value of about 40, although lower values have been suggested in the literature. Moreover, this critical number varies with conditions at the upper and lower boundaries, including whether flow occurs across these boundaries. Students are urged to examine Tritton (1988) and Phillips (1991) regarding these points. We will return to this topic in Chapter 16.

Geometrical configurations other than the Rayleigh–Bénard configuration also are of interest in many geological problems. Magma chambers typically are envisioned as having more complex shapes than the Rayleigh–Bénard configuration, and the possibility of buoyancy-driven flow in narrow openings is relevant to hydrothermal systems whose plumbing geometries involve narrow pipe-like conduits and open fracture systems. In general, convection tends to be suppressed in these more confined geometries because frictional influences of boundaries become more important than in an "open" Rayleigh–Bénard configuration. For example, a Hele–Shaw configuration (see Chapter 16) involves "very viscous" two-dimensional flow between parallel planar boundaries separated by a small aperture. In addition, we have neglected complexities such as variations in viscosity with varying temperature in the development above.

### 6.6.2 Gases

Consider now the case in which density $\rho(p, T)$ varies with both pressure and temperature, as in the atmosphere. Recall from Section 4.7 that as a parcel of gas is displaced upward it encounters a lower pressure and expands. In doing so work is performed such that, by the first law of thermodynamics (4.35), the temperature of the parcel decreases. Treated as an adiabatic process, the rate $\partial T/\partial z$ at which the temperature $T$ decreases with respect to vertical position $z$ is referred to as the *adiabatic lapse rate* $\Gamma$. Now, for a parcel of fluid that is cooling adiabatically to continue its ascent, it must locally possess a lower density, and therefore higher temperature, than that of surrounding fluid. For this to occur the temperature of the surrounding fluid must decrease faster vertically than the adiabatic rate $\Gamma$. That is, the *environmental lapse rate* $\Gamma_a$ must be less (more negative) than $\Gamma$.

The adiabatic lapse rate can be determined from the first law of thermodynamics. Written in the form,

$$dH = dU + dW \tag{6.59}$$

this law indicates that a small quantity of heat $dH$ added to a fluid system is partitioned between a change in its internal energy $dU$ and work performed by expansion $dW$. For a fluid with mass $m$ the change in internal energy $dU = c_V m dT$, where $c_V$ is the specific heat at constant volume. The small amount of work $dW$ is given by (4.36). Substituting these expressions into (6.59),

$$dH = c_V m dT + p dV \tag{6.60}$$

Observe that, using the chain rule,

$$d(pV) = p\,dV + V\,dp \tag{6.61}$$

giving $p\,dV = d(pV) - V\,dp$. From the ideal gas law (4.24), $d(pV) = mR\,dT$; and from (4.45), $R = c_p - c_V$. Substituting for $R$ and $d(pV)$, and then for $p\,dV$ in (6.60), and noting that $V = m/\rho$

$$dH = c_p\,dT - \frac{1}{\rho}dp \tag{6.62}$$

Since an adiabatic process is envisioned, no heat is added to the system, so $dH = 0$ and

$$c_p\rho\,dT = dp \tag{6.63}$$

From the equation of fluid statics (6.23), $dp = -\rho g\,dz$. Substituting and rearranging,

$$\frac{dT}{dz} = -\frac{g}{c_p} = \Gamma \tag{6.64}$$

The necessary condition for instability of an ideal gas, analogous to (6.55), is thus:

$$\frac{dT}{dz} < -\frac{g}{c_p} = \Gamma \tag{6.65}$$

If an ideal gas column has a stronger (more negative) lapse rate $\Gamma_a$ than the adiabatic rate $\Gamma$, a parcel of gas displaced upward will continue to ascend since it tracks an adiabatic cooling path whose temperature is higher than that of the surrounding gas (Figure 6.5). If the lapse rate $\Gamma_a$ is weaker than the adiabatic rate, a parcel momentarily displaced will tend to return to its original position, a stable behavior.

As a point of reference, $\Gamma = -0.0098\,°\text{C m}^{-1}$ for dry air. If during its ascent a gas parcel cools below its *dew point*, the adiabatic lapse rate decreases below the dry rate; condensation with cooling releases latent heat, so the rate at which the temperature decreases vertically is not as large given this internal source of heat (Figure 6.5).

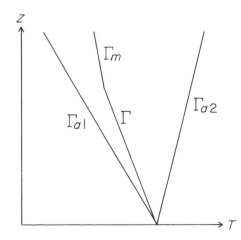

**Figure 6.5**   Graph of temperature $T$ versus vertical position $z$ depicting dry adiabatic lapse rate $\Gamma$ and moist adiabatic lapse rate $\Gamma_m$; adiabatically cooling air is mechanically metastable with the environmental lapse rate $\Gamma_{a1}$, and stable with the positive environmental lapse rate $T_{a2}$, or temperature "inversion."

As with the condition for instability (6.55) of a liquid column, the condition for instability (6.65) of a gas column may exist without onset of convective overturning. A useful criterion in this regard is the temperature gradient required for a gas column to spontaneously overturn, the *autoconvective lapse rate*. An approximate value for this rate can be obtained from the ideal gas law (4.25) by assuming that density varies negligibly with $z$. Differentiating with respect to $z$, $dp/dz = \rho R\, dT/dz$, and noting from the equation of fluid statics that $dp/dz = -\rho g$, substitution gives $dT/dz = -g/R$. According to this result a gas column will spontaneously overturn when

$$\frac{dT}{dz} < -\frac{g}{R} \tag{6.66}$$

The autoconvective lapse rate using (6.66) is $-0.034\,^{\circ}\mathrm{C}\ \mathrm{m}^{-1}$ for dry air. Note that this rate is smaller (of larger absolute magnitude) than the adiabatic rate for air. This difference reflects that the temperature gradient must be relatively strong to have a sufficiently destabilizing influence for spontaneous overturning in absence of an initial mechanical displacement of fluid. In contrast, the adiabatic gradient, in principle, is sufficient to produce overturning given an initial mechanical displacement of fluid.

## 6.7 EXAMPLE PROBLEMS

### 6.7.1 Buoyant force on a sediment particle: coordinate position of action

Consider a coarse sediment particle resting on a sloping stream bed (Figure 6.6). A classic problem in studies of sediment transport involves computing the forces that act on such a particle to determine the conditions under which it ought to move (the "initial motion" problem). Forces tending to move the particle arise from stresses imposed on it by fluid flow. These produce a resultant force $R$ that acts on an effective lever arm $l$ to pivot the particle about a fulcrum P out of its pocket (Figure 6.6). A small force arising from the downslope component of the buoyant weight of the particle also tends to rotate the particle. Resisting this motion is the buoyant weight of the particle. As asserted above, the buoyancy force on the particle may be considered as acting through the centroid of the particle. Establishing this assertion is important, because this centroid position defines a lever arm $m$, which, together with the buoyant weight $W$, produces a torque that resists motion (Figure 6.6).

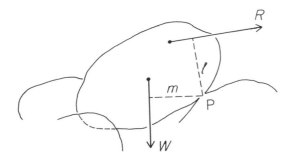

**Figure 6.6** Definition diagram for forces acting on a sediment particle resting on a rough bed.

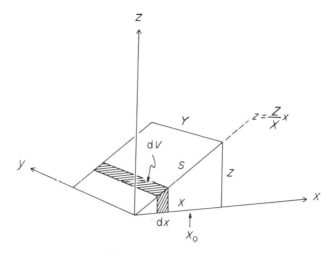

**Figure 6.7**   Definition diagram for forces acting on a finite wedge-shaped object immersed within a static fluid.

For illustration, consider an object whose centroid is not obvious: a wedge (Figure 6.7). From (6.52) the buoyancy force dF acting on the small volume dV is $\rho g\,dV$. Noting that $dV = Yz\,dx$, and that $z$ is given by (6.45),

$$dF = \rho g Yz\,dx = \rho g \frac{YZ}{X} x\,dx \tag{6.67}$$

Recall from general physics that the torque $d\tau$ associated with dV with respect to the origin equals the product of the force $dF$ and the length of the lever arm, here equal to $x$. Thus $d\tau = x\,dF$. The torque $\tau$ imposed by buoyancy on the entire wedge is then

$$\tau = \int_0^X x\,dF = \rho g \frac{YZ}{X} \int_0^X x^2\,dx \tag{6.68}$$

Further recall that the $x$-coordinate, $x_0$, coinciding with the position where the total torque (with respect to $x_0$) on the system equals zero is

$$x_0 = \frac{\displaystyle\int_0^X x\,dF}{\displaystyle\int_0^X dF} = \frac{\rho g \dfrac{YZ}{X} \displaystyle\int_0^X x^2\,dx}{\rho g \dfrac{YZ}{X} \displaystyle\int_0^X x\,dx} \tag{6.69}$$

The coordinate $x_0$ is the position where the total buoyancy force, if applied at a single point, gives the same value of torque (about the origin) as when the force is distributed over the wedge. Thus $x_0$ may be viewed as the coordinate position through which $F$ acts. As the constants in the last part of (6.69) cancel, this last expression should be recognized as the formula for the $x$-coordinate of the centroid of the wedge. Thus the position of action $x_0$ of the buoyant force is the same as that of the $x$-coordinate of the centroid. Evaluating the integrals in (6.69) gives $x_0 = 2X/3$. It is left as an exercise to deduce a similar result for the $y$-coordinate.

Returning to the case of a sediment particle, inasmuch as the weight of the particle acts through its centroid, the buoyant weight $W$ can be computed from (6.54).

The lever arm $m$ is then equal to the horizontal distance between $x_0$ and P, and the magnitude of the torque resisting motion is $mW$. Note that the particle weight in general acts through its centroid only if its density is constant throughout. Otherwise the particle weight and the buoyancy force may produce a couple that tends to rotate the particle.

### 6.7.2 Buoyant ascent of magma within a vertical dike

Although this chapter covers fluid statics, we nonetheless have gained the right kind of information to examine an interesting question regarding magma flow: How rapidly does magma ascend within the Earth's crust?

Let us choose a simple geometrical case: consider a lenticular blob of magma with density $\rho$, and assume that the vertical extent and breadth of the magma blob are much greater than its aperture (Figure 6.8). It is thought that in such cases the ascent of magma within the upper crust typically involves fracturing of the country rock at the top of the magma body, as the magma must part the country rock to intrude into it. In addition, the rate of ascent is in part governed by the viscous flow behavior of magma far below the leading tip. We shall examine this latter aspect of the problem, and in essence assume that viscous resistance to flow is more important in governing the rate of ascent than are mechanisms by which the country rock is fractured and parted to make room for the ascending magma.

Consider a rectangular control volume positioned within the center of the magma, which is bounded by parallel, vertical walls of country rock with aperture $2b$ (Figure 6.8). We will neglect any heat flow between the magma and country

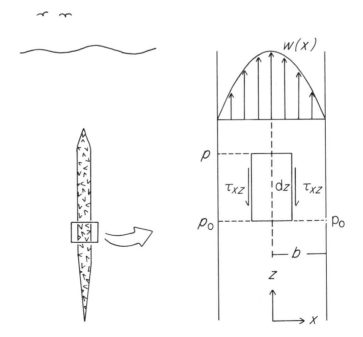

**Figure 6.8**  Lenticular magma blob ascending within crust, and definition diagram for forces acting on a control volume within magma.

rock, and assume that magma flows only in the vertical direction. Observe that the pressure $p_0$ at position $P_0$ on the dike boundary must equal the pressure at the same depth in the surrounding country rock. Further observe that, because magma flow is one-dimensional, the local flow velocity $w(x)$ within the magma must everywhere be normal to surfaces of equal pressure. It follows that the pressure $p_0$ at $P_0$ must be impressed throughout the magma at the level $z$; that is, $P_0$ lies on a horizontal surface of equal pressure $p_0$ (dashed line), of which the lower face of the control volume, with area $2x\,dy$, is a segment. A similar argument applies to the pressure $p_1$ at the top surface of the control volume.

It is left as an exercise to use a Taylor expansion about the lower face of the control volume, following the development in Section 6.2, to determine that a net vertical force equal to

$$-2x\frac{\partial p}{\partial z}dy\,dz \tag{6.70}$$

exists on the control volume. Assuming a density $\rho_c$ for the country rock, note from (6.23) that

$$\frac{\partial p}{\partial z} = -\rho_c g \tag{6.71}$$

Substituting into (6.70), the vertical force on the control volume is then

$$2\rho_c gx\,dy\,dz \tag{6.72}$$

In absence of acceleration (steady flow), this buoyancy force is balanced by the weight of the control volume, equal to $-2\rho gx\,dy\,dz$; and by a force equal to $-2\tau_{xz}dy\,dz$ arising from the shear stress $\tau_{xz}$ acting on the two faces $dy\,dz$ of the control volume. Assuming a Newtonian rheology, and noting the sense of fluid shear (Figure 6.8),

$$\tau_{xz} = -\mu\frac{dw}{dx} \tag{6.73}$$

Since forces on the control volume sum to zero,

$$2\rho_c g\,dy\,dz - 2\rho gx\,dy\,dz - 2\tau_{xz}\,dy\,dz = 0 \tag{6.74}$$

$$\rho_c gx - \rho gx + \mu\frac{dw}{dx} = 0 \tag{6.75}$$

Note that the velocity gradient $dw/dx$ is negative for positive $x$ (and positive for negative $x$), so the third term is implicitly negative, which is consistent with the idea that this viscous term acts in the direction of negative $z$. It is left as an exercise to separate the variables, integrate, then use the no-slip condition at $x = \pm b$ to evaluate the constant of integration, and demonstrate that

$$w = w(x) = \frac{1}{2\mu}(\rho_c - \rho)g(b^2 - x^2) \tag{6.76}$$

which is a parabolic velocity distribution.

It is useful to compare this equation for viscous flow within a vertical dike with equation (3.71) for viscous flow between parallel, horizontal boundaries. The forms of these equations are identical; the essential difference is that the pressure gradient $\Delta p/L$ in (3.71) that produces horizontal flow is replaced by the effect of a density contrast $(\rho_c - \rho)g$ in (6.76). It is left as an exercise to follow the development leading to (3.75) and demonstrate that the average, vertical velocity $\overline{w}$ of magma is given by

$$\overline{w} = \frac{1}{3\mu}(\rho_c - \rho)gb^2 \tag{6.77}$$

Thus, the average velocity $\overline{w}$ arises from a buoyant force associated with the density contrast between magma and country rock, and is proportional to the square of the aperture and inversely proportional to the magma viscosity. Assuming that the average velocity $\overline{w}$ roughly approximates the ascent speed, we note that this result is consistent with our result in Example Problem 1.4.2, that a big bubble immersed within a denser fluid will tend to rise faster than a small one.

As a point of reference, the dynamic viscosity of a diabase melt with 10 percent suspended crystals is about 150 Pa s at 1,200 °C, and the density is about 2,660 kg m$^{-3}$, depending on chemical composition. Assuming a density of 2,716 kg m$^{-3}$ for granodiorite country rock, $\overline{w}$ equals about 0.3 m s$^{-1}$ for $b = 0.5$ m. Does this value seem low or high? Keep in mind that we have neglected fracture phenomena and the related stress field in the surrounding country rock, and the possibility of non-Newtonian behavior. In addition, we have not considered what governs the half-aperture distance $b$. Equation (6.77), therefore, must be viewed only as an approximation of the ascent speed.

Now, how would these results change if the magma behaved as a Bingham substance with yield stress $\tau_0$? Suppose one considered fracturing at the leading tip. What are the relevant physical parameters involved in this process, and how would they influence the speed of ascent? What would be the effect of heat flow between the magma and surrounding country rock? What controls the value of $b$?

### 6.7.3 Growth and ascent of carbon dioxide bubbles in a liquid

Let us reconsider Example Problem 4.10.5, and specifically treat the buoyant ascent of bubbles in an isothermal liquid. We begin by rewriting equation (4.98) for the rate of change in bubble radius $r$ due to expansion as the bubble ascends and encounters decreasing hydrostatic pressure conditions, and due to a high partial pressure of dissolved gas within the liquid relative to that within the bubble, such that gas diffuses from the liquid into the bubble:

$$\frac{dr}{dt} = \frac{\alpha(p_{CO_2} - p_0 - \rho gh + \rho gz)r - \alpha 2\sigma + \dfrac{\rho g}{3RT}r^2\dfrac{dz}{dt}}{\dfrac{4\sigma}{3RT} + \dfrac{1}{RT}(p_0 + \rho gh - \rho gz)r} \tag{6.78}$$

We cannot directly integrate this formula to obtain an expression of how $r$ varies with time $t$, because it involves the vertical coordinate $z$ and the ascent velocity $dz/dt$, each of which depends on $r$. Additional expressions are required for a solution, and we first obtain one for $dz/dt$.

Let us assume that the buoyancy force acting on the bubble is at each instant essentially balanced by the drag force. This means that we are neglecting any acceleration of the bubble that is not associated with changes in bubble or liquid properties. Any acceleration of the bubble during its ascent therefore is related to a change in buoyancy or drag, or both, due to a change in the *size* of the bubble.

Treating the bubble as a rigid sphere, from Example Problem 1.4.2 the buoyancy force $F_B$ acting on a spherical bubble is

$$F_B \approx \frac{4}{3}\pi\rho g r^3 \tag{6.79}$$

where $\rho_b$ is neglected inasmuch as $\rho_b \ll \rho$. From Chapter 5 the drag force $F_D$ on the bubble is

$$F_D = -C_D \frac{1}{2}\pi\rho r^2 w^2 \tag{6.80}$$

where $w = dz/dt$, $C_D$ is a coefficient of drag, and the leading negative sign indicates that $F_D$ acts downward on an ascending bubble. Since $F_B$ and $F_D$ are balanced, $F_B + F_D = 0$; thus from (6.79) and (6.80),

$$w = \left(\frac{8gr}{3C_D}\right)^{1/2} \tag{6.81}$$

The coefficient $C_D$ is in general a function of the Reynolds number $Re$, defined here as $Re = 2\rho r w/\mu$.

An important case occurs when $Re < 1$. Then the result of Stokes can be used such that $C_D = 24/Re$ (Chapter 5), and $w = dz/dt$ becomes

$$w = \frac{2\rho g}{9\mu}r^2 \tag{6.82}$$

which may be substituted directly into (6.78). Unless the bubble is very small and ascends very slowly, however, the Reynolds number may be too large to use this result. Then, viscous forces no longer dominate, and motion of the bubble may involve the onset of turbulent flow around it (Chapter 14). We therefore must obtain another formula that is appropriate for larger Reynolds numbers.

Consider experimental results relating the drag coefficient $C_D$ to the Reynolds number $Re$ for a rigid sphere. Such results are available for seven orders of magnitude of $Re$ (Figure 6.9). Data for $0.01 < Re < 400$ are well fitted by a third-order polynomial in $\log_{10}$ coordinates:

$$\log C_D = A + B\log Re + C(\log Re)^2 + D(\log Re)^3 \tag{6.83}$$

where $A = 1.423$, $B = -0.8793$, $C = 0.07376$, and $D = 0.009431$. Letting $f(Re)$ denote the right side of (6.83), $C_D = 10^{f(Re)}$ may be substituted into (6.81) for the ascent velocity.

The acceleration $dw/dt$ due to changing bubble size, which influences both buoyancy and drag, is obtained from the derivative of (6.81). This is left as an exercise; start by rewriting (6.81) as $w^2 = 8gr/3C_D$, then use the chain rule to implicitly differentiate with respect to $t$, noting that both $r$ and $C_D$ may vary with respect to $t$. The first step is:

$$2w\frac{dw}{dt} = \frac{8g}{3C_D}\frac{dr}{dt} - \frac{8gr}{3C_D^2}\frac{dC_D}{dt} \tag{6.84}$$

Then note that

$$\frac{dC_D}{dt} = \ln 10 C_D \frac{d}{dt}f(Re) \tag{6.85}$$

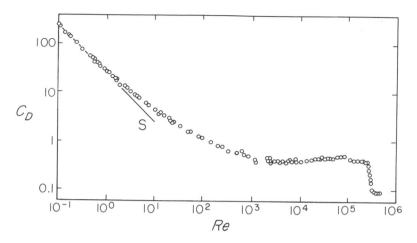

**Figure 6.9**    Graph of coefficient of drag $C_D$ for rigid spheres versus the Reynolds number $Re$; line $S$ coincides with theoretical relation due to Stokes; adapted from Schlichting (1979).

Now use the chain rule further to complete the differentiation of $f(Re)$, where it will be useful to observe that

$$\frac{dRe}{dt} = \frac{2\rho w}{\mu}\frac{dr}{dt} + \frac{2\rho r}{\mu}\frac{dw}{dt} \tag{6.86}$$

Substitute, then solve for $dw/dt$ and rearrange to obtain

$$\frac{dw}{dt} = \frac{w(1 - [B + 2C\log Re + 3D(\log Re)^2])}{r(2 + [B + 2C\log Re + 3D(\log Re)^2])}\frac{dr}{dt} \tag{6.87}$$

Note that as $Re$ becomes very small ($Re < 1$), $dw/dt$ in (6.87) approaches $(2w/r)dr/dt$. It is left as an exercise to demonstrate that this result is consistent with what one obtains using Stokes's law.

It is relatively straightforward to numerically solve equations (6.78), (6.81), (6.83), and (6.87) to predict how bubble radius $r$ and vertical position $z$ vary with time. Let us reconsider the case of beer bubbles (Example Problem 4.10.5). Assuming that $p_{CO_2} = 2 \times 10^5$ Pa (about 2 atm), $\sigma = 0.075$ N m$^{-1}$ (this has a minor effect), $\alpha = 8.2 \times 10^{-10}$ s m$^{-1}$, $\mu = 0.0013$ N s m$^{-2}$, $\rho = 1000$ kg m$^{-3}$, $T = 280$ K, $h = 0.16$ m, the initial velocity $w_0 = 0.025$ m s$^{-1}$, and the initial bubble radius $r_0 = 0.00018$ m, predicted curves of $r$ and $z$ reasonably match observed values (Figure 4.12b).

It is interesting that these results are based on the assumption that the drag on a bubble can be computed using data for rigid spheres. Actually, this is appropriate only for small $Re$; large, fast-moving bubbles tend to deform. In addition, circulation of fluid (gas) occurs within a bubble as it ascends due to viscous interaction between the gas and surrounding liquid at the bubble–liquid interface, and this circulation affects drag. Theory taking this circulation into account suggests that the drag on a bubble for small $Re$ is equal to $-4\pi\mu rw$. Observations of small, slowly ascending bubbles suggest, however, that the actual drag is closer to that of a rigid sphere, $-6\pi\mu rw$. This result is thought to be related to variations in surface tension over the bubble–liquid interface, which resist internal circulation.

Finally note that a bubble with finite initial radius may collapse if it is sufficiently small for given $p_{CO_2}$ and vertical position $z$. It also is possible for such a bubble to begin collapsing, but then start growing again after a short period. What are the conditions that would lead to this behavior?

Let us now turn to an interesting problem involving bubbles within a magma, for which the analysis above is relevant.

### 6.7.4 Advective overpressure within a magma chamber

Suppose that a magma charged with dissolved gases moves upward into a shallow chamber (1 km to 2 km depth) within the crust. The associated decrease in overburden pressure can lead to bubble formation if the partial pressure of dissolved gases is significantly greater than the hydrostatic pressure. The ascent of these bubbles within the chamber then can lead to pressurization of the magma, which may be sufficient to fracture overlying rock, leading to eruption of the magma. Let us consider a part of this problem; we will neglect initial bubble formation, and assume that a bubble has finite size compatible with pressure conditions within the magma.

Consider a rigid chamber with total volume $V_T$ that is filled entirely with an isothermal liquid, except for the volume $V_b$ occupied by a bubble at height $z$ from the bottom of the chamber (Figure 6.10). Let us momentarily neglect diffusion processes; the bubble does not react with the surrounding liquid. We therefore may envision the bubble as a small cylinder that is closed at one end, where the gas within the cylinder is separated from the surrounding liquid by a frictionless piston. Assuming equilibrium, the pressure $p_b$ within the bubble (exerted on the piston) therefore must equal the local hydrostatic pressure $p_l$. The liquid has an initial volume $V_{l0}$ and initial pressure $p_{l0} = p_0 + \rho g(h - z)$, where $p_0$ is the initial pressure at the top of the chamber at $z = h$. The bubble has initial volume $V_{b0}$, initial pressure $p_{b0}$, and initial position $z_0$. Also note that

$$V_T = V_l + V_b \qquad (6.88)$$

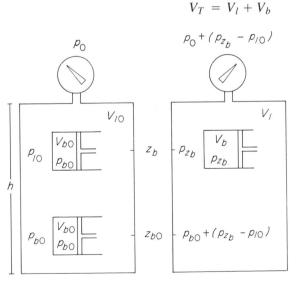

**Figure 6.10** Schematic of closed chamber with initial liquid volume $V_{l0}$ and initial bubble volume $V_{b0}(a)$, and final liquid volume $V_l$ and final bubble volume $V_b$ after bubble has ascended and expanded $(b)$.

Assuming the gas in the bubble behaves like an ideal gas, $V_b = mRT/p_b$. Then, with isothermal conditions and no exchange of mass between bubble and liquid, $mRT = V_{b0}p_{b0}$, so

$$V_b = \frac{V_{b0}p_{b0}}{p_b} \tag{6.89}$$

Let us also assume that the liquid obeys a linear equation of state,

$$V_l = V_{l0}[1 - \beta(p_l - p_{l0})] \tag{6.90}$$

where $\beta$ is liquid compressibility.

Suppose that, without piston motion, the bubble moves upward to a position where the liquid pressure is $p_{l0}$. At this time, the pressure in the bubble is the same as its initial value $p_{b0}$, which is greater than $p_{l0}$. Now suppose that the piston is allowed to move so that the bubble and liquid pressures equilibrate. The bubble expands, which leads to pressurization of the entire liquid column. This expansion, albeit very slight, is possible due to the finite compressibility of the liquid. The pressures equilibrate such that $p_b = p_l = p_z$ at the specified position $z$. Substituting (6.89) and (6.90) for $V_b$ and $V_l$ in (6.88), and with $p_b = p_z$, one obtains after rearranging:

$$\beta V_{l0}p_z^2 + (V_{b0} - \beta V_{l0}p_{l0})p_z - V_{b0}p_{b0} = 0 \tag{6.91}$$

Solving this quadratic equation, and making use of the hydrostatic equation $p_{l0} = p_0 + \rho g(h - z_b)$, where $z_b$ is used to emphasize that this is the position of the bubble,

$$p_z = -\frac{V_{b0}}{2\beta V_{l0}} + \frac{1}{2}[p_0 + \rho g(h - z_b)]$$
$$+ \frac{1}{2\beta V_{l0}}\left[(V_{b0} - \beta V_{l0}[p_0 + \rho g(h - z_b)])^2 + 4\beta V_{l0}V_{b0}p_{b0}\right]^{1/2} \tag{6.92}$$

When the bubble reaches the top of the chamber such that $z_b = h$, the new pressure $p_h$ at the top is

$$p_h = -\frac{V_{b0}}{2\beta V_{l0}} + \frac{1}{2}p_0 + \frac{1}{2\beta V_{l0}}[(V_{b0} - \beta V_{l0}p_0)^2 + 4\beta V_{l0}V_{b0}p_{b0}]^{1/2} \tag{6.93}$$

and the new pressure at any position $z$ is $p_h + \rho g(h - z)$.

These results more generally apply to a real bubble that ascends sufficiently slowly that a continuous equilibration of pressures occurs. Because the chamber is confined and the liquid is nearly incompressible, the bubble remains virtually the same size. This, in fact, is justification for assuming isothermal conditions for the bubble. Thus part of the initial bubble pressure is advected to the top of the chamber, and the pressure of the entire system increases.

Following the work of Steinberg et al. (1989), Sahagian and Proussevitch (1992) conducted a simple experiment to demonstrate this point. Consider a rigid cylinder with radius $R = 0.05$ m and height $h = 2.4$ m; then $V_T = 0.018850$ m$^3$. Suppose that an air-filled balloon is placed at the base of the cylinder, which is then filled with water and sealed at the top. The initial pressure $p_0 = 1 \times 10^5$ Pa. Neglecting effects of the curvature of the balloon, $p_{b0} = p_0 + \rho g h = 1.24 \times 10^5$ Pa. With an initial balloon radius $r = 0.04$ m, $V_{b0} = 0.000268$ m$^3$, and $V_{l0} = 0.018581$ m$^3$. Assume that $\beta = 4.8 \times 10^{-10}$ Pa$^{-1}$. Suppose now that the balloon is released and ascends to the top of the cylinder. Based on (6.93) $p_h$ equals, to three significant digits, 123,000 Pa, which is close to the measured value of 125,000 Pa.

In this experiment the balloon occupies a significant proportion of the chamber volume, and the liquid column is pressurized by an amount virtually equal to the initial bubble pressure. However, the effect of advective overpressurization by bubbles decreases with decreasing bubble size relative to the chamber size. To illustrate this, consider a bubble with $r = 0.005$ m at the base of a small magma chamber with volume $V_T = 1 \times 10^6$ m$^3$ and height $h = 100$ m. Let $p_0 = 3 \times 10^7$ Pa, so with $\rho = 3{,}000$ kg m$^{-3}$, $p_{b0} = 3.294 \times 10^7$ Pa. Assume that $\beta = 1 \times 10^{-11}$ Pa$^{-1}$. Then the magma is pressurized by an amount equal to $p_h - p_0 = 0.0063$ Pa, or about $2 \times 10^{-8}$ percent of its initial pressure. This mechanism must therefore involve many bubbles to produce significant overpressure in a magma chamber. Consider one percent bubbles by volume; then $V_{b0} = 1 \times 10^4$ m$^3$. (This is equivalent to about $1.9 \times 10^{10}$ bubbles, if each has radius $r = 0.005$ m.) Total pressurization then equals about $2.847 \times 10^6$ Pa, or about nine percent of the initial pressure.

Not all bubbles form at the bottom of a magma chamber; in fact, conditions for bubble growth are better near the top due to the lower magma pressure there. In addition, it is necessary to appeal to some mechanism, perhaps involving partial solidification, to confine the magma chamber. Sahagian and Proussevitch (1992) also point out other factors that influence the efficacy of advective overpressurization by bubbles. A complete treatment must consider bubbles that exchange mass with the surrounding liquid (Example Problem 8.4.4); students are urged to examine Sparks (1978), Steinberg et al. (1989a, b, c), and Proussevitch et al. (1993).

## 6.8 READING

Batchelor, G. K. 1967. *An introduction to fluid dynamics*. Cambridge: Cambridge University Press, 615 pp. This text includes a general treatment (pp. 229–38) of the drag on spherical drops and bubbles for low Reynolds number.

Middleton, G. V. and Southard, J. B. 1984. Mechanics of Sediment Motion. Lecture Notes for Short Course No. 3, Society of Economic Paleontologists and Mineralogists, Tulsa, Okla.: 401 pp. This monograph provides a clear description of the initial motion problem, with diagrams illustrating the forces (torques) that act on sediment grains immersed in a flow.

Oke, T. R. 1978. *Boundary layer climates*. London: Methuen, 372 pp. Appendix 1 (pp. 302–5) provides a clear discussion of atmospheric lapse rates and stability.

Phillips, O. M. 1991. *Flow and reactions in permeable rocks*. Cambridge: Cambridge University Press, 285 pp. This text contains a good discussion of how the critical Rayleigh number for convection in porous media depends on boundary conditions. We will follow parts of the treatment of convection in this text in Chapter 16.

Proussevitch, A. A., Sahagian, D. L., and Anderson, A. T. 1993. Dynamics of diffusive bubble growth in magmas: isothermal case. *Journal of Geophysical Research* 98:22, 283–307. This paper provides a comprehensive treatment of the growth of bubbles by diffusion during their ascent in a magma. Students should also consult the paper by Sparks listed below, and the Comment and Reply regarding the 1994 paper by Proussevitch, Sahagian, and Anderson that appear in the *Journal of Geophysical Research* 99:17, 827–32.

Ray, R. J., Krantz, W. B., Caine, T. N., and Gunn, R. D. 1983. A model for sorted patterned-ground regularity. *Journal of Glaciology* 29:317–37. This paper treats the saturated layer of soil water that seasonally occurs over permafrost (the active layer) in cold environments in terms of a Rayleigh–Bénard configuration, and derives a critical Rayleigh number of 27 for convection in the case of water between 0 °C and 4 °C. It is argued that the convective cells produce an undulatory water–ice interface that is a precursor to sorting of sediment grains into patterned ground.

Sahagian, D. L. and Proussevitch, A. A. 1992. Bubbles in volcanic systems. *Nature* 359(6395):485. This short paper reports on a simple experiment to demonstrate how advective overpressure arises from ascent of bubbles in a confined liquid chamber, and is the source of data used in Example Problem 6.7.4.

Schlichting, H. 1979. *Boundary-layer theory,* 7th ed. New York: McGraw-Hill, 817 pp. This is one of many texts containing data for the coefficient of drag $C_D$ in relation to the Reynolds number for spheres.

Sparks, R. S. J. 1978. The dynamics of bubble formation and growth in magmas: a review and analysis. *Journal of Volcanology and Geothermal Research* 3:1–38.

Steinberg, G. S., Steinberg, A. S., and Merzhanov, A. G. 1989a. Fluid mechanism of pressure growth in volcanic (magmatic) systems. *Modern Geology* 13:257–65. This is the first of three papers by these authors (see References) that provide an early theoretical and experimental treatment of the idea of advective overpressurization by bubbles.

Tritton, D. J. 1988. *Physical fluid dynamics,* 2nd ed. Oxford: Oxford University Press, 519 pp. This text provides a clear, qualitative description of convection and fluid–column instability (pp. 35–47) followed by theoretical treatments in a later chapter (pp. 162–88).

Turcott, D. L. 1990. On the role of laminar and turbulent flow in buoyancy driven magma fractures. In Ryan, M. P., ed., *Magma transport and storage.* Chichester: Wiley, pp. 103–11. This paper lends theoretical support to the idea that the resistance to magma flow in a dike due to mechanisms involving fracturing and parting of the country rock can be neglected relative to the viscous resistance to flow. It also points out that, whereas the density contrast between magma and country rock often provides the buoyancy to drive upward flow, this is not always the case, as evidenced by observations that dense magmas can ascend through relatively less dense country rock.

# CHAPTER 7

# Fluid Kinematics

Let us momentarily recall an elementary problem from physics: describing the motion of a ballistic particle. To do this in a formal way first required developing expressions for the geometry of motion, independently of any treatment of the forces producing the motion. One may recall that this task involved defining expressions for the speed, velocity, and acceleration of the particle with respect to a specified coordinate system. In fact, a clear geometrical description of particle motion was essential for understanding its cause.

We require similar, explicit descriptions for fluid motion to later relate this motion to the forces involved. *Fluid kinematics* thus involves the description of flow without explicit treatment of the forces producing motion. In this regard, our treatment of fluid kinematics actually is a pedagogical step toward the topic of *fluid dynamics,* wherein forces are explicitly treated.

Our essential objective is to derive an expression that describes how the velocity of a fluid is changing. A change in velocity implies that the fluid is accelerating and, therefore, that a net force is acting on the fluid. Consider Newton's second law, $\mathbf{F} = m\mathbf{a}$, which states that the net force $\mathbf{F}$ acting on a particle of mass $m$ equals the product of this mass and the acceleration $\mathbf{a}$. We thus seek an explicit expression for the acceleration $\mathbf{a}$. We will supplement this later with an expression for $\mathbf{F}$ to obtain a full dynamical description of fluid motion.

This chapter will introduce the idea of a *substantive derivative,* which will be used numerous times in subsequent chapters. In developing this idea, we will distinguish between *Eulerian* and *Lagrangian* views of fluid motion, and introduce the important concept of a convective acceleration. In most situations of interest in geology, convective accelerations arise when a real boundary induces a change in the direction or magnitude of the velocity of nearby flow; convective accelerations therefore typically involve *converging* or *diverging* flow. A good example is flow in a river channel whose bed is irregular due to bedform topography, for example, due to a point bar in a meander bend. Flow that shoals onto the bar is in part steered around the bar. As the flow is forced to change direction, it thereby undergoes acceleration.

The development of notation below requires care in keeping up with terms in the equations. For this reason, students who are not familiar with the notation are urged to re-derive each step in the development. In this regard the value of using the compact forms of vector notation will become particularly clear. Here we will use the definition of the gradient of a vector (Chapter 4), and introduce the definition of the inner product of vectors.

## 7.1 QUALITATIVE DESCRIPTIONS OF MOTION

It is convenient to speak of a *fluid particle* which, in one view, possesses a finite mass whose size is at least as large as a *physical point* as defined by the continuum concept (Chapter 2). Such a particle may undergo deformation during flow, thereby losing its distinct identity after a finite period. We may nonetheless consider its behavior for a brief interval before this occurs. Indeed, in developing the substantive derivative below, we will be concerned with the rate of change in properties of a fluid particle, for example, its temperature and velocity, in the limit as the interval of time goes to zero. Alternatively, it is sometimes useful to think of a fluid particle as a mathematical point (or a fluid molecule) whose motion mimics the behavior of a continuum.

A *pathline* is the trace formed by the set of points successively occupied by an individual fluid particle during its motion. A pathline can easily be envisioned as the trajectory of an identifiable particle embedded within a fluid, for example, a speck of dye or a pollen spore transported in groundwater. A *streakline* is the set of loci of all fluid particles that previously passed through a specific fixed position. A snapshot of a filament of smoke that has been continuously released from a fixed position within a moving fluid exhibits a streakline. A *streamline* is a continuous line that has the property that its tangent at each point coincides with the direction of the fluid velocity at that point. A streamline may or may not coincide with a pathline or streakline.

These are important distinctions in relation to techniques of flow visualization, of which the release of smoke within a flowing gas is an example. Such techniques generally provide either pathlines or streaklines, but not streamlines. This can be clarified with the ideas of *steady flow* and *unsteady flow*. Consider an arbitrary $xyz$-coordinate position within a fluid at which the velocity **u** is continuously measured. The flow is steady with respect to **u** if **u** does not vary with time (Figure 7.1), and it is unsteady if **u** varies with time. Similar remarks apply to other fluid properties, for example, temperature and pressure. In the case of steady flow, pathlines, streaklines, and streamlines coincide. For example, a particle starting at a given position on a streamline is at that instant moving tangentially to the streamline. If during an interval the particle changes to a new position, and velocities do not change, then the particle continues to move tangentially along the streamline, and its pathline coincides with the streamline. Moreover, all particles starting from a fixed position

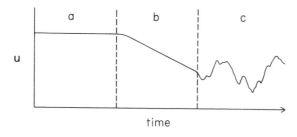

**Figure 7.1** Schematic diagram of steady (a) and unsteady (b and c) velocity signals; **u** may be unsteady in terms of direction or magnitude.

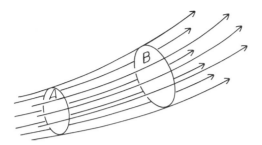

**Figure 7.2** Streamtube with control section bounded by areas $A$ and $B$.

on a streamline will track along the streamline, so the streakline formed coincides with the streamline. Conversely, with unsteady flow, streamlines may in general deform; a particle starting at a given position on a streamline may during a short interval take a path that does not track the changing streamline.

Streamlines possess an important property: at any given instant no two streamlines intersect, except at positions of zero velocity. Based on the definition of a streamline above, fluid at such an intersection would possess two velocities, which is nonsensical. A *streamtube* is a region within a flow that is bounded by an imaginary surface composed of a specified set of streamlines (Figure 7.2). As no streamlines intersect, it follows that the surface of a streamtube is an imaginary *no-flow boundary* in the sense that flow may not cross it. Streamlines within a streamtube at a given position remain within the streamtube throughout its length; all streamlines outside the streamtube remain outside.

Consider cross-sections of area $A$ and $B$ at two positions along a streamtube (Figure 7.2), and suppose that the streamtube is sufficiently small that fluid density and velocity do not vary over each cross section. The enclosed streamtube segment defines a control volume. The rate of mass flow into the segment through $A$ is $\rho_A v_A A$, with speed $v_A$ and density $\rho_A$; and the rate of mass flow out of the segment through $B$ is $\rho_B v_B B$, with speed $v_B$ and density $\rho_B$. In the case of steady flow *or* flow of an incompressible fluid, the mass in the streamtube segment remains constant, and therefore

$$\rho_A v_A A = \rho_B v_B B \qquad (7.1)$$

If the flow is incompressible, $\rho_A = \rho_B$, and

$$v_A A = v_B B \qquad (7.2)$$

Thus the rate of mass flow through all cross sections along a streamtube is constant in the case of steady flow or incompressible flow. Further, if streamlines defining the streamtube converge, an incompressible fluid within the streamtube must accelerate; if streamlines diverge, an incompressible fluid must decelerate.

A pathline is a good example of a Lagrangian view of motion. Describing the motion involves specifying the particle coordinates as a function of time $t$, and with respect to initial coordinate values, $x_0$, $y_0$, and $z_0$. Particle coordinates therefore are dependent variables, and time is the independent variable; thus $x = x(t; x_0, y_0, z_0)$, $y = y(t; x_0, y_0, z_0)$, and $z = z(t; x_0, y_0, z_0)$. Velocities and accelerations are then described in differential forms in terms of how these coordinates vary with respect to time. For example, the component of velocity parallel to the $x$-axis is $dx/dt$. This is the familiar approach adopted in particle mechanics, for example, in tracing the motion of a ballistic particle. In contrast, an Eulerian description of motion treats the $x$, $y$, and $z$ coordinates (and time) as independent variables. The solution to an Eulerian formulation therefore consists of a *velocity field*. Velocities are dependent quantities of position and time; thus $u = u(x, y, z, t)$, $v = v(x, y, z, t)$, and $w = w(x, y, z, t)$. An Eularian description of motion is normally adopted in fluid physics, although we will consider both views in this chapter.

## 7.2 SUBSTANTIVE DERIVATIVE

### 7.2.1 General expression

Consider the rate of change in a property associated with a specific fluid particle. It is convenient to begin with a scalar quantity, for example, the temperature $T$ of the particle. Suppose the particle is moving within a Cartesian coordinate system. A change in its temperature $dT$ as it moves a small distance whose components are $dx$, $dy$, and $dz$, during a small interval $dt$, may be written:

$$dT = \frac{\partial T}{\partial x}dx + \frac{\partial T}{\partial y}dy + \frac{\partial T}{\partial z}dz + \frac{\partial T}{\partial t}dt \qquad (7.3)$$

Each term in (7.3) is obtained as a first-order Taylor expansion (Chapter 6). A rate of change in $T$ is then formed by dividing by $dt$:

$$\frac{dT}{dt} = \frac{\partial T}{\partial x}\frac{dx}{dt} + \frac{\partial T}{\partial y}\frac{dy}{dt} + \frac{\partial T}{\partial z}\frac{dz}{dt} + \frac{\partial T}{\partial t} \qquad (7.4)$$

Because $dx$, $dy$, and $dz$ are chosen as the components of a small distance traveled by the fluid particle during $dt$, then in the limit as $dt$ goes to zero, $dx/dt$, $dy/dt$, and $dz/dt$ are the three components, $u$, $v$, and $w$, of the velocity of the particle. Thus,

$$\frac{DT}{Dt} = u\frac{\partial T}{\partial x} + v\frac{\partial T}{\partial y} + w\frac{\partial T}{\partial z} + \frac{\partial T}{\partial t} \qquad (7.5)$$

This is the rate of change of the temperature $T$ as viewed by an observer following the fluid particle. It is therefore a Lagrangian description of temperature change although it involves Eulerian coordinates.

To clarify this, let us derive (7.5) by a different procedure wherein the coordinates $x$, $y$, and $z$ are specifically associated with the fluid particle. Then the particle coordinates are functions $x(t)$, $y(t)$, and $z(t)$ of time $t$; and since the temperature in general is a function $T(x, y, z, t)$ of position and time,

$$T = T[x(t), y(t), z(t), t] \qquad (7.6)$$

Let $DT/Dt$ denote the rate of change in $T$ as viewed by an observer moving with the fluid particle. Thus,

$$\frac{DT}{Dt} = \frac{d}{dt}T[x(t), y(t), z(t), t] \qquad (7.7)$$

Applying the chain rule to (7.7) then leads to (7.4), from which we obtained the Lagrangian description (7.5).

The three terms on the right side of (7.5) involving velocities are referred to as *convective terms,* and the last term is referred to as a *local term.* These are to be interpreted as follows. The local term describes the effect on the particle associated with a change in the temperature of the entire field. Such a change in particle temperature can occur in absence of flow. Moreover, this effect is *local* in the sense that temperature locally changes independently of flow. Convective terms describe the effect on the particle temperature as it moves to a new position with different temperature. This effect can occur even if the temperature of the field is steady. The magnitude of this effect depends on the magnitude of variations in temperature over the flow field near the particle, and on the particle velocity, which determines how quickly the particle moves to a new position with different temperature. These terms are *convective* in the sense that they involve convection (or advection) of fluid mass.

The notation D/Dt denotes an operator, referred to as the *substantive derivative* (or *particle derivative*, or *total derivative*), in the same sense that d/dt denotes an operator, the ordinary derivative. The substantive derivative may be written as

$$\frac{D}{Dt} = u\frac{\partial}{\partial x} + v\frac{\partial}{\partial y} + w\frac{\partial}{\partial z} + \frac{\partial}{\partial t} \tag{7.8}$$

$$\frac{D}{Dt} = \mathbf{u} \cdot \nabla + \frac{\partial}{\partial t} \tag{7.9}$$

where $\mathbf{u} = \langle u, v, w \rangle$. Recall that $\nabla$ denotes the gradient operator (Chapter 4):

$$\nabla = \mathbf{i}\frac{\partial}{\partial x} + \mathbf{j}\frac{\partial}{\partial y} + \mathbf{k}\frac{\partial}{\partial z} \tag{7.10}$$

The notation $\mathbf{u} \cdot \nabla$ denotes the *inner product* of $\mathbf{u}$ and $\nabla$. In general, the inner product of two vectors $\mathbf{W} = \langle W_x, W_y, W_z \rangle$ and $\mathbf{V} = \langle V_x, V_y, V_z \rangle$ is

$$\mathbf{W} \cdot \mathbf{V} = \langle W_x, W_y, W_z \rangle \cdot \langle V_x, V_y, V_z \rangle = W_x V_x + W_y V_y + W_z V_z \tag{7.11}$$

It follows that if $\mathbf{u}$ is treated as $\mathbf{W}$, and $\nabla$ is treated as $\mathbf{V}$, taking the formal inner product leads to the first three terms on the right side of (7.8).

### 7.2.2 Convective accelerations

One of our main objectives is to obtain a kinematical expression for a change in fluid velocity, because such a change implies an acceleration of the fluid, and thus the action of a force. Recall that the rate of change in velocity $\mathbf{u}$ defines an acceleration $\mathbf{a}$,

$$\mathbf{a} = \lim_{\Delta t \to 0} \frac{\Delta \mathbf{u}}{\Delta t} = \frac{d\mathbf{u}}{dt} \tag{7.12}$$

and note that the direction of $\mathbf{a}$ may not coincide with that of $\mathbf{u}$ (Figure 7.3). To define the rate of change in velocity as viewed by an observer moving with the fluid, note that the substantive derivative (7.8) can operate on a vector as well as a scalar; thus

$$\frac{D\mathbf{u}}{Dt} = u\frac{\partial \mathbf{u}}{\partial x} + v\frac{\partial \mathbf{u}}{\partial y} + w\frac{\partial \mathbf{u}}{\partial z} + \frac{\partial \mathbf{u}}{\partial t} \tag{7.13}$$

The local term in (7.13) arises from unsteady flow. In the context of river flow, for example, this term is nonzero due to variations in discharge $Q(L^3 t^{-1})$ associated with the passage of a flood. As $Q$ increases on the rising limb of a flood wave,

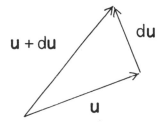

**Figure 7.3**  Vector diagram of components of acceleration $\mathbf{a} = d\mathbf{u}/dt$.

a given channel reach conveys an increasing volume of water per unit time. The velocity at a given coordinate position in the channel is therefore generally increasing, independently of convective effects. Conversely, the three convective terms in (7.13) may be significant in absence of changes in $Q$. Examination of these terms will reveal that each has the dimensions of acceleration. These accelerations arise when a real boundary induces a change in the direction or magnitude of the velocity of nearby flow. This idea is examined further below.

Note that because $\mathbf{u} = \mathbf{i}u + \mathbf{j}v + \mathbf{k}w$,

$$\frac{\partial \mathbf{u}}{\partial x} = \mathbf{i}\frac{\partial u}{\partial x} + \mathbf{j}\frac{\partial v}{\partial x} + \mathbf{k}\frac{\partial w}{\partial x} \tag{7.14}$$

$$\frac{\partial \mathbf{u}}{\partial y} = \mathbf{i}\frac{\partial u}{\partial y} + \mathbf{j}\frac{\partial v}{\partial y} + \mathbf{k}\frac{\partial w}{\partial y} \tag{7.15}$$

$$\frac{\partial \mathbf{u}}{\partial z} = \mathbf{i}\frac{\partial u}{\partial z} + \mathbf{j}\frac{\partial v}{\partial z} + \mathbf{k}\frac{\partial w}{\partial z} \tag{7.16}$$

$$\frac{\partial \mathbf{u}}{\partial t} = \mathbf{i}\frac{\partial u}{\partial t} + \mathbf{j}\frac{\partial v}{\partial t} + \mathbf{k}\frac{\partial w}{\partial t} \tag{7.17}$$

Substituting these into (7.13), expanding, then grouping terms in components of $\mathbf{i}$, $\mathbf{j}$, and $\mathbf{k}$ leads to

$$\mathbf{a} = \frac{D\mathbf{u}}{Dt} = \mathbf{i}\left(u\frac{\partial u}{\partial x} + v\frac{\partial u}{\partial y} + w\frac{\partial u}{\partial z} + \frac{\partial u}{\partial t}\right)$$

$$+ \mathbf{j}\left(u\frac{\partial v}{\partial x} + v\frac{\partial v}{\partial y} + w\frac{\partial v}{\partial z} + \frac{\partial v}{\partial t}\right)$$

$$+ \mathbf{k}\left(u\frac{\partial w}{\partial x} + v\frac{\partial w}{\partial y} + w\frac{\partial w}{\partial z} + \frac{\partial w}{\partial t}\right) \tag{7.18}$$

Because $\mathbf{a} = \langle a_x, a_y, a_z \rangle = \mathbf{i}a_x + \mathbf{j}a_y + \mathbf{k}a_z$, it follows that the three components of acceleration are

$$a_x = u\frac{\partial u}{\partial x} + v\frac{\partial u}{\partial y} + w\frac{\partial u}{\partial z} + \frac{\partial u}{\partial t} \tag{7.19}$$

$$a_y = u\frac{\partial v}{\partial x} + v\frac{\partial v}{\partial y} + w\frac{\partial v}{\partial z} + \frac{\partial v}{\partial t} \tag{7.20}$$

$$a_z = u\frac{\partial w}{\partial x} + v\frac{\partial w}{\partial y} + w\frac{\partial w}{\partial z} + \frac{\partial w}{\partial t} \tag{7.21}$$

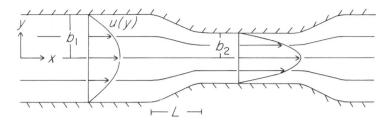

**Figure 7.4** Idealized velocity distributions $u(y)$ and streamlines for steady, incompressible fluid flow within a two-dimensional Venturi conduit.

which are kinematical descriptions of fluid acceleration; nothing as yet is explicitly stated regarding the forces that produce these accelerations.

Each component of acceleration thus has three convective terms and a local term. Each of the local terms has the same physical interpretation as that given above with regard to (7.13). Here we may examine the convective terms more carefully. Consider steady, incompressible fluid flow within a symmetrical two-dimensional Venturi conduit where the streamwise $x$-axis is parallel to the conduit axis (Figure 7.4). As $w$ everywhere equals zero in this two-dimensional conduit, the third convective term in (7.19) involving $w$ may be neglected. Conservation of mass requires that the flow must speed up as it converges within the conduit constriction. At any position along the conduit axis within the region of converging flow, for example, both $u$ and $\partial u / \partial x$ are positive. The first convective term of $a_x$ in (7.19)—the product $u \partial u / \partial x$—is therefore positive, and describes a streamwise acceleration induced by the conduit walls in satisfying continuity (Chapter 8). The velocity $v$ as well as the derivative $\partial u / \partial y$ equal zero at the conduit axis. Therefore the second convective term in (7.19) does not contribute to a streamwise acceleration of fluid. Near the wall, however, fluid is forced to flow toward the axis (Figure 7.4). The velocity $v$ therefore is nonzero and, due to the no-slip condition at the wall, the gradient $\partial u / \partial y$ is nonzero. What, then, is the significance of the product $v \partial u / \partial y$, which is nonzero and positive, and which evidently contributes to a positive streamwise acceleration of fluid?

The simplest answer to this question is that terms of the form $v \partial u / \partial y$ necessarily arise with an Eulerian formulation of fluid acceleration. Nonetheless, a geometrical interpretation of this term is possible. Notice that at a given position $y$ between the axis and wall, the magnitude of $\partial u / \partial y$ as well as the velocity component $u$, are increasing in a streamwise sense. Fluid with a component of velocity $v$ must displace fluid as it approaches the axis. Inasmuch as the displaced fluid moves with a streamwise component, it must be accelerated to make room for fluid arriving with component $v$ plus fluid arriving from upstream with component $u$. If $v$ is large, the streamwise acceleration must be large to compensate. In fact, for given $v$, $\partial u / \partial y$ is a measure of the local rate at which fluid must be displaced in a streamwise direction to compensate fluid arriving with component $v$. Note that the effect of this term is the same as that of the first term—an increasing streamwise velocity due to converging flow. The first convective term $u \partial u / \partial x$ is significant in absence of the no-slip condition, requiring only the convergence of flow due to the narrowing geometry of the conduit. The significance of the second convective term $v \partial u / \partial y$ is specifically due to wall effects: a nonzero $v$ arises because the wall geometry induces this component of flow; a nonzero gradient $\partial u / \partial y$ arises from the no-slip condition at the walls. Similar arguments apply to the two remaining kinematical expressions of acceleration, (7.20) and (7.21).

To anticipate our use of (7.18) in developing dynamical equations later, observe that $m\mathbf{D}\mathbf{u}/\mathbf{D}t$ is equal to a force acting on a fluid particle of mass $m$. In particular, by Newton's second law, this product of mass and acceleration is equal to the (net) sum of body and surface forces, $\mathbf{B} + \mathbf{S}$, acting on the fluid particle:

$$m\frac{\mathbf{D}\mathbf{u}}{\mathbf{D}t} = \mathbf{B} + \mathbf{S} \tag{7.22}$$

Noting that $\rho\mathbf{D}\mathbf{u}/\mathbf{D}t$ is then a force per unit volume,

$$\rho\frac{\mathbf{D}\mathbf{u}}{\mathbf{D}t} = \mathbf{B}_V + \mathbf{S}_V \tag{7.23}$$

where $\mathbf{B}_V$ and $\mathbf{S}_V$ are body and surface forces per unit volume. This is a simplified expression of the *equation of motion*, or the *dynamical equation*, of a fluid; it usually is referred to as the *momentum equation*. This name arises because $\rho\mathbf{D}\mathbf{u}/\mathbf{D}t$ is equivalent to the rate of change of momentum per unit volume of fluid. Equation (7.23) is a conservation law for momentum in the same sense that a continuity equation is a conservation law regarding mass (Chapter 8). Our task in the next several chapters is to develop explicit expressions for $\mathbf{B}_V$ and $\mathbf{S}_V$.

## 7.3 EXAMPLE PROBLEMS

### 7.3.1 Local and convective temperature changes within a dike

Let us choose a simplified problem to illustrate the significance of a substantive derivative of a scalar quantity. Consider the case of steady magma flow within a vertical dike (Example Problem 6.7.2). Magma ascending along the dike with velocity $w_1$ advects heat upward toward the surface. If the temperature of the magma is greater than the temperature of the surrounding country rock, part of this heat is conducted into the dike walls. Neglecting viscous generation of heat (Example Problem 12.5.3), we therefore may expect that the magma temperature decreases over the vertical distance $h$, from $T_m$ at the top of the magma chamber to $T_{s1}$ at the surface. Further suppose that steady temperature conditions exist. In this case $\partial T/\partial t = 0$, and with $u = v = 0$, (7.5) becomes

$$\frac{\mathbf{D}T}{\mathbf{D}t} = w\frac{\partial T}{\partial z} \tag{7.24}$$

If, for simplicity, $T$ varies linearly such that $\partial T/\partial z \approx (T_{s1} - T_m)/h$, then

$$\frac{\mathbf{D}T}{\mathbf{D}t} \approx w_1\frac{T_{s1} - T_m}{h} \tag{7.25}$$

This is the rate at which the temperature of a particle moving with the magma changes as it moves through stationary isotherms whose values decrease upward. This reflects the rate at which such a particle loses heat by conduction as it ascends.

Suppose that $w$ increases to $w_2$. Momentarily, the gradient $\partial T/\partial z$ remains unchanged over part of the distance $h$ such that isotherms are advected upward and a

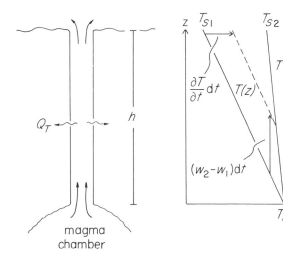

**Figure 7.5** Idealized steady, linear temperature distribution $T(z)$ associated with upward heat advection and lateral heat conduction $Q_T$ in a vertical dike, and transient upward migration of a "heat front" (*dashed line*) immediately following an increase in the ascent speed of the magma.

"heat front" migrates with speed $w_2 - w_1$ toward the surface (Figure 7.5). The local rate of change $\partial T/\partial t$ measured at the surface becomes nonzero, and the substantive rate of change $DT/Dt$ of a particle moving with the magma is then

$$\frac{DT}{Dt} \approx w_2 \frac{T_{s1} - T_m}{h} + \frac{\partial T}{\partial t} \qquad (7.26)$$

The first term on the right side of (7.26) describes the contribution to the total rate of change in temperature of the particle as it moves through the isotherms. This is a negative quantity. The last term describes the contribution to the total rate of change as the isotherms move upward and thereby tend to "overtake" the particle. This is a positive quantity. If new, steady temperature conditions compatible with $w_2$ are eventually reached, then $\partial T/\partial t = 0$, and the rate of change $DT/Dt$ of a particle moving with the magma is

$$\frac{DT}{Dt} \approx w_2 \frac{T_{s2} - T_m}{h} \qquad (7.27)$$

where $T_{s2}$ is the new (steady) surface temperature.

Let us suppose for simplicity that the rate of heat loss by conduction from a magma particle as it ascends is constant. (See Example Problem 9.3.3 for a more realistic treatment of this.) Then the substantive rate $DT/Dt$ is the same in all three cases above. Using the two cases of steady temperature, this would suggest that $T_m \approx (w_1 T_{s1} - w_2 T_{s2})/(w_1 - w_2)$, giving a rough estimate of $T_m$ based on measured values of $T_s$ and $w$.

## 7.3.2 Streamwise acceleration in a Venturi flow

Let us estimate the magnitude of the streamwise acceleration $a_x$ along the centerline ($y = 0$) of the Venturi conduit described in Section 7.2.2, and see how this magnitude varies with flow and Venturi geometry. Here we will assume that the change in aperture $b$ is small relative to the length of the constriction $L$ (Figure 7.4). First recall that $v = w = 0$ at the centerline. With steady flow, (7.19) therefore simplifies to

$$a_x = u \frac{\partial u}{\partial x} \tag{7.28}$$

Let $u_1$ and $u_2$ denote centerline velocities upstream and downstream of the constriction, which separates conduits with uniform half-apertures $b_1$ and $b_2$. Let us assume that $u$ is of the order of a characteristic velocity within the constriction, formed as the average of $u_1$ and $u_2$; thus $O(u) = (u_2 + u_1)/2$. Further assume that the velocity gradient $\partial u/\partial x$ is of the order $O(\partial u/\partial x) = (u_2 - u_1)/L$. Then from (7.28),

$$O(a_x) = \frac{u_2 + u_1}{2} \frac{u_2 - u_1}{L} = \frac{u_2^2 - u_1^2}{2L} \tag{7.29}$$

From (3.72),

$$u_1(0) = -\frac{1}{2\mu} \left( \frac{\partial p}{\partial x} \right)_1 b_1^2; \qquad u_2(0) = -\frac{1}{2\mu} \left( \frac{\partial p}{\partial x} \right)_2 b_2^2 \tag{7.30}$$

The total discharge $Q$ remains constant along the conduit; from (3.74) it therefore follows that

$$\left( \frac{\partial p}{\partial x} \right)_2 = \left( \frac{\partial p}{\partial x} \right)_1 \frac{b_1^3}{b_2^3} \tag{7.31}$$

Substituting (7.30) and (7.31) into (7.29) then rearranging,

$$O(a_x) = \frac{u_1^2}{2L} \left( \frac{b_1^2 - b_2^2}{b_2^2} \right) \tag{7.32}$$

Thus the magnitude of the centerline acceleration through the constriction is proportional to the square of the entrance velocity $u_1^2$ and to the ratio formed by the change in the squares of conduit apertures to the square of the downstream aperture, and inversely proportional to the constriction length $L$. Qualitatively, a large velocity requires a relatively strong acceleration to change the velocity a finite amount over the given distance $L$. Likewise, a large change in aperture over a short distance requires a strong acceleration of fluid to achieve a given change in velocity necessary to satisfy conservation of mass (constant $Q$). Also note that a rapid change in aperture size may lead to the onset of *flow separation,* a topic that we will examine in Chapters 11 and 14.

## 7.4 READING

Batchelor, G. K. 1967. *An introduction to fluid dynamics.* Cambridge: Cambridge University Press, 615 pp. This text contains a series of 24 plates that illustrate a variety of flow conditions as revealed by flow visualization techniques.

Nakayama, Y., Woods, W. A., and Clark, D. G., eds. 1988. *Visualized flow: fluid motion in basic and engineering situations.* Oxford: Pergamon, 137 pp. This book contains a descriptive summary of flow visualization techniques, followed by a comprehensive collection of photographs of flow as revealed by these techniques.

Tritton, D. J. 1988. *Physical fluid dynamics,* 2nd ed. Oxford: Oxford University Press, 519 pp. This text provides a clear discussion of streamlines, pathlines, and streaklines (pp. 73–75) in the context of flow visualization techniques (p. 425–28), and numerous photographs of flows involving these techniques.

# CHAPTER 8

# Conservation of Mass

The concept of *conservation of mass* holds a fundamental role in most problems in fluid physics. For a given problem this concept is cast in the form of an *equation of continuity*. Such an equation describes a condition—conservation of mass—that must be satisfied in any formal analysis of a problem. Thus an equation of continuity often is one of several complementary equations that are solved simultaneously to arrive at a solution to a flow problem, for example, the flow velocity as a function of coordinate position in a flow field. (Typically these complementary equations, as we will see in later chapters, involve conservation of momentum or energy, or both.) Although we did not explicitly use this idea in analyzing the one-dimensional flow problems at the end of Chapter 3, it turns out that continuity was implicitly satisfied in setting up each problem. We will return to these problems to illustrate this point.

We will develop equations of continuity for three general cases: purely fluid flow, saturated single-phase flow in porous media, and unsaturated flow in porous media. The most general of the three equations is that for unsaturated flow, where pores are partially filled with the fluid phase of interest, such that the degree of saturation with respect to that phase is less than one. We will then show that this equation reduces, in the special case in which the degree of saturation equals one, to a simpler form appropriate for saturated single-phase flow. Then, this equation for saturated flow could be reduced further, in the special case in which the porosity equals one, to a form appropriate for purely fluid flow. For pedagogical reasons, however, we shall reverse this order and consider purely fluid flow first.

In addition we will consider conservation of a solid or gas dissolved in a liquid, and take this opportunity to introduce *Fick's law* for molecular diffusion. For simplicity we will consider only species that do not react chemically with the liquid, nor with the solid phases of a porous medium.

Most of the derivations below are based on the idea of a small *control volume* of specified dimensions embedded within a fluid or porous medium. Continuity of mass in many problems, however, can be conveniently treated in terms of large control volumes defined by real boundaries. A lake is one example (Example Problem 8.4.5). It also is convenient to possess equations of continuity for different coordinate systems, as required by the geometry of different problems (Appendix 17.2). The simplest case involves a Cartesian coordinate system.

## 8.1 CONTINUITY IN CARTESIAN COORDINATES

### 8.1.1 Purely fluid flow

In the case of purely fluid flow (in absence of a porous medium), an equation of continuity states that mass flowing into a control volume must be compensated by mass flowing out of the control volume, or by a change in the density of the fluid within the volume, or some combination of both. Consider a small rectangular volume with edges having lengths $dx$, $dy$, and $dz$ embedded within a Cartesian coordinate system (Figure 8.1). Further, consider the amount of mass $m_{in}$ flowing into the volume through the face $dy\,dz$ during a small interval of time $dt$. Fluid entering the control volume travels a small distance $\delta$ during $dt$. The fluid velocity component $u = \delta/dt$, so the distance $\delta = u\,dt$. The mass of fluid that enters the volume during $dt$ is $\rho\delta\,dy\,dz$; that is, $m_{in} = \rho u\,dy\,dz\,dt$. The mass $m_{out}$ flowing out of the volume in the direction of positive $x$ during $dt$ is then specified using a Taylor expansion, in a manner similar to that used in Chapter 6 for specifying the static pressure acting on the surfaces of a control volume.

Recall Taylor's formula (6.10), with which we can approximate the value of a function $f(x)$ near the coordinate position $a$, where $f(a)$ is known:

$$f(x) = f(a) + \frac{f'(a)}{1!}(x - a) + \frac{f''(a)}{2!}(x - a)^2 + \cdots$$

$$+ \frac{f^{(n)}(a)}{n!}(x - a)^n + \frac{f^{(n+1)}(z)}{(n + 1)!}(x - a)^{n+1} \qquad a < z < x \quad (8.1)$$

Positioning $a$ at the entrance to the control volume, we wish to approximate the product $\rho u$ at position $x$, which may be denoted by $(\rho u)_x$. Note that either $\rho$ or $u$, or both, may change over the small distance $dx$. A change in $u$ would imply an acceleration or deceleration of fluid, a change in $\rho$ would imply a compression or decompression of fluid. Thus $f(a) = \rho u$, $x - a = dx$, and

$$(\rho u)_x = \rho u + \frac{\partial}{\partial x}(\rho u)\,dx + \frac{\partial^2}{\partial x^2}(\rho u)\frac{(\partial x)^2}{2!} + \cdots \qquad (8.2)$$

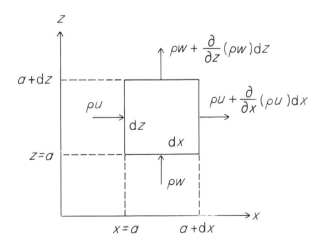

**Figure 8.1**  Definition diagram for mass flow quantities associated with an elementary control volume.

Then, the mass $m_{\text{out}}$ is to first order

$$\left[\rho u + \frac{\partial}{\partial x}(\rho u)\, dx\right] dy\, dz\, dt \tag{8.3}$$

The net flow of mass $N_x$ out of the control volume in the direction of positive $x$ is $N_x = m_{\text{out}} - m_{\text{in}}$. Thus

$$N_x = \left[\rho u + \frac{\partial}{\partial x}(\rho u) dx\right] dy\, dz\, dt - \rho u\, dy\, dz\, dt \tag{8.4}$$

$$N_x = \frac{\partial}{\partial x}(\rho u)\, dx\, dy\, dz\, dt \tag{8.5}$$

Making similar use of Taylor expansions, net flows $N_y$ and $N_z$ out of the control volume in directions of positive $y$ and $z$ are

$$N_y = \frac{\partial}{\partial y}(\rho v)\, dy\, dx\, dz\, dt \tag{8.6}$$

$$N_z = \frac{\partial}{\partial z}(\rho w)\, dz\, dx\, dy\, dt \tag{8.7}$$

The total net flow of mass $N$ out of the control volume is $N = N_x + N_y + N_z$. Thus

$$N = \left[\frac{\partial}{\partial x}(\rho u) + \frac{\partial}{\partial y}(\rho v) + \frac{\partial}{\partial z}(\rho w)\right] dx\, dy\, dz\, dt \tag{8.8}$$

According to the sign convention chosen, a positive value of $N$ implies that mass is exported from the control volume. Due to conservation of mass, this must be compensated by an equivalent decrease of mass within the volume. Since the volume $dx\, dy\, dz$ remains constant over the interval $dt$, the only way to decrease the mass is to alter the fluid density. The mass $m(t)$ within the control volume at time $t$ is $\rho\, dx\, dy\, dz$. Using a Taylor expansion with time as the independent variable, the mass $m(t + dt)$ within the volume after the brief interval of time $dt$ is to first order

$$m(t + dt) = \left(\rho + \frac{\partial \rho}{\partial t} dt\right) dx\, dy\, dz \tag{8.9}$$

The change in mass $dm = m(t + dt) - m(t)$. Thus

$$dm = \left(\rho + \frac{\partial \rho}{\partial t} dt\right) dx\, dy\, dz - \rho\, dx\, dy\, dz \tag{8.10}$$

$$dm = \frac{\partial \rho}{\partial t} dt\, dx\, dy\, dz \tag{8.11}$$

By conservation of mass, $N + dm = 0$, so summing (8.8) and (8.11) and dividing by $dx\, dy\, dz\, dt$,

$$\frac{\partial}{\partial x}(\rho u) + \frac{\partial}{\partial y}(\rho v) + \frac{\partial}{\partial z}(\rho w) = -\frac{\partial \rho}{\partial t} \tag{8.12}$$

This is an Eulerian equation of continuity in Cartesian coordinates. It is a condition that must be satisfied at each position within a flow field. As such, its principal value, as will be seen below, is to constrain the mathematics of a given flow problem.

Expanding (8.12) by taking derivatives of products using the chain rule,

$$u\frac{\partial \rho}{\partial x} + \rho\frac{\partial u}{\partial x} + v\frac{\partial \rho}{\partial y} + \rho\frac{\partial v}{\partial y} + w\frac{\partial \rho}{\partial z} + \rho\frac{\partial w}{\partial z} + \frac{\partial \rho}{\partial t} = 0 \tag{8.13}$$

Regrouping terms in $\rho$,

$$\rho\left(\frac{\partial u}{\partial x} + \frac{\partial v}{\partial y} + \frac{\partial w}{\partial z}\right) + u\frac{\partial \rho}{\partial x} + v\frac{\partial \rho}{\partial y} + w\frac{\partial \rho}{\partial z} + \frac{\partial \rho}{\partial t} = 0 \tag{8.14}$$

The four terms in (8.14) involving derivatives of density $\rho$ collectively constitute a substantive derivative denoted by $D\rho/Dt$, and we write

$$\frac{\partial u}{\partial x} + \frac{\partial v}{\partial y} + \frac{\partial w}{\partial z} = -\frac{1}{\rho}\frac{D\rho}{Dt} \tag{8.15}$$

This is referred to as a Lagrangian equation of continuity. The difference in the physical interpretations of the Eulerian form (8.12) and the Lagrangian form (8.15) follows from the discussion of the substantive derivative in Chapter 7. The Lagrangian form describes the total rate of change in fluid density as viewed by an observer moving with the fluid. The Eulerian form describes the local rate of change in density as viewed by an observer at a fixed position in the flow field. Both forms, however, involve Eulerian coordinates.

Recall that the density of a homogeneous fluid does not vary with respect to spatial coordinates. Thus $\partial \rho/\partial x = \partial \rho/\partial y = \partial \rho/\partial z = 0$, and (8.15) reduces to

$$\frac{\partial u}{\partial x} + \frac{\partial v}{\partial y} + \frac{\partial w}{\partial z} = -\frac{1}{\rho}\frac{\partial \rho}{\partial t} \tag{8.16}$$

Note that this statement leaves open the possibility that the fluid density $\rho$ may vary with time. In the case of steady flow of an incompressible fluid, or if the density $\rho$ is constant, $\partial \rho/\partial t = 0$ and (8.16) reduces to

$$\frac{\partial u}{\partial x} + \frac{\partial v}{\partial y} + \frac{\partial w}{\partial z} = 0 \tag{8.17}$$

This equation, implying that the substantive derivative of $\rho$ equals zero, is the appropriate statement of continuity for the case of incompressible fluid flow.

Equations (8.15), (8.16), and (8.17) involve the important mathematical concept of *divergence*. The divergence of a vector $\mathbf{V}$ is defined for Cartesian coordinates by

$$\text{div } \mathbf{V} = \frac{\partial V_x}{\partial x} + \frac{\partial V_y}{\partial y} + \frac{\partial V_z}{\partial z} \tag{8.18}$$

where $V_x$, $V_y$, and $V_z$ are the components of $\mathbf{V}$ in the $x$, $y$, and $z$ directions. Note that div $\mathbf{V}$ is a scalar quantity. Now, recall that the *vector differential operator* $\nabla$ in three dimensions is defined by (Chapter 4):

$$\nabla = \mathbf{i}\frac{\partial}{\partial x} + \mathbf{j}\frac{\partial}{\partial y} + \mathbf{k}\frac{\partial}{\partial z} \tag{8.19}$$

The vector $\mathbf{V}$ written in terms of unit vectors $\mathbf{i}$, $\mathbf{j}$, and $\mathbf{k}$ is

$$\mathbf{V} = \langle V_x, V_y, V_z \rangle = \mathbf{i}V_x + \mathbf{j}V_y + \mathbf{k}V_z \tag{8.20}$$

It follows that if $\mathbf{V}$ is the velocity $\mathbf{u} = \langle u, v, w \rangle$, taking the formal inner product $\nabla \cdot \mathbf{u}$ leads to

$$\nabla \cdot \mathbf{u} = \left( \mathbf{i} \frac{\partial}{\partial x} + \mathbf{j} \frac{\partial}{\partial y} + \mathbf{k} \frac{\partial}{\partial z} \right) \cdot (\mathbf{i}u + \mathbf{j}v + \mathbf{k}w) \qquad (8.21)$$

$$= \left\langle \frac{\partial}{\partial x}, \frac{\partial}{\partial y}, \frac{\partial}{\partial z} \right\rangle \cdot \langle u, v, w \rangle \qquad (8.22)$$

$$= \operatorname{div} \mathbf{u} \qquad (8.23)$$

Using these definitions, (8.15), (8.16), and (8.17) often are written in more compact forms:

$$\nabla \cdot \mathbf{u} = -\frac{1}{\rho} \frac{D\rho}{Dt} \qquad (8.24)$$

$$\nabla \cdot \mathbf{u} = -\frac{1}{\rho} \frac{\partial \rho}{\partial t} \qquad (8.25)$$

$$\nabla \cdot \mathbf{u} = 0 \qquad (8.26)$$

From statements above, when $\operatorname{div} \mathbf{u} = \nabla \cdot \mathbf{u} = 0$, the fluid is incompressible.

## 8.1.2  Flow in porous media

In the case of fluid flow within an unsaturated porous medium, at least two fluid phases are involved. Often we are concerned with a liquid phase and a gaseous phase—for example, liquid water and air. The development here pertains to flow of one of the phases, normally the liquid phase, wherein we neglect the other phase with the assumption that it can be readily displaced by the phase of interest. An equation of continuity then states that mass flowing into a control volume must be compensated by mass flowing out of the control volume, by a change in the density of the fluid within the volume, by a change in the pore volume occupied by fluid, or some combination of these. In addition, stresses acting on the solid matrix may change its porosity during a brief interval of time $dt$, thereby altering the volume of pores available to contain fluid; this must be taken into account. In the case of fluid flow within a saturated porous medium, the third of these possibilities is omitted since the pores remain fully occupied by the fluid.

Again consider a small, rectangular control volume with edges having lengths $dx$, $dy$, and $dz$ embedded within a Cartesian coordinate system (Figure 8.2). The amount of mass $m_{in}$ flowing into the volume through the face $dy\,dz$ during a small interval of time $dt$ is $\rho q_{hx}\,dy\,dz\,dt$, where $q_{hx}$ is the *specific discharge*. Note that $q_{hx}$ is a volumetric flux: a volume of water per unit time per unit area normal to the $x$-coordinate direction. Although $q_{hx}$ has dimensions of velocity, it does not correspond to the velocity of fluid molecules in the sense that the velocity $u$ does. The mass $m_{out}$ flowing out of the volume in the direction of positive $x$ during $dt$ is then specified using a Taylor expansion of the product $\rho q_{hx}$, as in the steps followed above for the product $\rho u$. Following these steps, the net flow of mass $N_x$ out of the control volume in the direction of positive $x$ is

$$N_x = \frac{\partial}{\partial x}(\rho q_{hx})\,dx\,dy\,dz\,dt \qquad (8.27)$$

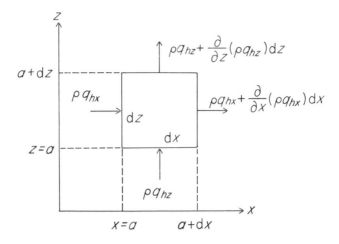

**Figure 8.2** Definition diagram for mass flow quantities associated with an elementary control volume embedded within a porous medium.

Making similar use of Taylor expansions, net flows of mass $N_y$ and $N_z$ out of the control volume in directions of positive $y$ and $z$ are obtained, giving for the total net flow of mass $N$ out of the volume:

$$N = \left[ \frac{\partial}{\partial x}(\rho q_{hx}) + \frac{\partial}{\partial y}(\rho q_{hy}) + \frac{\partial}{\partial z}(\rho q_{hz}) \right] dx\, dy\, dz\, dt \qquad (8.28)$$

where $q_{hy}$ and $q_{hz}$ denote specific discharges in the $y$ and $z$ directions. This is analogous to (8.8) for purely fluid flow.

According to the sign convention chosen, a positive value of $N$ implies that mass is exported from the control volume. Due to conservation of mass, this must be compensated by an equivalent decrease of mass within the volume. Since the volume $dx\, dy\, dz$ remains constant over the interval $dt$, the mass may be decreased by altering the fluid density, the proportion of the pores filled by fluid, or the pore space available for fluid. The mass $m(t)$ within the control volume at time $t$ is $n\theta_s \rho\, dx\, dy\, dz$, where $n$ is the porosity and $\theta_s$ is the degree of saturation. Using a Taylor expansion an expression analogous to (8.9) for the mass $m(t + dt)$ within the volume after the brief interval of time $dt$ is obtained. Then the change in mass $dm = m(t + dt) - m(t)$ is

$$dm = \frac{\partial}{\partial t}(n\theta_s \rho)\, dt\, dx\, dy\, dz \qquad (8.29)$$

By conservation of mass $N + dm = 0$, so summing (8.28) and (8.29) and dividing by $dx\, dy\, dz\, dt$,

$$\frac{\partial}{\partial x}(\rho q_{hx}) + \frac{\partial}{\partial y}(\rho q_{hy}) + \frac{\partial}{\partial z}(\rho q_{hz}) = -\frac{\partial}{\partial t}(n\rho\theta_s) \qquad (8.30)$$

which is an Eulerian equation of continuity, analogous to (8.12) for purely fluid flow.

In the special case in which the pores are entirely filled with the fluid phase of interest, $\theta_s = 1$, and (8.30) reduces to a form appropriate for saturated single-phase flow:

$$\frac{\partial}{\partial x}(\rho q_{hx}) + \frac{\partial}{\partial y}(\rho q_{hy}) + \frac{\partial}{\partial z}(\rho q_{hz}) = -\frac{\partial}{\partial t}(n\rho) \tag{8.31}$$

Further, in the special case in which the porosity $n = 1$, $q_{hx} = u$, $q_{hy} = v$, and $q_{hz} = w$, and (8.31) reduces to (8.12).

Expanding terms in (8.31) using the chain rule, regrouping terms involving derivatives of specific discharge and derivatives of density, then using the definitions of divergence and gradient (Chapter 4),

$$\rho(\nabla \cdot \mathbf{q}_h) + \mathbf{q}_h \cdot \nabla\rho = -n\theta_s\frac{\partial\rho}{\partial t} - n\rho\frac{\partial\theta_s}{\partial t} - \theta_s\rho\frac{\partial n}{\partial t} \tag{8.32}$$

where $\mathbf{q}_h = \langle q_{hx}, q_{hy}, q_{hz}\rangle$. This is a general equation of continuity for unsaturated single-phase flow.

With sufficiently thick, porous geological materials under unsaturated conditions, fluctuations in liquid content equate to fluctuations in overburden weight. This loading, if large enough, can lead to deformation of the medium at depth. In addition, many soils possess mineralogical components that expand and contract with varying amounts of moisture. In these circumstances, such a medium must be treated as deformable. Many problems of interest nonetheless involve essentially nondeformable media, in which case $n$ is constant, and (8.32) reduces to

$$\rho(\nabla \cdot \mathbf{q}_h) + \mathbf{q}_h \cdot \nabla\rho = -n\theta_s\frac{\partial\rho}{\partial t} - n\rho\frac{\partial\theta_s}{\partial t} \tag{8.33}$$

Moreover, liquids usually are incompressible over the range of pressures normally encountered in unsaturated flow. Thus derivatives involving density equal zero, and (8.33) reduces to

$$\nabla \cdot \mathbf{q}_h = -n\frac{\partial\theta_s}{\partial t} \tag{8.34}$$

Finally, in the case of steady unsaturated flow, this becomes

$$\nabla \cdot \mathbf{q}_h = 0 \tag{8.35}$$

Returning to (8.32), for saturated single-phase flow $\theta_s = 1$, so (8.32) becomes

$$\rho(\nabla \cdot \mathbf{q}_h) + \mathbf{q}_h \cdot \nabla\rho = -n\frac{\partial\rho}{\partial t} - \rho\frac{\partial n}{\partial t} \tag{8.36}$$

which is a general equation of continuity. Under many circumstances variations in density with respect to position are much smaller than local variations with respect to time. That is $\mathbf{q}_h \cdot \nabla\rho \ll n\partial\rho/\partial t$, so (8.36) reduces to

$$\rho(\nabla \cdot \mathbf{q}_h) = -n\frac{\partial\rho}{\partial t} - \rho\frac{\partial n}{\partial t} \tag{8.37}$$

This is the equation of mass continuity for transient flow typically developed in texts on groundwater hydrology and hydraulics. It is of considerable practical importance. In the case of steady saturated flow, (8.37) reduces to (8.35) above.

So far in our development we have not explicitly considered deformation of the porous matrix. But recall from Chapter 3 that, according to the relation $d\sigma = d\sigma_e + dp$, a change in fluid pressure $dp$ or total stress $d\sigma$ can lead to a change in the effective stress $d\sigma_e$, which may lead to deformation of the solid matrix.

We therefore ought to consider a more general statement of conservation of mass that takes this into account. Hereafter we will assume that the degree of saturation $\theta_s = 1$.

Consider a rectangular control volume embedded within a Cartesian system. The faces of this control volume are not to be associated with the solid phase, but rather are fixed with respect to the coordinate system. Here is it useful to consider the average fluid (molecular) velocity $\mathbf{u} = \langle u, v, w \rangle$ rather than the specific discharge $\mathbf{q}_h$. Now, for example, the fluid mass entering the control volume in a direction parallel to the $x$-axis during the interval $dt$ is $\rho n u \, dy \, dz \, dt$. To first order the fluid mass leaving the right face is $(\rho n u + [\partial(\rho n u)/\partial x] dx) \, dy \, dz \, dt$. The net flow of mass out of the control volume parallel to the $x$-axis is therefore $[\partial(\rho n u)/\partial x] dx \, dy \, dz \, dt$. After obtaining similar expressions for the other two coordinate directions involving the components $v$ and $w$ and summing these three expressions, then equating this result with the negative of the right side of (8.29) (with $\theta_s = 1$), we obtain a statement of conservation of fluid mass expressed in terms of the average fluid velocity $\mathbf{u}$:

$$\rho[\nabla \cdot (n\mathbf{u})] + (n\mathbf{u}) \cdot \nabla\rho = -n\frac{\partial\rho}{\partial t} - \rho\frac{\partial n}{\partial t} \tag{8.38}$$

Consider a second control volume that initially coincides with the fixed control volume, but which moves with the porous matrix as it undergoes deformation (Figure 8.3). Now envision deformation of the matrix in one dimension that involves the left face of the deformable control volume. During the interval $dt$, this face moves a distance $u_s dt$ into the fixed volume, where $u_s$ is the $x$-component of the matrix velocity $\mathbf{u}_s$. Thus a mass of solid equal to $\rho_s(1 - n)u_s \, dy \, dz \, dt$ moves into the fixed volume during $dt$, where $\rho_s$ is the density of the solid. Likewise, a mass of solid moves out of the right face of the fixed volume; to first order this is equal to $[\rho_s n u_s + (\partial[\rho_s(1 - n)u_s]/\partial x)dx] \, dy \, dz \, dt$. The net loss of solid mass out of the fixed volume associated with matrix deformation parallel to the $x$-axis is therefore $(\partial[\rho_s(1 - n)u_s]/\partial x) \, dx \, dy \, dz \, dt$. Obtaining similar expressions for the other two coordinate directions involving the matrix velocity components $v_s$ and $w_s$ and summing these quantities, the net loss of solid mass associated with matrix deformation is

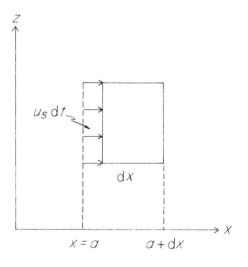

**Figure 8.3**  Definition diagram for mass flow quantities associated with a control volume that deforms with the porous matrix.

$$\nabla \cdot [\rho_s(1 - n)\mathbf{u}_s]dx\,dy\,dz\,dt \qquad (8.39)$$

The mass of solid in the fixed volume at time $t$ is $\rho_s(1 - n)dx\,dy\,dz$. The change in solid mass during the interval $dt$ is

$$\frac{\partial}{\partial t}[\rho_s(1 - n)]dx\,dy\,dz\,dt \qquad (8.40)$$

Setting the sum of (8.39) and (8.40) to zero and dividing by $dx\,dy\,dt$, we obtain a statement of conservation of solid mass:

$$\nabla \cdot [\rho_s(1 - n)\mathbf{u}_s] = -\frac{\partial}{\partial t}[\rho_s(1 - n)] \qquad (8.41)$$

Let us assume that any deformation of the matrix leading to a change in the porosity $n$ involves a change in the pore volume but not the solid volume. This is the same as assuming that the solid grains of the matrix are incompressible, in which case $\rho_s$ is a constant and may be removed from the differentials in (8.41). Then expanding and rearranging this equation,

$$\nabla \cdot \mathbf{u}_s = \frac{1}{1 - n}\mathbf{u}_s \cdot \nabla n + \frac{1}{1 - n}\frac{\partial n}{\partial t} \qquad (8.42)$$

Reconsider the deformation of the left face of the deformable control volume as above (Figure 8.3). In absence of relative motion between the fluid and matrix, $u = u_s$, and a mass of fluid equal to $\rho nu\,dy\,dz\,dt = \rho nu_s\,dy\,dz\,dt$ is imported into the fixed control volume during $dt$. Yet this does not involve a Darcy flux since fluid is not moving through the porous matrix. In fact, if we assert that Darcy's law describes fluid motion relative to the solid matrix, we must express the specific discharge $\mathbf{q}_h$ in terms of this relative motion:

$$\mathbf{q}_h = n(\mathbf{u} - \mathbf{u}_s) \qquad (8.43)$$

Solving for $\mathbf{u}$, substituting the result into (8.38), then expanding and rearranging,

$$\nabla \cdot (\rho\mathbf{q}_h) + \rho n\nabla \cdot \mathbf{u}_s + n\mathbf{u}_s \cdot \nabla\rho + \rho\mathbf{u}_s \cdot \nabla n = -n\frac{\partial\rho}{\partial t} - \rho\frac{\partial n}{\partial t} \qquad (8.44)$$

Substituting (8.42) into (8.44) then rearranging,

$$\nabla \cdot (\rho\mathbf{q}_h) + n\mathbf{u}_s \cdot \nabla\rho + \frac{\rho}{1 - n}\mathbf{u}_s \cdot \nabla n = -n\frac{\partial\rho}{\partial t} - \frac{\rho}{1 - n}\frac{\partial n}{\partial t} \qquad (8.45)$$

This is a general expression for conservation of total mass (fluid and solid). Notice that if spatial derivatives of $\rho$ and $n$ are small relative to local derivatives, (8.45) can be simplified to

$$\nabla \cdot (\rho\mathbf{q}_h) = -n\frac{\partial\rho}{\partial t} - \frac{\rho}{1 - n}\frac{\partial n}{\partial t} \qquad (8.46)$$

We will return to this expression in Chapter 13.

## 8.2 CONTINUITY OF SOLUTES

Many geological problems involve describing the movement of dissolved chemical species within a liquid flow. Recall, for example, that we examined the growth of $CO_2$ bubbles due to diffusion of $CO_2(aq)$ into them from surrounding liquid in

Example Problem 4.10.5. Likewise, the movement of dissolved species in groundwater is an important aspect of the geochemistry of dissolution of solid phases, leading to changes in media porosity, and precipitation of dissolved solids, for example, as interstitial cements. We will concentrate here on gas–liquid and solid–liquid systems, and refer to a dissolved gas or solid as the *solute*, and the liquid as the *solvent*. We will further restrict our treatment to simple systems, neglecting chemical interactions between solutes and solvents, and between solutes and solid media phases.

Solutes move by two mechanisms. First, a solute diffuses through a solvent in response to a spatial variation (gradient) in the concentration of the solute. Diffusion occurs from high to low concentrations. Second, a solute can be transported with a solvent, a process referred to as *advection*.

## 8.2.1 Simple solvent–solute flow

An equation of continuity for a solute states that the mass of solute entering a control volume must be compensated by mass leaving the control volume, or by a change in the concentration of the solute within the volume, or some combination of both. Let us start with the diffusion process.

Consider a small rectangular volume with edges having lengths $dx$, $dy$, and $dz$ embedded within a Cartesian coordinate system. Let $\mathbf{q}_c = \langle q_{cx}, q_{cy}, q_{cz} \rangle$ denote a mass flux density of solute ($M\,L^{-2}\,t^{-1}$). The mass of solute entering the control volume by diffusion in the direction of positive $x$ during $dt$ is $q_{cx}\,dy\,dz\,dt$. The mass of solute leaving the control volume by diffusion is to first order $[q_{cx} + (\partial q_{cx}/\partial x)\,dx]\,dy\,dz\,dt$. The net mass of solute diffusing out of the volume in the direction of $x$ during $dt$ is therefore $(\partial q_{cx}/\partial x)\,dx\,dy\,dz\,dt$. After obtaining similar expansions with respect to $y$ and $z$, the net mass diffusing out of the control volume during $dt$ is

$$\left( \frac{\partial q_{cx}}{\partial x} + \frac{\partial q_{cy}}{\partial y} + \frac{\partial q_{cz}}{\partial z} \right) dx\,dy\,dz\,dt \tag{8.47}$$

Recall from introductory chemistry and physics that simple diffusion of a solute is governed by *Fick's law:*

$$\mathbf{q}_c = -D_c \nabla c \tag{8.48}$$

where $c$ is the concentration (mass per unit volume) of the solute, and $D_c$ is a *coefficient of mass diffusion* ($L^2\,t^{-1}$) that is characteristic of the solute–solvent system. Note that this phenomenological law is homologous to Fourier's law (4.21). Now, assuming that $D_c$ is constant, substituting the components of $\mathbf{q}_c$ in (8.48) into (8.47) leads to

$$-D_c \left( \frac{\partial^2 c}{\partial x^2} + \frac{\partial^2 c}{\partial y^2} + \frac{\partial^2 c}{\partial z^2} \right) dx\,dy\,dz\,dt \tag{8.49}$$

Turning now to advection, assume that the solvent is incompressible. The volume of solvent entering the control volume in the direction of positive $x$ during $dt$ is $u\,dy\,dz\,dt$, so the mass of solute advected into the control volume is $uc\,dy\,dz\,dt$. The mass of solute advected out of the control volume is to first order $[uc + (\partial/\partial x)(uc)\,dx]\,dy\,dz$, and the net mass advected out of the control volume is

$(\partial/\partial x)(uc)\,dx\,dy\,dz$. After obtaining similar expansions with respect to $y$ and $z$, the net mass of solute advected out of the control volume during $dt$ is

$$\left[\frac{\partial}{\partial x}(uc) + \frac{\partial}{\partial y}(vc) + \frac{\partial}{\partial z}(wc)\right]dx\,dy\,dz\,dt \qquad (8.50)$$

The mass of solute in the control volume at time $t$ is $c\,dx\,dy\,dz$, the mass at time $t + dt$ is $[c + (\partial c/\partial t)dt]\,dx\,dy\,dz$, and the net change in mass during $dt$ is

$$\frac{\partial c}{\partial t}dx\,dy\,dz\,dt \qquad (8.51)$$

Setting the sum of (8.49), (8.50), and (8.51) to zero, expanding the convective terms by the chain rule and observing that $\nabla \cdot \mathbf{u} = 0$ for an incompressible fluid according to (8.17), then dividing by $dx\,dy\,dz\,dt$,

$$\mathbf{u} \cdot \nabla c + \frac{\partial c}{\partial t} = D_c \nabla^2 c \qquad (8.52)$$

$$\frac{Dc}{Dt} = D_c \nabla^2 c \qquad (8.53)$$

where

$$\nabla^2 c = \frac{\partial^2 c}{\partial x^2} + \frac{\partial^2 c}{\partial y^2} + \frac{\partial^2 c}{\partial z^2} \qquad (8.54)$$

Equation (8.52) (or [8.53]) is the continuity equation for a solute whose movement in a solvent is governed by diffusion and advection. When the velocity $\mathbf{u}$ equals zero, (8.52) reduces to

$$\frac{\partial c}{\partial t} = D_c \nabla^2 c \qquad (8.55)$$

This is the *diffusion equation* for solute in a static fluid.

Let us further consider a dissolved gas. Often it is convenient to describe the concentration of a gas in terms of its partial pressure, for example, in the case of a dissolved gaseous phase in a magma. Recall that at moderate pressures, *Henry's law* states that the concentration of a dissolved gas is proportional to its partial pressure $P$:

$$c = k_g P \qquad (8.56)$$

where $k_g$ is a coefficient that is characteristic of the particular gas–liquid system. (The coefficient $k_g$ can be related to the solubility or activity coefficient of the gas.) More generally, for a suitably restricted range of pressure,

$$dc = k_g\,dP \qquad (8.57)$$

Substituting (8.57) into (8.53), then assuming $k_g$ is constant such that it can be divided out,

$$\frac{DP}{Dt} = D_c \nabla^2 P \qquad (8.58)$$

Comparing (8.58) with (8.53), the continuity equation for a dissolved gas has the same form, whether expressed in terms of its concentration $c$ or partial pressure $P$, inasmuch as the gas obeys Henry's law in the form of (8.57). Finally, note that these statements of continuity can be expressed in terms of the salinity of the solution when several dissolved species are present.

## 8.2.2 Solutes in porous media flow

The results above must be modified for flow in a porous medium since fluid occupies only part of the control volume. We will assume here that the medium is homogeneous and isotropic with respect to porosity, such that the diffusion process is also—analogous to the solute–solvent system above—then consider the more general case of an anisotropic and heterogeneous medium in Chapter 13 when we examine the process of *mechanical dispersion.*

Diffusion occurs only over a fraction of the surface of the control volume; this fraction is equal to the porosity $n$. Diffusion also is influenced by pore geometry—the tortuosity of the interconnected pores—such that apparent diffusion rates are smaller than those observed in absence of the medium. Thus $D_c$ in (8.52) is replaced with an apparent diffusion coefficent $D_c^* = k_n D_c$, where $k_n$ is an empirical coefficient that is less than one.

Observe that the volume of fluid in the control volume is a fraction $n$ of the total volume. Thus the local term in (8.52) becomes $n \partial c / \partial t$. In addition the velocity $\mathbf{u} = \langle u, v, w \rangle$ is replaced by the specific discharge $\mathbf{q}_h = \langle q_{hx}, q_{hy}, q_{hz} \rangle$ and the advection term in (8.52) becomes $\mathbf{q}_h \cdot \nabla c$. Thus,

$$\mathbf{q}_h \cdot \nabla c + n \frac{\partial c}{\partial t} = n D_c^* \nabla^2 c \tag{8.59}$$

This statement of continuity for a nonreactive solute in a porous medium is referred to as the *advection–diffusion equation.* Notice that when $\mathbf{q}_h = 0$, (8.59) reduces to a diffusion equation for a static fluid, analogous to (8.55).

Equation (8.59) is applicable when solute transport by diffusion is of the same order as advection. This generally occurs when the advective terms are small, and $D_c^*$ and $n$ are large. The process of diffusion, however, normally does not fully account for observed rates of solute dispersion within the interstitial fluid of a porous medium. The reason for this is that fluid velocities vary at the pore scale. In analogy to Poiseuille flow, fluid "far" from solid boundaries generally moves faster than fluid next to boundaries. Moreover, fluid molecules that are at one instant in close proximity to each other are soon dispersed due to the individual tortuous paths that they take during flow. After an interval of time, some have traveled a significant distance whereas others have lagged behind; still others have become dispersed transversely with respect to the mean streamwise flow direction. These effects, referred to as *mechanical dispersion,* combine to give the appearance of Fick-like diffusion of solute at a macroscopic scale. It should be evident that, in absence of flow, this process of dispersion does not occur, and movement of solute is governed by molecular (Fickian) diffusion. In addition, the process of mechanical dispersion is anisotropic, despite presence of an isotropic medium. This means that a steady unidirectional flow, as defined by $q_{hx} \neq 0$ and $q_{hy} = q_{hz} = 0$, for example, produces dispersion in the $y$ and $z$ directions that generally is smaller than that parallel to $x$. We will examine this process of mechanical dispersion in Chapter 13.

## 8.3 CONTINUITY IN LARGE CONTROL VOLUMES

Many flow problems involve large control volumes that are, in principle, well defined by real boundaries. Examples include lakes, river reaches, plumbing systems of geysers, and casing and bore holes of wells. For some of these, developing a statement of continuity involves expanding mass flow quantities about the control volume, as in the developments above, but using quantities that are averaged over the volume surfaces as opposed to local quantities occurring at small elementary faces. In other cases, statements of mass continuity take the simple form of an equation relating inputs and outputs of mass to changes in the mass stored within a volume, subject to constraints imposed by its geometry. These points are illustrated below by the Saint-Venant shallow-water equation of continuity, and by the Dupuit approximation for flow in an unconfined aquifer.

### 8.3.1 Saint-Venant equation of continuity

Consider a short reach of open channel with flow in the direction of positive $s$ (Figure 8.4). The width $b$ and average depth $h$ of the channel, and the average streamwise velocity $u$, may change over the distance d$s$. (The velocity $u$ is obtained as a spatial average over the cross-sectional area of the channel; see Example Problem 1.4.3.) In addition, the width, depth, and velocity, and therefore the control volume, may vary with time if flow is unsteady. Expanding about an $s$-position centered within the volume, the mass flowing into the volume through its upstream face during a short interval of time d$t$ is to first order

$$\left[\rho ubh - \frac{\partial}{\partial s}(\rho ubh)\frac{\mathrm{d}s}{2}\right]\mathrm{d}t \tag{8.60}$$

and the mass flowing out of the volume through the downstream face is to first order

$$\left[\rho ubh + \frac{\partial}{\partial s}(\rho ubh)\frac{\mathrm{d}s}{2}\right]\mathrm{d}t \tag{8.61}$$

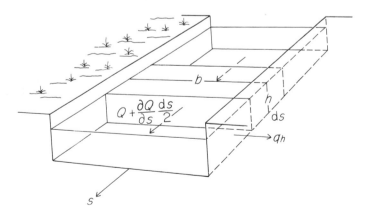

**Figure 8.4**  Definition diagram for average, mass flow quantities associated with a finite control volume within an open channel.

The net flow of mass $N_s$ out of the control volume is

$$N_s = \frac{\partial}{\partial s}(\rho ubh)\,ds\,dt \tag{8.62}$$

Water also may be exchanged between the channel and surrounding floodplain. In keeping with treatments of flow in porous media, let $q_h$ denote a volumetric flux density or specific discharge of water into or out of the channel through its bed and banks. Flow out of the channel, assigned a positive value according to our sign convention, moves through the *wetted perimeter* of the channel, so with an approximately rectangular channel, the flow of mass out of the control volume during $dt$ is

$$\rho q_h(b + 2h)\,ds\,dt \tag{8.63}$$

The volume of the control section is $bh\,ds$, so the mass $m(t)$ within it at time $t$ is $\rho bh\,ds$. The mass $m(t + dt)$ in the control section after the brief interval $dt$ is to first order

$$\left[\rho ubh + \frac{\partial}{\partial t}(\rho ubh)dt\right]ds \tag{8.64}$$

and the change in mass $dm$ is

$$dm = \frac{\partial}{\partial t}(\rho bh)\,ds\,dt \tag{8.65}$$

By conservation of mass, $N_s + \rho q_h(b + 2h)dsdt + dm = 0$, so summing (8.62), (8.63), and (8.65), dividing by $ds\,dt$, and noting that the density of water in an open channel is essentially constant,

$$\frac{\partial}{\partial s}(ubh) + q_h(b + 2h) = -\frac{\partial}{\partial t}(bh) \tag{8.66}$$

Here it is useful to observe that since the product $bh$ equals the area of the channel cross section, $ubh$ equals the volumetric rate of flow or *discharge,* usually denoted by $Q$ in fluvial hydrology. Substituting this into the first term, and expanding the term to the right of the equal sign

$$\frac{\partial Q}{\partial s} + q_h(b + 2h) = -b\frac{\partial h}{\partial t} - h\frac{\partial b}{\partial t} \tag{8.67}$$

For a channel with nearly vertical banks, width does not vary significantly with time insofar as flow does not exceed *bankfull* and spill onto the floodplain. Then $\partial b/\partial t = 0$ and

$$\frac{\partial Q}{\partial s} + q_h(b + 2h) = -b\frac{\partial h}{\partial t} \tag{8.68}$$

The quantity $b + 2h$ equals the wetted perimeter. Typically $b \gg h$ for large channels, and the wetted perimeter is approximately constant. In this case, the term $q_h(b + 2h)$ may be denoted by $q_P$, which equals the volumetric flux per unit length of channel, and

$$\frac{\partial Q}{\partial s} + q_P = -b\frac{\partial h}{\partial t} \tag{8.69}$$

If the channel bed is impermeable, for example, because the river flows over bedrock, the term $q_h(b + 2h)$ reduces to $2q_h h$, which equals the volumetric flux per unit length of channel bank. Equation (8.69) typically is used in describing the movement of a flood wave down a channel.

## 8.3.2  Dupuit flow

Consider a homogeneous, unconfined aquifer where the lower boundary is impermeable and horizontal, and the water table is gently sloping in the direction of the $x$-coordinate (Figure 8.5). The saturated thickness $h$ thus varies with $x$, and it is assumed that $h \ll L$, where $L$ is the horizontal dimension of the aquifer. Two piezometers at different $x$-coordinate positions whose bottoms are at the same vertical position $z$ register different water levels that essentially coincide with the local water table elevation. We may infer from the hydrostatic equation that pressure varies with the local height of the water table, and thus flow is induced by a pressure gradient associated with the sloping water table. Now, a water table normally involves a vertical component of flow in the presence of recharge from the land surface above, and in general a sloping water table requires a vertical component of flow at the water table. (Why?) Nonetheless if we assume that the slope of the water table is very small, flow essentially is horizontal and therefore one-dimensional in the direction of the $x$-coordinate. Then, because flow at any given vertical position occurs along a horizontal line, it is sufficient to consider the flow to be induced by a pressure gradient $\partial p / \partial x$ proportional to the water table slope $\partial h / \partial x$. These assumptions of horizontal flow and a pressure gradient proportional to the water table slope are referred to as the *Dupuit approximations*. (Students are urged to reexamine this section once the idea of hydraulic head is formalized in the next chapter.)

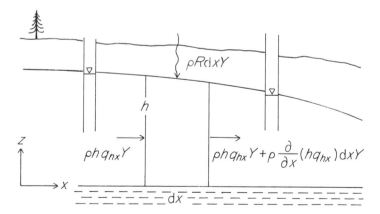

**Figure 8.5** Definition diagram for mass flow quantities associated with a finite control volume embedded within an unconfined aquifer; the bases of the two piezometers are at the same level $z$.

Assuming steady flow in the $x$ direction, consider a control volume whose edges are of length $dx$ and $Y$ (Figure 8.5). The inflow face of the volume has a vertical height $h$ defined by the position of the water table. The mass $m_{in}$ flowing into the control volume during the small interval $dt$ is $\rho q_{hx} Y dt$, where $q_{hx}$ is the specific discharge over the inflow face $Yh$. Note that both $q_{hx}$ and $h$ may change over the distance $dx$. Expanding the product $hq_{hx}$ about the inflow position, the mass $m_{out}$ flowing out of the control volume during the interval $dt$ is to first order

$$\rho h q_{hx} Y dt + \rho \frac{\partial}{\partial x}(h q_{hx}) dx\, Y\, dt \qquad (8.70)$$

The net flow of mass out of the volume is then given by $m_{out} - m_{in}$:

$$\rho \frac{\partial}{\partial x}(h q_{hx}) dx\, Y\, dt \qquad (8.71)$$

Suppose that mass is recharged to the control volume from above at a rate $R$ (with dimensions of specific discharge). By continuity $m_{out} = m_{in} + \rho R Y dx\, dt$; that is, under steady conditions the quantity of mass discharged through the outflow face must equal that which is delivered from the inflow face plus that which is added over the intervening distance $dx$. Thus $m_{out} - m_{in} = \rho R Y dx\, dt$, so from (8.71)

$$\rho \frac{\partial}{\partial x}(h q_{hx}) dx\, Y\, dt = \rho R\, dx\, Y\, dt \qquad (8.72)$$

$$\frac{\partial}{\partial x}(h q_{hx}) = R \qquad (8.73)$$

This is a statement of mass continuity under conditions imposed by the Dupuit assumptions.

Instead of expanding (8.73), let us substitute Darcy's law for horizontal flow, $q_{hx} = -K_h \partial h/\partial x$, where $K_h$ is the hydraulic conductivity (Chapter 1):

$$K_h \frac{\partial}{\partial x}\left(h \frac{\partial h}{\partial x}\right) = -R \qquad (8.74)$$

Observe that $\partial h^2/\partial x = 2h\partial h/\partial x$; thus

$$\frac{K_h}{2} \frac{\partial}{\partial x}\left(\frac{\partial h^2}{\partial x}\right) = -R \qquad (8.75)$$

$$\frac{\partial^2 h^2}{\partial x^2} = -\frac{2R}{K_h} \qquad (8.76)$$

This is referred to as *Forchheimer's equation*. Let us substitute $v = h^2$ then observe that, because the second derivative of $v$ with respect to $x$ is a constant, the first derivative $\partial v/\partial x$ ought to be a linear function of $x$. Further, because $\partial v/\partial x$ is a linear function, $v$ ought to vary as a second-order polynomial of $x$. That is, integrating (8.76) twice with respect to $x$,

$$v = h^2 = -\frac{R}{K_h}x^2 + ax + b \qquad (8.77)$$

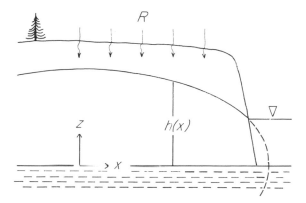

**Figure 8.6** Elliptical ($R > 0$) solution $h(x)$ for steady Dupuit flow in an unconfined aquifer bounded by a free-surface reservoir; adapted from Bear (1972).

where $a$ and $b$ are constants of integration that may be determined from boundary conditions. For $R > 0$, $h$ has the form of an ellipse (Figure 8.6). For $R < 0$, implying a loss of mass, $h$ has the form of a hyperbola. Equation (8.77) has considerable practical value, although it is subject to errors that arise from the Dupuit assumptions.

## 8.4 EXAMPLE PROBLEMS

### 8.4.1 Steady conduit flow: inferences from continuity

Reconsider the problem of flow within a horizontal conduit with parallel walls (Example Problem 3.7.1; Figure 3.18) in view of the idea of conservation of mass covered above. Although this is an elementary problem, it will serve to illustrate how one may sometimes use the physical constraint provided by continuity to anticipate certain information about the solution to a problem.

Assuming steady flow of an incompressible fluid, the relevant equation of continuity is

$$\frac{\partial u}{\partial x} + \frac{\partial v}{\partial y} + \frac{\partial w}{\partial z} = 0 \tag{8.78}$$

Our first task is to simplify this. Observe from the geometry of the problem that $v = 0$ at the walls ($y = \pm b$). One may be tempted to conclude that $\partial v/\partial y$ therefore also must equal zero. The fact that $v = 0$ at the walls does not generally imply, however, that $\partial v/\partial y$ is necessarily zero at the wall or elsewhere. One can envision an oscillatory flow within the conduit for which $v$ and $\partial v/\partial y$ locally are nonzero. (Let us momentarily recall an analogous problem in elementary mechanics. Whereas the vertical velocity component $w$ of a ballistic particle at the peak of its parabolic trajectory equals zero, its acceleration, $dw/dt = -g$, is nonzero.) If, however, $v = 0$ throughout the flow, then the term $\partial v/\partial y$ vanishes. Likewise, if $w = 0$ everywhere, then the term $\partial w/\partial z$ vanishes. In this situation (8.78) reduces to

$$\frac{\partial u}{\partial x} = 0 \tag{8.79}$$

In defining the problem, we asserted one-dimensional flow such that the pressure gradient has a component parallel to the $x$-axis only. Thus $\partial p/\partial y = \partial p/\partial z = 0$, so no pressure gradients capable of inducing flow in either the direction of $y$ or $z$ exist. It follows that $v = w = 0$ throughout the flow and therefore (8.79) indeed is generally applicable.

The condition $\partial u/\partial x = 0$ implies that $u = $ constant for given $(y, z)$ coordinate position. Formally, integrating (8.78) with respect to $x$,

$$u = u(y, z) \tag{8.80}$$

Therefore $u$ is independent of $x$, which means that a solution possibly has the form $u(y, z)$, but not $u(x, y, z)$. Further, because surfaces normal to the $z$-axis are assumed to be far from any parallel real boundary, fluid on either side of such a surface is uninfluenced by a no-slip condition and therefore possesses the same velocity with zero shear stress on the surface. We may thus conclude that $u$ does not vary with $z$, and anticipate a solution of the form $u(y)$. This in fact provides the justification for the balance of forces (not involving $x$) used to obtain (3.71) for $u(y)$. Taking the derivative of (3.71) with respect to $x$ illustrates that this solution indeed is compatible with the condition $\partial u/\partial x = 0$. Thus continuity is automatically satisfied by the geometry of this problem with specification of one-dimensional flow. We will examine several more difficult problems in later chapters.

## 8.4.2 Depth-integrated equation of continuity for open-channel flow

When describing flow within an open channel, sometimes it is convenient to work with *depth-integrated* or *depth-averaged* quantities. Then, for example, instead of explicitly considering how velocity varies with vertical position $z$ in a channel (Figure 8.7), one considers the depth-averaged velocity. In this case we require a depth-averaged version of the equation of continuity.

Consider a straight channel with uniform width (Figure 8.8). Let $x$ denote a streamwise coordinate that is positive downstream; let $y$ denote a transverse coordinate that is positive toward the left bank; and let $z$ denote a nearly vertical coordinate that is positive upward and normal to the bed. Assuming steady incompressible fluid flow, the appropriate equation of continuity is (8.17):

$$\frac{\partial u}{\partial x} + \frac{\partial v}{\partial y} + \frac{\partial w}{\partial z} = 0 \tag{8.81}$$

In streams and rivers, flow normally is turbulent, so in this case $u$, $v$, and $w$ in (8.81) refer to time-averaged values at a given coordinate position (see Chapter 14). This distinction, however, is not essential to the development here, which applies equally to viscous (nonturbulent) flows.

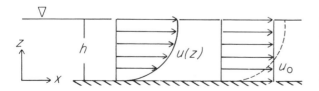

**Figure 8.7** Local, vertical velocity distribution $u(z)$ in an open channel, and imaginary uniform distribution associated with depth-averaged velocity $u_0$ obtained from mean-value theorem.

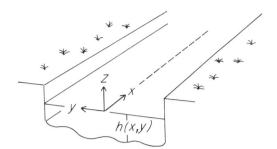

**Figure 8.8** Definition diagram for coordinate system used in depth-integration of velocity components in an open channel.

Elevations of the bed and water surface vary with position in a natural channel, so local depth $h$ is a function $h(x, y)$ of coordinates $x$ and $y$. Further, $u$, $v$, and $w$ vary as functions $u(x, y, z)$, $v(x, y, z)$, and $w(x, y, z)$ of coordinate position $(x, y, z)$. Thus (8.81) may be written

$$\frac{\partial}{\partial x} u(x, y, z) + \frac{\partial}{\partial y} v(x, y, z) + \frac{\partial}{\partial z} w(x, y, z) = 0 \qquad (8.82)$$

Now, we wish to integrate each term over the local flow depth, from the bed ($z = \eta$) to the water surface ($z = \zeta$), where $h = \zeta - \eta$, to obtain a depth-averaged form:

$$\int_{\eta}^{\zeta} \frac{\partial}{\partial x} u(x, y, z)\, dz + \int_{\eta}^{\zeta} \frac{\partial}{\partial y} v(x, y, z)\, dz + \int_{\eta}^{\zeta} \frac{\partial}{\partial z} w(x, y, z)\, dz = 0 \quad (8.83)$$

Let us start with the third term in (8.83). Observe that this term involves the antiderivative of the derivative of the function $w$ with respect to $z$, an operation that retrieves the original function $w$. Evaluating this between the limits $\eta$ and $\zeta$ therefore leads to

$$\int_{\eta}^{\zeta} \frac{\partial}{\partial z} w(x, y, z)\, dz = w(x, y, z)\Big|_{\eta}^{\zeta} = w(x, y, \zeta) - w(x, y, \eta) \qquad (8.84)$$

Consider the term $w(x, y, \zeta)$, which equals the vertical component of velocity at the water surface, and envision the motion of a fluid particle at the surface with coordinates $(x, y, \zeta)$. Here we will observe the particle motion in a Lagrangian sense; thus the coordinates $x$, $y$, and $\zeta$ are to be momentarily associated with the particle, and are therefore dependent quantities. Further, note that the water surface is a (steady) function $\zeta(x, y)$ of $x$ and $y$. Now, because $x$ and $y$ are functions $x(t)$ and $y(t)$ of time $t$, the coordinate $\zeta$ associated with the particle is a function of time also. Thus,

$$\zeta = \zeta[x(t), y(t)] \qquad (8.85)$$

The vertical component of the (Lagrangian) particle velocity is $d\zeta/dt$, which must also equal the Eulerian component $w(x, y, \zeta)$; thus

$$w(x, y, \zeta) = \frac{d}{dt} \zeta[x(t), y(t)] \qquad (8.86)$$

We now apply the chain rule to obtain:

$$w(x, y, \zeta) = \frac{\partial \zeta}{\partial x} \frac{dx}{dt} + \frac{\partial \zeta}{\partial y} \frac{dy}{dt} \qquad (8.87)$$

Observe that $dx/dt = u(x, y, \zeta)$ and $dy/dt = v(x, y, \zeta)$, from which it follows that

$$w(x, y, \zeta) = u(x, y, \zeta)\frac{\partial \zeta}{\partial x} + v(x, y, \zeta)\frac{\partial \zeta}{\partial y} \qquad (8.88)$$

Substituting this into (8.84), then substituting the result into (8.83), and noting that the no-slip condition leads to $w(x, y, \eta) = 0$, (8.83) becomes

$$\int_{\eta}^{\zeta} \frac{\partial}{\partial x} u(x, y, z)\,dz + \int_{\eta}^{\zeta} \frac{\partial}{\partial y} v(x, y, z)\,dz + u(x, y, \zeta)\frac{\partial \zeta}{\partial x} + v(x, y, \zeta)\frac{\partial \zeta}{\partial y} = 0 \qquad (8.89)$$

Let us now turn to the first two terms in (8.89), and make use of *Leibnitz's rule* for differentiation of an integral. Notice that each of these terms involves at most two variables (coordinates) of integration and differentiation. For example, the first term involves differentiation with respect to $x$ and integration with respect to $z$; the coordinate $y$ is treated as a constant. Also note that the limits of integration $\eta$ and $\zeta$ are, in general, functions of $x$ and $y$ (but not $z$). Writing Leibnitz's rule in its simplest form,

$$\frac{d}{ds}\int_{a(s)}^{b(s)} f(s, t)\,dt = \int_{a(s)}^{b(s)} \frac{\partial f}{\partial s}\,dt + f(s, b)\frac{db}{ds} - f(s, a)\frac{da}{ds} \qquad (8.90)$$

where $a$ and $b$ are limits of integration that are, in general, functions of $s$, where $s$ and $t$ may denote any two spatial coordinates (or time). For example, with $f(s, t) = u(x, z)$, $a(s) = \eta(x)$, and $b(s) = \zeta(x)$ (where $y$ is constant),

$$\frac{d}{dx}\int_{\eta(x,y)}^{\zeta(x,y)} u(x, y, z)\,dz = \int_{\eta(x,y)}^{\zeta(x,y)} \frac{\partial u}{\partial x}\,dz + u(x, y, \zeta)\frac{d\zeta}{dx} - u(x, y, \eta)\frac{d\eta}{dx} \qquad (8.91)$$

Solving for the first term on the right side of (8.91), and noting that the no-slip condition leads to $u(x, y, \eta) = 0$, the first term in (8.89) becomes

$$\int_{\eta(x,y)}^{\zeta(x,y)} \frac{\partial u}{\partial x}\,dz = \frac{d}{dx}\int_{\eta(x,y)}^{\zeta(x,y)} u(x, y, z)\,dz - u(x, y, \zeta)\frac{d\zeta}{dx} \qquad (8.92)$$

We similarly obtain for the second term in (8.89):

$$\int_{\eta(x,y)}^{\zeta(x,y)} \frac{\partial v}{\partial y}\,dz = \frac{d}{dy}\int_{\eta(x,y)}^{\zeta(x,y)} v(x, y, z)\,dz - v(x, y, \zeta)\frac{d\zeta}{dy} \qquad (8.93)$$

Substituting (8.92) and (8.93) into (8.89), and noting that terms involving $\partial\zeta/\partial x$ and $\partial\zeta/\partial y$ cancel, (8.89) becomes

$$\frac{\partial}{\partial x}\int_{\eta}^{\zeta} u(x, y, z)\,dz + \frac{\partial}{\partial y}\int_{\eta}^{\zeta} v(x, y, z)\,dz = 0 \qquad (8.94)$$

Now, regardless of how $u$ and $v$ vary with $z$, the integral quantities in (8.94) may be replaced by the product of the depth $h = (\zeta - \eta)$ and the vertically averaged velocities $u_0$ and $v_0$. For example, the first integral quantity is geometrically represented by the area behind the $u(z)$ curve (Figure 8.7). By the mean-value theorem, this area is equal to $hu_0$, where $u_0$ is the depth-averaged value of $u(z)$. Thus (8.94) reduces to

$$\frac{\partial}{\partial x}(hu_0) + \frac{\partial}{\partial y}(hv_0) = 0 \tag{8.95}$$

which is a depth-integrated equation of continuity. This equation physically implies that the divergence of the product $h\mathbf{u}_0$ equals zero, where $\mathbf{u}_0 = \langle u_0, v_0 \rangle$. In the special case in which no transverse flow occurs, $v_0 = 0$, and (8.95) reduces to

$$\frac{\partial}{\partial x}(hu_0) = 0 \tag{8.96}$$

This implies that the product $hu_0$ is constant; as depth increases, the average streamwise velocity decreases, and vice versa.

In treatments of open-channel flow, (8.95) normally is used in conjunction with equations of conservation of momentum (Chapters 12 and 15) that are vertically integrated in a manner similar to that above. This set of equations is then solved simultaneously for specified initial and boundary conditions to obtain formulae that describe how $u_0$, $v_0$, and $h$ vary with $x$ and $y$. This set of equations, however, is nonlinear and difficult to solve directly. A technique referred to as *linearization* commonly is used to solve them. This technique involves specifying values of $u_0$, $v_0$, and $h$ that are averaged over a channel reach, then working with a linear set of equations that describe deviations in $u_0$, $v_0$, and $h$ from their reach-averaged values. Students are encouraged to examine this technique of linearization in the reading suggested below, and in Example Problem 15.8.3.

### 8.4.3 Continuity of radial flow at a well with steady pumping

Let us consider flow within the vicinity of a well that penetrates an aquifer, making use of cylindrical coordinates. We will start by briefly introducing the equation governing flow, then use conservation of mass to obtain a boundary condition of significant practical importance when used in solving this equation.

Consider for simplicity a confined horizontal aquifer that is homogeneous and isotropic with respect to hydraulic conductivity $K_h$. A well with radius $R$ fully penetrates the aquifer over its thickness $b$. We will neglect "skin effects"; these arise from local alterations to the natural hydraulic conductivity in the immediate vicinity of the well due to such things as presence of a "sand pack" around the well screen, or presence of drilling mud in pores, a remnant from the drilling operation. Let $r$ denote a horizontal radial coordinate with $r = 0$ at the well center, and let $z$ denote a vertical coordinate. From Appendix 17.2 the relevant equation of continuity is

$$\frac{1}{r}\frac{\partial}{\partial r}(rq_{hr}) + \frac{1}{r}\frac{\partial q_{h\theta}}{\partial \theta} + \frac{\partial q_{hz}}{\partial z} = -n\frac{\partial \rho}{\partial t} - \rho\frac{\partial n}{\partial t} \tag{8.97}$$

where $\theta$ denotes the azimuthal angle about the $z$-axis. With well pumpage, flow occurs radially, parallel to $r$ (neglecting regional flow). Therefore $q_{h\theta}$ everywhere equals zero. The aquifer in the vicinity of the well compresses in response to a lowering of fluid pressure in the well associated with pumping (Section 3.5.2). Thus $b$ varies, in principle, inducing a component $q_{hz}$. However, compression normally is so slight that the change in $b$ is negligible. Thus $q_{hz}$ effectively equals zero and (8.97) reduces to

$$\frac{1}{r}\frac{\partial}{\partial r}(rq_{hr}) = -n\frac{\partial \rho}{\partial t} - \rho\frac{\partial n}{\partial t} \tag{8.98}$$

The appropriate form of Darcy's law in this case is $q_{hr} = -K_h\partial h/\partial r$. Substituting, then expanding using the chain rule,

$$\frac{\partial^2 h}{\partial r^2} + \frac{1}{r}\frac{\partial h}{\partial r} = \frac{n}{K_h}\frac{\partial \rho}{\partial t} + \frac{\rho}{K_h}\frac{\partial n}{\partial t} \tag{8.99}$$

The right side of (8.99) is equal to $(S/T)\partial h/\partial t$, where $T = K_h b$ is the aquifer *transmissivity*; $S$ is the *storativity,* a measure of the water released from the aquifer per unit decline in hydraulic head $h$ due to expansion of the water, associated with a change in $\rho$, and due to compression of the aquifer matrix, leading to a change in $n$. (Domenico and Schwartz [1990] provide a delightful historical perspective (pp. 109–15) of the derivation of this expression.) Then

$$\frac{\partial^2 h}{\partial r^2} + \frac{1}{r}\frac{\partial h}{\partial r} = \frac{S}{T}\frac{\partial h}{\partial t} \tag{8.100}$$

Obtaining a solution $h(r, t)$ of (8.100) requires specifying initial and boundary conditions. A convenient initial condition is given by $h(r, 0) = h_0$, which indicates that the head $h$ initially ($t = 0$) equals $h_0$ at all radial distances $r$. One suitable boundary condition is given by $h(\infty, t) = h_0$, which indicates that the head $h$ equals $h_0$ infinitely far from the well for all times $t$. Implicit in this is the idea that the aquifer has infinite extent, which is merely a mathematical artifice rather than a realistic physical condition. A second boundary condition of significant practical importance is obtained from a consideration of continuity.

Suppose that water is pumped from the well at a steady volumetric rate $Q$ (a condition that can be practicably obtained). The discharge $Q$ must equal the volumetric flow rate into the well. More generally $Q$ must equal the product of the specific discharge $q_{hr}$ and the surface area $2\pi r b$ through which flow occurs at any radial distance $r$. Applying Darcy's law again for $q_{hr}$, $Q = 2\pi K_h b r(\partial h/\partial r) = 2\pi T r(\partial h/\partial r)$. (Why is the negative sign convention omitted here?) The boundary condition is therefore $\partial h/\partial r = Q/2\pi T R$. Since $R$ is small relative to the flow field, however, this is normally expressed for mathematical convenience as

$$\lim_{r \to 0}\left(r\frac{\partial h}{\partial r}\right) = \frac{Q}{2\pi T} \tag{8.101}$$

Then the well is referred to as a *line sink* ($R \to 0$).

The solution of (8.100) for the case of a well that fully penetrates an isotropic aquifer was originally provided by C. V. Theis in 1935. It is now referred to as the *Theis nonequilibrium solution,* and can be found in introductory groundwater texts. The Theis solution forms the basis of a conventional *pumping test,* wherein hydraulic parameters $K_h$ and $S$ are estimated based on the observed drawdown, $h_0 - h(r, t)$, within an observation well at distance $r$ from the pumping well, after the onset of pumping. Numerous variations on the Theis solution also exist; these attempt to take into account the effects of different boundary conditions, unconfined conditions, partial well penetration, layered rock units with different hydraulic properties, and anisotropic conditions.

### 8.4.4  Diffusion of carbon dioxide into a slowly ascending bubble

Let us reconsider diffusion of $CO_2(aq)$ toward a bubble ascending within a liquid, as in Example Problem 4.10.5. This is a good example in which dimensional analysis suggests the need to modify our initial treatment of this problem. Based on conservation of mass during gas diffusion, let us assume that the rate at which mass is added to the bubble, $dm/dt$, is a function of bubble radius $r_b$, the coefficient of mass diffusion $D_c$, and the $CO_2$ partial-pressure difference $\Delta P = P_\infty - P_b$, where $P_\infty$ is the partial pressure of $CO_2(aq)$ in the solvent far from the bubble and $P_b$ is the partial pressure in the bubble. Now, since we are using $D_c$ to characterize diffusion, it is useful to assume that the functional relation involves the quantity $k_g \Delta P$, rather than merely $\Delta P$, where $k_g$ is the constant in Henry's law. Notice that $k_g \Delta P$ has units of concentration, which is consistent with our using $D_c$ in Fick's law. Thus,

$$\frac{dm}{dt} = f(r_b, D_c, k_g \Delta P) \tag{8.102}$$

It is left as an exercise to determine by dimensional analysis that this takes the form

$$\frac{dm}{dt} = k k_g D_c (P_\infty - P_b) r_b \tag{8.103}$$

where $k$ is a dimensionless constant. Notice that this result suggests that $dm/dt$ is proportional to $r_b$, rather than to $r_b^2$, as assumed in (4.95) of Example Problem 4.10.5. Which is correct? The answer resides in the choice of coordinate systems, and with an assumption of steady diffusion.

Let us choose a spherical $r\theta\phi$-coordinate system whose origin is centered on the bubble. Because we are assuming that the $CO_2$ obeys Henry's law, from Appendix 17.2 the relevant continuity equation for $CO_2$ in terms of its partial pressure $P$ is

$$u_r \frac{\partial P}{\partial r} + \frac{v_\theta}{r} \frac{\partial P}{\partial \theta} + \frac{w_\phi}{r \sin \theta} \frac{\partial P}{\partial \phi} + \frac{\partial P}{\partial t}$$

$$= D_c \left[ \frac{1}{r^2} \frac{\partial}{\partial r} \left( r^2 \frac{\partial P}{\partial r} \right) + \frac{1}{r^2 \sin \theta} \frac{\partial}{\partial \theta} \left( \sin \theta \frac{\partial P}{\partial \theta} \right) + \frac{1}{r^2 \sin^2 \theta} \frac{\partial^2 P}{\partial \phi^2} \right] \tag{8.104}$$

which may be written in the compact form

$$\mathbf{u} \cdot \nabla P + \frac{\partial P}{\partial t} = D_c \nabla^2 P \tag{8.105}$$

where $r$ is the radial distance from the center of the bubble, $\theta$ is the azimuthal angle about the axis $\phi = 0$, and $u_r$, $v_\theta$, and $w_\phi$ are the components of the velocity $\mathbf{u}$ in the three coordinate directions.

Two boundary conditions are relevant. Observe that $P_b$ coincides with the $CO_2$ partial pressure in the liquid at the bubble–liquid interface. Then

$$P = P_b, \quad r = r_b; \qquad P = P_\infty, \quad r \to \infty \tag{8.106}$$

Note that $P_\infty$ changes with time due to $CO_2$ diffusion into the bubble. Nonetheless, assume that the rate of extraction of $CO_2(aq)$ from the liquid is small such that $P_\infty$ varies so slowly with time that it is essentially constant.

Further assume that the rate of growth of the bubble is small such that

$$\left| \frac{dr_b}{dt} \right| \ll \left( w_b + \frac{D_c}{r_b} \right) \tag{8.107}$$

where $w_b$ is the ascent speed of the bubble. If (8.107) is satisfied, then the local velocity field and the distribution of $CO_2(aq)$ around the bubble are negligibly influenced by bubble growth. With the assumptions of (8.107) and constant $P_\infty$, $CO_2$ diffusion toward the bubble is steady with respect to the coordinate system moving with the bubble; then (8.105) reduces to

$$\mathbf{u} \cdot \nabla P = D_c \nabla^2 P \qquad (8.108)$$

Now consider the case where $w_b \ll D_c/r_b$; then the distribution of $CO_2(aq)$ around the bubble is negligibly influenced by the local velocity field, and the advective terms in (8.108) may be omitted. (The effect of nonnegligible $CO_2$ advection is considered below.) Then (8.108) reduces to the *Laplacian* of $P$:

$$\nabla^2 P = 0 \qquad (8.109)$$

Because diffusion effectively occurs only parallel to $r$ with small $w_b$, only terms involving derivatives of $P$ with respect to $r$ in (8.109), from (8.104), are retained, and it becomes after expansion

$$\frac{\partial^2 P}{\partial r^2} + \frac{2}{r}\frac{\partial P}{\partial r} = 0 \qquad (8.110)$$

Integrating twice, then evaluating the two constants of integration using the boundary conditions (8.106),

$$P = -\frac{(P_\infty - P_b)r_b}{r} + P_\infty \qquad (8.111)$$

which is a steady solution for $P$. It is worth pondering the idea that this solution is steady with respect to $r \geq r_b$, where $r_b$ is changing.

According to Fick's law the rate at which $CO_2$ mass is added to the bubble is

$$\frac{dm}{dt} = 4\pi k_g D_c r_b^2 \left(\frac{dP}{dr}\right)_{r_b} \qquad (8.112)$$

where $k_g$ is the coefficient in Henry's law (8.56), and $(dP/dr)_{r_b}$ is the partial pressure gradient at the bubble–liquid interface. Differentiating (8.111) with respect to $r$ and evaluating the result at $r = r_b$, then substituting this expression for $(dP/dr)_{r_b}$ into (8.112),

$$\frac{dm}{dt} = 4\pi k_g D_c (P_\infty - P_b) r_b \qquad (8.113)$$

Evidently the dimensionless constant $k$ in (8.103) equals $4\pi$. It is left as an exercise to follow the analysis in Example Problem 4.10.5 to obtain an expression for the rate of bubble growth based on (8.113).

With increasing $w_b$, advection of $CO_2$ near the bubble becomes important, and the rate at which $CO_2$ mass is added to the bubble is enhanced by a factor equal to

$$1 + 0.624 \left(\frac{r_b w_b}{D_c}\right)^{1/3} \qquad (8.114)$$

This factor is discussed in Levich (1962) and Loper (1992).

### 8.4.5 Lake mass balance

Inputs of mass to a lake include rain, condensation, stream flow, and seepage from the banks and bed. Outputs of mass involve evaporation, transpiration, stream flow,

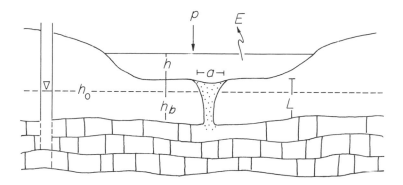

**Figure 8.9**   Definition diagram for components of lake mass balance.

and seepage through the lake bed. For the purpose here, consider an internally drained lake; examples include many of the karstic lakes of Florida. In some of these lakes, input effectively is limited to rain and intermittent stream flow; condensation and seepage are negligible. Output chiefly involves evaporation and seepage through the lake bed and through active sinkhole features (ponors) that are hydraulically connected with the underlying sedimentary layers (Figure 8.9).

The flow of mass into the lake during the interval of time $dt$ is

$$\rho P A \, dt + \rho Q \, dt \tag{8.115}$$

where $P$ is the rate of precipitation (L t$^{-1}$), $A$ is the surface area of the lake, and $Q$ is the rate of stream discharge (L$^3$ t$^{-1}$). The flow of mass out due to evaporation is $\rho E A dt$, where $E$ is the rate of evaporation (L t$^{-1}$). With reference to Figure 8.9, the flow of mass out of the lake also involves seepage through its bed, given by

$$\rho A K_A \left( \frac{h + h_b - h_0}{L} \right) dt \tag{8.116}$$

and may involve seepage through a ponor, given by

$$\rho a K_p \left( \frac{h + h_b - h_0}{L} \right) dt \tag{8.117}$$

In these formulae, $h$ is the height of the water surface above the lake bed, $h_b$ is the height of the lake bed above the underlying aquifer, $h_0$ is the height of the potentiometric surface of the aquifer, $L$ is the thickness of the sediment separating the lake from the aquifer, $a$ is the effective cross-sectional area of the ponor, and $K_A$ and $K_p$ are the effective hydraulic conductivities of the sediment and ponor. It is assumed that $A \gg a$, and that flow downward is essentially one-dimensional.

Let $A(h)$ denote a hypsometric function, which relates the surface area of the lake to its level $h$. The volume $V$ of the lake is then

$$V(h) = \int_0^h A(z) \, dz \tag{8.118}$$

and the mass within it at time $t$ is $\rho V(h)$. The mass within the lake after a short interval $dt$ is to first order

$$\rho V + \rho \frac{\partial V}{\partial t}\, dt \tag{8.119}$$

so the change in mass $dm$ is

$$\rho \frac{\partial V}{\partial t}\, dt \tag{8.120}$$

Here it is useful to observe that

$$\frac{\partial V}{\partial t} = \frac{\partial V}{\partial h}\frac{\partial h}{\partial t} = A\frac{\partial h}{\partial t} \tag{8.121}$$

Then by conservation of mass

$$\rho PA\, dt + \rho Q\, dt - \rho EA\, dt - (\rho AK_A + \rho aK_p)\left(\frac{h + h_b - h_0}{L}\right)dt = \rho A\frac{\partial h}{\partial t}\, dt \tag{8.122}$$

Assuming constant density, and noting that $h_b = L$,

$$PA + Q - EA - \left(\frac{K_A A}{L} + \frac{K_p a}{L}\right)h + \frac{K_A h_0 A}{L} - K_A A + \frac{K_p a h_0}{L} - K_p a = A\frac{\partial h}{\partial t} \tag{8.123}$$

Between storms, after stream flow has subsided, this reduces to

$$-E - \left(\frac{K_A}{L} + \frac{K_p a}{LA}\right)h + \frac{K_A h_0}{L} - K_A + \frac{K_p a h_0}{LA} - \frac{K_p a}{A} = \frac{\partial h}{\partial t} \tag{8.124}$$

This suggests that when the surface area $A$ is very large, terms involving ponor properties contribute little to the lake balance. In absence of ponor seepage,

$$-\frac{K_A}{L}h + \frac{K_A h_0}{L} - K_A - E = \frac{\partial h}{\partial t} \tag{8.125}$$

indicating that the rate of lake-level decline is linearly related to the lake level. This suggests that, if good estimates of evaporation are available, one can estimate seepage from observed rates of decline in lake level. Seepage must be of the same order as evaporation or larger to obtain sufficient precision to do this. Also note that the hydraulic conductivity $K_A$ is not likely to be uniform over the lake bed, and is therefore a function of $A$. In addition, the potentiometric level $h_0$ is likely to fluctuate with time.

### 8.4.6 Flow within a two-dimensional fracture network

An important problem in groundwater hydrology is describing flow in dual-porosity systems. One example is a rock unit that contains a primary porosity composed of intergranular pores and a secondary porosity composed of open fractures. Flow can occur simultaneously in both sets of pores. Flow within the intergranular pores is normally envisioned as being diffuse as embodied in Darcy's law, whereas flow within interconnected fractures is more like that in a channel network. Furthermore, flow through interconnected fractures, under certain circumstances, can account for most of the transport of mass—water and solutes—in dual-porosity systems. The reason for this resides in the cubic law (3.74), which suggests that for a given pressure gradient the volumetric discharge $Q$ is proportional to the third power of pore aperture. Thus, to describe transport in these systems requires understanding how open fractures may preferentially channelize flow.

For a given geologic setting, however, specifying the geometry of a fracture network is problematical, and can be achieved only in a probabilistic way based in part on field sampling of the orientations and spacing of exposed fractures. The problem then is reduced to this: Can transport be predicted in a probabilistic sense based on an artificially generated fracture geometry whose statistical properties match those of the field data? See the relevant papers listed at the end of this chapter to obtain an idea of the many aspects of this question. Let us consider here one part: how conservation of mass is applied in computing flow through a simple fracture network.

Envision a two-dimensional network composed of vertical fractures that transect a horizontal rock unit of constant thickness. Consider a portion of this unit that is bounded by two vertical planes (Figure 8.10$a$) containing five fractures labeled 1 to 5, only two of which completely transect the area. There are 12 nodes: A, B, C, J, K, and L are boundary nodes formed by intersections of fractures and bounding planes; D, E, F, G, H, and I are interior nodes formed by intersections of two fractures. Some parts of fractures are "dead ends" that do not involve flow. These fracture segments may be "removed" to obtain a view of the effective flow system (Figure 8.10$b$). Nodes D, F, and I are composed of four connecting fracture segments, E and G are composed of three, and H is composed of two segments. Consider for illustration node E created by the intersection of fractures 2 and 3 (Figure 8.10$c$).

Neglect diffuse flow through rock elements bounded by fractures, and assume steady flow between nodes. Then choose as a positive quantity the volumetric discharge $Q$ delivered to node E. By conservation of mass,

$$Q_{DE} + Q_{FE} + Q_{JE} = 0 \qquad (8.126)$$

where each subscript pair denotes the fracture segment defined by neighboring nodes. This equation simply states that the sum of positive terms must equal the sum of negative terms such that the flow of mass toward E is exactly compensated

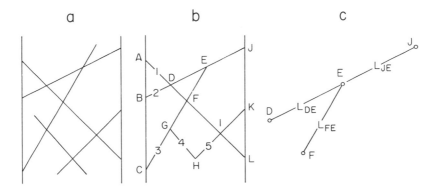

**Figure 8.10** Portion of two-dimensional fracture network bounded by two vertical planes ($a$); identification of fractures (1-5), exterior nodes (A, B, C, J, K, L) and interior nodes (D, E, F, G, H, I) ($b$); and isolation of one fracture-node subset for computation of mass flow quantities associated with central node E ($c$).

by flow of mass away from E per unit time. Now assume momentarily that flow toward E is proportional to the difference in the fluid pressure $p$ of the neighboring node and the pressure at E, divided by the segment length $L$. Thus

$$k_2 \frac{p_D - p_E}{L_{DE}} + k_3 \frac{p_F - p_E}{L_{FE}} + k_2 \frac{p_J - p_E}{L_{JE}} = 0 \tag{8.127}$$

where the subscript on $p$ denotes the node. Each coefficient of proportionality $k$ is a property of the fracture, as denoted by the subscript. One physical interpretation of $k$ is this: Assume that each fracture can be approximated as a conduit of uniform aperture $2b$ and vertical breadth $dz$. Comparing each term in (8.127) with the cubic law (3.74), evidently $k = (2b^3/3\mu)\,dz$. Substituting this expression into (8.127), then expanding and solving for $p_E$,

$$p_E = \frac{b_2^3 L_{FE} L_{JE} p_D + b_3^3 L_{DE} L_{JE} p_F + b_2^3 L_{DE} L_{FE} p_J}{b_2^3 L_{FE} L_{JE} + b_3^3 L_{DE} L_{JE} + b_2^3 L_{DE} L_{FE}} \tag{8.128}$$

This actually is a weighted average; to illustrate this, momentarily assume that $b_1 = b_2 = b_3 = b$, and $L_{DE} = L_{FE} = L_{JE} = L$. Then $p_E$ is simply equal to the average pressure of the three neighboring nodes: $p_E = (p_D + p_F + p_J)/3$.

Formula (8.128) describes a condition that must be satisfied at node E. Similar formulae can readily be obtained for all remaining interior nodes. For example, the pressure $p_H$ at node H is given by

$$p_H = \frac{b_4^3 L_{IH} p_G + b_5^3 L_{GH} p_I}{b_4^3 L_{IH} + b_5^3 L_{GH}} \tag{8.129}$$

Now, assume that values of pressure at boundary nodes are specified. There are then several ways to compute the flow through the network. A numerically simple way involves *successive relaxation,* a technique, which, in this example, is relatively easy to program on a computer. The steps include the following. (1) Obtain the rest of the pressure formulae for interior nodes. (2) Assign initial values of $p$ to interior nodes. (Initial values may be arbitrary, although the average of known boundary nodes is a good choice, as this normally will speed the convergence to a solution.) (3) Systematically compute $p$ for interior nodes using the pressure formulae and initial values. These "new" values of $p$ are reassigned to the nodes. Specified values of $p$ at boundary nodes are, of course, retained. (4) Successively recompute $p$ for interior nodes, each time reassigning new values to be used in the next iteration. (5) Continue this procedure until changes in $p$ with successive iterations are negligible. (One may specify this change to a desired precision, as well as apply several techniques to speed the convergence.) The final values of $p$ represent a steady-state solution. These values may then be substituted into the cubic law to obtain the discharge $Q$ associated with each fracture segment. Obtaining this type of solution for arbitrarily large networks is, in principle, limited only by computer memory and speed. It is left as an exercise to compute the total water discharge passing nodes J, K, and L based on the following parametric values: $b_1 = 0.0030$, $b_2 = 0.0015$, $b_3 = 0.0020$, $b_4 = 0.0015$, $b_5 = 0.0010$ (all in m); $L_{AD} = 1.41$, $L_{DF} = 0.92$, $L_{FI} = 1.91$, $L_{IL} = 1.41$, $L_{BD} = 1.12$, $L_{DE} = 1.68$, $L_{EJ} = 1.68$, $L_{CG} = 1.97$, $L_{GF} = 1.32$, $L_{FE} = 1.64$, $L_{GH} = 1.56$, $L_{HI} = 1.41$, $L_{IK} = 1.41$ (all in m); and $p_A = p_B = p_C = 20$, $p_J = p_K = p_L = 10.5$ (in kPa).

The problem is more complicated if fracture flow is coupled with diffuse flow within porous rock elements bounded by fractures. Then pressure values at nodes cannot be computed just based on continuity at nodes; one must simultaneously consider flow through the elements, whose boundary conditions involve variations in pressure along bounding fractures. In addition, the treatment of fractures as conduits of uniform aperture is only a rough approximation. Students should examine relevant papers listed below to see how this problem has been treated.

## 8.5 READING

Bear, J. 1972. *Dynamics of fluids in porous media.* New York: Elsevier (Dover edition, 1988), 764 pp. This text provides a thorough discussion of the Dupuit approximations, the Forchheimer modification of the Dupuit model, and errors that arise from the assumptions involved.

Bras, R. L. 1990. *Hydrology: an introduction to hydrologic science.* Reading, Mass.: Addison-Wesley, 643 pp. This text provides (pp. 478–91) a straightforward introductory treatment of the Saint-Venant equations for open-channel flow in the context of the flood-routing problem.

Domenico, P. A. and Schwartz, F. W. 1990. *Physical and chemical hydrogeology.* New York: Wiley, 824 pp. This text provides a clear presentation of how the concept of effective stress and aquifer compressibility are translated into the idea of aquifer storativity then incorporated into the equation of continuity. The development involves a delightful historical commentary on the original treatment of this problem by Jacob who, in 1940, obtained the correct result from incorrect reasoning. This text also provides a good introductory treatment of the advection–diffusion equation.

Jacob, C. E. 1940. On the flow of water in an elastic artesian aquifer. *Transactions of the American Geophysical Union* 22:574–86. This is the paper in which Jacob originally derived an expression for aquifer storativity $S$.

Levich, V. G. 1962. *Physicochemical hydrodynamics.* Englewood Cliffs, N.J.: Prentice-Hall, 700 pp. This text contains a discussion of the influence of advection on bubble growth embodied in (8.114).

Long, J. C. S. and Billaux, D. M. 1987. From field data to fracture network modeling: an example incorporating spatial structure. *Water Resources Research* 23:1201–16. This paper examines the problem of predicting flow in a probabilistic sense based on an artificially generated fracture network whose statistical properties match those of field data. In this regard, Jane Long and her colleagues have pioneered recent work on this problem. Students interested in pursuing this problem should consult the reference section of this paper, particularly the 1982 article by Long and others regarding the idea of porous-media equivalents for fracture networks, and the 1980 article by Witherspoon and others. The 1970 dissertation by C. R. Wilson provides an alternative numerical approach for computing flow within fracture networks.

Loper, D. E. 1992. A nonequilibrium theory of a slurry. *Continuum Mechanics and Thermodynamics* 4:213–45. This paper examines growth and melting of crystals in a slurry due to diffusion of mass and energy between the crystals and surrounding liquid, including effects of advection.

Nelson, J. M. and Smith, J. D. 1989. Evolution and stability of erodible channel beds. In Ikeda, S. and Parker, G., eds., River Meandering. American Geophysical Union, Water Resources Monograph 12, Washington, D.C., 321–77. This work and the paper by Struiksma and Crosato are good examples of how depth-integrated equations of conservation of mass and momentum are linearized, then used to examine flow and bedform development in an alluvial channel.

Struiksma, N. and Crosato, A. 1989. Analysis of a 2-D bed topography model for rivers. In Ikeda, S. and Parker, G., eds., River Meandering. American Geophysical Union, Water Resources Monograph 12, Washington, D.C., 153–80.

Theis, C. V. 1935. The relation between the lowering of the piezometric surface and rate and duration of discharge of a well using groundwater storage. *Transactions of the American Geophysical Union* 2:519–24. This is the paper in which Theis originally presented the solution of (8.100) for the case of a well that fully penetrates an isotropic aquifer. Readers will note the structural homology with radial heat flow toward a line sink.

Tsang, Y. W. 1984. The effect of tortuosity on fluid flow through a single fracture. *Water Resources Research* 20:1209–15. This paper examines the influence of variable aperture width and surface roughness on flow within fractures, and how these lead to a departure from the cubic law for flow between parallel walls; it also provides a good list of references concerning related work on this problem.

# CHAPTER 9

# Conservation of Energy

We briefly considered in Chapter 6 a geologically important class of flows—buoyancy driven flows—in which thermal effects hold an essential part in creating the forces that induce flow. We also should recall that fluid properties such as viscosity can vary with temperature. As fluids are heat-conducting media, taking into account the thermal energy conditions and heat flow within a fluid therefore is often an essential part of describing a flow field. Thermal energy, however, is not transported merely by conduction; in a moving fluid, thermal energy also is advected from position to position. In addition, recall that the thermal energy of a fluid, according to the first law of thermodynamics, is inextricably coupled with energy in the form of work performed between a fluid element and its surroundings. It is therefore important to consider the mechanical energy of a fluid when describing its thermal conditions. To this end, the developments below concern *conservation of energy* in both thermal and mechanical forms. Analogous to our treatment of conservation of mass, we will derive equations that describe a condition—conservation of energy—which must be satisfied at each coordinate position in a fluid.

An important outcome of our development of expressions for conservation of energy is a set of dynamical equations for the special case of an ideal fluid. In component form, these are referred to as *Euler's equations,* and arise from conservation of purely mechanical energy, neglecting thermal forms. (We will derive Euler's equations again in Chapter 10 using an explicit treatment of the forces involved in fluid motion.) Conservation of energy also applies to flow in porous media; the relevant expressions are similar to those for purely fluid flow, but with several important differences that arise from the two-phase character (solid and fluid) of flow in porous media. In relation to this topic, we also will develop the idea of *Hubbert's potential,* and the relation of this potential to *piezometric head.* This is a cornerstone of the theory of flow in porous media. Our objective is to illustrate how Hubbert's potential, and head, are obtained from applying the idea of conservation of mechanical energy to a fluid.

## 9.1 ENERGY EQUATION

### 9.1.1 Conservation in absence of viscous dissipation

An equation of energy conservation states that energy flowing into a control volume must be compensated by energy flowing out of the volume, by a change in the total energy within the volume, or some combination of both. We will consider potential, kinetic, and thermal forms of energy. In addition, we must consider energy in the form of work performed on the fluid within the control volume, as this must be compensated by a change in the internal energy of the fluid according to the first law of thermodynamics. Work also is performed against friction arising from the viscosity of real fluids; we will neglect this topic for now, however, deferring it to a later chapter when we consider viscous forces. The development below therefore pertains to inviscid flows.

A statement of conservation of energy can be formulated in a straightforward way, involving only a few steps, if use is made of the substantive derivative (Chapter 7). Nonetheless, we will adopt here an approach involving expansions of energy quantities about the faces of a control volume, as used in the previous chapter to derive equations of mass continuity. Although this approach involves a greater number of steps, it is particularly well suited to illustrating how each form of energy is imported to a particular site, or exported from it. This is an Eulerian approach. When we examine the energy equation again in Chapter 12, and include effects of viscous forces, we will begin with the substantive derivative of the energy of a fluid parcel, a Lagrangian approach.

Consider a small rectangular control volume with edges of lengths d$x$, d$y$, and d$z$ embedded within a local Cartesian coordinate system (Figure 9.1). This local system has an arbitrary orientation with respect to the Earth coordinate system; the $x$-axis is inclined at an angle $\alpha$ measured from the horizontal. Acceleration due to gravity acts vertically, and the centroid of the control volume is at a height $h$ above a horizontal datum. The height $h$ provides a measure of the position of the fluid within the gravitational field. The task now consists of examining the fluxes of each form of energy into and out of the control volume. We must adopt a consistent sign convention; herein a net flow of energy out of the control volume during a

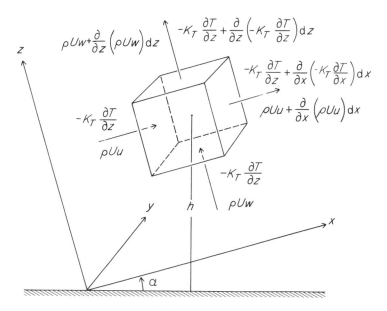

**Figure 9.1**  Definition diagram for heat flow by conduction, and advection of internal fluid energy, associated with an elementary control volume.

small interval d$t$ is assigned a positive value. Let us begin with thermal energy, neglecting radiation.

According to Fourier's law (4.21), the heat flux density by conduction in the direction of positive $x$ is $-K_T \partial T/\partial x$, where $K_T$ is the thermal conductivity of the fluid and $T$ is temperature. Therefore the total quantity of heat flowing into the control volume through the left face d$y$d$z$ during a small interval d$t$ is $-K_T(\partial T/\partial x)\,\mathrm{d}y\,\mathrm{d}z\,\mathrm{d}t$. Using a Taylor expansion to first order about the left face, the total quantity of heat flowing out of the control volume through the right face during d$t$ is $-K_T(\partial T/\partial x)\,\mathrm{d}y\,\mathrm{d}z\,\mathrm{d}t + (\partial/\partial x)[-K_T(\partial T/\partial x)]\,\mathrm{d}x\,\mathrm{d}y\,\mathrm{d}z\,\mathrm{d}t$, so the net quantity of heat flowing out of the volume in the direction of positive $x$ is

$$-\frac{\partial}{\partial x}\left(K_T \frac{\partial T}{\partial x}\right)\mathrm{d}x\,\mathrm{d}y\,\mathrm{d}z\,\mathrm{d}t \tag{9.1}$$

Using similar expansions, the net quantities of heat flowing out of the control volume in the directions of positive $y$ and $z$ are

$$-\frac{\partial}{\partial y}\left(K_T \frac{\partial T}{\partial y}\right)\mathrm{d}x\,\mathrm{d}y\,\mathrm{d}z\,\mathrm{d}t \tag{9.2}$$

$$-\frac{\partial}{\partial z}\left(K_T \frac{\partial T}{\partial z}\right)\mathrm{d}x\,\mathrm{d}y\,\mathrm{d}z\,\mathrm{d}t \tag{9.3}$$

Consider now the internal (thermal) energy of the fluid. The quantity of fluid mass flowing into the control volume through the left face d$y$d$z$ in the direction of positive $x$ during d$t$ is $\rho u\,\mathrm{d}y\,\mathrm{d}z\,\mathrm{d}t$. This mass carries a quantity of internal energy with it; the quantity of internal energy flowing into the volume during d$t$ is therefore $\rho U u\,\mathrm{d}y\,\mathrm{d}z\,\mathrm{d}t$, where $U$ is the internal energy per unit mass of fluid. Defining similar expressions for the flow of internal energy in the directions of $y$ and $z$, then using Taylor expansions of these quantities, the net quantities of internal energy flowing out of the control volume in the three coordinate directions are

$$\frac{\partial}{\partial x}(\rho U u)\,\mathrm{d}x\,\mathrm{d}y\,\mathrm{d}z\,\mathrm{d}t \tag{9.4}$$

$$\frac{\partial}{\partial y}(\rho U v)\,\mathrm{d}x\,\mathrm{d}y\,\mathrm{d}z\,\mathrm{d}t \tag{9.5}$$

$$\frac{\partial}{\partial z}(\rho U w)\,\mathrm{d}x\,\mathrm{d}y\,\mathrm{d}z\,\mathrm{d}t \tag{9.6}$$

The fluid mass within the control volume at time $t$ is $\rho\,\mathrm{d}x\,\mathrm{d}y\,\mathrm{d}z$, so the internal energy of the fluid within the volume at time $t$ is $\rho U\mathrm{d}x\,\mathrm{d}y\,\mathrm{d}z$. Using a Taylor expansion with time as the independent variable, the change in internal energy after the small interval d$t$ is

$$\frac{\partial}{\partial t}(\rho U)\,\mathrm{d}x\,\mathrm{d}y\,\mathrm{d}z\,\mathrm{d}t \tag{9.7}$$

Let us now turn to mechanical energy: potential and kinetic. The potential energy per unit volume of fluid at the centroid of the control volume is $\rho g h$. This suggests that the flux of potential energy in the direction of positive $x$ through a surface $dy\,dz$ positioned at the centroid is $\rho g h u\,dy\,dz$. The potential energy of fluid moving into the volume through the left face $dy\,dz$ during $dt$ is therefore $g[\rho hu - (\partial/\partial x)(\rho hu)(dx/2)]\,dy\,dz\,dt$, and the potential energy of fluid moving out of the volume is $g[\rho hu + (\partial/\partial x)(\rho hu)(dx/2)]\,dy\,dz\,dt$. The height $h$ is included within the parentheses because fluid arriving at the left face is on average lower than the volume centroid, and fluid leaving the right face is higher (Figure 9.1). The net quantity of potential energy flowing out of the control volume during $dt$ is

$$g\frac{\partial}{\partial x}\left(\rho hu\right) dx\,dy\,dz\,dt \tag{9.8}$$

Using similar expansions, the net quantities of potential energy flowing out of the control volume in the directions of positive $y$ and $z$ are

$$g\frac{\partial}{\partial y}(\rho hv) dx\,dy\,dz\,dt \tag{9.9}$$

$$g\frac{\partial}{\partial z}(\rho hw) dx\,dy\,dz\,dt \tag{9.10}$$

The potential energy of fluid within the control volume at time $t$ is $\rho g h\,dx\,dy\,dz$. Expanding with respect to time, the change in potential energy after $dt$ is

$$gh\frac{\partial \rho}{\partial t}dx\,dy\,dz\,dt \tag{9.11}$$

Here $h$ is not included as part of the differential because its value is a fixed quantity associated with the stationary control volume.

Consider the flux of kinetic energy into and out of the control volume. The quantity of fluid mass $m$ flowing into the control volume through the left face $dy\,dz$ in the direction of positive $x$ during $dt$ is $\rho u\,dy\,dz\,dt$. This mass carries a quantity of kinetic energy with it. One may be tempted to assume that this kinetic energy is given by $\frac{1}{2}mu^2 = \frac{1}{2}\rho u\,u^2\,dy\,dz\,dt$. However, the mass entering the control volume in general possesses a velocity $\mathbf{u} = \langle u, v, w\rangle$ whose magnitude is $\sqrt{(u^2 + v^2 + w^2)}$, a scalar quantity. The total kinetic energy carried by fluid entering the control volume with a component velocity $u$ is therefore $\frac{1}{2}\rho u(u^2 + v^2 + w^2)\,dy\,dz\,dt$. Defining similar expressions for the flow of kinetic energy in the directions of $y$ and $z$, then using Taylor expansions of these quantities, the net quantities of kinetic energy flowing out of the control volume in the three coordinate directions are

$$\frac{1}{2}\frac{\partial}{\partial x}[\rho u(u^2 + v^2 + w^2)] dx\,dy\,dz\,dt \tag{9.12}$$

$$\frac{1}{2}\frac{\partial}{\partial y}[\rho v(u^2 + v^2 + w^2)] dx\,dy\,dz\,dt \tag{9.13}$$

$$\frac{1}{2}\frac{\partial}{\partial z}[\rho w(u^2 + v^2 + w^2)] dx\,dy\,dz\,dt \tag{9.14}$$

The kinetic energy of fluid within the control volume at time $t$ is $\frac{1}{2}\rho(u^2 + v^2 + w^2)\,dx\,dy\,dz$. Expanding with respect to time, the change in kinetic energy after $dt$ is

$$\frac{1}{2}\frac{\partial}{\partial t}[\rho(u^2 + v^2 + w^2)] dx\,dy\,dz\,dt \tag{9.15}$$

Finally consider energy in the form of work performed on the fluid within the control volume. This is necessary because, according to the first law of thermodynamics (4.35), any net flux of heat into the control volume is partitioned between a change in its internal energy, work performed by the fluid on its surroundings, or both. (Conversely, any net flux of heat out of the control volume is compensated by a change in its internal energy, work performed on the fluid by its surroundings, or both.) The force associated with fluid pressure $p$ acting on the left face $dy\,dz$ is $p\,dy\,dz$. According to (4.36), the work performed equals the product of this force and the displacement of fluid in the direction of the force. The rate at which work is performed is therefore $pu\,dy\,dz$, and the total work performed during $dt$ is $pu\,dy\,dz\,dt$. This may be considered a quantity of energy (in the form of work) imparted to fluid within the control volume during $dt$. The work performed at the right face is obtained to first order using a Taylor expansion. This is the quantity of energy (in the form of work) exported from the control volume during $dt$. The net work performed on the fluid in the direction of positive $x$ is

$$\frac{\partial}{\partial x}(pu)\,dx\,dy\,dz\,dt \tag{9.16}$$

This may be considered a net change in the energy of fluid within the control volume. Using similar expansions, the net changes in energy of fluid within the control volume with respect to work performed in the directions of positive $y$ and $z$ are

$$\frac{\partial}{\partial y}(pv)\,dx\,dy\,dz\,dt \tag{9.17}$$

$$\frac{\partial}{\partial z}(pw)\,dx\,dy\,dz\,dt \tag{9.18}$$

By conservation of energy, the quantities (9.1) through (9.18) must sum to zero. Assuming that $K_T$ is constant, then expanding terms, dividing throughout by $dx\,dy\,dz\,dt$, and rearranging,

$$-K_T\left(\frac{\partial^2 T}{\partial x^2} + \frac{\partial^2 T}{\partial y^2} + \frac{\partial^2 T}{\partial z^2}\right)$$

$$+\rho\left(u\frac{\partial U}{\partial x} + v\frac{\partial U}{\partial y} + w\frac{\partial U}{\partial z} + \frac{\partial U}{\partial t}\right) + U\left[\frac{\partial}{\partial x}(\rho u) + \frac{\partial}{\partial y}(\rho v) + \frac{\partial}{\partial z}(\rho w) + \frac{\partial \rho}{\partial t}\right]$$

$$+\rho g\left(u\frac{\partial h}{\partial x} + v\frac{\partial h}{\partial y} + w\frac{\partial h}{\partial z}\right) + gh\left[\frac{\partial}{\partial x}(\rho u) + \frac{\partial}{\partial y}(\rho v) + \frac{\partial}{\partial z}(\rho w) + \frac{\partial \rho}{\partial t}\right]$$

$$+\rho u\left(u\frac{\partial u}{\partial x} + v\frac{\partial u}{\partial y} + w\frac{\partial u}{\partial z} + \frac{\partial u}{\partial t}\right) + \frac{1}{2}u^2\left[\frac{\partial}{\partial x}(\rho u) + \frac{\partial}{\partial y}(\rho v) + \frac{\partial}{\partial z}(\rho w) + \frac{\partial \rho}{\partial t}\right]$$

$$+\rho v\left(u\frac{\partial v}{\partial x} + v\frac{\partial v}{\partial y} + w\frac{\partial v}{\partial z} + \frac{\partial v}{\partial t}\right) + \frac{1}{2}v^2\left[\frac{\partial}{\partial x}(\rho u) + \frac{\partial}{\partial y}(\rho v) + \frac{\partial}{\partial z}(\rho w) + \frac{\partial \rho}{\partial t}\right]$$

$$+\rho w\left(u\frac{\partial w}{\partial x} + v\frac{\partial w}{\partial y} + w\frac{\partial w}{\partial z} + \frac{\partial w}{\partial t}\right) + \frac{1}{2}w^2\left[\frac{\partial}{\partial x}(\rho u) + \frac{\partial}{\partial y}(\rho v) + \frac{\partial}{\partial z}(\rho w) + \frac{\partial \rho}{\partial t}\right]$$

$$+u\frac{\partial p}{\partial x} + v\frac{\partial p}{\partial y} + w\frac{\partial p}{\partial z} + p\left(\frac{\partial u}{\partial x} + \frac{\partial v}{\partial y} + \frac{\partial w}{\partial z}\right) = 0 \tag{9.19}$$

The first line in (9.19) concerns heat flow by conduction, the second concerns internal (thermal) energy, the third concerns potential energy, the fourth, fifth, and sixth lines concern kinetic energy, and the seventh concerns energy in the form of work. This can immediately be simplified; each of the terms involving brackets in lines two through six equals zero based on the equation of continuity (8.12). Moreover, using definitions of the substantive derivative and the vector gradient operator (Chapter 7), (9.19) may be written

$$\rho\frac{DU}{Dt} + \rho u\frac{Du}{Dt} + \rho v\frac{Dv}{Dt} + \rho w\frac{Dw}{Dt} = K_T\nabla^2 T - \rho g\mathbf{u}\cdot\nabla h - \mathbf{u}\cdot\nabla p - p\nabla\cdot\mathbf{u}$$

(9.20)

or further simplified to

$$\rho\frac{DU}{Dt} + \rho\mathbf{u}\cdot\frac{D\mathbf{u}}{Dt} = K_T\nabla^2 T - \rho g\mathbf{u}\cdot\nabla h - \mathbf{u}\cdot\nabla p - p\nabla\cdot\mathbf{u} \qquad (9.21)$$

This is referred to as the *energy equation* for inviscid flow. Like an equation of mass continuity, (9.21) defines a condition—conservation of energy—that must be satisfied at each coordinate position in a flow. In the case of an incompressible fluid, the last term equals zero and (9.21) reduces to

$$\rho\frac{DU}{Dt} + \rho\mathbf{u}\cdot\frac{D\mathbf{u}}{Dt} = K_T\nabla^2 T - \rho g\mathbf{u}\cdot\nabla h - \mathbf{u}\cdot\nabla p \qquad (9.22)$$

The derivation leading to (9.8) through (9.11) above is based on an arbitrary orientation of the $xyz$-coordinate system within an Earth reference frame. In the case where $z$ is vertical (in a gravitational field), $\partial h/\partial x = \partial h/\partial y = 0$ and $\partial h/\partial z = 1$ in (9.19); then (9.21) becomes

$$\rho\frac{DU}{Dt} + \rho\mathbf{u}\cdot\frac{D\mathbf{u}}{Dt} = K_T\nabla^2 T - \rho g w\mathbf{k} - \mathbf{u}\cdot\nabla p - p\nabla\cdot\mathbf{u} \qquad (9.23)$$

and (9.22) becomes

$$\rho\frac{DU}{Dt} + \rho\mathbf{u}\cdot\frac{D\mathbf{u}}{Dt} = K_T\nabla^2 T - \rho g w\mathbf{k} - \mathbf{u}\cdot\nabla p \qquad (9.24)$$

where the unit vector $\mathbf{k} = \langle 0, 0, 1\rangle$. Additional terms must be added to (9.21) through (9.24) when viscous effects are considered (Chapter 12).

### 9.1.2 Mechanical energy and Euler's equation

Let us momentarily consider a flow for which thermal and viscous effects are unimportant, and further assume that the fluid is incompressible. Neglecting terms involving heat flow and internal energy in (9.22), then dividing remaining terms by $\rho$ and rearranging:

$$\mathbf{u}\cdot\left(\frac{D\mathbf{u}}{Dt} + g\nabla h + \frac{1}{\rho}\nabla p\right) = 0 \qquad (9.25)$$

For nonzero $\mathbf{u}$, the parenthetical part in (9.25) generally must equal zero. Isolating this parenthetical part and writing it in component form,

$$u\frac{\partial u}{\partial x} + v\frac{\partial u}{\partial y} + w\frac{\partial u}{\partial z} + \frac{\partial u}{\partial t} = -\frac{1}{\rho}\frac{\partial p}{\partial x} - g\frac{\partial h}{\partial x} \tag{9.26}$$

$$u\frac{\partial v}{\partial x} + v\frac{\partial v}{\partial y} + w\frac{\partial v}{\partial z} + \frac{\partial v}{\partial t} = -\frac{1}{\rho}\frac{\partial p}{\partial y} - g\frac{\partial h}{\partial y} \tag{9.27}$$

$$u\frac{\partial w}{\partial x} + v\frac{\partial w}{\partial y} + w\frac{\partial w}{\partial z} + \frac{\partial w}{\partial t} = -\frac{1}{\rho}\frac{\partial p}{\partial z} - g\frac{\partial h}{\partial z} \tag{9.28}$$

Observe that the left sides of these equations are identical to the kinematical expressions for fluid acceleration, (7.19), (7.20), and (7.21). Thus, (9.26), (9.27),
and (9.28) are dynamical equations that relate fluid accelerations to the forces
involved—in this case, forces associated with pressure and gravitational potentials.
These are known as *Euler's equations* or the *momentum equations* for an inviscid,
incompressible fluid (ideal fluid). We will derive them again in Chapter 10 using
an explicit treatment of the forces involved in fluid motion.

The result (9.25) can be used to simplify the energy equation (9.22). We will examine this point in Example Problem 9.3.3, and then again in Chapter 10. We may
meanwhile note that, according to (9.25), a statement of conservation of mechanical energy is obtained as the inner product of the velocity **u** and the momentum
equation, a result that we will return to in Chapter 14.

### 9.1.3 Porous Media

To obtain a statement of conservation of energy for flow in a porous medium, let us
assume that the medium is saturated with an incompressible liquid, and that local
thermal equilibrium exists between the fluid and solid phases. Under these circumstances it is convenient to define an *effective thermal conductivity* $K_{Te}$, which
characterizes how the liquid and solid phases act together as a thermal conductor.
Then $K_{Te}$ replaces $K_T$ in (9.1), (9.2), and (9.3).

Consider the local term (9.7). This must be modified since the internal energy of
the control volume is partitioned between liquid and solid. The volume of the liquid
is $n\,dx\,dy\,dz$, and the volume of the solid is $(1 - n)\,dx\,dy\,dz$, where $n$ is porosity.
Thus the change in the internal energy of the control volume (liquid plus solid)
after a small interval $dt$ is

$$\frac{\partial}{\partial t}(\rho U_f)n\,dx\,dy\,dz\,dt + \frac{\partial}{\partial t}(\rho_s U_s)(1 - n)\,dx\,dy\,dz\,dt \tag{9.29}$$

where $U_f$ is the internal energy per unit mass of liquid, $U_s$ is the internal energy
per unit mass of solid, and $\rho_s$ is the density of the solid. Assuming that both $\rho$ and
$\rho_s$ are constant, and using the relations $dU_f = c\,dT$ and $dU_s = c_s dT$,

$$[n\rho c + (1 - n)\rho_s c_s]\frac{\partial T}{\partial t}dx\,dy\,dz\,dt \tag{9.30}$$

Letting $c_e$ denote the quantity in brackets, the *effective specific heat*, the local rate
of change in the internal energy per unit volume is

$$c_e\frac{\partial T}{\partial t} \tag{9.31}$$

Because the fluid is incompressible, observe that the second term on the left of (9.22), according to (9.25), equals the last two terms on the right side of (9.22). Then, replacing the velocities $u$, $v$, and $w$ with the specific discharges $q_{hx}$, $q_{hy}$, and $q_{hz}$ in the substantive derivative of internal energy in (9.22), making further use of the relation $dU_f = cdT$, using the local term (9.31), and replacing $K_T$ with $K_{Te}$,

$$q_{hx}\frac{\partial T}{\partial x} + q_{hy}\frac{\partial T}{\partial y} + q_{hz}\frac{\partial T}{\partial z} + \frac{c_e}{\rho c}\frac{\partial T}{\partial t} = \frac{K_{Te}}{\rho c}\nabla^2 T \tag{9.32}$$

$$\mathbf{q}_h \cdot \nabla T + \frac{c_e}{\rho c}\frac{\partial T}{\partial t} = \kappa_e \nabla^2 T \tag{9.33}$$

where $\mathbf{q}_h = \langle q_{hx}, q_{hy}, q_{hz}\rangle$, and $\kappa_e = K_{Te}/\rho c$ is the *effective thermal diffusivity*. This is the energy equation normally adopted for incompressible liquid flows within porous media. It states that the rate at which the temperature changes at a given coordinate position equals the net rate at which heat is conducted and advected to (or away from) the site.

It is interesting that (9.32) and (9.33) follow from a development involving inviscid flow. Yet flow in a porous medium is clearly viscous at the pore scale since it involves the shearing of fluid around grains and through pores. Keep in mind, however, that the specific discharge $\mathbf{q}_h$ is a macroscopic quantity that does not describe the details of fluid motion at the pore scale. Indeed, recall our observation in Chapter 3 with reference to Figure 3.1, that flow in a porous medium next to a geologic boundary has, at a macroscopic scale, the appearance of an inviscid flow. Like inviscid flow, mechanical energy is apparently conserved, so the balance of energy in (9.33) only involves thermal conduction and advection, and the local term $(c_e/\rho c)\partial T/\partial t$. We will return to this below in developing the ideas of Hubbert's potential, hydraulic head, and Darcy's law. In addition, we will reconsider the ideas of effective conductivity and specific heat in Chapter 13.

## 9.2 HUBBERT'S POTENTIAL

Consider an inclined tube filled with a porous medium, whose ends allow for fluid flow into and out of the tube (Figure 9.2). Two manometers whose ends are embedded within the tube wall register different levels of fluid. Judging from these different levels, and the hydrostatic equation (6.29), the fluid within the tube at the base of the higher manometer is at a lower pressure than fluid at the base of the lower manometer. Based on pressure alone, one would infer that fluid flows upward through the tube. Nonetheless, intuition suggests the correct result, that flow occurs from higher to lower positions in this example, despite the adverse pressure conditions. Knowledge of the pressure variation between the two manometer positions evidently is not sufficient to predict the direction of flow. Indeed, hydrologists refer to *head* as the quantity whose variation with respect to position induces fluid flow within a porous medium. This is embodied, for example, in the $x$-component form of Darcy's law (Chapter 1),

$$q_{hx} = -K_h\frac{\partial h}{\partial x} \tag{9.34}$$

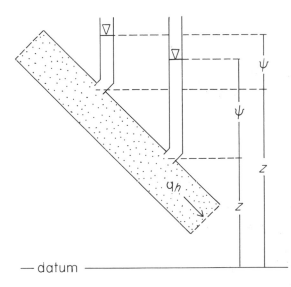

**Figure 9.2** Inclined tube containing a porous medium through which saturated one-dimensional flow occurs; water levels in manometers reflect fluid pressure conditions at the bases of the manometers according to the hydrostatic equation; adapted from Hubbert (1940).

where $q_{hx}$ is specific discharge (volumetric flux density), $K_h$ is hydraulic conductivity, and $h$ is *hydraulic head*. (The use of $h$ here to denote head should not be confused with the use of $h$ above to denote the arbitrary elevation of a fluid control volume.) Let us examine the idea of head, and show how it involves conservation of mechanical energy. The development here follows that of M. K. Hubbert, who in 1940, formalized the idea of head using an ingenious artifice, a simple thermodynamic pump.

The *potential* $\Phi$ of a fluid may be defined as its mechanical energy per unit mass. Recall that mechanical energy is a relative quantity in the sense that it is measurable as the amount of work performed in bringing a given mass from its initial (datum) state to a specified final state. This final state is specified in terms of the elevation $z$, pressure $p$, density $\rho$, and speed $u$ of the fluid. The datum state is specified as $z = z_0$, $p = p_0$, $\rho = \rho_0$, and $u = 0$. The actual choice of $z_0$ is arbitrary, since only a relative vertical position in Earth's gravitational field is important. The datum pressure state is envisioned as a purely hydrostatic pressure—like that which occurs beneath a static liquid with a *free surface* that does not involve effects of surface tension due to surface curvature (Chapter 3). In practice, $p_0$ is set to atmospheric pressure; then $p - p_0$ is referred to as *gauge pressure* (in allusion to its actual measurement). Note also that the volume $V_0$ of a unit mass of fluid ($m = 1$) is $V_0 = 1/\rho_0$.

Consider a pump consisting of a cylinder with two one-way valves (inlet and outlet) and a frictionless piston (Figure 9.3). The pressure behind the piston is set to zero. (It is left as an exercise to show that this choice, albeit convenient, is not essential.) The cylinder inlet valve is placed in contact with fluid at the datum state. The piston is displaced in a reversible, quasi-static manner, and the cylinder fills with a unit mass ($m = 1$) of fluid. A quantity of work $W_1$ is performed:

$$W_1 = -p_0 V_0 \qquad (9.35)$$

where the negative sign indicates that work is performed on the fluid entering the cylinder by the fluid at datum state. The cylinder is then raised to an elevation $z$.

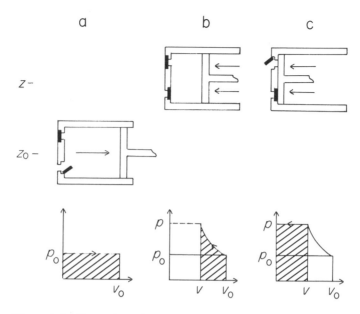

**Figure 9.3** Thermodynamic pump as envisioned by Hubbert (1940); shaded areas in pressure-volume graphs represent quantities of work performed, as described in the text: $-p_0 V_0$ (a), $\int p\,dV$ (b), and $pV$ (c).

The work $W_2$ performed is

$$W_2 = g(z - z_0) + m_c g(z - z_0) \qquad (9.36)$$

where $m_c$ is the mass of the cylinder and piston. The fluid within the cylinder is then injected into a (porous medium) system at pressure $p$, again in a reversible, quasi-static manner. The work $W_3$ performed is

$$W_3 = \int_V^{V_0} p\,dV + pV \qquad (9.37)$$

The term involving the integral, according to (4.37), describes the work involved in pressurizing the fluid from $p_0$ to $p$, with a decrease in volume from $V_0$ to $V$. The last term describes the work involved in injecting the fluid with final volume $V$ into the system against the pressure $p$. The positive signs on these terms indicate that work is performed by the fluid on its surroundings (the porous medium system). The fluid is then accelerated to a speed $u$; the work $W_4$ performed is

$$W_4 = \frac{u^2}{2} \qquad (9.38)$$

Finally the cylinder is lowered to the datum elevation $z_0$; the work $W_5$ performed is

$$W_5 = -m_c g(z - z_0) \qquad (9.39)$$

The total energy per unit mass of fluid $\Phi = W_1 + W_2 + W_3 + W_4 + W_5$. Summing (9.35) through (9.39) leads to

$$\Phi = gz - p_0 V_0 + \int_V^{V_0} p\,dV + pV + \frac{u^2}{2} \qquad (9.40)$$

Note that

$$\int_{pV}^{p_0V_0} d(pV) = \int_{V}^{V_0} p \, dV + \int_{p}^{p_0} V \, dp \tag{9.41}$$

Rearranging and substituting into (9.40),

$$\Phi = gz - p_0V_0 + \int_{pV}^{p_0V_0} d(pV) - \int_{p}^{p_0} V \, dp + pV + \frac{u^2}{2} \tag{9.42}$$

Substituting $1/\rho$ for $V$ in the second integral, then evaluating the first integral and simplifying,

$$\Phi = gz + \int_{p_0}^{p} \frac{1}{\rho} dp + \frac{u^2}{2} \tag{9.43}$$

Here it must be noted that $\rho$ can be a function $\rho(p)$ of pressure only. This will generally be true under isothermal or adiabatic conditions (Chapter 4). Equation (9.43) is a form of Bernoulli's equation for compressible fluid flow (Chapter 10). In the case of an incompressible fluid, $\rho = $ constant, and (9.43) becomes

$$\Phi = gz + \frac{p - p_0}{\rho} + \frac{u^2}{2} \tag{9.44}$$

Fluid velocities in groundwater flow normally are sufficiently small that the kinetic energy term in (9.44) can be neglected. Further, since it may be assumed that $g$ is constant for most groundwater flow problems, it is conventional to work with a quantity $h$ equal to the mechanical energy per unit weight, which is obtained by dividing each term in (9.44) by $g$. Thus

$$h = z + \frac{p - p_0}{\rho g} \tag{9.45}$$

The quantity $h$ is the *hydraulic head* or *piezometric head* of the fluid. Referring to Figure 9.2, according to the hydrostatic equation (6.26), $p = \rho g \psi + p_0$, where $\psi = h - z$, and $p_0$ is taken as atmospheric pressure. Substituting this into (9.45),

$$h = z + \psi \tag{9.46}$$

which is simply the height of the fluid surface in the manometer relative to the datum elevation.

It now remains to be illustrated that the head gradient, as in (9.34), is a measure of the force that induces flow. Consider a rectangular fluid element within a porous medium (Figure 9.4). The element has edges of lengths $dx$, $dy$, and $dz$ parallel to the axes of a Cartesian coordinate system. The net force acting on the fluid element is the sum of the gravitational force and forces due to pressure acting on the faces of the element. Recalling the development leading to the equation of hydrostatics (6.23), no net components of force due to pressure exist in either the $x$ or $y$ direction, and the net vertical force due to pressure (buoyant force) is exactly balanced by the gravitational force in the case of a static fluid. As the present problem involves a moving fluid, however, we must now allow for the possibility that a net force due to pressure acts in a direction other than the vertical (due to horizontal as well as vertical variations in pressure). Following the development involving Taylor

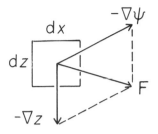

**Figure 9.4** Elementary rectangular control volume embedded within porous medium, acted on by normal pressures and gravity; adapted from Hubbert (1940).

expansions leading to the equation of hydrostatics, the components of force per unit volume due to pressure, $P_x$, $P_y$, and $P_z$, in directions of positive $x$, $y$, and $z$ are

$$P_x = -\frac{\partial p}{\partial x}; \qquad P_y = -\frac{\partial p}{\partial y}; \qquad P_z = -\frac{\partial p}{\partial z} \tag{9.47}$$

where the negative signs indicate that a decrease in pressure in a positive coordinate direction leads to a positive component of force. Using the definition of $\nabla p$ (Chapter 4), the net force per unit mass $\mathbf{P}$ due to pressure is

$$\mathbf{P} = \frac{1}{\rho}\langle P_x, P_y, P_z \rangle = -\frac{1}{\rho}\nabla p = -\frac{1}{\rho}\left(\mathbf{i}\frac{\partial p}{\partial x} + \mathbf{j}\frac{\partial p}{\partial y} + \mathbf{k}\frac{\partial p}{\partial z}\right) \tag{9.48}$$

Substituting $\rho g\psi + p_0$ (from above) for $p$ and expanding

$$\mathbf{P} = -g\left(\mathbf{i}\frac{\partial \psi}{\partial x} + \mathbf{j}\frac{\partial \psi}{\partial y} + \mathbf{k}\frac{\partial \psi}{\partial z}\right) \tag{9.49}$$

The force per unit mass due to gravity is $\mathbf{g}$, which may be written

$$\mathbf{g} = -g\nabla z = -g\left(\mathbf{i}\frac{\partial z}{\partial x} + \mathbf{j}\frac{\partial z}{\partial y} + \mathbf{k}\frac{\partial z}{\partial z}\right) \tag{9.50}$$

With $h = z + \psi$, the net force $\mathbf{F}$ per unit weight acting on the fluid element is then obtained by summing (9.49) and (9.50) and dividing by $g$:

$$\mathbf{F} = -(\nabla z + \nabla \psi) = -\left(\mathbf{i}\frac{\partial h}{\partial x} + \mathbf{j}\frac{\partial h}{\partial y} + \mathbf{k}\frac{\partial h}{\partial z}\right) = -\nabla h \tag{9.51}$$

Assuming that specific discharge $\mathbf{q}_h$ is proportional to $\mathbf{F}$ then gives Darcy's law in three dimensions

$$\mathbf{q}_h = K_h \mathbf{F} = -K_h \nabla h \tag{9.52}$$

of which (9.34) is the $x$-component. Note that this development depends on the assumption of an incompressible fluid and negligible kinetic energy of flow.

It becomes apparent that the hydraulic conductivity $K_h$ is merely a coefficient of proportionality. Also noteworthy is the structural homology that exists between Darcy's law (9.52) and Fourier's law (4.21). Like thermal conductivity $K_T$, the hydraulic conductivity $K_h$ may, in general, be a function of $K_h(x, y, z)$ of coordinate position, reflecting spatial variations in pore size and structure related to natural heterogeneities in the composition of porous media. In addition, $K_h$ may vary with coordinate direction; for example, the hydraulic conductivity $K_{hx}$ measured parallel to the $x$-axis may be greater than the hydraulic conductivity $K'_{hz}$ measured parallel to the $z$-axis. This situation reflects *anisotropy* with respect to hydraulic conductivity, a topic to which we will return in Chapter 13 when we generalize Darcy's law (9.52) for arbitrary orientation of the coordinate axes.

## 9.3 EXAMPLE PROBLEMS

### 9.3.1 Velocity measurements in open channel flow using a meter stick

A variety of instruments and techniques have been developed for measuring flow velocities within an open channel. These can be as simple as determining the time required for a float to travel a specified distance along the channel, but more typically involve using electromechanical current meters. The designs of these meters involve cup anemometers, propellers, pitot tubes, and magnetic induction systems. More sophisticated techniques involve hot-film and laser-Doppler anemometry. Average velocities also can be obtained by timing the advection of a solute (dye or salt) along a channel. (These techniques are summarized in the publication by the U.S. Geological Survey [1977].) The choice of instrument and technique depends on the objectives, and required precision, of the measurements.

Suppose that one requires only a quick estimate of the velocity in a small channel for a rough computation of discharge. In this case a surprisingly simple technique involving only a meter stick can be obtained from a straightforward consideration of conservation of mechanical energy. Consider placing a flat meter stick into the flow such that the length divisions face upstream. Flow collides with the meter stick and some of the water "climbs" the upstream face of the stick (Figure 9.5). Flow colliding with the stick possesses kinetic energy; a portion of this is converted to potential energy, manifest as water climbing the stick. As the local velocity fluctuates (due to turbulence in open channel flow), the height of climb fluctuates, but has an average value $h$.

The kinetic energy per unit mass of flow approaching the meter stick is $\frac{1}{2}u^2$, and the maximum potential energy per unit mass of water that has climbed the meter stick is $gh$. Momentarily assuming this conversion is conservative, equating these

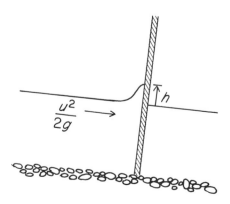

**Figure 9.5** Definition diagram for meter-stick technique for measuring velocity in an open channel.

two expressions and solving for $u$ gives

$$u = \sqrt{2gh} \tag{9.53}$$

This is the essential basis of "impact-pressure" devices for estimating velocity. In actuality, friction is involved, which has the effect of reducing the height of climb $h$, therefore leading to an underestimation of $u$. On the other hand, this technique obtains an estimate of the surface velocity, which is an overestimate of the average velocity in the fluid column. Simple semi-empirical corrections can be derived. (How?) Also note that, to use this simple technique, velocities must be sufficiently large that $h$ can be measured with reasonable precision.

Consider a very different problem: estimating the speed $u$ of a water–steam mixture as it exits the vent of an erupting geyser. It is left as an exercise to present an argument that formula (9.53) is an estimate of this speed, where $h$ is the height of the eruption (Example Problem 10.3.2).

### 9.3.2 Heat flow within glacier ice

Let us reexamine the problem of heat flow within glacier ice, recalling that ice temperature strongly influences the coefficient $A$ in Glen's law and therefore the rate of ice deformation for given stress. In contrast to our simple treatment in Example Problem 4.10.1, we shall here consider how heat is redistributed within the ice as its boundary is heated or cooled. With negligible ice velocity the energy equation (9.22) reduces to

$$K_T \left( \frac{\partial^2 T}{\partial x^2} + \frac{\partial^2 T}{\partial y^2} + \frac{\partial^2 T}{\partial z^2} \right) = \rho \frac{\partial U}{\partial t} \tag{9.54}$$

where it is assumed that the ice is homogeneous and isotropic with respect to thermal conductivity $K_T$.

Consider the case in which a broad, horizontal ice surface is heated uniformly; thus heat flow occurs only in the vertical direction. For convenience, set $z$ positive downward with $z = 0$ at the ice surface. Then (9.54) reduces to

$$K_T \frac{\partial^2 T}{\partial z^2} = \rho \frac{\partial U}{\partial t} \tag{9.55}$$

The specific internal energy $dU = c\,dT$ where $c$ is the specific heat of ice; substituting for $dU$,

$$\frac{\partial T}{\partial t} = \kappa \frac{\partial^2 T}{\partial z^2} \tag{9.56}$$

where $\kappa = K_T / \rho c$ is the *thermal diffusivity*. This one-dimensional heat flow equation appears in many physics problems, and therefore has a long history of study. A solution of (9.56) has the form $T(z, t)$, denoting that ice temperature $T$ is a function of position $z$ and time $t$, and requires specifying initial and boundary conditions.

It is convenient to consider the case in which the ice is initially of uniform temperature $T_0$. This may be specified as

$$T(z, 0) = T_0; \qquad z \geq 0 \tag{9.57}$$

In words, the temperature equals $T_0$ for all positions $z$ within the ice at time zero. The lower boundary condition is determined by the temperature at the base of the glacier. Rather than treating this explicitly, it is convenient to envision the ice as being sufficiently thick that the influence of the lower boundary on near-surface temperatures, which are of interest here, is not important. Justification for this will become clear below. This is equivalent to treating the ice as an "infinite half-space." Then the lower boundary condition may be specified as

$$T(\infty, t) = T_0; \qquad t \geq 0 \qquad (9.58)$$

In words, the temperature far from the surface ($z \to \infty$) equals $T_0$ for all times $t$.

The temperature at the ice surface ($z = 0$) is denoted by $T_s(0, t)$. The function $T_s$ is a complex result of solar radiation, long-wave radiation from the atmosphere, and ice albedo (reflectivity). In addition, $T_s$ is influenced by energy advected to or from the ice surface, which depends on thermal conditions of the atmosphere, wind, and ice-surface roughness. $T_s$ also is influenced by energy transferred between the ice surface and atmosphere through sublimation or accumulation of ice. The function $T_s$ therefore represents the net effect of numerous processes. In general, $T_s$ will exhibit diurnal variations chiefly related to solar input, annual variations related to solar and atmospheric conditions, and possibly longer variations related to climatic changes.

The clear cyclical nature of the first two of these processes suggests that a reasonable form for $T_s$ is that of a sinusoid; for example,

$$T_s = T_A \sin\left(\frac{2\pi}{\lambda} t\right) = T_A \sin(\omega t) \qquad (9.59)$$

where $T_A$ is the amplitude and $\lambda$ is the period of the temperature signal. For example, $\lambda = 24$ hrs for a diurnal signal. The ratio $\omega = 2\pi/\lambda$ is referred to as the angular frequency. Note that the product $\omega t$ transforms a real time $t$ into a dimensionless quantity measured in radians.

Recall that $K_T$ and $c$, and therefore $\kappa$, vary with $T$ (Chapter 4); however, let us momentarily assume that $T_A$ is sufficiently small that $\kappa$ is essentially constant. With the initial and boundary conditions specified above, the solution of (9.56) is well known. It involves a transient part related to the uniform initial temperature; this part dissipates after a finite period. The steady (dynamical) solution is then

$$T(z, t; \omega) = T_A \exp\left(-z\sqrt{\frac{\omega}{2\kappa}}\right) \sin\left(\omega t - z\sqrt{\frac{\omega}{2\kappa}}\right) \qquad (9.60)$$

where the notation $T(z, t; \omega)$ indicates that $\omega$ is treated as a parameter; the importance of this will become clear below. This solution looks like a series of waveforms that propagate downward (Figure 9.6). The wave speed is given by $\sqrt{2\kappa\omega}$; thus high-frequency waveforms travel downward faster than low-frequency waveforms. Moreover, the amplitude of the waveform decreases with distance $z$, as given by the first part in (9.60) involving the product of $T_A$ and the exponential function. Thus high-frequency temperature waveforms (short wavelengths) are attenuated with distance $z$ more rapidly than low-frequency waveforms, and therefore do not propagate far below the ice surface. The importance of this is that a temperature distribution $T(z, t)$ at any instant normally does not possess small bumps related

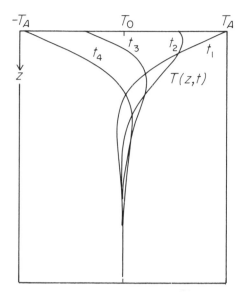

**Figure 9.6** Instantaneous configurations of temperature waveforms $T(z, t)$ propagating downward in glacier ice, where wave amplitude decreases with depth.

to short-lived fluctuations in boundary temperature, except near the surface. Rapid fluctuations in $T$ therefore do not significantly alter the overall thermal conditions of a thick ice body. For example, variations related to the diurnal solar cycle normally are not detectable below depths on the order of centimeters to decimeters. Variations related to annual cycles normally are not detectable below about 10 to 20 meters. Slower fluctuations of large amplitude, however, for example, those related to climatic changes, can be propagated significant distances downward and therefore can influence the overall thermal conditions of glacier ice. This also means that the vertical temperature distribution of thick ice can exhibit a temperature signal that is relict from ancient surface conditions.

The mechanisms of wave propagation and attenuation reside in the effects of specific heat and waveform geometry. Recall that, with steady conditions (Chapter 1), heat flowing from any given coordinate position is continuously replaced by heat arriving from up-gradient, so the temperature at the coordinate position does not vary. In contrast, consider the leading side of a waveform just in front of the temperature peak. The temperature gradient is not uniform; it increases in magnitude with depth, so at any instant the heat flowing from a given coordinate position is not replaced by heat from above at the same rate. Thus the temperature at any instant is decreasing over this part of the waveform. By a similar argument, the temperature is increasing over the leading side of a waveform just behind the temperature trough. Moreover, a finite quantity of heat is involved with the motion of a waveform. As this heat is "consumed" in raising the temperature of the ice, depending on the specific heat, it is dispersed with depth. A large specific heat therefore is manifest as a rapid attenuation of the waveform as heat is dissipated. On the leading side of a waveform, the temperature gradient at any instant induces downward flow of heat. On the trailing side, the gradient induces upward flow. But the waveform is asymmetrical; the magnitude of the temperature gradient on the leading side of a temperature peak is on average less than the gradient on the trailing side of the peak. (How can this be demonstrated qualitatively, then quantitatively?) It is left as an exercise to demonstrate on geometrical grounds that this asymmetry leads to downward propagation of the waveform. (Begin by considering what should happen to a symmetrical waveform that is not attenuated with depth.)

The solution (9.60) possesses a mathematical property with far-reaching implications. Suppose that the surface temperature signal $T_s(0, t)$ is composed of multiple frequencies, for example, a diurnal signal superimposed on the annual signal. Then the net solution $T(z, t)$ consists of the superposition of individual solutions $T(z, t; \omega)$ associated with each frequency present. Students familiar with Fourier series know that an arbitrarily complex signal can be formed by adding many sinusoids with different frequencies and amplitudes. Thus the property of superposition can be used to more carefully examine real signals involving multiple frequencies.

With ice flow, consideration must be given to differential velocities with depth and generation of heat due to viscous dissipation. In addition, the character of the surface of a glacier normally is changing due to snowfall, metamorphosis of snow to firn, and ablation, so thermal properties are not constant with depth. For example, the thermal diffusivity of snow is less than that of ice, in which case attenuation of a temperature signal near the surface is stronger.

### 9.3.3 Solidification and melting along dikes with Newtonian magma flow

Consider the upward flow of magma within a dike (Figure 9.7). Magma entering the dike with temperature $T_m$ continuously advects heat upward. The temperature of the country rock is less than that of the magma, and the associated temperature gradient induces flow of heat by conduction from the magma into the country rock. The magma therefore cools during its ascent. If this loss of heat by conduction

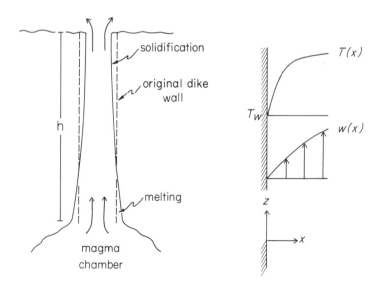

**Figure 9.7** Definition diagram for magma flow in a vertical dike, with velocity profile $w(x)$ and temperature profile $T(x)$ near the dike wall; after Bruce and Huppert (1990).

is sufficiently large, solidification of magma on the dike walls can occur. (This releases latent heat, which tends to counter solidification.) With solidification, the dike becomes narrower. A feedback occurs: As the dike becomes narrower the flow of magma decreases, the upward rate of heat advection decreases, and solidification is enhanced. The dike eventually can close. In contrast, if flow is large, the upward advection of heat may be sufficient to counter solidification, eventually leading to melting of the country rock and widening of the dike. (This involves consumption of latent heat.) Again, a feedback occurs: As the dike widens, the flow of magma increases, the upward rate of heat advection increases, and melting of the dike walls is enhanced. To predict which of these two behaviors is more likely therefore requires determining the importance of heat conduction relative to advection in terms of extant physical and thermal quantities.

These ideas follow the theoretical work of Bruce and Huppert (1990). Although a full treatment of the problem is beyond the scope of this example, essential results can be obtained from a simple dimensional analysis of the governing equations. (The development here departs slightly from that of Bruce and Huppert.) In addition, the example illustrates the important idea of a moving boundary problem.

Let us consider only thermal effects, and neglect heating due to viscous dissipation. The dike may possess a variable half-width $b$ due to irregularity in the initial dike aperture, and due to the possibility of vertical variations in narrowing (solidification) or widening (melting) of the dike walls. The dike has a vertical length $h$, and its horizontal breadth is much greater than its half-width $b$. The temperature of the country rock far from the dike is denoted by $T_c$. The density of the magma is denoted by $\rho$, and that of the country rock by $\rho_c$. The compositions of the magma and country rock are the same; a difference in density is therefore due to thermal expansion alone. The thermal conductivity $K_T$ and specific heat $c$ are the same for the magma and country rock. The magma is incompressible; flow is induced buoyantly due to a constant pressure drop between the magma chamber and surface over the distance $h$. The magma is assumed to behave like a Newtonian fluid with viscosity $\mu$.

Consider heat flow within the magma. Under the conditions defined above, $v = 0$, $\partial^2 T/\partial y^2 = 0$, and the energy equation (9.22) reduces to

$$-K_T\left(\frac{\partial^2 T}{\partial x^2} + \frac{\partial^2 T}{\partial z^2}\right) + \rho\left(u\frac{\partial U}{\partial x} + w\frac{\partial U}{\partial z} + \frac{\partial U}{\partial t}\right) = 0 \qquad (9.61)$$

Making use of the relation $dU = c\,dT$, and thermal diffusivity $\kappa = K_T/\rho c$, (9.61) becomes

$$-\kappa\left(\frac{\partial^2 T}{\partial x^2} + \frac{\partial^2 T}{\partial z^2}\right) + u\frac{\partial T}{\partial x} + w\frac{\partial T}{\partial z} + \frac{\partial T}{\partial t} = 0 \qquad (9.62)$$

This may be simplified if we can show that upward conduction of heat along the dike is insignificant relative to the upward advection of heat accompanying magma flow, and that local horizontal velocities related to variations in dike width are insignificant relative to streamwise (vertical) velocities. To do this, let us define several characteristic quantities of the flow. A characteristic temperature difference is given by $\Delta T = T_m - T_c$, which represents the largest possible drop in temperature capable of inducing heat flow by conduction over a given distance. Characteristic linear dimensions include the half-width $b$ and the dike length $h$. A convenient characteristic vertical velocity is the average velocity $w_0$ across the dike (although another velocity, for example, the maximum velocity, could be selected). With $x = 0$ at the left wall, and momentarily assuming a uniform dike width, it is left as an exercise to follow the development of Example Problem 6.7.2 and demonstrate that the vertical velocity distribution $w(x)$ for a Newtonian buoyancy-driven flow is:

$$w = \frac{1}{\mu}(\rho_c - \rho)g\left(bx - \frac{x^2}{2}\right) \tag{9.63}$$

for which the average velocity $w_0$ is

$$w_0 = \frac{1}{3\mu}(\rho_c - \rho)gb^2 \tag{9.64}$$

If the half-width varies, a local horizontal component of magma velocity must exist. A characteristic horizontal velocity $u_0$ is given by $w_0 db/dz$, which represents the magnitude of the largest conceivable horizontal velocity. Finally, a characteristic time scale $t_0$ is given by the average interval required for magma to travel the distance $h$, equal to $t_0 = h/w_0$.

Let us now use these characteristic quantities to estimate the magnitudes of the first four terms in (9.62). Since cooling of the magma occurs near the walls, the temperature there approaches $T_c$, whereas magma at the center of the dike possesses a temperature approaching $T_m$. It follows that the temperature gradient $\partial T/\partial x$ is of the order $\Delta T/b$. The lowest magma temperatures, approaching $T_c$, also occur at the surface, so the gradient $\partial T/\partial z$ is of the order $\Delta T/h$. Derivatives of temperature gradients $\partial^2 T/\partial x^2$ and $\partial^2 T/\partial z^2$ are therefore of the orders $\Delta T/b^2$ and $\Delta T/h^2$. With $b$ on the order of 1 m and $h$ on the order of 1,000 m, it follows that $\partial^2 T/\partial z^2 \ll \partial^2 T/\partial x^2$. Neglecting the term $\kappa\partial^2 T/\partial z^2$ in (9.62) therefore would involve an insignificant error relative to the magnitude of $\kappa\partial^2 T/\partial x^2$.

Except near the walls, the velocity $w$ is of the order $w_0$, and the velocity $u$ is of the order $w_0 db/dz$. Further, the spatial rate of variation in dike width $db/dz$ is of the order $b/h$, which geometrically coincides with the average rate of variation. This implies that the advection terms $u\partial T/\partial x$ and $w\partial T/\partial z$ are of the order $w_0\Delta T/h$ within the interior of the flow. However, the temperature gradient $\partial T/\partial x$ is most important near the walls where $u$ goes to zero (no-slip condition). In addition, if we insist that $db/dz$ is very small, at least initially, the advection term $u\partial T/\partial x$ is much smaller than the term $w\partial T/\partial z$.

Finally, it is useful to consider the conditions under which the vertical conduction term is insignificant relative to the vertical advection term. Since $t_0$ is the average time required to replace magma at the surface, it is the longest period over which we need to consider the effectiveness of conduction relative to vertical advection of heat. The dimensions of $\kappa$ are $L^2 t^{-1}$; taking the square root of the product $\kappa t_0$ retrieves a length scale $L_c$ over which conduction is effective during $t_0$. That is, $L_c = \sqrt{\kappa t_0} = \sqrt{\kappa h/w_0}$. With $w_0$ on the order of 1 m s$^{-1}$, $\kappa$ on the order of $10^{-6}$ m$^2$ s$^{-1}$, and $h$ on the order of 1,000 m, $L_c$ is about 0.08 m. Thus vertical heat conduction can be neglected relative to upward advection of heat so long as $w_0$ is much greater than $L_c/t_0$, or $\kappa/h \ll w_0$. In this regard, the ratio $w_0 h/\kappa = Pe_T$ is referred to as a *Peclet number* for heat transport. The Peclet number may be regarded as the ratio of heat transported by advection and heat transported by conduction.

Vertical conduction in this problem therefore can be neglected when $Pe_T \gg 1$. The small magnitude of the length scale $L_c$ also implies that the horizontal thermal boundary layer—that part of the magma next to the wall over which the temperature varies—is significantly less than the half-width $b$. The importance of this will become clear below.

These simple estimates are sufficient to justify neglecting the terms $\kappa \partial^2 T/\partial z^2$ and $u \partial T/\partial x$, in which case (9.62) reduces to

$$-\kappa \frac{\partial^2 T}{\partial x^2} + w \frac{\partial T}{\partial z} + \frac{\partial T}{\partial t} = 0 \qquad (9.65)$$

The problem is thus reduced to a consideration of conservation of thermal energy involving one-dimensional heat conduction into the walls, vertical advection of heat, and possible unsteady effects.

Now consider heat flow within the country rock. Velocities here are zero, so the energy equation reduces to

$$-\kappa \frac{\partial^2 T}{\partial x^2} + \frac{\partial T}{\partial t} = 0 \qquad (9.66)$$

which is the one-dimensional heat flow equation examined in Example Problem 9.3.2 above.

Note that with solidification onto the wall, or melting of the wall, the $xyz$-coordinate system (Figure 9.7) must move to maintain the position $x = 0$ at the wall–magma interface. This constitutes a moving-boundary problem, and (9.65) and (9.66) must be modified. Let $u_w$ denote the rate of horizontal migration of the wall-magma interface ($x = 0$) due to solidification ($u_w > 0$) or melting ($u_w < 0$). Then (9.65) and (9.66) become

$$-\kappa \frac{\partial^2 T}{\partial x^2} + w \frac{\partial T}{\partial z} + \frac{\partial T}{\partial t} - u_w \frac{\partial T}{\partial x} = 0 \qquad (9.67)$$

$$-\kappa \frac{\partial^2 T}{\partial x^2} + \frac{\partial T}{\partial t} - u_w \frac{\partial T}{\partial x} = 0 \qquad (9.68)$$

The additional term, $u_w \partial T/\partial x$, has a simple geometrical interpretation. Let us write $u_w$ as $dx/dt$, where $dx$ is the small distance that the interface (and coordinate system) move during $dt$. Assume that $dx$ is sufficiently small that $\partial T/\partial x$, the change in $T$ with position $x$ at any instant, is essentially constant. Then $dx \partial T/\partial x$ is the change in $T$ over the distance $dx$, and $(dx/dt)\partial T/\partial x = u_w \partial T/\partial x$ is the local rate at which $T$ changes, for specified (moving) coordinate position $x$, when $u_w$ is nonzero. (Why does this term have a negative sign?)

For a solution of the form $T(x, z, t)$, initial and boundary conditions for the magma fluid are specified by

$$T(x, z, 0) = T_m; \qquad T(x, 0, t) = T_m; \qquad T(0, z, t) = T_w; \qquad T(\infty, z, t) = T_m \qquad (9.69)$$

where $T_w$ is the temperature at the wall–magma interface. The last of these indicates that the fluid magma is treated as an infinite half-space, despite its finite width. This is justified inasmuch as the thermal boundary layer is small relative to the half-width $b$. For the solid rock,

$$T(x, z, 0) = T_c; \qquad T(0, z, t) = T_w; \qquad T(-\infty, z, t) = T_c \qquad (9.70)$$

The last of these indicates that the solid also is treated as an infinite half-space, with the temperature far from the dike (the *far-field* temperature) equal to $T_c$ for all times $t$.

Using the initial and boundary conditions (9.69) and (9.70), (9.67) and (9.68) are solved numerically. Note that $u_w$ is proportional to the difference in conductive heat flow across the wall–magma interface, and requires specifying the latent heat of the magma–solid composition. The velocity $w$ is given by (9.63), for example. Such a solution for the case of basaltic magma leads to the results introduced at the beginning of this Example Problem.

## 9.4 READING

Anderson, D. A., Tannehill, J. C., and Pletcher, R. H. 1984. *Computational fluid mechanics and heat transfer.* New York: McGraw-Hill, 599 pp. This text provides a comprehensive treatment of numerical methods used to compute heat flow. Students familiar with numerical methods will find that the one-dimensional case is straightforward.

Bruce, P. M. and Huppert, H. E. 1990. Solidification and melting along dykes by the laminar flow of basaltic magma. In Ryan, M. P., ed., *Magma transport and storage.* Chichester: Wiley, pp. 87–101. This article forms the basis for Example Problem 9.4.3, and presents the idea that a critical dike width determines whether dikes are likely to undergo narrowing (solidification onto walls) or widening (melting of walls). The article also discusses how these alternate behaviors may influence dike evolution in three dimensions.

Carslaw, H. S. and Jaeger, J. C. 1959. *Conduction of heat in solids,* 2nd ed. Oxford: Oxford University Press, 510 pp. This classic text is a compendium of analytical solutions to heat flow problems for a wide variety of geometrical configurations and initial and boundary conditions.

Hubbert, M. K. 1940. The theory of groundwater motion. *Journal of Geology* 48:785-944. This classic paper formalizes the idea of groundwater potential, based on considerations of conservation of energy, which has become the cornerstone of groundwater hydraulics.

Paterson, W. S. B. 1981. *The physics of glaciers,* 2nd ed. Oxford: Pergamon, 380 pp. Chapter 10 covers heat flow in glaciers, including the idea that ancient low-frequency temperature signals are preserved as relicts in modern distributions of temperature.

Ryan, M. P., ed. 1990. *Magma transport and storage.* Chichester: Wiley, 420 pp. This well-edited book is a collection of articles given to recent advances in the physics of magmatic processes. Topics include magma generation, storage, and flow, and volcanic eruption dynamics. The articles offer a nice balance of theoretical, experimental, and field studies, and the accompanying color photographs are spectacular.

U. S. Geological Survey, 1977. National Handbook of Recommended Methods for Water-Data Acquisition. Reston, Va. Chapter 1 describes methods and instruments used in measuring flow velocities in open channels, including the principle of impact-pressure devices.

# CHAPTER 10

# Inviscid Flows

This chapter covers an important step toward our development of dynamical equations of fluid motion. Herein we will develop explicit expressions for the forces that produce the fluid accelerations that we described kinematically in Chapter 7. In particular, we will consider the behavior of inviscid fluids. Viscous forces therefore are not involved; accelerations are wholly due to body forces and normal surface forces associated with fluid pressure. The results of our development are *Euler's equations*, or the momentum equations for inviscid flow.

One consequence of the inviscid assumption is that slip flow may occur at real boundaries, in contrast to the no-slip condition that occurs with real fluids. This is unrealistic for the viscous flows of interest in many geological problems. Nonetheless, situations exist in which viscous fluids can be treated as inviscid. Examples include fluids having small viscosity, and flows far from boundaries. The study of inviscid flow therefore is justified in its own right. A particularly important example involves the consideration of how velocity and pressure vary along a streamline, which leads to *Bernoulli's equation*.

## 10.1 EULER'S EQUATIONS

Consider a rectangular control volume with edges of length $dx$, $dy$, and $dz$ embedded within a local Cartesian coordinate system (Figure 10.1). This local system has an arbitrary orientation with respect to the Earth coordinate system; the $x$-axis is inclined at an angle $\alpha$ measured from the horizontal. Acceleration due to gravity $g$ acts vertically, and the centroid of the control volume is at height $h$ above a horizontal datum. The height $h$ provides a measure of the position of the fluid within the gravitational field. Consider, now, forces acting on the control volume parallel to the $x$-axis.

The weight $W$ of fluid within the control volume possesses a component $W_x$ parallel to the $x$-axis:

$$W_x = -\rho g \sin \alpha \, dx \, dy \, dz \tag{10.1}$$

where $\rho$ is the fluid density, and the negative sign indicates that $W_x$ acts in the

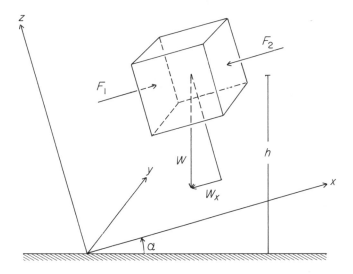

**Figure 10.1** Definition diagram for forces acting on an elementary control volume.

direction of negative $x$. On geometrical grounds, $\sin\alpha = \partial h/\partial x$, so

$$W_x = -\rho g \frac{\partial h}{\partial x} \, dx \, dy \, dz \tag{10.2}$$

As in our treatment of fluid statics (Chapter 6), we may identify the normal stresses acting on the faces of the control volume with the ordinary scalar, fluid pressure $p$. (This generally will not be the case when viscous forces are present; we will reexamine the relation between normal stresses and pressure when we distinguish between the *mechanical pressure* and the *thermodynamical pressure* in Chapter 12.) Then, setting fluid pressure $p = p_0$ at the centroid, surface (normal) forces $F_1$ and $F_2$ associated with fluid pressure acting on faces $dy \, dz$ are obtained as first-order Taylor expansions about the centroid:

$$F_1 = \left(p_0 - \frac{\partial p}{\partial x}\frac{dx}{2}\right) dy \, dz \tag{10.3}$$

$$F_2 = -\left(p_0 + \frac{\partial p}{\partial x}\frac{dx}{2}\right) dy \, dz \tag{10.4}$$

where the negative sign indicates that $F_2$ acts in the direction of negative $x$. The net component force $F_x$ acting on the control volume parallel to the $x$-axis is $F_x = F_1 + F_2 + W_x$. From (10.2), (10.3), and (10.4),

$$F_x = -\frac{\partial p}{\partial x} \, dx \, dy \, dz - \rho g \frac{\partial h}{\partial x} \, dx \, dy \, dz \tag{10.5}$$

From Newton's second law,

$$F_x = m a_x = \rho a_x \, dx \, dy \, dz \tag{10.6}$$

where $a_x$ is the component of acceleration of the fluid mass $m$ parallel to the $x$-axis.

Comparing (10.5) with (10.6),

$$a_x = -\frac{1}{\rho}\frac{\partial p}{\partial x} - g\frac{\partial h}{\partial x} \tag{10.7}$$

This equals the component of acceleration $a_x$ defined by (7.19) in Chapter 7. Using similar developments for $a_y$ and $a_z$, and the kinematical expressions (7.19), (7.20), and (7.21),

$$u\frac{\partial u}{\partial x} + v\frac{\partial u}{\partial y} + w\frac{\partial u}{\partial z} + \frac{\partial u}{\partial t} = -\frac{1}{\rho}\frac{\partial p}{\partial x} - g\frac{\partial h}{\partial x} \tag{10.8}$$

$$u\frac{\partial v}{\partial x} + v\frac{\partial v}{\partial y} + w\frac{\partial v}{\partial z} + \frac{\partial v}{\partial t} = -\frac{1}{\rho}\frac{\partial p}{\partial y} - g\frac{\partial h}{\partial y} \tag{10.9}$$

$$u\frac{\partial w}{\partial x} + v\frac{\partial w}{\partial y} + w\frac{\partial w}{\partial z} + \frac{\partial w}{\partial t} = -\frac{1}{\rho}\frac{\partial p}{\partial z} - g\frac{\partial h}{\partial z} \tag{10.10}$$

These are referred to as Euler's equations, or the *momentum equations,* for inviscid flow. These dynamical equations indicate that each component of fluid acceleration arises from a pressure gradient and a potential energy gradient (gravitational potential) acting in that component direction.

Recall that these equations were obtained by considering conservation of mechanical energy, (9.26), (9.27), and (9.28), in the case of an ideal fluid (incompressible and inviscid). Making use of the substantive derivative (Chapter 7), Euler's equations are written in vector form as

$$\frac{D\mathbf{u}}{Dt} = -\frac{1}{\rho}\nabla p - g\nabla h \tag{10.11}$$

where $\mathbf{u} = \langle u,v,w\rangle$.

The equations above pertain to an arbitrary orientation of the $xyz$-coordinate axes. For the special case where $z$ is vertical, $\partial h/\partial x = \partial h/\partial y = 0$, and $\partial h/\partial z = 1$. The component forms of Euler's equations then become

$$u\frac{\partial u}{\partial x} + v\frac{\partial u}{\partial y} + w\frac{\partial u}{\partial z} + \frac{\partial u}{\partial t} = -\frac{1}{\rho}\frac{\partial p}{\partial x} \tag{10.12}$$

$$u\frac{\partial v}{\partial x} + v\frac{\partial v}{\partial y} + w\frac{\partial v}{\partial z} + \frac{\partial v}{\partial t} = -\frac{1}{\rho}\frac{\partial p}{\partial y} \tag{10.13}$$

$$u\frac{\partial w}{\partial x} + v\frac{\partial w}{\partial y} + w\frac{\partial w}{\partial z} + \frac{\partial w}{\partial t} = -\frac{1}{\rho}\frac{\partial p}{\partial z} - g \tag{10.14}$$

and the vector form becomes

$$\frac{D\mathbf{u}}{Dt} = -\frac{1}{\rho}\nabla p - \mathbf{k}g \tag{10.15}$$

where $\mathbf{k}$ is the unit vector $\langle 0,0,1\rangle$.

Let us briefly return to the energy equations (9.21) and (9.22). Observe that the second term on the left side of (9.21) equals the second and third terms on the right side according to (10.11). Using the relation $dU = c_V\,dT$ for an inviscid, compressible gas flow,

$$\rho c_V \frac{DT}{Dt} = K_T \nabla^2 T - p \nabla \cdot \mathbf{u} \tag{10.16}$$

where $c_V$ is the specific heat at constant volume. For an incompressible fluid, the last term in (10.16) equals zero, $c_V$ is replaced with $c$, and

$$\frac{DT}{Dt} = \kappa \nabla^2 T \tag{10.17}$$

where $\kappa = K_T / \rho c$ is the *thermal diffusivity*.

## 10.2 BERNOULLI'S EQUATION

Envision a curvilinear *snz*-coordinate system embedded within an inviscid flow such that the *s*-axis coincides with a streamline. By the definition of a streamline (Section 7.1), $v_n = w_z = 0$. It is left as an exercise to show that, by assuming steady flow and using a development virtually identical to that above, the equation of motion along the streamline is

$$u_s \frac{\partial u_s}{\partial s} = -\frac{1}{\rho} \frac{\partial p}{\partial s} - g \frac{\partial h}{\partial s} \tag{10.18}$$

Observe that the first term $u_s \partial u_s / \partial s$ may be written as $\frac{1}{2} \partial (u_s^2)/\partial s$. Then assuming an incompressible fluid (constant $\rho$), (10.18) can be written as

$$\frac{\partial}{\partial s}\left(\frac{u_s^2}{2} + \frac{p}{\rho} + gh\right) = 0 \tag{10.19}$$

This implies that

$$\frac{u_s^2}{2} + \frac{p}{\rho} + gh = \Phi \tag{10.20}$$

where the quantity $\Phi$ is constant along the streamline. This is referred to as *Bernoulli's equation*.

Each term in (10.20) has the dimensions of mechanical energy per unit mass; that in $u_s^2$ is a kinetic energy term, that in $h$ is a potential energy term, and that in $p$ describes work performed against pressure. Thus $\Phi$ is identical to Hubbert's potential (Section 9.2). Bernoulli's equation therefore is a statement of conservation of mechanical energy along a streamline, in the following sense. Flow in the direction of an adverse pressure gradient performs work against the gradient and therefore loses kinetic energy and decelerates. Likewise, flow in a direction of increasing elevation involves a conversion of kinetic energy to potential energy and therefore it decelerates. Conversely, flow in the direction of a negative pressure gradient or decreasing elevation extracts energy from these forms and accelerates.

Neglecting gravity momentarily, (10.20) may be written as

$$\frac{1}{2}\rho u_s^2 + p = p_T \tag{10.21}$$

where $p_T = \rho \Phi$ is referred to as the *total pressure* or *stagnation pressure*, and $\frac{1}{2}\rho u_s^2$ is referred to as the *dynamic pressure*. Also note that if $h$ in (10.20) is replaced by

$z$, then dividing (10.20) by $g$ leads to an expression for total head, $h_T = \Phi/g = h + u_s^2/2g$, where now $h = z + \psi$ is the *hydraulic head* as defined with reference to Hubbert's potential in Section 9.2, and $u_s^2/2g$ is the *dynamic head*.

Bernoulli's equation in the form of (10.20) or (10.21) is valid for incompressible liquid flows, or subsonic gaseous flows where effects of compressibility are small such that changes in density are negligible. The pressure $p$ is referred to as the *static pressure,* although it is not the same as the hydrostatic pressure (Chapter 6) inasmuch as it responds to changes in velocity, but is not a thermodynamic variable. In the case of supersonic gaseous flows, where the Mach number $M > 1$, the effect of compressibility becomes important. Then (10.20) must be modified to take into account work performed in fluid compression, and it takes the form of (9.43), as derived by Hubbert:

$$\frac{u_s^2}{2} + \int \frac{1}{\rho}\,dp + gh = \Phi \qquad (10.22)$$

Note that (10.22) is valid for isothermal or adiabatic conditions. Here, the pressure $p$ takes on a thermodynamic role as well as a dynamic one. In using (10.22), an equation of state relating density and pressure must be specified. In addition, the relevant speed of sound $c$ in computing the Mach number ($M = u_s/c$) is that for the flowing fluid. For example, a thermal geyser jet can involve supersonic flow, not because of high flow velocities, but because the speed of sound in a steam–water mixture can be low. Similar circumstances can occur with volcanic jets. These ideas are examined in Kieffer and Sturtevant (1984).

## 10.3 EXAMPLE PROBLEMS

### 10.3.1 Pressure conditions in a Venturi flow

Let us reexamine the steady Venturi flow of Example Problem 7.3.2 assuming, for simplicity, that flow is horizontal. We will consider flow along a streamline near the center of the conduit, away from the boundaries, and assume that flow here approximates the behavior of an inviscid fluid. Indeed, shear stresses vanish at the centerline since $\partial u/\partial y = 0$, due to the symmetry of the flow. Alternatively, we may consider a high speed, but subsonic, flow such that the boundary layers are compressed and the streamwise velocity $u$ is approximately uniform across the conduit. With $x$ denoting the streamwise coordinate, the relevant form of Bernoulli's equation is

$$\frac{1}{2}\rho u^2 + p = p_T \qquad (10.23)$$

Since $p_T$ is constant along a streamline, this implies that

$$\frac{1}{2}\rho u_1^2 + p_1 = \frac{1}{2}\rho u_2^2 + p_2 \qquad (10.24)$$

where the subscripts 1 and 2, respectively, refer to quantities at the upstream and constricted parts of the conduit.

Because the discharge through the conduit is constant, (10.24) implies that flow accelerates into the constriction such that $u_2 > u_1$, and thus $p_2 < p_1$. Likewise,

flow decelerates out of the constriction, whereas the pressure increases. It is easy
to construct a simple lab apparatus to demonstrate this by placing manometers at
the three sections of a Venturi conduit. Fluid levels in the manometers, which
measure local pressure conditions, reflect the predicted changes in $p$ along the
conduit.

If flow is truly inviscid, the pressure at the downstream section should return to
a value equal to that at the upstream section. In actuality, however, this does not
occur. Viscous losses occur due to boundary effects. Since this consists of work
performed against friction, energy of the flow is consumed and dissipated as heat,
and the pressure at the downstream section is lower than at the upstream section.
This suggests that in practice an "energy loss" term can be added to (10.24),

$$\frac{1}{2}\rho u_1^2 + p_1 = \frac{1}{2}\rho u_2^2 + p_2 + \Delta E \tag{10.25}$$

such that $\Delta E$ is a measure of the viscous dissipation within the conduit, and is a
function of both fluid properties and Venturi geometry.

Let us now write the differential form of Bernoulli's equation:

$$u\frac{\partial u}{\partial x} = -\frac{1}{\rho}\frac{\partial p}{\partial x} \tag{10.26}$$

This suggests that the term in $u$, which equals the streamwise acceleration $a_x$, must
be of the same order as the term in $p$. This latter term is of order $-(p_2 - p_1)/\rho L$,
where $L$ is the distance over which the aperture changes. Rewriting (10.24) and
dividing by $L$,

$$O(a_x) = \frac{u_2^2 - u_1^2}{2L} = -\frac{p_2 - p_1}{\rho L} \tag{10.27}$$

which is equivalent to (7.29), obtained with regard to viscous flow along the centerline of a Venturi conduit. This provides some justification for the assumption that
flow of a real fluid along a streamline near the center of the conduit approximates
the behavior of an inviscid fluid.

### 10.3.2 Near-vent velocity of a subsonic geyser eruption

Let us reconsider the geyser problem briefly introduced at the end of Example Problem 9.3.1. Assume for simplicity that the steam–water mixture exits the geyser vent
as a jet with subsonic velocity ($\rho \approx$ constant). Then the pressure within this mixture throughout the erupting column is nearly equal to atmospheric pressure $p_0$.
Further assume that the height of geyser play is sufficiently small that $p_0$ is essentially constant. (In contrast, this is not necessarily true for large volcanic eruptions
where erupted material is lofted high into the atmosphere.) With $w$ denoting the
velocity along a vertical streamline within the core of the geyser jet, the relevant
form of Bernoulli's equation is

$$\frac{w^2}{2} + \frac{p}{\rho} + gh = \text{const} \tag{10.28}$$

With $p \approx p_0 =$ constant, this reduces to

$$\frac{w^2}{2} + gh = \text{const} \tag{10.29}$$

This implies that

$$\frac{1}{2}w_1^2 + gh_1 = \frac{1}{2}w_2^2 + gh_2 \tag{10.30}$$

where the subscripts 1 and 2, respectively, denote positions near the geyser vent and at the top of the erupting fluid column. Setting the elevation datum $h = 0 = h_1$ at the geyser vent, and noting that $w$ decreases to $w_2 = 0$ at the top of the column, (10.30) leads to

$$w_1 = \sqrt{2gh_2} \tag{10.31}$$

We thus retrieve the simple result suggested in Example Problem 9.3.1, that the exit velocity $w_1$ can be roughly estimated from the height of geyser play $h_2$. Note, however, that flow may actually be supersonic depending on the proportions of water and steam in the fluid mixture.

The height $h_2$ of the steam–water eruption column of Old Faithful geyser in Yellowstone National Park, Wyoming, is about 30 m, during a steady-flow interval that lasts about 30 s. Based on (10.31), this suggests a near-vent velocity of about 24 m s$^{-1}$. The discharge of liquid water from Old Faithful during its initial phase of eruption has been estimated to be about 6.8 m$^3$ s$^{-1}$. This is a conservative value of the total fluid discharge since it does not account for the steam fraction. The cross-sectional area of the vent is about 0.88 m$^2$. Thus a conservative estimate of the exit velocity is 7.7 m s$^{-1}$. Both of these estimates of velocity are less than the speed of sound $c$ in the steam–water mixture (about 50 m s$^{-1}$) estimated by Kieffer (1984), suggesting that flow is subsonic. Nonetheless, $c$ can be significantly reduced depending on the steam fraction and associated pressure and temperature conditions.

## 10.3.3 Speed of shallow waves

In studies of near-shore waves in lakes and oceans, a useful parameter is the speed $c$ at which the waves propagate. The motion of these surface waves may be initiated, for example, by stresses that winds impose on the water surface. Once set in motion, the waves then continue to move more or less independently of the wind. The energy to maintain motion resides in the gravitational potential energy of the wave crests; thus these surface waves are one type of *gravity wave*. Here we will consider two-dimensional *shallow waves*. A shallow wave is one in which the depth of liquid $z_*$ is much less than the wavelength $\lambda$. We will further restrict our development to waves whose amplitude $a$ is much smaller than the wavelength (Figure 10.2). To obtain an estimate of wave speed, let us start by developing a general description of wave motion based on kinematics.

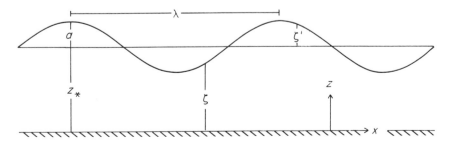

**Figure 10.2** Definition diagram for dynamical description of motion of shallow-water wave.

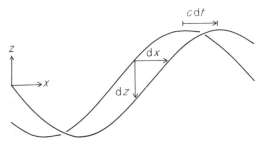

**Figure 10.3** Definition diagram for kinematical description of wave motion.

Suppose that a surface wave is moving in the direction of positive $x$. Now, it is important to concentrate on the motion of the wave and, at least momentarily, not on the motion of fluid particles composing the wave. Thus we consider vertical motions of wave particles, and neglect any horizontal fluid motion. In this view a wave particle is like a particle on a string that has been jerked with a transverse (vertical) motion such that a wave moves along it.

Consider a small interval $dt$, during which the wave travels between two successive positions (Figure 10.3). At specified position $x$ a particle has an instantaneous position $z$ and vertical velocity $w(x, t) = dz/dt$. The slope of the wave at position $x$ is $dz/dx$. During $dt$, the wave moves a small distance $dx = c\,dt$, where $c$ is the wave speed, and the particle at $x$ moves downward a distance $dz$. On geometrical grounds, $dz/dx = -(1/c)dz/dt$, and in the limit as $dt \to 0$,

$$\frac{\partial z}{\partial x} = -\frac{1}{c}\frac{\partial z}{\partial t} \tag{10.32}$$

Alternatively, consider a coordinate system that moves with the wave at the same speed; let $x = \chi + ct$, or $\chi = x - ct$. Here $\chi$ is the moving coordinate position of a specified wave particle whose position is fixed in a stationary coordinate system, and $x$ is the coordinate (with fixed value in the moving system) of a geometrically similar position on the wave, moving at the same speed as the wave. By the chain rule, $\partial z/\partial x = (dz/d\chi)(\partial\chi/\partial x) = dz/d\chi$ since $\partial\chi/\partial x = 1$. Further, $\partial z/\partial t = (dz/d\chi)(\partial\chi/\partial t) = -c\,dz/d\chi$ since $\partial\chi/\partial t = -c$. Equating expressions for $dz/d\chi$ then retrieves (10.32). Similarly, for a wave moving in the direction of negative $x$, we can demonstrate that

$$\frac{\partial z}{\partial x} = \frac{1}{c}\frac{\partial z}{\partial t} \tag{10.33}$$

Equations (10.32) and (10.33) are kinematical expressions of the velocity $\partial z/\partial t$, formulated without regard to any forces involved. Let us next obtain an expression for the acceleration $\partial^2 z/\partial t^2$, with the eventual aim of relating this acceleration to the forces that produce it. Observe that (10.32) states that the derivative of $z$ with respect to $x$ is equal to $-1/c$ times the derivative of $z$ with respect to $t$. Therefore it follows that

$$\frac{\partial}{\partial x}\left(\frac{\partial z}{\partial x}\right) = -\frac{1}{c}\frac{\partial}{\partial t}\left(-\frac{1}{c}\frac{\partial z}{\partial t}\right) \tag{10.34}$$

(Why?) This becomes

$$\frac{\partial^2 z}{\partial x^2} = \frac{1}{c^2}\frac{\partial^2 z}{\partial t^2} \tag{10.35}$$

The same result is obtained starting with (10.33); thus (10.35) is more general than either (10.32) or (10.33) since it is valid for a wave moving in the direction of positive or negative $x$. If a wave is approximately sinusoidal, (10.32) and (10.33) suggest that the largest positive and negative vertical velocities ($\partial z/\partial t$) occur at positions where the water-surface slope ($\partial z/\partial x$) is steepest, halfway between the crest and trough of the wave. Equation (10.35) suggests that the largest positive and negative vertical accelerations ($\partial^2 z/\partial t^2$) occur at positions where the water-surface possesses its greatest curvature (proportional to $\partial^2 z/\partial x^2$), at the crest and trough.

Equation (10.35) is referred to as the one-dimensional *wave equation*, since it appears in many different problems involving wave behavior. In fact, the mathematical description leading to (10.35) can be readily translated to the problem of describing the motion of longitudinal waves where particle displacements are parallel to the wave motion. Shock waves, including sound waves, are an example. As a kinematical expression, however, (10.35) reveals little about the forces producing surface-wave motion. Let us now turn to the dynamical equations.

For two-dimensional shallow waves, the governing equations include the $x$-component and $z$-component of Euler's equations, and the continuity equation:

$$u\frac{\partial u}{\partial x} + w\frac{\partial u}{\partial z} + \frac{\partial u}{\partial t} = -\frac{1}{\rho}\frac{\partial p}{\partial x} \tag{10.36}$$

$$u\frac{\partial w}{\partial x} + w\frac{\partial w}{\partial z} + \frac{\partial w}{\partial t} = -\frac{1}{\rho}\frac{\partial p}{\partial z} - g \tag{10.37}$$

$$\frac{\partial u}{\partial x} + \frac{\partial w}{\partial z} = 0 \tag{10.38}$$

These can be simplifed under certain conditions and assumptions. Let us start by defining a set of characteristic quantities, treating magnitudes without regard to signs. It is convenient to define the interval $t_*$ over which velocities $u$ and $w$ change from zero to their maximum values $u_*$ and $w_*$. If $t_0 = \lambda/c$ is the wave period, then $t_* = t_0/4 = \lambda/4c$. Similarly, the horizontal distance $\lambda_*$ over which $u$ varies from zero to $u_*$ is $\lambda_* = \lambda/4$. Letting $z = \zeta$ at the wave surface, then from (10.32), $w_* = \partial\zeta/\partial t = c\,\partial\zeta/\partial x$. Since $\partial\zeta/\partial x$ is of order $4a/\lambda$, $w_*$ at the surface is of order $4ac/\lambda$. Now, $w$ varies from zero at the bed to its maximum $w_*$ at the surface over the depth $z_*$; thus $\partial w/\partial z$ is of order $w_*/z_* = 4ac/\lambda z_*$. Further, $\partial u/\partial x$ is of order $u_*/\lambda_* = 4u_*/\lambda$. Then based on the continuity equation (10.38), $u_*$ is of order $u_* = ac/z_*$, or $w_* = 4z_*u_*/\lambda$. Therefore, if $4z_* \ll \lambda$, $w_* \ll u_*$, and it follows that $\partial w/\partial t \ll \partial u/\partial t$. Thus, under these conditions we may neglect the local term $\partial w/\partial t$ relative to $\partial u/\partial t$ in (10.36) and (10.37). Herein is the significance of the assumption of a shallow wave—that the depth is much less than the wavelength.

Now consider the acceleration terms in (10.36). Using characteristic quantities defined above, the term $u\,\partial u/\partial x$ is of order $u_*^2/\lambda_* = 4a^2c^2/\lambda z_*^2$, the term $w\,\partial u/\partial z$ is of order $w_*u_*/z_* = 4a^2c^2/\lambda z_*^2$, and the local term $\partial u/\partial t$ is of order $u_*/t_* = 4ac^2/\lambda z_*$. If $a \ll \lambda$, it follows that the nonlinear (convective) terms are much smaller than the local term. For example, if $a$ is of order unity, $a^2$ is much less than unity. Moreover, since $z_*$ is greater than unity, division by $z_*^2$ further reduces the magnitude of the nonlinear terms. A similar argument leads to the conclusion that nonlinear terms in (10.37) are much smaller than the local term. Herein is the significance of the assumption that the wave amplitude is small relative to the wavelength.

With these simplifications of (10.36) and (10.37), the approximate equations of motion are

$$\frac{\partial u}{\partial t} = -\frac{1}{\rho}\frac{\partial p}{\partial x} \qquad (10.39)$$

$$0 = -\frac{1}{\rho}\frac{\partial p}{\partial z} - g \qquad (10.40)$$

Observe that the second of these, when rearranged, is equivalent to the equation of hydrostatics: $\partial p/\partial z = -\rho g$. Thus the assumption of negligible vertical accelerations, leading to (10.40), is equivalent to assuming that the fluid is under hydrostatic conditions. The vertical position of the surface $\zeta = z_* + \zeta'$, where $\zeta'$ is a deviation from $z_*$ (Figure 10.2). The pressure at the surface is equal to atmospheric pressure $p_0$, so the fluid pressure $p = p_0 + \rho g(\zeta - z) = p_0 + \rho g(z_* - z + \zeta')$. Since $\zeta'$ varies with $x$ (but not with $z_*$ or $z$), $\partial p/\partial x = \rho g \partial \zeta'/\partial x$, so (10.39) becomes

$$\frac{\partial u}{\partial t} = -g\frac{\partial \zeta'}{\partial x} \qquad (10.41)$$

Differentiating this with respect to $x$,

$$\frac{\partial}{\partial x}\left(\frac{\partial u}{\partial t}\right) = -g\frac{\partial^2 \zeta'}{\partial x^2} \qquad (10.42)$$

which will be used below.

Consider the continuity equation (10.38), and assume that the velocity gradient $\partial u/\partial x$ is essentially independent of $z$. (How is this justified? Consider the distance over which $u$ varies.) Then integrating with respect to $z$ between the bed ($z = 0$) and $z_*$,

$$w = -\frac{\partial u}{\partial x}\int_0^{z_*} \mathrm{d}z = -z_*\frac{\partial u}{\partial x} \qquad (10.43)$$

At the surface, $w = \partial \zeta'/\partial t$; thus

$$w_{\zeta'} = \frac{\partial \zeta'}{\partial t} = -z_*\frac{\partial u}{\partial x} \qquad (10.44)$$

This has a simple physical interpretation: a nonzero gradient $\partial u/\partial x$ implies that a net quantity of mass is imported to (or exported from) a fixed control volume with edge $\mathrm{d}x$ during a small interval $\mathrm{d}t$. By continuity, this must be compensated by a change in the height of the water surface during $\mathrm{d}t$. Now, differentiating (10.44) with respect to $t$,

$$\frac{\partial^2 \zeta'}{\partial t^2} = -z_*\frac{\partial}{\partial t}\left(\frac{\partial u}{\partial x}\right) \qquad (10.45)$$

Solving for $(\partial/\partial t)(\partial u/\partial x)$, noting that the order of partial differentiation with respect to $x$ and $t$ is immaterial, then equating (10.42) and (10.45),

$$\frac{\partial^2 \zeta'}{\partial x^2} = \frac{1}{g z_*}\frac{\partial^2 \zeta'}{\partial t^2} \qquad (10.46)$$

This has the form of the wave equation (10.35) obtained on kinematical grounds above. Comparing the constants in these equations, evidently

$$c = \sqrt{gz_*} \qquad (10.47)$$

So the wave speed is equal to the square root of the product of $g$ and the average water depth $z_*$. This is qualitatively consistent with the simple observation that the speed of ocean waves decreases as they shoal toward the beach—before they reach the surf zone, where other factors, including bed friction, become important. A complete analysis of more general validity must involve nonlinear terms and effects of friction.

## 10.4 READING

Batchelor, G. K. 1967. *An introduction to fluid dynamics.* Cambridge: Cambridge University Press, 615 pp. This text provides a general derivation of Bernoulli's equation (pp. 156–64), including a treatment of the pressure term for the case of compressible flow, based on the second law of thermodynamics.

Kieffer, S. W. 1977. Sound speed in liquid-gas mixtures: water-air and water-steam. *Journal of Geophysical Research* 82:2895–904. This paper develops, on thermodynamical grounds, expressions for the speed of sound in two-phase fluids with reference to geyser and volcanic systems. In mixtures such as water and air or water and steam, the speed of sound varies with the proportions of the phases, temperature and pressure, and the extent of thermodynamical equilibration between phases. The speed of sound in two-phase mixtures can be very different from the speed of sound in each phase taken separately.

Kieffer, S. W. 1981. Blast dynamics at Mount St Helens on 18 May 1980. *Nature* 291:568–70. This article examines the energetics of the 1980 eruption of Mount Saint Helens in part using Bernoulli's equation, and decribes why much of the flow involved supersonic velocities. To obtain an equation of state for use in the pressure term, it treats the erupted material, gases and cntrained solid particles, as a pseudo-gas characterized by the ideal gas law. This involves taking into account the transfer of heat between the gaseous and solid phases.

Kieffer, S. W. 1984. Seismicity at Old Faithful geyser: an isolated source of geothermal noise and possible analogue of volcanic seismicity. *Journal of Volcanology and Geothermal Research* 22:59–95. This paper provides a clear description of the several phases of eruption of Old Faithful geyser. An interesting aspect of this involves using seismicity associated with shock waves released during collapse of bubbles to infer aspects of eruption dynamics. The paper also explains the pulsed behavior of the eruption jet in terms of resonance of shock waves over the subsurface length of the geyser fluid column—in the same sense that an ordinary cylinder closed at the bottom resonates a specific frequency of sound wave.

Kieffer, S. W. and Sturtevant, B. 1984. Laboratory studies of volcanic jets. *Journal of Geophysical Research* 89(B10):8253–68. This article provides a systematic discussion of the use of Bernoulli's equation in describing geyser and volcanic jets, and includes a description of conditions that lead to high subsonic and supersonic flows when the jet fluid consists of a mixture of phases, steam and water, for example, that possesses a low speed of sound. The article examines the behavior of a series of laboratory-scale jets (including a delightful set of photographs) in the context of volcanic eruptions.

Young, H. D. 1964. *Fundamentals of mechanics and heat.* New York: McGraw-Hill, 638 pp. This text provides a lucid, introductory treatment of the kinematics and dynamics of waves (Chapter 20). I recommend it as "essential" reading for students who are beginning their study of waves.

# CHAPTER 11

# Vorticity and Fluid Strain

Many flows involve a sense of rotation. Clear examples include cyclones, whirlpools, and eddies. Less apparent, perhaps, is the interesting result that a one-dimensional shearing flow—for example, Couette flow—possesses a rotational component. As we will see below, the idea of fluid *vorticity* provides a way to characterize the rotational qualities of such flows. In addition, our treatment of vorticity will provide a way to distinguish between *simple shear* and *pure shear* of a fluid. Because shearing motions involve viscous dissipation of energy in real fluids, our descriptions of vorticity and shear will form an important part of the development of dynamical equations for flows that involve viscous forces (Chapter 12). The idea of vorticity also is useful in visualizing the onset of flow separation (Example Problem 11.4.2), very viscous flow behavior (Example Problem 12.6.5), and certain aspects of turbulence (Chapters 14 and 15). Beyond this, our treatment of vorticity is not emphasized.

## 11.1 FLOW WITH ROTATION

Let us envision a *vorticity meter* made of two small orthogonal vanes, with the end of one vane marked for easy identification (Figure 11.1). (Such meters can readily be constructed and used as described next.) Consider placing this meter at some position within a fluid that is rotating like a rigid body (Figure 11.2). The vorticity meter in this case rotates with the fluid in such a way that its orientation relative to the axis of rotation of the fluid remains fixed. As we will see below, the fluid possesses a definite vorticity that is reflected in the observation that the vorticity meter rotates with respect to its own axis. In this regard, we also may observe that the angular velocity of this *local* rotation of the meter is the same regardless of its distance from the fluid axis.

Now consider a one-dimensional (Couette) shear flow (Figure 11.3). A vorticity meter placed at any position within this flow also rotates about its center due to the streamwise velocity differential over the span of the meter. Judging from the behavior of the meter, this flow also possesses a definite vorticity. We also may envision that the rate of rotation varies directly with the velocity gradient $du/dy$.

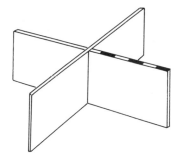

**Figure 11.1**   Vorticity meter.

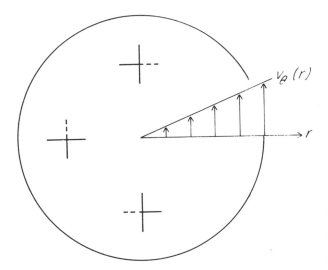

**Figure 11.2** Tangential velocity distribution $v_\theta(r)$, and
successive positions of a vorticity meter within a fluid that is
rotating like a rigid body.

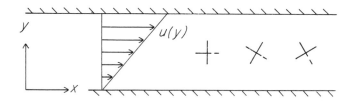

**Figure 11.3**   Velocity distribution $u(y)$, and successive positions of a
vorticity meter within Couette flow.

Finally consider a *free vortex* (Figure 11.4), of which a whirlpool is an example.
A vorticity meter placed within the vortex revolves around its core, yet the meter
does not rotate about its own axis. Flow within a free vortex therefore possesses
zero vorticity, and is said to be *irrotational*. (This is not actually true near the vortex
core.) Such irrotational behavior can be observed in other settings. For example, a

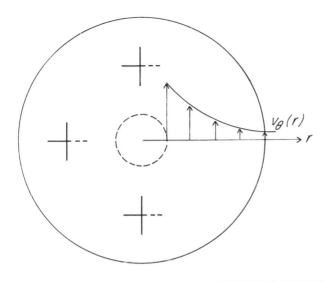

**Figure 11.4** Tangential velocity distributions $v_\theta(r)$ outside the core (*dashed line*) of an irrotational vortex, and successive positions of a vorticity meter within the vortex.

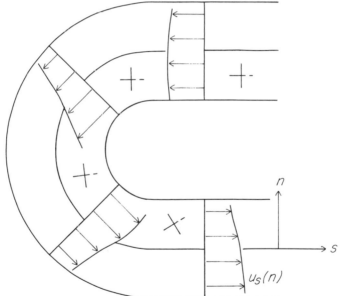

**Figure 11.5** Schematic diagram of near-surface streamwise velocity distributions $u_s(n)$ in a curved, open channel in absence of bed topography, and successive positions of a vorticity meter showing approximate irrotational behavior within upstream part of bend; adapted from Grishanin (1979).

vorticity meter placed on the water surface within a curved channel whose bed is flat exhibits very little rotation about its axis as it moves with the flow through the channel (Figure 11.5).

Let us now examine the idea of vorticity more closely, then return to these three examples once we have defined vorticity in mathematical terms.

## 11.2 VORTICITY

Consider a local Cartesian coordinate system positioned within a flow such that the flow is one-dimensional in the direction of positive $x$ (Figure 11.6). The flow velocity $u$ varies as a function $u(y)$ of position $y$. Assume that the region of flow being considered is sufficiently small that the velocity gradient $\partial u/\partial y$ may be treated as constant near the origin of the coordinate system. The function $u(y)$ is then given by the formula for a straight line,

$$u(y) = u_0 + \frac{\partial u}{\partial y} y \tag{11.1}$$

where $u_0$ denotes the fluid velocity at the origin. This implies that the local coordinate system we have selected is moving with velocity $u_0$ in the direction of positive $x$.

Consider fluid that is instantaneously at a point P whose polar coordinates are $r$ and $\theta$ relative to the origin of the moving coordinate system (Figure 11.6). The fluid velocity $u(y)$ at P has a tangential component (normal to the direction of $r$) equal in magnitude to $u(y) \sin \theta$. This implies that fluid at P possesses an angular velocity $\omega$ about the local origin. Recall that

$$\omega = \frac{d\theta}{dt} = \frac{1}{r} v_\theta \tag{11.2}$$

where $v_\theta$ is the tangential velocity. Since the origin moves with velocity $u_0$, the tangential fluid velocity at P relative to the origin is $v_\theta = [u(y) - u_0] \sin \theta$. In polar coordinates, the angular velocity of fluid at P, using (11.1), is therefore

$$\omega(r, \theta) = -\frac{1}{r}[u(y) - u_0] \sin \theta \tag{11.3}$$

$$= -\frac{1}{r}\frac{\partial u}{\partial y} y \sin \theta \tag{11.4}$$

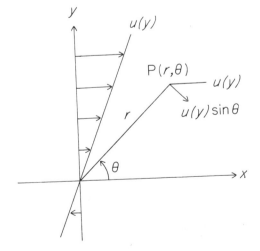

**Figure 11.6**  Definition diagram for vorticity associated with a one-dimensional flow $u(y)$ with velocity gradient $\partial u/\partial y$; adapted from Williams and Elder (1989).

where the negative sign arises from the convention that clockwise motion is assigned a negative value, whereas counterclockwise motion is assigned a positive value. Using the fact that $y = r \sin \theta$, (11.4) becomes

$$\omega(\theta) = -\frac{\partial u}{\partial y} \sin^2 \theta \tag{11.5}$$

where omission of $r$ in the functional notation indicates that $\omega$ is independent of $r$. This function indicates that $\omega$ has its greatest magnitude when $\theta = \pi/2$ (along the $y$-axis), and $\omega$ equals zero when $\theta = 0$ (along the $x$-axis).

Let us now characterize the rotation of fluid about the origin. Here it is useful to define an ensemble average $\omega_1$ of $\omega(\theta)$, which essentially defines the average angular velocity of fluid in the vicinity of the origin over all possible angles $\theta$:

$$\omega_1 = -\frac{1}{2\pi} \int_0^{2\pi} \omega(\theta) d\theta \tag{11.6}$$

Substituting (11.5) for $\omega(\theta)$ and evaluating the integral,

$$\omega_1 = -\frac{1}{2\pi} \frac{\partial u}{\partial y} \int_0^{2\pi} \sin^2 \theta d\theta = -\frac{1}{2} \frac{\partial u}{\partial y} \tag{11.7}$$

We may envision this averaging process as being applied to an infinitesimal area about the origin such that, in the limit as this area goes to zero, (11.7) defines the angular velocity of the fluid at a point.

An expression for the angular velocity associated with one-dimensional flow parallel to the $y$-axis can be developed in a similar manner. The velocity $v$ is then specified as a function $v(x)$ of position $x$, with gradient $\partial v/\partial x$ (Figure 11.7). It is left as an exercise to demonstrate that the ensemble average $\omega_2$ analogous to (11.7) is given by

$$\omega_2 = \frac{1}{2} \frac{\partial v}{\partial x} \tag{11.8}$$

which is, in this example, a negative quantity since the sign of $\partial v/\partial x$ is negative (Figure 11.7).

Both senses of shear flow (Figures 11.6 and 11.7) lead to the same sense of fluid rotation. For a fluid with both components of shear, the angular velocity $\omega_z$ is the sum of the two components of angular velocity:

$$\omega_z = \frac{1}{2} \left( \frac{\partial v}{\partial x} - \frac{\partial u}{\partial y} \right) \tag{11.9}$$

where the subscript $z$ denotes rotation about the $z$-axis. Similar expressions are obtained for $\omega_x$ and $\omega_y$:

$$\omega_x = \frac{1}{2} \left( \frac{\partial w}{\partial y} - \frac{\partial v}{\partial z} \right) \tag{11.10}$$

$$\omega_y = \frac{1}{2} \left( \frac{\partial u}{\partial z} - \frac{\partial w}{\partial x} \right) \tag{11.11}$$

Note that (11.9), (11.10), and (11.11) define the three components of an angular velocity vector. Here it is useful to introduce a mathematical quantity, the *curl* of

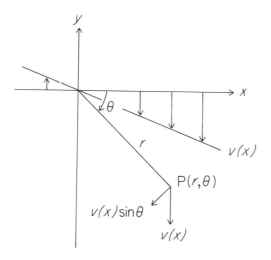

**Figure 11.7** Definition diagram for vorticity associated with a one-dimensional flow $v(x)$ with velocity gradient $\partial v/\partial x$.

the velocity **u**. We start by defining the cross product of two vectors $\mathbf{U} = U_1\mathbf{i} + U_2\mathbf{j} + U_3\mathbf{k}$ and $\mathbf{V} = V_1\mathbf{i} + V_2\mathbf{j} + V_3\mathbf{k}$ in terms of a determinant expansion:

$$\mathbf{U} \times \mathbf{V} = \begin{vmatrix} \mathbf{i} & \mathbf{j} & \mathbf{k} \\ U_1 & U_2 & U_3 \\ V_1 & V_2 & V_3 \end{vmatrix}$$

$$= (U_2V_3 - V_2U_3)\mathbf{i} + (U_3V_1 - V_3U_1)\mathbf{j} + (U_1V_2 - V_1U_2)\mathbf{k} \qquad (11.12)$$

Suppose that the vector function $\mathbf{V}(x, y, z) = V_1(x, y, z)\mathbf{i} + V_2(x, y, z)\mathbf{j} + V_3(x, y, z)$, where $V_1$, $V_2$, and $V_3$ have partial derivatives in some region. Then the curl of **V**, denoted by curl **V** or $\nabla \times \mathbf{V}$, is obtained as the formal cross product:

$$\nabla \times \mathbf{V} = \begin{vmatrix} \mathbf{i} & \mathbf{j} & \mathbf{k} \\ \dfrac{\partial}{\partial x} & \dfrac{\partial}{\partial y} & \dfrac{\partial}{\partial z} \\ V_1 & V_2 & V_3 \end{vmatrix} = \left(\frac{\partial V_3}{\partial y} - \frac{\partial V_2}{\partial z}\right)\mathbf{i} + \left(\frac{\partial V_1}{\partial z} - \frac{\partial V_3}{\partial x}\right)\mathbf{j} + \left(\frac{\partial V_2}{\partial x} - \frac{\partial V_1}{\partial y}\right)\mathbf{k}$$

$$(11.13)$$

The curl of the velocity **u** is therefore

$$\text{curl } \mathbf{u} = \nabla \times \mathbf{u} = \left(\frac{\partial w}{\partial y} - \frac{\partial v}{\partial z}\right)\mathbf{i} + \left(\frac{\partial u}{\partial z} - \frac{\partial w}{\partial x}\right)\mathbf{j} + \left(\frac{\partial v}{\partial x} - \frac{\partial u}{\partial y}\right)\mathbf{k} \qquad (11.14)$$

Comparing the components of (11.14) with (11.9), (11.10), and (11.11), it is conventional to define the *vorticity* **ω** of fluid motion as twice the angular velocity of fluid rotation; that is **ω** = curl **u**. Vorticity is therefore a vector quantity that may be defined at all coordinate positions in a flow, just as we may assign a velocity to all positions in a flow. Denoting $\boldsymbol{\omega} = \langle \xi, \eta, \zeta \rangle = \mathbf{i}\xi + \mathbf{j}\eta + \mathbf{k}\zeta = \text{curl } \mathbf{u}$,

$$\xi = \frac{\partial w}{\partial y} - \frac{\partial v}{\partial z} \qquad (11.15)$$

$$\eta = \frac{\partial u}{\partial z} - \frac{\partial w}{\partial x} \qquad (11.16)$$

$$\zeta = \frac{\partial v}{\partial x} - \frac{\partial u}{\partial y} \qquad (11.17)$$

To examine the physical significance of vorticity, let us reconsider the three examples of flow with rotation introduced in Section 11.1. In each of these, only rotation about the $z$-axis is possible, and therefore we will consider only the possibility that the $z$-component of vorticity is nonzero.

Consider a fluid that is rotating like a rigid body with angular velocity $\omega$. Positioning the $z$-axis at the center of rotation, the tangential velocity $v_\theta$ at a radial distance $r$ equals $r\omega$. The (Cartesian) component velocities $u$ and $v$ of $v_\theta$ are

$$u = -v_\theta \sin \theta = -r\omega \sin \theta \qquad (11.18)$$

$$v = v_\theta \cos \theta = r\omega \cos \theta \qquad (11.19)$$

Noting that $x = r\cos \theta$ and $y = r \sin \theta$, substitution gives $u = -\omega y$ and $v = \omega x$. Then,

$$\frac{\partial u}{\partial y} = -\omega; \qquad \frac{\partial v}{\partial x} = \omega \qquad (11.20)$$

It follows from (11.17) that the component of vorticity $\zeta$ equals twice the angular velocity; that is $\zeta = 2\omega$.

In the case of the simple shear flow illustrated in Figure 11.3, the velocity field is specified by

$$u = \frac{U}{b} y; \qquad v = w = 0 \qquad (11.21)$$

where $U$ is the maximum streamwise velocity at the upper boundary, and $b$ is the distance between boundaries. With $\partial v/\partial x = 0$, it follows from (11.17) that the $z$-component of vorticity $\zeta = -\partial u/\partial y = -U/b$, which is consistent with our having envisioned that a vorticity meter within this shear flow rotates in a clockwise manner. More generally, for a one-dimensional flow where $u = u(y)$ and $v = w = 0$, the vorticity $\zeta = -\partial u/\partial y$. With parabolic Poiseuille flow in a conduit, for example, the vorticity is zero at the conduit axis, it is at a maximum at the walls, and is of opposite sign on either side of the axis (Figure 11.8). It is left as an exercise to derive an expression for vorticity as a local function $\zeta(y)$ of position $y$ in terms of conduit dimensions and physical properties of the fluid, starting with the velocity profile $u(y)$ (Example Problem 11.4.1).

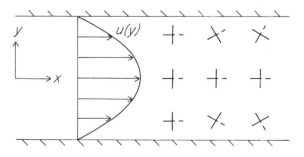

**Figure 11.8** Successive positions of vorticity meters within Poiseuille flow.

Now reconsider flow within a free vortex, which we asserted above is irrotational. Positioning the $z$-axis at the center of the vortex, the tangential velocity field, without proof, is given by $v_\theta = C/r$ (except close to the vortex core), where $C$ is a constant that characterizes the intensity of the vortex (Example Problem 12.5.7). Using (11.18) and (11.19) together with $x = r\cos\theta$, $y = r\sin\theta$, and $r^2 = x^2 + y^2$,

$$u = -\frac{Cy}{x^2 + y^2} \tag{11.22}$$

$$v = \frac{Cx}{x^2 + y^2} \tag{11.23}$$

Differentiating $u$ with respect to $y$, and $v$ with respect to $x$,

$$\frac{\partial u}{\partial y} = \frac{\partial v}{\partial x} = C\frac{(y^2 - x^2)}{(x^2 + y^2)^2} \tag{11.24}$$

According to (11.17), $\zeta = 0$, which agrees with our assertion that flow within a free vortex is irrotational. In general, an irrotational flow possesses zero vorticity, $\boldsymbol{\omega} = 0$, or equally, from (11.15), (11.16), and (11.17),

$$\frac{\partial w}{\partial y} = \frac{\partial v}{\partial z} \tag{11.25}$$

$$\frac{\partial u}{\partial z} = \frac{\partial w}{\partial x} \tag{11.26}$$

$$\frac{\partial v}{\partial x} = \frac{\partial u}{\partial y} \tag{11.27}$$

This free vortex problem suggests that, rather than using Cartesian expressions of vorticity, a more direct approach would be to express vorticity in polar coordinates. For a velocity vector $\mathbf{u}$, where

$$\mathbf{u} = \mathbf{e}_r u_r + \mathbf{e}_\theta v_\theta + \mathbf{e}_z w_z \tag{11.28}$$

the three components of vorticity, $\xi_r$, $\eta_\theta$, and $\zeta_z$, are (Appendix 17.2)

$$\xi_r = \frac{1}{r}\frac{\partial w_z}{\partial \theta} - \frac{\partial v_\theta}{\partial z} \tag{11.29}$$

$$\eta_\theta = \frac{\partial u_r}{\partial z} - \frac{\partial w_z}{\partial r} \tag{11.30}$$

$$\zeta_z = \frac{1}{r}v_\theta + \frac{\partial v_\theta}{\partial r} - \frac{1}{r}\frac{\partial u_r}{\partial \theta} \tag{11.31}$$

where $u_r$, $v_\theta$, and $w_z$ are the three components of velocity, and $\mathbf{e}_r$, $\mathbf{e}_\theta$, and $\mathbf{e}_z$ are orthogonal unit vectors in the coordinate directions $r$, $\theta$, and $z$. With $v_\theta = C/r$ for a free vortex, $\partial v_\theta/\partial r = -C/r^2$ (and $\partial u_r/\partial\theta = 0$). Substituting into (11.31) gives $\zeta_z = 0$, which again indicates that the flow is irrotational.

As vorticity is a vector quantity that can be defined at all positions in a flow, it is possible to define the rate of change in vorticity with respect to position and time in the same manner that we defined the substantive derivative of velocity (Chapter 7), then further cast this into the form of a dynamical equation. The resulting *vorticity equation* can, under certain circumstances, provide significant insight to flow problems. Although this treatment of vorticity is beyond the scope of this text, interested students are urged to consider the references on this important topic listed at the end of the chapter.

## 11.3 FLUID STRAIN

Fluid flow involves strain that in general is manifest as changes in the positions of fluid particles relative to each other. As the motion of a fluid is entirely specified by a velocity field $\mathbf{u}(x, y, z)$, the instantaneous rate of strain can be defined in terms of the relative velocities of fluid points. Consider two neighboring points A and B within a moving fluid (Figure 11.9). The instantaneous position of A is given by the coordinates $x$, $y$, and $z$, and the instantaneous velocity at A is specified by the components $u$, $v$, and $w$. The position of B at the same instant is given by the coordinates $x + dx$, $y + dy$, and $z + dz$. The velocity components at B then may be obtained from Taylor expansions to first order (Chapter 6), where small changes in velocity $du$, $dv$, and $dw$ that occur over the small distances $dx$, $dy$, and $dz$ are added to the velocities $u$, $v$, and $w$:

$$u + du = u + \frac{\partial u}{\partial x}dx + \frac{\partial u}{\partial y}dy + \frac{\partial u}{\partial z}dz \qquad (11.32)$$

$$v + dv = v + \frac{\partial v}{\partial x}dx + \frac{\partial v}{\partial y}dy + \frac{\partial v}{\partial z}dz \qquad (11.33)$$

$$w + dw = w + \frac{\partial w}{\partial x}dx + \frac{\partial w}{\partial y}dy + \frac{\partial w}{\partial z}dz \qquad (11.34)$$

Subtracting $u$, $v$, and $w$ from each side, the three components $du$, $dv$, and $dw$ may then be written as a vector that is equal to the product of a $3 \times 3$ matrix and a $1 \times 3$ column vector of the small distances $dx$, $dy$, and $dz$:

$$\langle du, dv, dw \rangle \leftrightarrow \begin{bmatrix} du \\ dv \\ dw \end{bmatrix} = \begin{bmatrix} \dfrac{\partial u}{\partial x} & \dfrac{\partial u}{\partial y} & \dfrac{\partial u}{\partial z} \\ \dfrac{\partial v}{\partial x} & \dfrac{\partial v}{\partial y} & \dfrac{\partial v}{\partial z} \\ \dfrac{\partial w}{\partial x} & \dfrac{\partial w}{\partial y} & \dfrac{\partial w}{\partial z} \end{bmatrix} \begin{bmatrix} dx \\ dy \\ dz \end{bmatrix} \qquad (11.35)$$

Now, because the relative motion between A and B is represented by $du$, $dv$, and $dw$, and because $dx$, $dy$, and $dz$ are at any instant fixed infinitesimal distances, the instantaneous relative motion between A and B is entirely characterized by the $3 \times 3$ matrix in (11.35). This matrix is referred to as the *velocity gradient tensor*, and describes all possible components of the rate at which a fluid may undergo strain. Note that each diagonal element in this tensor describes a change in the velocity component with respect to the coordinate direction of that component. Thus a nonzero diagonal element is associated with a *linear dilation* (or contraction) of the distance between neighboring fluid points along a given coordinate axis, a point we will examine below. Each off-diagonal element describes a change in the velocity component with respect to a coordinate direction normal to the direction of the

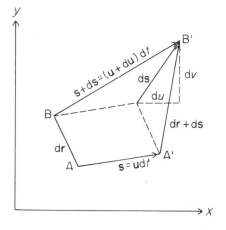

**Figure 11.9**   Definition diagram for two-dimensional infinitesimal strain associated with relative motion between fluid points A and B initially separated by distance **dr**; during $dt$ A moves along displacement vector **s** to A', and B moves along **s** + d**s** to B', where $\mathbf{u} = \langle u, v \rangle$.

velocity component. Thus a nonzero off-diagonal element is associated with shearing motion, which may involve either *pure shear* or rotation, or some combination of pure shear and rotation.

Consider as examples the off-diagonal gradients $\partial u/\partial y$ and $\partial v/\partial x$, and expand them as follows:

$$\frac{\partial u}{\partial y} = \frac{1}{2}\left(\frac{\partial u}{\partial y} + \frac{\partial v}{\partial x}\right) - \frac{1}{2}\left(\frac{\partial v}{\partial x} - \frac{\partial u}{\partial y}\right) \tag{11.36}$$

$$\frac{\partial v}{\partial x} = \frac{1}{2}\left(\frac{\partial v}{\partial x} + \frac{\partial u}{\partial y}\right) + \frac{1}{2}\left(\frac{\partial v}{\partial x} - \frac{\partial u}{\partial y}\right) \tag{11.37}$$

Comparing these with (11.9), the last term in each is equal to the angular velocity component $\omega_z$. Thus (11.36) and (11.37) may be written as

$$\frac{\partial u}{\partial y} = \frac{1}{2}\left(\frac{\partial u}{\partial y} + \frac{\partial v}{\partial x}\right) - \omega_z \tag{11.38}$$

$$\frac{\partial v}{\partial x} = \frac{1}{2}\left(\frac{\partial v}{\partial x} + \frac{\partial u}{\partial y}\right) + \omega_z \tag{11.39}$$

In general each velocity gradient in (11.35) can be expressed as the sum of two terms; one describes a rate of rotation and, as we will see below, the other describes a rate of shear strain involving no rotation. Using similar expansions for the remaining off-diagonal terms, the velocity gradient tensor in (11.35) can be written in the equivalent form

$$\begin{bmatrix} \dfrac{\partial u}{\partial x} & \dfrac{1}{2}\left(\dfrac{\partial u}{\partial y} + \dfrac{\partial v}{\partial x}\right) - \omega_z & \dfrac{1}{2}\left(\dfrac{\partial u}{\partial z} + \dfrac{\partial w}{\partial x}\right) + \omega_y \\[3mm] \dfrac{1}{2}\left(\dfrac{\partial v}{\partial x} + \dfrac{\partial u}{\partial y}\right) + \omega_z & \dfrac{\partial v}{\partial y} & \dfrac{1}{2}\left(\dfrac{\partial v}{\partial z} + \dfrac{\partial w}{\partial y}\right) - \omega_x \\[3mm] \dfrac{1}{2}\left(\dfrac{\partial w}{\partial x} + \dfrac{\partial u}{\partial z}\right) - \omega_y & \dfrac{1}{2}\left(\dfrac{\partial w}{\partial y} + \dfrac{\partial v}{\partial z}\right) + \omega_x & \dfrac{\partial w}{\partial z} \end{bmatrix} \tag{11.40}$$

Let us now examine each type of element in this matrix, beginning with diagonal elements.

Consider a small, rectangular fluid element whose edges have lengths $dx$, $dy$, and $dz$, where one corner is positioned at the origin of a Cartesian coordinate system (Figure 11.10). Suppose that all elements in (11.40) are zero except the leading diagonal element and, for illustration, $\partial u/\partial x > 0$. The velocity of any point on the right face relative to the left face is $du = (\partial u/\partial x)dx$. The fluid element thus undergoes uniform linear extension in the direction of the $x$-axis, the right face moving away from the left face with increasing velocity. The rate of this strain, denoted by $\dot{\varepsilon}_x = \partial u/\partial x$, thus describes a *rate of linear dilation* of the fluid element. Similarly, $\dot{\varepsilon}_y = \partial v/\partial y$ and $\dot{\varepsilon}_z = \partial w/\partial z$ describe rates of linear dilation parallel to the $y$-axis and the $z$-axis. Note that these dilations involve no distortions of angular relations among faces of the fluid element.

Now envision the case in which all three diagonal elements in (11.40) are nonzero and off-diagonal elements are zero. A small linear dilation of the fluid element in the direction of the $x$-axis during a small interval $dt$ is given by $(\partial u/\partial x)dxdt$. Therefore a small change in the volume of the fluid element due to dilation in the direction of the $x$-axis during $dt$ is $(\partial u/\partial x)dxdydzdt$. Likewise, small changes in volume of the fluid element due to dilation in the directions of the $y$-axis and $z$-axis during $dt$ are given by $(\partial v/\partial y)dxdydzdt$ and $(\partial w/\partial z)dxdydzdt$. The total infinitesimal change in volume $dV$ during $dt$ is the sum of these. Recalling that volumetric strain is defined by the ratio $dV/V$ (Chapter 3), the *volumetric dilation rate* $\dot{\varepsilon}$ is

$$\dot{\varepsilon} = \frac{1}{V}\frac{dV}{dt} = \frac{\left(\dfrac{\partial u}{\partial x} + \dfrac{\partial v}{\partial y} + \dfrac{\partial w}{\partial z}\right)dx\,dy\,dz\,dt}{dx\,dy\,dz\,dt} = \nabla \cdot \mathbf{u} \qquad (11.41)$$

Thus the sum of the diagonals in the tensor (11.40) defines a volumetric rate of expansion or contraction that involves no distortion of angular relations among faces

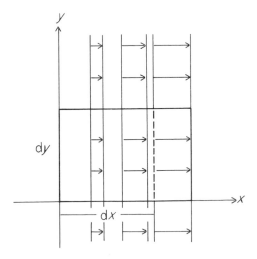

**Figure 11.10** Infinitesimal linear dilation of a fluid element, where $\partial u/\partial x > 0$.

of the fluid element. By conservation of mass (Chapter 8), $\dot{\varepsilon} = -(D\rho/Dt)/\rho$ when the fluid is compressible, and $\dot{\varepsilon} = 0$ when the fluid is incompressible.

Again consider a small, rectangular fluid element whose edges have lengths $dx$, $dy$, and $dz$, where one corner is positioned at the origin of a Cartesian coordinate system (Figure 11.11). Suppose for illustration that the off-diagonal element $\partial u/\partial y = \frac{1}{2}(\partial u/\partial y + \partial v/\partial x) - \omega_z$ in (11.40) is nonzero. (This implies that the element $\partial v/\partial x = \frac{1}{2}(\partial v/\partial x + \partial u/\partial y) + \omega_z$ also is nonzero.) In the first case of interest, both gradients $\partial u/\partial y$ and $\partial v/\partial x$ are nonzero and equal in sign and magnitude. Thus $\omega_z = \frac{1}{2}(\partial v/\partial x - \partial u/\partial y) = 0$. The fluid element is deformed to a parallelogram represented by compression along one diagonal and extension along the other, and the angle at the origin deforms at twice the rate given by $\frac{1}{2}(\partial u/\partial y + \partial v/\partial x)$; that is, at a rate equal in magnitude to $\partial u/\partial y = \partial v/\partial x$. Consistent with the fact that $\omega_z = 0$, the motion is irrotational (zero vorticity), and describes the case of *pure shear strain*, which involves distortion of angular relations among faces of the fluid element. However, absence of any change in the angular relation between the two fluid-element diagonals (Figure 11.11) reflects the irrotational quality of this strain.

Consider again the nonzero element $\partial u/\partial y = \frac{1}{2}(\partial u/\partial y + \partial v/\partial x) - \omega_z$ in (11.40) when $\partial u/\partial y = -\partial v/\partial x$. In this case the first term of this element equals zero, and $\omega_z = \frac{1}{2}(\partial v/\partial x - \partial u/\partial y) = \partial v/\partial x = -\partial u/\partial y$. As described in Section 11.2 above, this defines the case of pure rotation, involving no distortion of angular relations among faces of the fluid element. Note that $\omega_z$ leads to the counterclockwise (positive) rotation of the fluid element (Figure 11.12).

Let us now return to the case of simple shear. Again consider a small, rectangular fluid element whose edges have lengths $dx$, $dy$, and $dz$, where one corner is positioned at the origin of a Cartesian coordinate system (Figure 11.13). Assume that flow is one-dimensional such that $u = u(y)$ and $\partial u/\partial y > 0$. Consider the term $\partial u/\partial y = \frac{1}{2}(\partial u/\partial y + \partial v/\partial x) - \omega_z$ in (11.40). With $\partial v/\partial x = 0$, $\omega_z = \frac{1}{2}(\partial v/\partial x - \partial u/\partial y) = -\frac{1}{2}\partial u/\partial y$. The fluid element is distorted into a parallelogram after a short interval $dt$, such that faces originally normal to the $x$-axis are inclined. Point D on the left-hand face has a new position D$'$ with coordinates

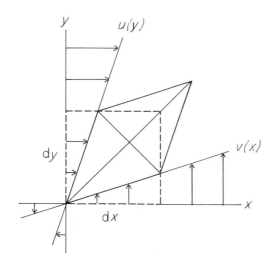

**Figure 11.11** Infinitesimal pure-shear strain of a fluid element, where $\partial u/\partial y = \partial v/\partial x$.

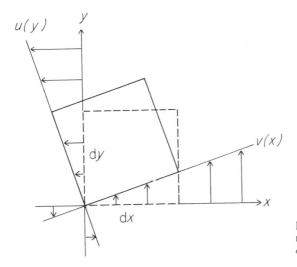

**Figure 11.12** Infinitesimal pure rotation of a fluid element, where $\partial u/\partial y = -\partial v/\partial x$.

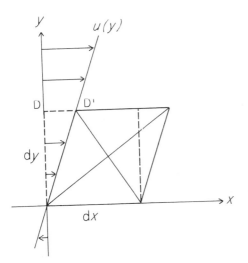

**Figure 11.13** Infinitesimal simple-shear strain of a fluid element, where $\partial u/\partial y > 0$ and $\partial v/\partial x = 0$.

$[(\partial u/\partial y)\mathrm{d}y\mathrm{d}t, \mathrm{d}y]$. Note that the diagonals of the fluid element (Figure 11.13) have rotated during $\mathrm{d}t$, reflecting the finite vorticity involved. The fluid deformation in this case is referred to as *simple shear strain* and involves distortion of angular relations among faces of the fluid element.

This leads to an interesting, approximate geometrical interpretation of simple shear motion described by the two terms of the element $\frac{1}{2}(\partial u/\partial y + \partial v/\partial x) - \omega_z$ in the example where $\partial u/\partial y > 0$ and $\partial v/\partial x = 0$ (Figure 11.13). The first term gives a shear-rate component equal to $\frac{1}{2}\partial u/\partial y$, and the second term, including the leading negative sign, gives a rotational component equal to $-\omega_z = \frac{1}{2}\partial u/\partial y$. This positive value, however, may seem at odds with our conclusion that this case of simple shear involves clockwise (negative) rotation. The simplest explanation is that this term describes a clockwise rotation whose effect is numerically equivalent to a positive component of shear equal to $\frac{1}{2}\partial u/\partial y$. To further clarify this, let us envision decomposing the motion of the fluid into its two components (Figure 11.14), where it is important to keep in mind that any such geometrical interpretation, which necessarily involves depicting finite strains, will not be entirely adequate since we actually are dealing with infinitesimal strains. The first component involves pure strain of the initially rectangular fluid element by an amount equal to $(\frac{1}{2}\partial u/\partial y)\mathrm{d}y\mathrm{d}t$. As clockwise rotation is occurring simultaneously, we must envision this pure strain in the context of a coordinate system that is rotating clockwise, where the rate of rotation keeps up with the rate of shear strain. Thus the second component involves rotating the deforming fluid element (including the coordinate system) clockwise by an amount equal to $\omega_z\mathrm{d}y\mathrm{d}t = (-\frac{1}{2}\partial u/\partial y)\mathrm{d}y\mathrm{d}t$. To then produce a deformed fluid element whose appearance coincides with simple shear strain (Figure 11.13) thus requires rotating the coordinate system counterclockwise by an amount equal to $(\frac{1}{2}\partial u/\partial y)\mathrm{d}y\mathrm{d}t$, that is, at a positive rate $-\omega_z = \frac{1}{2}\partial u/\partial y$ during the interval $\mathrm{d}t$. Thus the matrix element $\frac{1}{2}(\partial u/\partial y + \partial v/\partial x) - \omega_z$ describes the rates of two motions (involving positive shear and counterclockwise rotation of the coordinate system) that would be required to produce simple shear strain, given that this strain involves clockwise rotation. Our geometrical interpretation must stop here; the final orientation and shape of a fluid element involving finite strain, as is necessarily depicted in Figure 11.14, will in general depend on the order in which each component of strain is added.

In addition to pure shear and pure rotation, the case of simple shear illustrates that combinations of shear strain and rotation occur when the magnitudes of velocity gradients defined in a single plane are not equal, for example when $|\partial u/\partial y| \neq |\partial v/\partial x|$. In general, we may interpret terms of the form $\frac{1}{2}(\partial u/\partial y + \partial v/\partial x)$ in off-diagonal elements in (11.40) as describing the rate at which the shape of a fluid element is distorted without rotation, and terms of the form $\omega_z = \frac{1}{2}(\partial v/\partial x - \partial u/\partial y)$ as describing the rate of rigid-body rotation. It is left as an exercise to provide a geometrical interpretation similar to that above for the case in which, for example, $\partial v/\partial x = 2\partial u/\partial y$.

Thus, fluid motion, in general, involves four components. The first is pure translation, which is specified by the velocity components $u$, $v$, and $w$. The second is rigid-body rotation, specified by the angular velocity components $\omega_x$, $\omega_y$, and $\omega_z$, or equivalently, by the vorticity components $\xi$, $\eta$, and $\zeta$. The third is linear dilation, specified by the diagonal elements of (11.40) which, taken together, define the volumetric dilation rate, $\nabla \cdot \mathbf{u}$. The fourth is shearing strain, specified by the

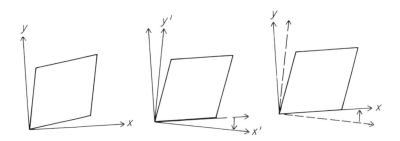

**Figure 11.14** Simple shear strain decomposed into pure-shear strain and rotation components.

parenthetical parts of the off-diagonal elements in (11.40). The motion of a fluid element consists of the superposition of these four types of motion, one or more of which may be zero at any instant or position in the flow. Whereas the first two components describe changes in the location of a fluid element, the last two describe deformation in the shape of a fluid element.

The velocity gradient tensor can be written as the sum of two tensors, one describing pure strain and one describing rotation:

$$
\begin{bmatrix}
\dfrac{\partial u}{\partial x} & \dfrac{1}{2}\left(\dfrac{\partial u}{\partial y}+\dfrac{\partial v}{\partial x}\right) & \dfrac{1}{2}\left(\dfrac{\partial u}{\partial z}+\dfrac{\partial w}{\partial x}\right) \\[2ex]
\dfrac{1}{2}\left(\dfrac{\partial v}{\partial x}+\dfrac{\partial u}{\partial y}\right) & \dfrac{\partial v}{\partial y} & \dfrac{1}{2}\left(\dfrac{\partial v}{\partial z}+\dfrac{\partial w}{\partial y}\right) \\[2ex]
\dfrac{1}{2}\left(\dfrac{\partial w}{\partial x}+\dfrac{\partial u}{\partial z}\right) & \dfrac{1}{2}\left(\dfrac{\partial w}{\partial y}+\dfrac{\partial v}{\partial z}\right) & \dfrac{\partial w}{\partial z}
\end{bmatrix}
+
\begin{bmatrix}
0 & -\omega_z & \omega_y \\[1ex]
\omega_z & 0 & -\omega_x \\[1ex]
-\omega_y & \omega_x & 0
\end{bmatrix}
\tag{11.42}
$$

The first, the *rate-of-strain tensor*, is symmetric. The second, the *rotation tensor*, is skew-symmetric. Consider now the rate-of-strain tensor, and write it as

$$
\begin{bmatrix}
\dot{\varepsilon}_x & \dot{\varepsilon}_{xy} & \dot{\varepsilon}_{xz} \\
\dot{\varepsilon}_{yx} & \dot{\varepsilon}_y & \dot{\varepsilon}_{yz} \\
\dot{\varepsilon}_{zx} & \dot{\varepsilon}_{zy} & \dot{\varepsilon}_z
\end{bmatrix}
\equiv
\begin{bmatrix}
\dfrac{\partial u}{\partial x} & \dfrac{1}{2}\left(\dfrac{\partial u}{\partial y}+\dfrac{\partial v}{\partial x}\right) & \dfrac{1}{2}\left(\dfrac{\partial u}{\partial z}+\dfrac{\partial w}{\partial x}\right) \\[2ex]
\dfrac{1}{2}\left(\dfrac{\partial v}{\partial x}+\dfrac{\partial u}{\partial y}\right) & \dfrac{\partial v}{\partial y} & \dfrac{1}{2}\left(\dfrac{\partial v}{\partial z}+\dfrac{\partial w}{\partial y}\right) \\[2ex]
\dfrac{1}{2}\left(\dfrac{\partial w}{\partial x}+\dfrac{\partial u}{\partial z}\right) & \dfrac{1}{2}\left(\dfrac{\partial w}{\partial y}+\dfrac{\partial v}{\partial z}\right) & \dfrac{\partial w}{\partial z}
\end{bmatrix}
\tag{11.43}
$$

where elements symbolized by $\dot{\varepsilon}_{ij}$ in the first matrix correspond to elements in similar row-column positions in the second matrix. The symmetry of this tensor should be evident, such that

$$
\dot{\varepsilon}_{xy} = \dot{\varepsilon}_{yx}; \qquad \dot{\varepsilon}_{xz} = \dot{\varepsilon}_{zx}; \qquad \dot{\varepsilon}_{yz} = \dot{\varepsilon}_{zy}
\tag{11.44}
$$

We will make use of these ideas and definitions of fluid strain in treating viscous forces in the next chapter. In particular, we will relate the rates of strain in the tensor (11.43) to normal and tangential stresses acting on a fluid element.

## 11.4 EXAMPLE PROBLEMS

### 11.4.1 Local vorticity and grain rotation in a dike or sill

Suppose that crystal grains are suspended within a magma as it flows within a dike or sill. We may envision the grains as small, crude vorticity meters. With steady Newtonian flow, the velocity distribution is known (Example Problem 3.7.1), so we can immediately compute the vorticity as a local function of position. In this problem, the only possible component of vorticity is $\zeta = \partial v/\partial x - \partial u/\partial y$; and since $\partial v/\partial x$ vanishes for this one-dimensional flow, $\zeta = -du/dy$. This suggests that vorticity $\zeta$ is a local function $\zeta(y)$ of position $y$. Taking the derivative of the velocity distribution (3.71) with respect to $y$,

$$\frac{\mathrm{d}u}{\mathrm{d}y} = \frac{\nabla p}{\mu}y \qquad (11.45)$$

It follows that

$$\zeta = -\frac{\nabla p}{\mu}y \qquad (11.46)$$

Thus a negative pressure gradient $\nabla p = \partial p/\partial x$ leads to positive vorticity (counterclockwise rotation) over $y > 0$, and negative vorticity (clockwise rotation) over $y < 0$. The vorticity is zero at the centerline ($y = 0$). Inasmuch as a crystal behaves like a vorticity meter, its angular velocity $\omega_z$ is

$$\omega_z = -\frac{\nabla p}{2\mu}y \qquad (11.47)$$

which suggests that grains near the dike walls rotate more rapidly than grains near the centerline. But this raises an interesting question: Does a grain rotate with the same angular velocity as computed from the local vorticity, and what is the effect of this rotation?

Suppose that a crystal grain is approximately spherical. It is well known that a spinning cylinder or sphere that moves through a stationary fluid experiences a net force acting normal to the axis of rotation and to the direction of motion. This is referred to as the *Magnus effect* in the case of a cylinder and the *Robins effect* in the case of a sphere. A partial explanation of these effects is obtained from Bernoulli's theorem, as viewed from a frame of reference moving with the cylinder or sphere. On the side of the object where the tangential motion is in the same direction as the *relative* motion of the surrounding fluid, the fluid speed is increased, so the pressure is reduced. On the side of the object where the tangential motion is in the opposite direction as the relative motion of the surrounding fluid, the fluid speed is decreased, so the pressure is increased. A force (*lift*) associated with this pressure difference thus acts on the object. (We will consider a similar situation in the next Example Problem.)

But does this effect occur with a particle that "passively" rotates in a moving fluid with nonuniform velocity distribution, as in the case of a crystal grain in a dike? Evidently, a small Robins effect occurs, which tends to induce motion of the grain toward the center of the dike. This effect, however, is negligible relative to other forces that induce motion toward the center. These include *grain dispersive pressures* that arise from hydrodynamic interactions among grains, and which do not necessarily involve grain-to-grain collisions. This interaction involves interference between the fluid velocity fields surrounding grains that are near each other. (An idea of similar hydrodynamic interactions can readily be gained by watching the motions of bubbles that approach each other while ascending within a bottle containing viscous shampoo or syrup.) In addition, there is a *wall effect* due to the hydrodynamic interaction between a grain and the nearby wall. This mechanism is effective only near the dike wall. Both the Robbins effect and the wall effect vary with the lag velocity, the difference between the local fluid velocity and the speed of the grain. Komar (1972) provides a general discussion of these effects in the context of flow differentiation in dikes and sills.

Finally, note that with absence of vorticity at the centerline, fluid here is effectively inviscid (in the limit as $y \to 0$), a point used in Example Problem 7.3.2.

## 11.4.2 Generation of vorticity in viscous boundary flow: precursor to turbulence

Let us consider another boundary flow with the aim of learning one way in which vorticity associated with a velocity gradient can be transformed into distinct vorticies (eddies), as a precursor to turbulence. Consider steady two-dimensional flow of an incompressible fluid around a cylinder, neglecting gravity. It will suffice for our purpose to provide a qualitative description of fluid motion, combining ideas of inviscid flow with ideas regarding vorticity within viscous boundary flow. Momentarily assume that the flow has a small Reynolds number defined by $Re = \rho R U / \mu$ (Chapter 5), where $R$ is the cylinder radius and $U$ is the free-stream velocity far from the cylinder. The no-slip condition leads to a boundary layer next to the cylinder (Figure 11.15). Flow outside this boundary layer is essentially inviscid since velocity gradients vanish; and if $U$ is small such that velocity gradients within the boundary layer are small, flow here also is approximately inviscid.

Streamlines are compressed around the cylinder. In general they converge between A and B and diverge between B and C (Figure 11.15), such that relatively high streamline velocities occur at B, just outside the boundary layer. Inasmuch as flow is inviscid, the fluid pressure along a streamline therefore decreases between A and B, then increases between B and C. The negative pressure gradient between A and B is transformed to kinetic energy and flow accelerates; then the kinetic energy of the flow performs work against the adverse pressure gradient between B and C and flow decelerates. Moreover, the pressure at C recovers to the value at A. Now, let us restate this in a way that will be particularly useful below: Since the flow is conservative with respect to mechanical energy (inviscid), fluid arriving at B has gained precisely the kinetic energy required to overcome the adverse pressure gradient between B and C such that, when this gained quantity of kinetic energy is consumed, the kinetic energy of the fluid arriving at C has returned to its initial (upstream) value, the pressure has returned to its upstream value, and a constant total pressure (static plus dynamic) along the streamline has been maintained.

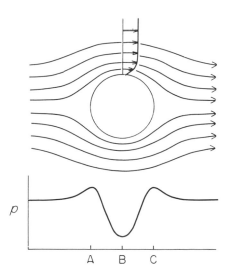

$p$

A    B    C

**Figure 11.15**  Schematic diagram of two-dimensional viscous flow around a cylinder, and associated variation in pressure $p$ along a streamline between A and C; adapted from Schlichting (1979).

This is, so far, reminiscent of inviscid flow in a Venturi conduit (Example Problem 10.3.1), with the difference that flow is unbounded away from the cylinder. As $Re$ increases, streamlines are increasingly compressed around the cylinder, and velocity gradients next to the cylinder strengthen. Viscous resistance to shear within the boundary layer therefore becomes increasingly significant. Like the Venturi problem involving real fluids, work is performed against this viscous resistance such that energy is extracted from the flow and dissipated as heat. Now, the static pressure outside the boundary layer (within the inviscid region) is impressed throughout the boundary layer. Thus the adverse pressure gradient between B and C is the same within the boundary layer. Along a streamline within the boundary layer, however, kinetic energy is consumed by friction such that fluid arriving at B has not gained sufficient kinetic energy to entirely overcome the adverse pressure gradient between B and C, so fluid close to the cylinder stalls at some point between B and C. The adverse pressure gradient at this point is unbalanced by the extant, dynamic pressure gradient. The adverse pressure gradient is therefore capable of inducing flow upstream close to the cylinder, which provides the onset of concentrated vorticity in the lee of the cylinder (Figure 11.16).

The inviscid portion of the flow is displaced outward. This is the same as stating that flow "separates" around the site of concentrated vorticity. The sense of shear associated with this separation enhances the rotational motion of fluid next to the cylinder. This region may then become a more or less distinct parcel of rotating fluid, an eddy. Once initiated, eddy motion is further enhanced by shear of fluid

a

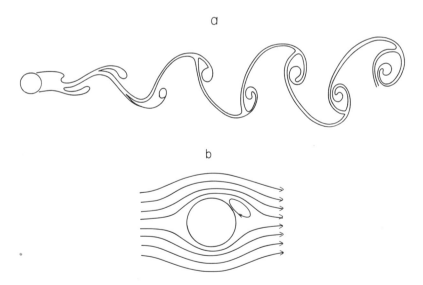

b

**Figure 11.16** Simplified sketch of streaklines associated with von Kármán vortex street (*a*) for $Re = 67$ (sketch based on photograph from Tritton [1988], by permission of Oxford University Press), and schematic diagram of two-dimensional flow around a cylinder with separation (*b*).

around it. Thus the rotational kinetic energy of an eddy reflects further extraction of energy from the mean flow.

These eddies are unstable in the sense that they detach, are shed from the cylinder, and then move downstream with the mean flow. Shedding occurs in a regular pattern, with a regular frequency, at low to moderate $Re$. Eddies grow and then are shed alternately from one side of the cylinder, then the other. This pattern of shedding, together with the wake of regularly spaced eddies moving downstream, is referred to as a *Von Kármán vortex street*. This behavior of flow is a precursor to turbulence in the sense that, with a further increase of $Re$, shedding becomes irregular, the eddy wake becomes disorganized, and the complexity of its velocity field becomes a dominant, characteristic feature of the flow (Chapter 14).

Half the cylinder in Figure 11.16 is like a bump on a flat surface—for example, a pebble or a small bedform on a stream bed. Like the cylinder, these bumps tend to be sites where eddies develop. These eddies, after detaching from the bumps, then move up into the fluid column where they contribute to turbulent velocity fluctuations (Chapter 15).

## 11.5 READING

Bagnold, R. A. 1954. Experiments on a gravity-free dispersion of large solid spheres in a Newtonian fluid under shear. *Proceedings of the Royal Society of London* Series A, 255:49–63. Bagnold describes in this classic paper the results of careful measurements of the phenomenon of grain-dispersive pressures. The idea of dispersive pressures subsequently has been used to describe a variety of fluid-flow processes, including the segregation of different grain sizes in mud flows.

Johnson, A. M. 1970. *Physical processes in geology.* San Francisco: Freeman, Cooper & Co., 577 pp. This text provides a clear discussion of stress, strain, infinitesimal strain, and finite strain, in the context of simple and pure shear, and rotation. It also examines the idea of superposition of infinitesimal versus finite strains.

Komar, P. D. 1972. Mechanical interactions of phenocrysts and flow differentiation of igneous dikes and sills. *Geological Society of America Bulletin* 83:973–88. This paper provides a general discussion of the Magnus effect, grain-dispersive pressures, and wall effects, in the context of grains suspended within a magma in a dike or sill.

Nielson, J. E. and Nakata, J. K. 1994. Mantle origin and flow sorting of megacryst-xenolith inclusions in mafic dikes of Black Canyon, Arizona. U. S. Geological Survey Professional Paper 1541, Washington, D.C., 41 pp. This paper provides extensive field evidence that axial concentrations of xenoliths and xenocrysts within dikes can derive from particle sorting associated with the intrusive flow of magmas.

Schlichting, H. 1979. *Boundary-layer theory,* 7th ed. New York: McGraw-Hill, 817 pp. This text provides a set of figures that clearly illustrate the components of infinitesimal fluid strain. It also includes a section on the two-dimensional vorticity transport equation.

Shapiro, A. H. Vorticity. Encyclopaedia Britannica Educational Corporation. This delightful film provides a demonstration of vorticity meters, and several clear visual examples of vorticity.

Tritton, D. J. 1988. *Physical fluid dynamics,* 2nd ed. Oxford: Oxford University Press, 519 pp. This text provides (pp. 81–85) a clear treatment of vorticity in relation to strain. It also has a nice summary (pp. 159–61) of the Magnus and Robins effects, including the idea that the direction of these forces can be reversed under certain circumstances.

# CHAPTER 12

# Viscous Flows

Most flow problems in geology involve viscous fluids which exhibit resistance to shearing motions. Mechanical treatments of such flows therefore must involve a consideration of frictional forces associated with this viscous behavior. Our first objective is to obtain a general equation of motion based on Newton's second law that involves body and surface forces acting on a fluid element, regardless of the specific fluid involved. The steps in this development are similar to those leading to Euler's equation for inviscid flows (Chapter 10); the difference is that tangential stresses acting on a fluid element, in addition to normal stresses, are included in the description of surface forces. It will be necessary, when describing normal forces, to distinguish between the thermodynamic pressure $p$, as used in treating inviscid flows, and a *mechanical pressure* $\sigma_0$, which arises in treating viscous effects associated with compressible flows, and flows that simultaneously involve chemical reactions and possibly other phenomena.

The second step in obtaining an equation of motion is determined by the specific fluid involved. Here we require a supplemental set of equations that describe the relation between surface forces and rates of fluid strain, as defined by the rate-of-strain tensor examined in Chapter 11. This set of equations, referred to as the *constitutive equations* of a fluid, varies with fluid rheology. The emphasis of this chapter is on Newtonian fluids. The set of constitutive equations in this case, when coupled with the general equation of motion obtained in the first step, lead to the well known *Navier–Stokes equations*. In addition, we will briefly examine the case of glacier ice as an example of a non-Newtonian fluid whose rheology is described by Glen's law (Chapter 3).

Treatments of viscous flows that incorporate conservation of energy similarly must involve a consideration of work performed against the frictional effects of viscosity. This performance of work against friction is nonconservative. Its effect therefore is to continuously extract mechanical energy from the main fluid motion, dissipating it in the form of heat. Our treatment of this irreversible conversion of energy will lead to a *dissipation function* that is added to the energy equation developed in Chapter 9.

This chapter will complete the formal developments necessary to examine viscous flows. We will reexamine at the end of this chapter several problems introduced in the preceding chapters.

## 12.1  VISCOUS FORCES

Consider a rectangular control volume with edges of lengths $dx$, $dy$, and $dz$ embedded within a Cartesian coordinate system (Figure 12.1). Tangential and normal stresses may act on each face. Those acting on faces normal to the $x$-axis include $\tau_{xy}$, $\tau_{xz}$, and $\sigma_x$. Taking $\tau_{xy}$ as an example, the first subscript $x$ denotes that the stress $\tau_{xy}$ acts on a face normal to the $x$-axis; the second subscript $y$ denotes that $\tau_{xy}$ acts parallel to the $y$-axis. Similar remarks pertain to $\tau_{xz}$. The single subscript on $\sigma_x$ is sufficient to indicate that this normal stress acts parallel to the $x$-axis. With this convention, stresses acting on faces normal to the $y$-axis include $\tau_{yx}$, $\tau_{yz}$, and $\sigma_y$, and stresses acting on faces normal to the $z$-axis include $\tau_{zx}$, $\tau_{zy}$, and $\sigma_z$.

Tangential stresses are to be identified with viscous forces that arise during shearing motions of a real fluid. The normal stresses $\sigma_x$, $\sigma_y$, and $\sigma_z$ are to be identified with three types of forces: viscous forces associated with linear and volumetric dilations of a fluid element; apparent viscous forces associated with dissipation of mechanical energy during fluid dilation, which arise from the bulk viscosity of a compressible fluid (Chapter 3); and surface forces associated with the thermodynamic pressure $p$ as used, for example, in our treatment of fluid statics (Chapter 6) and inviscid flows (Chapter 10). In the special cases of inviscid fluids and viscous incompressible fluids, we will see below that the normal stresses $\sigma_x$, $\sigma_y$, and $\sigma_z$ are in practice equated to the thermodynamic pressure.

At this point we must choose a sign convention regarding the directions of the stress components acting on the faces of a control volume (Figure 12.1). Students should be aware that this convention varies in the literature, and hinges in part on the sign convention adopted in defining the constitutive equations of a fluid (Sections 12.2.1 and 12.4.1). Here we will adopt a convention that follows the kinetic interpretation of tangential stresses presented in Section 3.4.1: that such a stress is to be identified with a flux of molecular momentum (Section 3.4.1). Then, for example, a positive tangential stress $\tau_{yx}$ acting on a face normal to the $y$-axis from below (at lesser values of $y$) has the interpretation that it is equal to the flux of $x$-component momentum through the face in the positive direction of the $y$-axis. (Students should consult the text by Bird, Stewart, and Lightfoot [1960], pp. 76–81, regarding this point.) With reference to Figure 12.1, the component $\tau_{yx}$ acts in the positive $x$-direction on the lower face of the control volume and in the negative $x$-direction on the upper face. Similarly, the component $\tau_{xy}$ acts in the positive

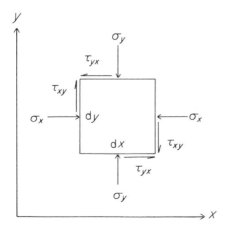

**Figure 12.1**  Definition diagram of normal and tangential surface stresses acting on the faces of a fluid control volume.

$y$-direction on the left face and in the negative $y$-direction on the right face. Similar remarks apply to the other tangential stress components. The normal stresses $\sigma_x$, $\sigma_y$, and $\sigma_z$ act in positive coordinate directions on faces having lower coordinate positions and in negative directions on faces having higher coordinate positions. For example, the normal stress $\sigma_x$ acts in the positive $x$-direction on the left face, and in the negative $x$-direction on the right face (Figure 12.1). This is equivalent to assigning a positive sign to a compressive state.

A net force on the control volume due to one of these stresses exists if the stress varies from one face to another; that is, if a gradient in the stress exists. Consider stresses acting parallel to the $x$-axis: $\sigma_x$, $\tau_{yx}$, and $\tau_{zx}$ (Figure 12.1). The force associated with the normal stress $\sigma_x$ acting on the left face with area $dy\,dz$ equals $\sigma_x\,dy\,dz$. Using a Taylor expansion about the left face, the force associated with $\sigma_x$ acting in the right face with area $dy\,dx$ is $-[\sigma_x + (\partial\sigma_x/\partial x)dx]dy\,dz$, where the negative sign indicates that this force acts in the negative direction of the $x$-axis. The net component of force parallel to the $x$-axis equals the sum of the normal forces acting on the two faces, and is therefore

$$\sigma_x\,dy\,dz - \left(\sigma_x + \frac{\partial\sigma_x}{\partial x}dx\right)dy\,dz = -\frac{\partial\sigma_x}{\partial x}dx\,dy\,dz \qquad (12.1)$$

Thus a positive component of force is associated with a decreasing normal stress $\sigma_x$ in the direction of positive $x$, which is analogous to the term in (10.5) that describes the $x$-component of force associated with the fluid pressure $p$ in Euler's equations.

Turning to tangential forces, fluid external to the control volume exerts a stress $\tau_{zx}$ on the lower face over its area $dx\,dy$ (Figure 12.2); the associated force equals $\tau_{zx}\,dx\,dy$. As a point of reference, momentarily suppose that the control volume is embedded within a one-dimensional flow involving a Newtonian fluid such that $v = w = 0$, and the velocity gradient $\partial u/\partial z$ varies with the $z$-coordinate position. Recall the definition of shear stress (3.8), which is appropriate for describing the stress exerted on a fluid surface from below (at lesser values of $z$). In the present context,

$$\tau_{zx} = -\mu\frac{\partial u}{\partial z} \qquad (12.2)$$

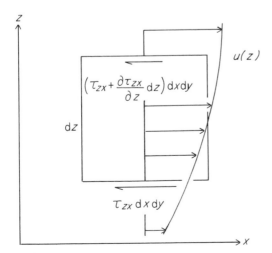

**Figure 12.2** Definition diagram for a variation in shear stress $\tau_{zx}$ across a control volume in presence of a velocity gradient $\partial u/\partial z$.

Since the velocity gradient $\partial u/\partial z$ at the lower face is positive in this example, the force $\tau_{zx}\, dx\, dy$ is a negative quantity, indicating that this force acts in the direction of negative $x$. Now, fluid within the control volume similarly exerts a tangential stress on the upper face. The force associated with this stress may be specified in terms of $\tau_{zx} dx dy$ using a Taylor expansion to first order about the bottom face:

$$\left(\tau_{zx} + \frac{\partial \tau_{zx}}{\partial z} dz\right) dx\, dy \tag{12.3}$$

Noting the sense of shear, this tangential force is a negative quantity based on (12.2). Further, it is equal in magnitude, and of opposite sign, to a tangential force exerted on the upper face from above by fluid external to the control volume. This external force is therefore

$$-\left(\tau_{zx} + \frac{\partial \tau_{zx}}{\partial z} dz\right) dx\, dy \tag{12.4}$$

and the net shearing force associated with $\tau_{zx}$ parallel to the $x$-axis is

$$\tau_{zx}\, dx\, dy - \left(\tau_{zx} + \frac{\partial \tau_{zx}}{\partial z} dz\right) dx\, dy = -\frac{\partial \tau_{zx}}{\partial z} dx\, dy\, dz \tag{12.5}$$

Thus a net force on the control volume exists in the presence of a nonzero gradient $\partial \tau_{zx}/\partial z$.

In providing a physical interpretation of the expansion (12.3), we appealed to the case of a Newtonian fluid using the definition of shear stress involving a negative sign (12.2). The significance of this sign convention will become clear when we generalize the problem to three-dimensional flows (Section 12.2.1). Meanwhile, note that the expression (12.5) for the shearing force on a fluid element is generally valid, regardless of the specific fluid involved.

A similar treatment involving the tangential stress $\tau_{yx}$ leads to the expression $-(\partial \tau_{yx}/\partial dx)\, dy\, dz$ for the net shearing force parallel to the $x$-axis associated with this stress. Summing this expression with (12.1) and (12.5) then gives the net force acting in the direction of $x$:

$$-\left(\frac{\partial \sigma_x}{\partial x} + \frac{\partial \tau_{yx}}{\partial y} + \frac{\partial \tau_{zx}}{\partial z}\right) dx\, dy\, dz \tag{12.6}$$

Dividing by $dx dy dz$, the force per unit volume $\tau_x$ acting in the direction of $x$ is

$$\tau_x = -\left(\frac{\partial \sigma_x}{\partial x} + \frac{\partial \tau_{yx}}{\partial y} + \frac{\partial \tau_{zx}}{\partial z}\right) \tag{12.7}$$

Similar results are obtained for the net forces per unit volume, $\tau_y$ and $\tau_z$, acting in the directions of $y$ and $z$:

$$\tau_y = -\left(\frac{\partial \tau_{xy}}{\partial x} + \frac{\partial \sigma_y}{\partial y} + \frac{\partial \tau_{zy}}{\partial z}\right) \tag{12.8}$$

$$\tau_z = -\left(\frac{\partial \tau_{xz}}{\partial x} + \frac{\partial \tau_{yz}}{\partial y} + \frac{\partial \sigma_z}{\partial z}\right) \tag{12.9}$$

The quantities $\tau_x$, $\tau_y$, and $\tau_z$ are the three components of a vector $\mathbf{S}_V = \langle \tau_x, \tau_y, \tau_z \rangle$ that describes the viscous force per unit volume acting on the control volume.

It is convenient to organize the normal and tangential stresses in the form of a $3 \times 3$ matrix or *stress tensor* $\boldsymbol{\tau}$:

$$\boldsymbol{\tau} = \begin{bmatrix} \sigma_x & \tau_{xy} & \tau_{xz} \\ \tau_{yx} & \sigma_y & \tau_{yz} \\ \tau_{zx} & \tau_{zy} & \sigma_z \end{bmatrix} \tag{12.10}$$

The net force per unit volume $\mathbf{S}_V$ associated with normal and tangential stresses is then denoted as the negative of the divergence of $\boldsymbol{\tau}$, that is $-\nabla \cdot \boldsymbol{\tau}$. To illustrate this, the operator $\nabla$ may be written formally as a $1 \times 3$ matrix, whence

$$-\nabla \cdot \boldsymbol{\tau} = -\begin{bmatrix} \dfrac{\partial}{\partial x} & \dfrac{\partial}{\partial y} & \dfrac{\partial}{\partial z} \end{bmatrix} \begin{bmatrix} \sigma_x & \tau_{xy} & \tau_{xz} \\ \tau_{yx} & \sigma_y & \tau_{yz} \\ \tau_{zx} & \tau_{zy} & \sigma_z \end{bmatrix} \tag{12.11}$$

Students familiar with matrix multiplication will recognize that since the number of columns in the first matrix equals the number of rows in the second, the result is a $1 \times 3$ matrix:

$$-\nabla \cdot \boldsymbol{\tau} = -\left[ \left( \frac{\partial \sigma_x}{\partial x} + \frac{\partial \tau_{yx}}{\partial y} + \frac{\partial \tau_{zx}}{\partial z} \right) \left( \frac{\partial \tau_{xy}}{\partial x} + \frac{\partial \sigma_y}{\partial y} + \frac{\partial \tau_{zy}}{\partial z} \right) \left( \frac{\partial \tau_{xz}}{\partial x} + \frac{\partial \tau_{yz}}{\partial y} + \frac{\partial \sigma_z}{\partial z} \right) \right]$$

$$= \mathbf{S}_V \tag{12.12}$$

The operation $\nabla \cdot \boldsymbol{\tau}$ above is referred to as a *vector–dyadic* dot product.

The development above departs from the convention usually found in fluid mechanics texts: that the viscous force per unit volume equals the positive of the divergence of the stress tensor. The convention used here, that this viscous force equals $-\nabla \cdot \boldsymbol{\tau}$, is the one adopted in the thermodynamics literature, as presented in de Groot and Mazur (1984). (Also see Bird, Stewart, and Lightfoot [1960].) This convention is consistent with our previous use of Taylor expansions to specify surface forces in Chapters 6 and 7. Additional reasons for using this convention will become clear in the next section; these involve the choice of a negative sign in defining shear stress, as in (12.2), and the choice of sign in defining thermodynamic work. It is left as an exercise to develop the necessary geometrical arguments to obtain the equally correct result that the viscous force per unit volume equals the positive divergence $\nabla \cdot \boldsymbol{\tau}$.

Recall from Chapter 11 that the rate-of-strain tensor (11.43) is symmetric such that $\dot{\varepsilon}_{xy} = \dot{\varepsilon}_{yx}$, $\dot{\varepsilon}_{xz} = \dot{\varepsilon}_{zx}$, and $\dot{\varepsilon}_{yz} = \dot{\varepsilon}_{zy}$. Since we wish to relate these strain rates to the stress tensor (12.10), it is reasonable to consider whether $\boldsymbol{\tau}$ possesses a similar symmetry. (There are, in fact, several ways to demonstrate that this symmetry indeed exists.) Consider the off-diagonal stress pair $\tau_{xy}$ and $\tau_{yx}$ acting on a rectangular fluid element (Figure 12.3). We may anticipate that if $\tau_{xy} = \tau_{yx}$, the net torque on the element about the $z$-axis (and thus the angular acceleration $\dot{\omega}_z = d\omega_z/dt$) must equal zero. The force due to $\tau_{yx}$ on face $dx\,dz$ is $\tau_{yx}\,dx\,dz$, and the associated lever arm with respect to the $z$-axis is equal to $dy$. The torque on the fluid element due to $\tau_{yx}$ is therefore $(\tau_{yx}\,dx\,dz)\,dy$; similarly the torque due to $\tau_{xy}$ is $(\tau_{xy}\,dy\,dz)\,dx$. The net torque $\tau$ on the fluid element with volume $dV = dx\,dy\,dz$ is therefore $\tau = (\tau_{xy}\,dy\,dz)\,dx - (\tau_{yx}\,dx\,dz)\,dy = (\tau_{xy} - \tau_{yx})dV$. By Newton's second law, $\tau = \omega_z dI$, where $dI$ is the elementary moment of inertia of the fluid element

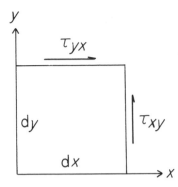

**Figure 12.3** Definition diagram of the shear stress pair, $\tau_{xy}$ and $\tau_{yx}$, exerting torque on a fluid element.

about the $z$-axis. Equating these expressions for $\tau$ and dividing by $dI$,

$$\dot{\omega}_z = (\tau_{xy} - \tau_{yx})\frac{dV}{dI} \tag{12.13}$$

The volume $dV$ is proportional to the third power of the linear dimensions of the fluid element. The moment of inertia $dI$ is proportional to the fifth power of the linear dimensions. As the dimensions of the fluid element approach zero, $dI$ therefore approaches zero faster than $dV$, and the ratio $dV/dI$ becomes infinitely large in the limit. If we insist on physical grounds that $\dot{\omega}_z$ cannot become infinitely large with this contraction of the fluid element, we also must insist that $\tau_{xy} - \tau_{yx} = 0$. It follows that $\tau_{xy} = \tau_{yx}$. Similar results are obtained for the other off-diagonal stress elements in (12.10); thus the tensor $\boldsymbol{\tau}$ is symmetric such that

$$\tau_{xy} = \tau_{yx}; \qquad \tau_{xz} = \tau_{zx}; \qquad \tau_{yz} = \tau_{zy} \tag{12.14}$$

This property of symmetry means that, instead of considering nine stress components, we need to specify only six—three diagonal normal stresses and three off-diagonal tangential stresses—when characterizing the surface force acting on a fluid element.

Recall from Chapter 7 that we developed a kinematical expression of fluid acceleration in the form of a substantive derivative of fluid velocity, $D\mathbf{u}/dt$. Then based on Newton's second law, we equated the product $\rho D\mathbf{u}/Dt$ with the sum of body and surface forces per unit volume, $\mathbf{B}_V + \mathbf{S}_V$, to arrive at a general equation of motion:

$$\rho\frac{D\mathbf{u}}{Dt} = \mathbf{B}_V + \mathbf{S}_V \tag{12.15}$$

The body force per unit volume $\mathbf{B}_V$ is that associated with gravity, which we obtained in developing Euler's equations (Chapter 10). Namely, $\mathbf{B}_V = -\rho g\nabla h$, where $h$ denotes the height (relative to a specified datum) within the gravitational field when the Cartesian coordinate axes have an arbitrary orientation with respect to the gravitational field. Based on the development above, the surface force per unit volume $\mathbf{S}_V = -\nabla \cdot \boldsymbol{\tau}$. Substituting these expressions,

$$\rho\frac{D\mathbf{u}}{Dt} = -\rho g\nabla h - \nabla \cdot \boldsymbol{\tau} \tag{12.16}$$

Resolving this vector form of the equation of motion into components,

$$\rho\frac{Du}{Dt} = -\rho g\frac{\partial h}{\partial x} - \left(\frac{\partial\sigma_x}{\partial x} + \frac{\partial\tau_{yx}}{\partial y} + \frac{\partial\tau_{zx}}{\partial z}\right) \tag{12.17}$$

$$\rho \frac{Dv}{Dt} = -\rho g \frac{\partial h}{\partial y} - \left( \frac{\partial \tau_{xy}}{\partial x} + \frac{\partial \sigma_y}{\partial y} + \frac{\partial \tau_{zy}}{\partial z} \right) \tag{12.18}$$

$$\rho \frac{Dw}{Dt} = -\rho g \frac{\partial h}{\partial z} - \left( \frac{\partial \tau_{xz}}{\partial x} + \frac{\partial \tau_{yz}}{\partial y} + \frac{\partial \sigma_z}{\partial z} \right) \tag{12.19}$$

We now require expressions that relate the six stresses appearing in these equations to rates of fluid strain. These expressions in general depend on the type of fluid; we will start with Newtonian fluids.

## 12.2 NEWTONIAN FLUIDS

### 12.2.1 Constitutive equations

It is convenient to express the normal stresses $\sigma_x$, $\sigma_y$, and $\sigma_z$ in terms of the thermodynamic, or equilibrium, pressure $p$ and *deviatoric stresses* $\tau_{xx}$, $\tau_{yy}$, and $\tau_{zz}$:

$$\sigma_x = p + \tau_{xx}; \qquad \sigma_y = p + \tau_{yy}; \qquad \sigma_z = p + \tau_{zz} \tag{12.20}$$

The subscripts on $\tau$ follow the same convention as that used for off-diagonal elements in (12.10). These deviatoric stresses are to be interpreted as follows. When a fluid is at rest a hydrostatic stress field develops. Deviatoric stresses vanish at all points in the fluid, and the stress on any fluid element equals the fluid pressure $p$, as defined by a thermodynamic equation of state. This may be stated as $\sigma_x = \sigma_y = \sigma_z = p$. Equivalently, the fluid pressure may be taken as the arithmetic average of the three normal stresses:

$$p = \frac{1}{3}(\sigma_x + \sigma_y + \sigma_z) \tag{12.21}$$

It then can be verified by substituting (12.20) into (12.21) that the sum of the three normal deviatoric stresses equals zero. Moreover, (12.21) is valid when fluid motion is like that of a rigid body, since the hydrostatic stress field that develops is not associated with fluid deformation.

More generally, deviatoric components of the normal stresses $\sigma_x$, $\sigma_y$, and $\sigma_z$ are nonzero, and are associated with the rate of fluid deformation during motion. Because real fluids exhibit viscous resistance to deforming motions, these deviatoric stresses therefore must be associated with dissipation of mechanical energy in the form of work performed against friction. It is therefore convenient to define a mechanical pressure $\sigma_0$, in a manner analogous to (12.21), for a viscous fluid in motion:

$$\sigma_0 = \frac{1}{3}(\sigma_x + \sigma_y + \sigma_z) \tag{12.22}$$

The mechanical pressure then does not necessarily coincide in magnitude with $\sigma_x$, $\sigma_y$, or $\sigma_z$, as does the thermodynamic pressure in the case of a static fluid, nor does it generally equal the thermodynamic pressure, except in special cases.

To clarify these points, consider a fluid element in the shape of a sphere. Suppose, as in Section 4.9, that the element is uniformly subjected to a normal stress $\sigma_0$. In absence of motion, $\sigma_0$ is equal in magnitude to the thermodynamic pressure $p$ of fluid within the element. If the stress $\sigma_0$ is slowly increased in a quasi-static, reversible manner, work is performed on the fluid element and, according to the first law of thermodynamics (4.35), this work must involve an increase in the internal energy of the fluid. With slow compression, this energy is readily partitioned among all possible modes of molecular motion—translational, rotational, and vibrational—compatible with the equilibrium (thermodynamic) pressure state. From (4.36) the rate at which work is performed is

$$\frac{dW}{dt} = p\frac{dV}{dt} = \sigma_0\frac{dV}{dt} \tag{12.23}$$

which, based on (11.41), may be written in terms of the rate at which work is performed per unit volume:

$$\frac{dW_V}{dt} = \sigma_0 \nabla \cdot \mathbf{u} = \sigma_0 \dot{\varepsilon} \tag{12.24}$$

Further recall from Section 4.9 that, if the fluid element is rapidly compressed, the mechanical work performed is instantaneously transferred to the translational mode of molecular motion. But this rapid compression also may involve a lag associated with relaxation, as energy is repartitioned among other possible modes compatible with the equilibrium pressure state that otherwise would exist for the same quantity of energy added under quasi-static conditions. Since the translational mode is responsible for exerting a pressure that opposes $\sigma_0$, this pressure is momentarily higher (until relaxation is complete) than would occur in quasi-static compression. The effect is that more work is performed beyond that associated with the resistance that would otherwise be provided by the thermodynamic pressure during quasi-static compression. This effect is characterized in terms of the bulk viscosity $\mu_b$ (Section 4.9). Further note that this effect vanishes with a monatomic gas since, in this case, the only mode of molecular energy is translational.

Let us isolate the elements of this effect in the stress tensor $\boldsymbol{\tau}$. First, substituting (12.20) into (12.22),

$$\sigma_0 = p + \frac{1}{3}(\tau_{xx} + \tau_{yy} + \tau_{zz}) = p + \tau_0 \tag{12.25}$$

The quantity $\tau_0$ is the average deviatoric normal stress. Substituting (12.25) into (12.24) then gives

$$\frac{dW_V}{dt} = p\dot{\varepsilon} + \tau_0\dot{\varepsilon} \tag{12.26}$$

Comparing with (12.24), and based on (12.25), this reveals that $\tau_0$ may be identified as that part of the mechanical pressure that is associated with the relaxation process during which energy is repartitioned among nontranslational modes of molecular motion. Now, using (12.20) the stress tensor $\boldsymbol{\tau}$ may be written as the sum of a *hydrostatic stress tensor* and a *deviatoric stress tensor:*

$$\boldsymbol{\tau} = \begin{bmatrix} p & 0 & 0 \\ 0 & p & 0 \\ 0 & 0 & p \end{bmatrix} + \begin{bmatrix} \tau_{xx} & \tau_{xy} & \tau_{xz} \\ \tau_{yx} & \tau_{yy} & \tau_{yz} \\ \tau_{zx} & \tau_{zy} & \tau_{zz} \end{bmatrix} \tag{12.27}$$

For reasons that will become clear below, let us further decompose the deviatoric stress tensor as follows:

$$
\boldsymbol{\tau} = \begin{bmatrix} p & 0 & 0 \\ 0 & p & 0 \\ 0 & 0 & p \end{bmatrix} + \begin{bmatrix} \tau_0 & 0 & 0 \\ 0 & \tau_0 & 0 \\ 0 & 0 & \tau_0 \end{bmatrix} + \begin{bmatrix} \tau_{xx} - \tau_0 & \tau_{xy} & \tau_{xz} \\ \tau_{yx} & \tau_{yy} - \tau_0 & \tau_{yz} \\ \tau_{zx} & \tau_{zy} & \tau_{zz} - \tau_0 \end{bmatrix} \tag{12.28}
$$

This may be written in the more compact form

$$
\boldsymbol{\tau} = p\mathbf{I} + \tau_0\mathbf{I} + \begin{bmatrix} \tau_{xx} - \tau_0 & \tau_{xy} & \tau_{xz} \\ \tau_{yx} & \tau_{yy} - \tau_0 & \tau_{yz} \\ \tau_{zx} & \tau_{zy} & \tau_{zz} - \tau_0 \end{bmatrix} \tag{12.29}
$$

where $\mathbf{I}$ is the identity matrix:

$$
\mathbf{I} = \begin{bmatrix} 1 & 0 & 0 \\ 0 & 1 & 0 \\ 0 & 0 & 1 \end{bmatrix} \tag{12.30}
$$

Let us similarly decompose the rate-of-strain tensor (11.43):

$$
\begin{bmatrix} \dot{\varepsilon}_x & \dot{\varepsilon}_{xy} & \dot{\varepsilon}_{xz} \\ \dot{\varepsilon}_{yx} & \dot{\varepsilon}_y & \dot{\varepsilon}_{yz} \\ \dot{\varepsilon}_{zx} & \dot{\varepsilon}_{zy} & \dot{\varepsilon}_z \end{bmatrix} = \frac{1}{3}\dot{\varepsilon}\mathbf{I} + \begin{bmatrix} \dot{\varepsilon}_x - \frac{1}{3}\dot{\varepsilon} & \dot{\varepsilon}_{xy} & \dot{\varepsilon}_{xz} \\ \dot{\varepsilon}_{yx} & \dot{\varepsilon}_y - \frac{1}{3}\dot{\varepsilon} & \dot{\varepsilon}_{yz} \\ \dot{\varepsilon}_{zx} & \dot{\varepsilon}_{zy} & \dot{\varepsilon}_z - \frac{1}{3}\dot{\varepsilon} \end{bmatrix} \tag{12.31}
$$

where, based on (11.41) and (11.43), $\dot{\varepsilon} = \nabla \cdot \mathbf{u} = \dot{\varepsilon}_x + \dot{\varepsilon}_y + \dot{\varepsilon}_z$. The last matrix in each of (12.29) and (12.31) possesses the property that its *trace*—the sum of the diagonal elements—equals zero.

Now, for an isotropic Newtonian fluid it is assumed that elements of the rate-of-strain tensor with zero trace are linearly related to corresponding elements of the deviatoric stress tensor with zero trace. The fluid (thermodynamic) pressure $p$, and elements of the rotation tensor (11.42), are not involved since these are not associated with fluid deformation and therefore viscous forces. These linear relations are:

$$
\tau_0 = -\mu_b\dot{\varepsilon} \tag{12.32}
$$

$$
\tau_{xx} - \tau_0 = -2\mu\left(\dot{\varepsilon}_x - \frac{1}{3}\dot{\varepsilon}\right) \tag{12.33}
$$

$$
\tau_{yy} - \tau_0 = -2\mu\left(\dot{\varepsilon}_y - \frac{1}{3}\dot{\varepsilon}\right) \tag{12.34}
$$

$$
\tau_{zz} - \tau_0 = -2\mu\left(\dot{\varepsilon}_z - \frac{1}{3}\dot{\varepsilon}\right) \tag{12.35}
$$

$$
\tau_{xy} = \tau_{yx} = -2\mu\dot{\varepsilon}_{xy} = -2\mu\dot{\varepsilon}_{yx} \tag{12.36}
$$

$$
\tau_{xz} = \tau_{zx} = -2\mu\dot{\varepsilon}_{xz} = -2\mu\dot{\varepsilon}_{zx} \tag{12.37}
$$

$$
\tau_{yz} = \tau_{zy} = -2\mu\dot{\varepsilon}_{yz} = -2\mu\dot{\varepsilon}_{zy} \tag{12.38}
$$

These are, in the language of thermodynamics, *phenomenological relations*. The leading negative signs are a conscious choice, and are consistent with the convention adopted with what are probably more familiar examples of phenomenological relations: Darcy's law, Fourier's law, and Fick's law. Recall that by this convention a positive flux arises from a negative gradient. The fluxes involved here are of molecular momentum. (Recall the kinetic interpretation of shear stress covered in Chapter 3.) Coefficients of proportionality—$\mu$ and $\mu_b$ in the relations above—are referred to as *phenomenological coefficients*. The factor 2 appearing in each relation, although not essential, is convenient, as we will see below.

The first of these relations is analogous to the definition of the bulk modulus (3.43) of an elastic substance; the (bulk) modulus of elasticity $E$ is replaced by the bulk viscosity $\mu_b$, and the volumetric strain $dV/V$ is replaced by the volumetric strain rate $\dot{\varepsilon} = (1/V)dV/dt$. The last three describe relations between shear stresses and rates of shear strain. These are anologous to the definition of the shear modulus of an elastic substance; the shear modulus is replaced by the shear viscosity $\mu$, and the shear strain is replaced by the shear strain rate. Returning to the second, third, and fourth, these describe relations between normal deviatoric stresses and fluid dilation, accounting for the effect of bulk viscosity. For example, the quantity $\tau_{xx} - \tau_0$ may be regarded as that component of the mechanical pressure associated with the purely translational mode of molecular motion, and therefore with molecular momentum. Qualitatively, if $\tau_0$ is associated with an apparent resistance to strain involving relaxation, the difference $\tau_{xx} - \tau_0$ is associated with a "true" viscous resistance to strain. That is, this viscous normal stress arises due to interaction among molecules whose average translational motion (the fluid velocity) varies in the x-direction. This produces a resistance to fluid dilation, which, analogous to the kinetic interpretation of viscous shear stress (Section 3.4.1), is proportional to the net flux of molecular momentum across planes normal to the x-axis.

This raises a question: Why are elements of the rate-of-strain tensor with zero trace used in forming the phenomenological relations, rather than proposing a direct relation between elements of the deviatoric stress tensor and the rate-of-strain tensor with nonzero trace? The result of the former choice is that, whereas the shear elements are unaffected, the volumetric dilation rate $\dot{\varepsilon}$ appears in each of the normal deviatoric stress relations, (12.33) through (12.35). For example, the normal stress $\tau_{xx} - \tau_0$ is proportional to the linear dilation rate $\dot{\varepsilon}_x$ minus one-third the volumetric dilation rate $\dot{\varepsilon}$. The full answer to this question resides in the tensor analysis leading to these relations; students familiar with tensor analysis are urged to examine Batchelor (1970). We will momentarily defer answering this question, and return to it when we examine properties of a modified version of the phenomenological relations (12.32) through (12.38). We may anticipate, meanwhile, that the answer resides in the definition of mechanical pressure, and the fact that a fluid element cannot undergo dilation in one dimension without concomitant changes in the normal stresses acting in the other two dimensions.

Substituting (12.32) for $\tau_0$ in (12.33) through (12.35), then making use of (11.41), (11.43), and (12.20),

$$\sigma_x = p + \left(\frac{2}{3}\mu - \mu_b\right)\nabla \cdot \mathbf{u} - 2\mu\frac{\partial u}{\partial x} \tag{12.39}$$

$$\sigma_y = p + \left(\frac{2}{3}\mu - \mu_b\right)\nabla \cdot \mathbf{u} - 2\mu\frac{\partial v}{\partial y} \tag{12.40}$$

$$\sigma_z = p + \left(\frac{2}{3}\mu - \mu_b\right)\nabla \cdot \mathbf{u} - 2\mu\frac{\partial w}{\partial z} \tag{12.41}$$

$$\tau_{xy} = \tau_{yx} = -\mu\left(\frac{\partial u}{\partial y} + \frac{\partial v}{\partial x}\right) \tag{12.42}$$

$$\tau_{xz} = \tau_{zx} = -\mu\left(\frac{\partial u}{\partial z} + \frac{\partial w}{\partial x}\right) \tag{12.43}$$

$$\tau_{yz} = \tau_{zy} = -\mu\left(\frac{\partial v}{\partial z} + \frac{\partial w}{\partial y}\right) \tag{12.44}$$

These relations require specifying two fluid properties, the shear viscosity $\mu$ and the bulk viscosity $\mu_b$. Kinetic theory suggests that, with monatomic gases, the quantity $\sqrt{\mu_b/\mu}$ is proportional to the ratio of the volume occupied by molecules to the total volume of gas; therefore with low-density gases involving simple binary collisions, $\mu_b \to 0$. Apart from this theoretical conclusion, direct measurement of $\mu_b$ is difficult. With dense gases and liquids, $\mu_b$ evidently is very small, and usually is neglected. Bulk viscosity, however, may become important in flows involving rapid pressure fluctuations (for example, shock waves) and when flows are influenced by chemical reactions or magnetic phenomena.

The assumption that $\mu_b$ is negligible for Newtonian fluids is referred to as *Stokes's hypothesis* (wherein the idea that $\mu_b \to 0$ is obtained in a different manner; see, for example, Schlichting [1979]). With $\mu_b = 0$ in (12.39), (12.40), and (12.41), substitution of these equations into (12.22) then reveals that this hypothesis is equivalent to assuming that $\sigma_0 = p$; that is, the mechanical pressure is equal to the thermodynamic pressure. This also means that, according to (12.25), the average of the normal deviatoric stresses $\tau_0$ equals zero. Stokes's hypothesis leads to a "working set" of phenomenological relations:

$$\sigma_x = p + \frac{2}{3}\mu\nabla \cdot \mathbf{u} - 2\mu\frac{\partial u}{\partial x} \tag{12.45}$$

$$\sigma_y = p + \frac{2}{3}\mu\nabla \cdot \mathbf{u} - 2\mu\frac{\partial v}{\partial y} \tag{12.46}$$

$$\sigma_z = p + \frac{2}{3}\mu\nabla \cdot \mathbf{u} - 2\mu\frac{\partial w}{\partial z} \tag{12.47}$$

$$\tau_{xy} = \tau_{yx} = -\mu\left(\frac{\partial u}{\partial y} + \frac{\partial v}{\partial x}\right) \tag{12.48}$$

$$\tau_{xz} = \tau_{zx} = -\mu\left(\frac{\partial u}{\partial z} + \frac{\partial w}{\partial x}\right) \tag{12.49}$$

$$\tau_{yz} = \tau_{zy} = -\mu\left(\frac{\partial v}{\partial z} + \frac{\partial w}{\partial y}\right) \tag{12.50}$$

These are referred to as the constitutive equations for isotropic Newtonian fluids. Let us consider several important properties of these equations and, in doing so, further clarify their significance.

The relation between stress and strain rate, as characterized by the constitutive equations, must be independent of the orientation of axes in an isotropic fluid.

Consider a particular flow field $\mathbf{u}(x, y, z)$ relative to a fixed $xyz$-coordinate system. The stress field at any coordinate position is entirely characterized by the extant velocity gradients appearing in the constitutive equations, such that the net viscous force $(-\nabla \cdot \boldsymbol{\tau})_{xyz}$ acting on a fluid element at a given position P is specified. Keeping the flow field the same, envision a new $x'y'z'$-coordinate system with arbitrary orientation relative to the original one. The stress field at the same position P is now characterized by a new set of extant velocity gradients in the constitutive equations, defined relative to the new coordinate system, such that the net viscous force $(-\nabla \cdot \boldsymbol{\tau})_{x'y'z'}$ acting on a fluid element at the specified position P coincides globally with $(-\nabla \cdot \boldsymbol{\tau})_{xyz}$. The constitutive equations must be such that they are independent of the particular coordinate axes; that is, they are *invariant* with a rotation of coordinate axes. This property of invariance is manifest in a simple way. Any interchange of the axis pairs $(x, u)$, $(y, v)$, and $(z, w)$ in (12.45) through (12.50), for example, $x \to y$, $u \to v$, $y \to z$, $v \to w$, $z \to x$, $w \to u$, retrieves the same set of equations. In this regard the coefficients $2\mu/3$ and $\mu$ must be the same in each equation for the invariant property to hold. Further note that the expression (12.49) for the shear stress $\tau_{zx}$ reduces to the definition (12.2) first introduced in Chapter 3 for the special case of simple shear involving one-dimensional flow, such that $\partial w/\partial x = 0$.

Let us now reexamine the appearance of the volumetric dilation rate $\dot{\varepsilon}$ in the normal deviatoric stress relations (12.39) through (12.41), and (12.45) through (12.47). It is sufficient to consider the latter equations since the conclusions we obtain for these also can be obtained for the former, more general relations involving the bulk viscosity $\mu_b$. Envision a fluid element whose motion consists of purely one-dimensional dilation parallel to the $x$-axis. Thus $v = w = 0$, and the deviatoric parts of (12.45), (12.46), and (12.47) become

$$\tau_{xx} = \frac{2}{3}\mu \frac{\partial u}{\partial x} - 2\mu \frac{\partial u}{\partial x} \tag{12.51}$$

$$\tau_{yy} = \frac{2}{3}\mu \frac{\partial u}{\partial x} \tag{12.52}$$

$$\tau_{zz} = \frac{2}{3}\mu \frac{\partial u}{\partial x} \tag{12.53}$$

Suppose that $\partial u/\partial x > 0$, which implies extension of the fluid element in the $x$-direction. One may envision that a viscous drag $\tau_{xx}$ is exerted on the element as the fluid is stretched due to the gradient $\partial u/\partial x$. Since the stress $\tau_{xx}$ on the element is negative in this example, the normal stress $\sigma_x$ on the element in the $x$-direction is reduced according to (12.45). Moreover, since $\mu_b = 0$ in this case, by definition the mechanical pressure $\sigma_0$ must equal the fluid pressure $p$; this means that, according to (12.25), the average deviatoric normal stress $\tau_0 = 0$, and therefore $\tau_{xx} + \tau_{yy} + \tau_{zz} = 0$. With nonzero $\tau_{xx}$, this, in turn, requires that $\tau_{yy}$ and $\tau_{zz}$ become nonzero due to the gradient $\partial u/\partial x$, despite zero gradients $\partial v/\partial y$ and $\partial w/\partial z$; $\tau_{yy}$ and $\tau_{zz}$ increase by an amount given by (12.52) and (12.53). The condition that $\tau_0 = \frac{1}{3}(\tau_{xx} + \tau_{yy} + \tau_{zz}) = 0$ is therefore satisfied only if the expression for $\tau_{xx}$ contains the term $(2\mu/3)\partial u/\partial x$. A similar result occurs with flows that are more complicated than one-dimensional dilation. Thus the constitutive equations are constructed to account for this interaction—that a fluid element cannot undergo dilation in one dimension without concomitant changes in the normal stresses acting in the other two dimensions—requiring the appearance of $\ddot{\varepsilon}$ in each normal stress relation.

## 12.2.2 Navier–Stokes equations

Substituting the constitutive equations, (12.45) through (12.50), into the equations of motion, (12.17), (12.18), and (12.19), then rearranging terms,

$$\rho \frac{Du}{Dt} = -\rho g \frac{\partial h}{\partial x} - \frac{\partial p}{\partial x}$$

$$+ \frac{\partial}{\partial x}\left[\mu\left(2\frac{\partial u}{\partial x} - \frac{2}{3}\nabla \cdot \mathbf{u}\right)\right] + \frac{\partial}{\partial y}\left[\mu\left(\frac{\partial u}{\partial y} + \frac{\partial v}{\partial x}\right)\right] + \frac{\partial}{\partial z}\left[\mu\left(\frac{\partial w}{\partial x} + \frac{\partial u}{\partial z}\right)\right]$$

$$(12.54)$$

$$\rho \frac{Dv}{Dt} = -\rho g \frac{\partial h}{\partial y} - \frac{\partial p}{\partial y}$$

$$+ \frac{\partial}{\partial y}\left[\mu\left(2\frac{\partial v}{\partial y} - \frac{2}{3}\nabla \cdot \mathbf{u}\right)\right] + \frac{\partial}{\partial z}\left[\mu\left(\frac{\partial v}{\partial z} + \frac{\partial w}{\partial y}\right)\right] + \frac{\partial}{\partial x}\left[\mu\left(\frac{\partial u}{\partial y} + \frac{\partial v}{\partial x}\right)\right]$$

$$(12.55)$$

$$\rho \frac{Dw}{Dt} = -\rho g \frac{\partial h}{\partial z} - \frac{\partial p}{\partial z}$$

$$+ \frac{\partial}{\partial z}\left[\mu\left(2\frac{\partial w}{\partial z} - \frac{2}{3}\nabla \cdot \mathbf{u}\right)\right] + \frac{\partial}{\partial x}\left[\mu\left(\frac{\partial w}{\partial x} + \frac{\partial u}{\partial z}\right)\right] + \frac{\partial}{\partial y}\left[\mu\left(\frac{\partial v}{\partial z} + \frac{\partial w}{\partial y}\right)\right]$$

$$(12.56)$$

These equations express Newton's second law for the motion of a compressible Newtonian fluid, and form the cornerstone of all fluid mechanics dealing with Newtonian fluids; they are referred to as the *Navier–Stokes equations*.

The Navier–Stokes equations can be greatly simplified for the case of an incompressible fluid. Then, by conservation of mass, $\nabla \cdot \mathbf{u} = 0$ based on (8.17), and the terms $(2/3)\nabla \cdot \mathbf{u}$ vanish. If it is further assumed that temperature variations are small, the dynamic viscosity $\mu$ is essentially constant, so $\mu$ is taken outside the differentials. Using the $x$-component (12.54) for illustration,

$$\rho \frac{Du}{Dt} = -\rho g \frac{\partial h}{\partial x} - \frac{\partial p}{\partial x} + 2\mu\frac{\partial}{\partial x}\left(\frac{\partial u}{\partial x}\right) + \mu\frac{\partial}{\partial y}\left(\frac{\partial u}{\partial y} + \frac{\partial v}{\partial x}\right) + \mu\frac{\partial}{\partial z}\left(\frac{\partial w}{\partial x} + \frac{\partial u}{\partial z}\right)$$

$$(12.57)$$

Let us expand this and regroup terms:

$$\rho \frac{Du}{Dt} = -\rho g \frac{\partial h}{\partial x} - \frac{\partial p}{\partial x} + \mu\frac{\partial^2 u}{\partial x^2} + \mu\frac{\partial^2 u}{\partial y^2} + \mu\frac{\partial^2 u}{\partial z^2}$$

$$+ \mu\frac{\partial}{\partial x}\left(\frac{\partial u}{\partial x}\right) + \mu\frac{\partial}{\partial y}\left(\frac{\partial v}{\partial x}\right) + \mu\frac{\partial}{\partial z}\left(\frac{\partial w}{\partial x}\right) \qquad (12.58)$$

Recall that the order of partial differentiation in the last two terms is unimportant; for example,

$$\frac{\partial}{\partial y}\left(\frac{\partial v}{\partial x}\right) = \frac{\partial^2 v}{\partial y \partial x} = \frac{\partial^2 v}{\partial x \partial y} = \frac{\partial}{\partial x}\left(\frac{\partial v}{\partial y}\right) \tag{12.59}$$

Thus, after rearranging the last two terms in (12.58), this equation becomes

$$\rho\frac{Du}{Dt} = -\rho g\frac{\partial h}{\partial x} - \frac{\partial p}{\partial x} + \mu\nabla^2 u + \mu\frac{\partial}{\partial x}\left(\frac{\partial u}{\partial x} + \frac{\partial v}{\partial y} + \frac{\partial w}{\partial z}\right) \tag{12.60}$$

The parenthetical part of (12.60) equals $\nabla \cdot \mathbf{u}$, which, as above, is zero for an incompressible fluid. With similar simplifications of (12.55) and (12.56), the Navier–Stokes equations for incompressible fluid flow are:

$$\rho\frac{Du}{Dt} = -\rho g\frac{\partial h}{\partial x} - \frac{\partial p}{\partial x} + \mu\nabla^2 u \tag{12.61}$$

$$\rho\frac{Dv}{Dt} = -\rho g\frac{\partial h}{\partial y} - \frac{\partial p}{\partial y} + \mu\nabla^2 v \tag{12.62}$$

$$\rho\frac{Dw}{Dt} = -\rho g\frac{\partial h}{\partial z} - \frac{\partial p}{\partial z} + \mu\nabla^2 w \tag{12.63}$$

These have the compact vector form:

$$\frac{D\mathbf{u}}{Dt} = -g\nabla h - \frac{1}{\rho}\nabla p + \nu\nabla^2\mathbf{u} \tag{12.64}$$

where the kinematic viscosity $\nu = \mu/\rho$.

In the case of inviscid flow, the viscous term vanishes and we retrieve Euler's equation (10.11):

$$\frac{D\mathbf{u}}{Dt} = -g\nabla h - \frac{1}{\rho}\nabla p \tag{12.65}$$

## 12.2.3 Energy equation

We are now in a position to modify the energy equation (9.21) so that it includes a description of the work performed against the frictional effects of viscosity. Rather than considering conservation of energy associated with a control volume that is fixed in space, as we did in developing (9.21), let us consider here an elementary volume of fluid as it moves with a flow. This elementary volume may be treated as a thermodynamic system, for which we begin by writing a general form of the first law of thermodynamics:

$$\frac{dQ}{dt} = \frac{DE}{Dt} + \frac{dW}{dt} \tag{12.66}$$

The first law in this form states that the quantity of heat $dQ$ added to the fluid element during the interval $dt$ is partitioned between a change in the total energy $DE$ of the element and a quantity of work $dW$ performed by the element on surrounding fluid. The change in total energy $DE$ involves both thermal and mechanical forms. The substantive derivative $DE/Dt$ describes the rate of change of total energy as viewed by an observer moving with the fluid element, and therefore includes local and convective parts (Chapter 7). The notation d/d$t$ used with $Q$ and $W$ indicates that these are not state variables.

Consider a fluid element of mass $m$ within a flow. At the instant we observe the element, it is rectangular with edges of length d$x$, d$y$, and d$z$, has volume d$V = $ d$x$d$y$d$z$, and possesses density $\rho$. The mass of the element is thus $m = \rho$d$V$. At the same instant the element is positioned at height $h$ in a gravitational field, possesses internal energy per unit mass $U$, and velocity $\mathbf{u} = \langle u, v, w \rangle$. While observing the element over the interval d$t$, its shape and volume may change, and therefore its density may change since its mass remains the same. However, since we wish to compute the change in the energy state of the element in the limit as d$t$ approaches zero, we may specify the density without considering its rate of change, unlike an Eulerian view when the control volume is fixed in space. The change in total energy D$E$ of the element during a small interval d$t$ in general consists of a change in its internal energy $m$D$U$, a change in its potential energy $mg$D$h$, and a change in its kinetic energy $\frac{1}{2}m$D$(u^2 + v^2 + w^2)$. Thus

$$\frac{DE}{Dt} = \rho\frac{DU}{Dt}dV + \rho g\frac{Dh}{Dt}dV + \frac{1}{2}\rho\frac{D}{Dt}(u^2 + v^2 + w^2)\,dV \qquad (12.67)$$

Since the height $h$ serves merely as a vertical coordinate, the local part of D$h$/D$t$ equals zero; that is, $\partial h/\partial t = 0$, so D$h$/D$t = \mathbf{u} \cdot \nabla h$. Using this fact, and rewriting the kinetic energy term in vector form,

$$\frac{DE}{Dt} = \rho\frac{DU}{Dt}dV + \rho g\mathbf{u} \cdot \nabla h\,dV + \rho\mathbf{u} \cdot \frac{D\mathbf{u}}{Dt}\,dV \qquad (12.68)$$

Consider the quantity of heat d$Q$ in (12.66). According to the sign convention of (12.66) a positive value of d$Q$ indicates that this quantity is added to the fluid element during d$t$. Recall that in deriving our first version of the energy equation (9.21) we obtained the expression $-K_T\nabla^2 T$ for the flux of heat per unit volume by conduction out of a fluid element. It follows that by changing the sign of this expression,

$$\frac{dQ}{dt} = K_T\nabla^2 T\,dV \qquad (12.69)$$

Here we have neglected radiation, and assumed that the thermal conductivity $K_T$ is constant.

Turning to the quantity of work d$W$ in (12.66), a positive value of d$W$ indicates that the fluid element performs work on its surroundings. This work is associated with elements of the stress tensor $\boldsymbol{\tau}$. Consider first the normal stress $\sigma_x$ (Figure 12.4). The force on the left face associated with $\sigma_x$ is $\sigma_x$d$y$d$z$. Suppose that this force displaces the left face by a small distance $u$d$t$ during d$t$. Then the rate at which work is performed on the left face is $\sigma_x u$d$y$d$z$. Now, since this quantity of work is performed on the element by surrounding fluid, it is therefore assigned a negative value according to the sign convention of (12.66):

$$-\sigma_x u\,dy\,dz \qquad (12.70)$$

The stress $\sigma_x$ and the velocity component $u$ may vary between the left and right faces. The rate at which work is performed on the right face to first order is therefore equal to $-[\sigma_x u + (\partial/\partial x)(\sigma_x u)dx]dy\,dz$, where the negative sign arises because the normal stress at the right face acts in the direction of negative $x$. Now, since

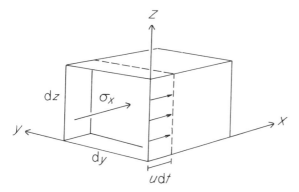

**Figure 12.4**   Control volume where the normal stress $\sigma_x$ performs work on the face $dy\,dz$, equal to $\sigma_x u\,dy\,dz\,dt$.

this quantity of work is performed on the element by surrounding fluid, it also is assigned a negative value according to (12.66):

$$- \left\{ - \left[ \sigma_x u + \frac{\partial}{\partial x}(\sigma_x u)\,dx \right] dy\,dz \right\} \tag{12.71}$$

The net rate at which work is performed by fluid within the element on surrounding fluid is therefore the sum of (12.70) and (12.71):

$$\frac{dW_{\sigma_x}}{dt} = \frac{\partial}{\partial x}(\sigma_x u)\,dV \tag{12.72}$$

As a point of reference, (12.72) may be expanded to $(\sigma_x \partial u/\partial x + u\partial\sigma_x/\partial x)\,dV$. A positive value of the first term in this expansion implies a net (infinitesimal) dilation of the fluid element during $dt$, which is consistent with the idea embodied in (4.36) and Figure 4.7, that work is performed by the fluid element on its surroundings.

Now consider work associated with the tangential stress components in $\tau$, for example $\tau_{zx}$. The force associated with $\tau_{zx}$ acting on the bottom face of the fluid element is $\tau_{zx}\,dx\,dy$, and the velocity parallel to $\tau_{zx}$ at the bottom face is $u$. The rate at which work is performed on the bottom face is therefore $\tau_{zx} u\,dx\,dy$. Since this quantity of work is performed on the element by surrounding fluid, it is therefore assigned a negative value according to (12.66). The stress $\tau_{zx}$ and the velocity component $u$ may vary between the bottom and top faces. Following a similar development to that used above for $\sigma_x$, and maintaining the appropriate sign convention, the rate at which work is performed by fluid within the element on surrounding fluid is

$$\frac{dW_{\tau_{zx}}}{dt} = \frac{\partial}{\partial z}(\tau_{zx} u)\,dV \tag{12.73}$$

After treating the remaining normal and tangential stresses in a similar manner, the rate at which work is performed by fluid within the element on surrounding fluid is

$$\frac{dW}{dt} = \left[ \frac{\partial}{\partial x}(\sigma_x u + \tau_{xy} v + \tau_{xz} w) \right.$$

$$\left. + \frac{\partial}{\partial y}(\tau_{yx} u + \sigma_y v + \tau_{yz} w) + \frac{\partial}{\partial z}(\tau_{zx} u + \tau_{zy} v + \sigma_z w) \right] dV \tag{12.74}$$

This has the compact vector form:

$$\frac{dW}{dt} = \nabla \cdot (\mathbf{u} \cdot \boldsymbol{\tau}) dV \tag{12.75}$$

Now, substituting (12.68), (12.69), and (12.75) into the first law (12.66), dividing through by $dV$ and rearranging,

$$\rho \frac{DU}{Dt} + \rho \mathbf{u} \cdot \frac{D\mathbf{u}}{Dt} = K_T \nabla^2 T - \rho g \mathbf{u} \cdot \nabla h - \nabla \cdot (\mathbf{u} \cdot \boldsymbol{\tau}) \tag{12.76}$$

This is a general form of the energy equation for viscous flow.

Let us now specifically consider a Newtonian flow. We begin by expanding (12.74) using the product rule, then selectively substitute the relations (12.20) and rearrange terms. For example, the first term on the right side of the equal sign in (12.74) becomes

$$p\frac{\partial u}{\partial x} + u\frac{\partial \sigma_x}{\partial x} + v\frac{\partial \tau_{xy}}{\partial x} + w\frac{\partial \tau_{xz}}{\partial x} + \tau_{xx}\frac{\partial u}{\partial x} + \tau_{xy}\frac{\partial v}{\partial x} + \tau_{xz}\frac{\partial w}{\partial x} \tag{12.77}$$

After expanding the remaining parts of (12.74), we may identify three groups of terms. The first group includes three terms, each of which involves the product of pressure $p$ and a velocity gradient. This group may be written as $p\nabla \cdot \mathbf{u}$. A second group includes nine terms, each of which involves the product of a velocity and a stress gradient. Using (12.12), this group may be written as $\mathbf{u} \cdot (\nabla \cdot \boldsymbol{\tau})$. A third group includes nine terms, each of which involves the product of a stress and a velocity gradient. Substituting the constitutive equations, (12.45) through (12.50), into the terms of this third group and rearranging leads to the expression:

$$\begin{aligned}
\Phi = {} & 2\mu\left[\left(\frac{\partial u}{\partial x}\right)^2 + \left(\frac{\partial v}{\partial y}\right)^2 + \left(\frac{\partial w}{\partial z}\right)^2\right] \\
& + \mu\left[\left(\frac{\partial u}{\partial y} + \frac{\partial v}{\partial x}\right)^2 + \left(\frac{\partial u}{\partial z} + \frac{\partial w}{\partial x}\right)^2 + \left(\frac{\partial v}{\partial z} + \frac{\partial w}{\partial y}\right)^2\right] \\
& - \frac{2}{3}\mu(\nabla \cdot \mathbf{u})^2
\end{aligned} \tag{12.78}$$

This is referred to as the *dissipation function* for a Newtonian fluid. Thus (12.75) may now be written as

$$\frac{dW}{dt} = [p\nabla \cdot \mathbf{u} + \mathbf{u} \cdot (\nabla \cdot \boldsymbol{\tau}) - \Phi] dV \tag{12.79}$$

Observe that the first term on the right side of (12.79) describes the rate at which work is performed during volumetric dilation. The term involving the divergence of the stress tensor describes the rate at which work is performed in translating and deforming a fluid element. The dissipation function describes the rate at which mechanical energy is extracted from the main fluid motion and irreversibly transformed into heat energy due to the effect of viscosity.

Substituting (12.68), (12.69), and (12.79) into the first law (12.66), dividing through by $dV$, and rearranging,

$$\rho\frac{DU}{Dt} + \rho\mathbf{u} \cdot \frac{D\mathbf{u}}{Dt} = K_T\nabla^2 T - \rho g\mathbf{u} \cdot \nabla h - p\nabla \cdot \mathbf{u} - \mathbf{u} \cdot (\nabla \cdot \boldsymbol{\tau}) + \Phi$$

(12.80)

$$\rho\frac{DU}{Dt} + \mathbf{u} \cdot \left(\rho\frac{D\mathbf{u}}{Dt} + \rho g\nabla h + \nabla \cdot \boldsymbol{\tau}\right) = K_T\nabla^2 T - p\nabla \cdot \mathbf{u} + \Phi$$

(12.81)

Observe that, according to the equation of motion (12.16), $\rho D\mathbf{u}/dt + \rho g\nabla h + \nabla \cdot \boldsymbol{\tau} = 0$. Thus, the second term in (12.81) equals zero, so this equation is simplified to

$$\rho\frac{DU}{Dt} = K_T\nabla^2 T - p\nabla \cdot \mathbf{u} + \Phi$$

(12.82)

This is the energy equation for a Newtonian flow. For an ideal gas, using $dU = c_V dT$,

$$\rho c_V \frac{DT}{Dt} = K_T\nabla^2 T - p\nabla \cdot \mathbf{u} + \Phi$$

(12.83)

For a gas that is treated as incompressible, $\nabla \cdot \mathbf{u} = 0$, and

$$\rho c_V \frac{DT}{Dt} = K_T\nabla^2 T + \Phi$$

(12.84)

where the dissipation function reduces to

$$\Phi = 2\mu\left[\left(\frac{\partial u}{\partial x}\right)^2 + \left(\frac{\partial v}{\partial y}\right)^2 + \left(\frac{\partial w}{\partial z}\right)^2\right]$$

$$+ \mu\left[\left(\frac{\partial u}{\partial y} + \frac{\partial v}{\partial x}\right)^2 + \left(\frac{\partial u}{\partial z} + \frac{\partial w}{\partial x}\right)^2 + \left(\frac{\partial v}{\partial z} + \frac{\partial w}{\partial y}\right)^2\right] \quad (12.85)$$

For an incompressible liquid, $c_V$ is normally replaced by $c$ in (12.84).

Note that for the special case of inviscid flow, the dissipation function in (12.83) and (12.84) vanishes and these equations reduce to (10.16) and (10.17).

## 12.3 INCOMPRESSIBLE NEWTONIAN FLOWS

Because flows involving incompressible Newtonian fluids are so important in geology, let us collect here the set of equations governing these flows:

$$\frac{D\mathbf{u}}{Dt} = -g\nabla h - \frac{1}{\rho}\nabla p + \nu\nabla^2\mathbf{u}$$

(12.86)

$$\nabla \cdot \mathbf{u} = 0$$

(12.87)

$$\rho c_V \frac{DT}{Dt} = K_T \nabla^2 T + \Phi \qquad (12.88)$$

$$\rho = \rho(T, p) \qquad (12.89)$$

$$\mu = \mu(T) \qquad (12.90)$$

The first, second, and third, respectively, describe conservation of momentum, mass, and energy. The fourth is an equation of state, in which normally the dependence of density on pressure can be neglected. The fifth describes viscosity as a function of temperature. These five equations, together with initial and boundary conditions, completely specify the flow field of an incompressible Newtonian fluid.

## 12.4  NON-NEWTONIAN FLUIDS:
## THE EXAMPLE OF GLACIER ICE

Equations (12.17), (12.18), and (12.19) are general expressions of Newton's second law for the motion of a viscous fluid. They are therefore valid for non-Newtonian fluids as well as Newtonian fluids. Any differences in the treatment of these two types of fluid thus reside in differences between the respective constitutive equations relating stresses and strain rates. The variety of non-Newtonian rheologies is great. So rather than attempt to systematically treat these, let us briefly examine glacier ice as an example of a non-Newtonian fluid.

### 12.4.1  Constitutive equations

The constitutive equations for ice are obtained from a general form of Glen's law (3.40), under the condition that these equations must be invariant with a rotation of coordinate axes. We will not go through their full derivation, but rather only cover key ingredients to illustrate the flavor of the problem.

As with an incompressible Newtonian fluid, it is assumed that elements of the rate-of-strain tensor are related to corresponding elements of the deviatoric stress tensor. For example,

$$\dot{\varepsilon}_x = \lambda \tau_{xx}; \qquad \dot{\varepsilon}_{yx} = \lambda \tau_{yx} \qquad (12.91)$$

Similar expressions are written for the remaining elements in the two tensors. To evaluate the factor $\lambda$, it is useful to define the effective strain rate $\dot{\varepsilon}$ and the effective shear stress $\tau$:

$$2\dot{\varepsilon}^2 = \dot{\varepsilon}_x^2 + \dot{\varepsilon}_y^2 + \dot{\varepsilon}_z^2 + 2(\dot{\varepsilon}_{xy}^2 + \dot{\varepsilon}_{yz}^2 + \dot{\varepsilon}_{zx}^2) \qquad (12.92)$$

$$2\tau^2 = \tau_{xx}^2 + \tau_{yy}^2 + \tau_{zz}^2 + 2(\tau_{xy}^2 + \tau_{yz}^2 + \tau_{zx}^2) \qquad (12.93)$$

where $\dot{\varepsilon}$ and $\tau$ are taken as positive quantities. Substituting the six expressions represented by (12.91) into (12.92), then comparing the result with (12.93), we obtain

$$\dot{\varepsilon} = \lambda \tau \qquad (12.94)$$

The quantities $\dot{\varepsilon}$ and $\tau$ are defined such that they are invariant with a rotation of axes. It is assumed that these quantities are related according to a general form of

Glen's law:

$$2\dot{\varepsilon} = -A\tau^m \tag{12.95}$$

The negative sign follows the convention adopted above for Newtonian fluids. The factor 2 is convenient, but not essential. Recall from Chapter 3 that $A$ and $m$ are determined empirically, normally for the case of polycrystalline ice whose bulk behavior is that of an isotropic fluid. The parameter $A$ varies with ice temperature, crystal size, and concentration of impurities. A value of $m = 3$ normally is adopted in glacier mechanics. Equating (12.94) with (12.95) and solving for $\lambda$,

$$\lambda = -\frac{1}{2}A\tau^{m-1} \tag{12.96}$$

Substituting this into the six expressions represented by (12.91), and making use of (11.43) and (11.44), the constitutive equations for polycrystalline ice are:

$$2\frac{\partial u}{\partial x} = -A\tau^{m-1}\tau_{xx} \tag{12.97}$$

$$2\frac{\partial v}{\partial y} = -A\tau^{m-1}\tau_{yy} \tag{12.98}$$

$$2\frac{\partial w}{\partial z} = -A\tau^{m-1}\tau_{zz} \tag{12.99}$$

$$\frac{\partial u}{\partial y} + \frac{\partial v}{\partial x} = -A\tau^{m-1}\tau_{xy} = -A\tau^{m-1}\tau_{yx} \tag{12.100}$$

$$\frac{\partial u}{\partial z} + \frac{\partial w}{\partial x} = -A\tau^{m-1}\tau_{xz} = -A\tau^{m-1}\tau_{zx} \tag{12.101}$$

$$\frac{\partial v}{\partial z} + \frac{\partial w}{\partial y} = -A\tau^{m-1}\tau_{yz} = -A\tau^{m-1}\tau_{zy} \tag{12.102}$$

As a point of reference, the expression (12.101) involving $\tau_{zx}$ reduces to the definition of Glen's law first introduced in Chapter 3 in the case of simple shear where $\partial u/\partial z$ is the only nonzero part of (12.92) and $\tau_{zx}$ is the only nonzero part of (12.93).

## 12.4.2 Equations of motion

Glacier flow by ice deformation is sufficiently slow that acceleration terms in the equations of motion, (12.17), (12.18), and (12.19), can be neglected. These equations thus reduce to expressions that describe a balance between body and surface forces:

$$\rho g\frac{\partial h}{\partial x} = -\frac{\partial p}{\partial x} - \left(\frac{\partial \tau_{xx}}{\partial x} + \frac{\partial \tau_{yx}}{\partial y} + \frac{\partial \tau_{zx}}{\partial z}\right) \tag{12.103}$$

$$\rho g\frac{\partial h}{\partial y} = -\frac{\partial p}{\partial y} - \left(\frac{\partial \tau_{xy}}{\partial x} + \frac{\partial \tau_{yy}}{\partial y} + \frac{\partial \tau_{zy}}{\partial z}\right) \tag{12.104}$$

$$\rho g\frac{\partial h}{\partial z} = -\frac{\partial p}{\partial z} - \left(\frac{\partial \tau_{xz}}{\partial x} + \frac{\partial \tau_{yz}}{\partial y} + \frac{\partial \tau_{zz}}{\partial z}\right) \tag{12.105}$$

where the expressions in (12.20) have been substituted for the normal stresses $\sigma_x$, $\sigma_y$, and $\sigma_z$. These together with the constitutive equations, (12.97) through (12.102), and the continuity equation (8.17), completely specify the velocity field in flowing ice for given thermal state and boundary conditions. We shall leave (12.103), (12.104), and (12.105) in their present form rather than substitute the constitutive equations, as we did above for a Newtonian fluid.

### 12.4.3 Energy equation

Internal production of heat in glacier ice is due to viscous dissipation associated with ice deformation, and to release of latent heat with refreezing of meltwater near the glacier surface. Neglecting this latter contribution of heat, the energy equation for glacier flow has the form of (12.84), where the dissipation function $\Phi$ has the general form:

$$
\Phi = \sigma_x \frac{\partial u}{\partial x} + \sigma_y \frac{\partial v}{\partial y} + \sigma_z \frac{\partial w}{\partial z}
$$

$$
+ \tau_{xy}\left(\frac{\partial v}{\partial x} + \frac{\partial u}{\partial y}\right) + \tau_{xz}\left(\frac{\partial w}{\partial x} + \frac{\partial u}{\partial z}\right) + \tau_{yz}\left(\frac{\partial w}{\partial y} + \frac{\partial v}{\partial z}\right)
$$

(12.106)

Solving (12.84) using (12.106) requires specifying initial and boundary conditions together with knowledge of the velocity field. Analytical solutions are available for only a few simple flow geometries. Boundary conditions at the surface are determined by weather (Example Problem 9.3.2), including the contribution of thermal energy contained in precipitation. Boundary conditions at the base depend on whether the glacier is frozen to the bed, or basal ice is at the pressure-melting point. In addition, the geothermal flux can add heat to the base of a glacier. Students are urged to examine Paterson (1981, Chapter 10).

---

## 12.5 EXAMPLE PROBLEMS

### 12.5.1 Buoyant ascent of magma in a vertical dike

Let us use the results of this chapter to reexamine Example Problem 6.7.2. Recall that this problem involved determining the velocity distribution associated with a Newtonian magma ascending a vertical dike of constant half-width $b$ (Figure 6.8). Let us treat this problem as a purely mechanical one, neglecting heat flow. Assume that flow is steady, and that the magma is incompressible with density $\rho$ and constant viscosity $\mu$. The surrounding country rock has density $\rho_c$. For this unidirectional flow, $u = v = 0$, and the $x$-component and $y$-component of the momentum equation can be neglected. For completeness, let us write out the $z$-component equation together with the continuity equation for an incompressible fluid:

$$
\rho\left(u\frac{\partial w}{\partial x} + v\frac{\partial w}{\partial y} + w\frac{\partial w}{\partial z} + \frac{\partial w}{\partial t}\right) = -\rho g\frac{\partial h}{\partial z} - \frac{\partial p}{\partial z} + \mu\left(\frac{\partial^2 w}{\partial x^2} + \frac{\partial^2 w}{\partial y^2} + \frac{\partial^2 w}{\partial z^2}\right)
$$

(12.107)

$$
\frac{\partial u}{\partial x} + \frac{\partial v}{\partial y} + \frac{\partial w}{\partial z} = 0
$$

(12.108)

With $u = v = 0$ everywhere, the first two convective terms vanish, and the continuity equation reduces to $\partial w/\partial z = 0$. With steady unidirectional flow the convective term $w\partial w/\partial z$, the local term $\partial w/\partial t$, and the last two viscous terms vanish. Based on our positioning of the coordinate axes relative to the gravitational field, $\partial h/\partial z = 1$. Thus (12.107) reduces to

$$\frac{\partial^2 w}{\partial x^2} = \frac{\rho g}{\mu} + \frac{1}{\mu}\frac{\partial p}{\partial z} \tag{12.109}$$

Following Example Problem 6.7.2, we shall assume that the vertical pressure gradient arises from lithostatic overburden, and is given by the equation of hydrostatics: $\partial p/\partial z = -\rho_c g$. Substituting this into (12.109) and integrating,

$$\frac{dw}{dx} = \frac{1}{\mu}(\rho - \rho_c)gx + C_1 \tag{12.110}$$

where partial notation is no longer necessary. The constant of integration $C_1$ is obtained from the symmetry of the problem; that is, $dw/dx = 0$ at $x = 0$, so $C_1 = 0$. This is equivalent to stating that the shear stress $\tau_{xz} = \mu\partial w/\partial x$ vanishes at $x = 0$. Integrating (12.110) a second time (with $C_1 = 0$),

$$w = \frac{1}{2\mu}(\rho - \rho_c)gx^2 + C_2 \tag{12.111}$$

The constant of integration $C_2$ is evaluated using the no-slip condition that $w = 0$ at $x = \pm b$; thus $C_2 = -(\rho - \rho_c)gb^2/2\mu$, so (12.111) becomes

$$w(x) = \frac{1}{2\mu}(\rho_c - \rho)g(b^2 - x^2) \tag{12.112}$$

which is the answer obtained in Example Problem 6.7.2. It is left as an exercise to consider the analogous problem of flow in a horizontal sill. Assume that flow is induced by a constant pressure gradient $\partial p/\partial x$.

## 12.5.2 Steady free-surface flow down an inclined bed

Let us now reexamine Example Problem 3.7.1. Recall that this problem involved determining the velocity distribution associated with a steady free-surface flow of constant depth $\zeta$ down a bed inclined at an angle $\alpha$ (Figure 3.19). The $x$-axis is parallel to the inclined bed and positive downstream, and the $z$-axis is normal to the bed and positive upward, with origin at the bed. Assume in this first example that the liquid is Newtonian with constant density $\rho$ and viscosity $\mu$. For this unidirectional flow, $v = w = 0$, and the $y$-component and $z$-component of the momentum equation can be neglected. For completeness, let us write out the $x$-component equation together with the continuity equation:

$$\rho\left(u\frac{\partial u}{\partial x} + v\frac{\partial u}{\partial y} + w\frac{\partial u}{\partial z} + \frac{\partial u}{\partial t}\right) = -\rho g\frac{\partial h}{\partial x} - \frac{\partial p}{\partial x} + \mu\left(\frac{\partial^2 u}{\partial x^2} + \frac{\partial^2 u}{\partial y^2} + \frac{\partial^2 u}{\partial z^2}\right) \tag{12.113}$$

$$\frac{\partial u}{\partial x} + \frac{\partial v}{\partial y} + \frac{\partial w}{\partial z} = 0 \tag{12.114}$$

With $v = w = 0$ everywhere, the second and third convective terms vanish, and the continuity equation reduces to $\partial u/\partial x = 0$. With steady unidirectional flow the convective term $u\partial u/\partial x$, the local term $\partial u/\partial t$, and the first and second viscous terms vanish. Based on our positioning of the $xyz$-coordinate axes relative to the gravitational field, $\partial h/\partial x = \sin\alpha$, where it is important to note that $\alpha$ is a negative angle since it involves a clockwise rotation of the $xyz$-coordinate axes relative to the Earth reference frame. Finally, at all points on any surface defined by $z =$ constant, the pressure $p$ is constant. An example is the free surface, where $p$ equals atmospheric pressure at all points on the surface. Another example is the surface defined by the bed. Pressure $p$ therefore is independent of $x$, so $\partial p/\partial x = 0$. Thus (12.113) reduces to

$$\frac{d^2 u}{dz^2} = \frac{\rho g}{\mu}\sin\alpha \qquad (12.115)$$

where partial notation is no longer necessary. Integrating once with respect to $z$,

$$\frac{du}{dz} = \frac{\rho g}{\mu}\sin\alpha z + C_1 \qquad (12.116)$$

The constant of integration $C_1$ is evaluated using the stress condition at the free surface, that $\tau_{zx} = \mu\partial u/\partial z = 0$, whence $du/dz = 0$ at $z = \zeta$. Here we are neglecting any shear stress that the atmosphere exerts on the liquid surface. Thus $C_1 = -(\rho g/\mu)\sin\alpha\zeta$ and

$$\frac{du}{dz} = \frac{\rho g}{\mu}\sin\alpha z - \frac{\rho g}{\mu}\sin\alpha\zeta \qquad (12.117)$$

Integrating a second time,

$$u = \frac{\rho g}{2\mu}\sin\alpha z^2 - \frac{\rho g}{\mu}\sin\alpha\zeta z + C_2 \qquad (12.118)$$

The constant of integration $C_2$ is evaluated using the no-slip condition at the bed, that $u = 0$ at $z = 0$. Thus $C_2 = 0$ and

$$u(z) = -\frac{\rho g}{\mu}\sin\alpha\left(\zeta z - \frac{1}{2}z^2\right) \qquad (12.119)$$

This may be transformed to the form of (3.83) by noting that $y = \zeta - z$.

Now turn to the case of steady one-dimensional flow of glacier ice down an inclined bed. Momentarily assume zero basal sliding (the ice is frozen to the bed), and begin with the $x$-component of the equation of motion:

$$\rho g\frac{\partial h}{\partial x} = -\frac{\partial p}{\partial x} - \left(\frac{\partial\tau_{xx}}{\partial x} + \frac{\partial\tau_{yx}}{\partial y} + \frac{\partial\tau_{zx}}{\partial z}\right) \qquad (12.120)$$

The quantity $\partial h/\partial x$ is again given by $\sin\alpha$, where $\alpha$ is a negative angle. The pressure gradient $\partial p/\partial x$ vanishes for the same reasons as described in the preceding problem. Further, the only nonzero shear-stress term involves $\tau_{zx}$. Thus (12.120) reduces to

$$\frac{\partial\tau_{zx}}{\partial z} = -\rho g\sin\alpha \qquad (12.121)$$

Integrating with respect to $z$,

$$\tau_{zx} = -\rho g \sin \alpha z + C_1 \tag{12.122}$$

The constant of integration $C_1$ is evaluated using the stress condition at the free surface, that $\tau_{zx} = 0$ at $z = \zeta$. Thus $C_1 = \rho g \sin \alpha \zeta$ and

$$\tau_{zx} = \rho g \sin \alpha (\zeta - z) \tag{12.123}$$

To this point $\alpha$ is a negative angle. Thus $\tau_{zx}$ is a negative value according to (12.123), which is consistent with the interpretation that this stress is exerted on surfaces normal to the $z$-axis from beneath (where we anticipate that $\partial u / \partial z$ is positive). Now, for reasons that will become clear below, we must insist in this problem that $\alpha$ be measured as a positive angle. In this case $\tau_{zx}$ in (12.123) is a positive value, which is consistent with the interpretation that this stress is exerted on surfaces normal to the $z$-axis from above. Glen's law under the one-dimensional flow conditions is $\partial u / \partial z = -A \tau_{zx}^m$. This form involving a negative sign, however, conforms with the interpretation that $\tau_{zx}$ is exerted on surfaces from below. Since we wish to work with positive $\tau_{zx}$, the appropriate form of Glen's law is therefore

$$\frac{\partial u}{\partial z} = A \tau_{zx}^m \tag{12.124}$$

Substituting (12.123) into (12.124),

$$\frac{\partial u}{\partial z} = A[\rho g \sin \alpha (\zeta - z)]^m \tag{12.125}$$

It now becomes clear that, by insisting that $\alpha$ (and $\tau_{zx}$) be positive, we obviate dealing with exponents of negative quantities.

Integrating (12.125) with respect to $z$, and evaluating the constant of integration using the no-slip condition that $u = 0$ at $x = 0$, we obtain

$$u(z) = \frac{A}{m+1}(\rho g \sin \alpha)^m [\zeta^{m+1} - (\zeta - z)^{m+1}] \tag{12.126}$$

When the ice is not frozen to the bed, the ice can possess a finite velocity at $z = 0$, denoted by $u_b$; this is added to (12.126) to give

$$u(z) = u_b + \frac{A}{m+1}(\rho g \sin \alpha)^m [\zeta^{m+1} - (\zeta - z)^{m+1}] \tag{12.127}$$

When a value of $m = 3$ is adopted,

$$u(z) = u_b + \frac{A}{4}(\rho g \sin \alpha)^3 [\zeta^4 - (\zeta - z)^4] \tag{12.128}$$

which suggests that ice velocity due to deformation is proportional to $\sin^3 \alpha$ and $\zeta^4$. Ice velocity is therefore strongly sensitive to both slope and thickness, particularly to the latter.

### 12.5.3 Viscous generation of heat in the vertical dike problem

Recall that we neglected the generation of heat due to viscous friction in Example Problem 9.3.3, assuming that this contribution of heat was negligible relative to heat advected upward by hot magma. Let us now examine this assumption by estimating the quantity of heat generated by friction and comparing it with advected heat. First, consider a slightly different version of Example Problem 9.3.3.

Envision a vertical dike of uniform half-width $b$ and height $h$ (Figure 6.8). Assume steady flow of a Newtonian magma with constant viscosity $\mu$, density $\rho$, specific heat $c$, and thermal conductivity $K_T$. The surrounding country rock has density $\rho_c$. Further assume that the temperature $T_w$ at the dike walls is constant; the significance of this simple case will become clear below. Because, under these conditions, fluid properties are independent of heat flow, this is a problem of *forced convection* (or *forced advection*), which means that the velocity field does not depend on the temperature field, whereas the temperature field does depend on the velocity field. We may therefore use the velocity distribution $w(x)$ obtained in Example Problem 12.5.1.

With steady unidirectional flow, $u = v = 0$, and the energy equation (12.84) reduces to

$$\rho c w \frac{\partial T}{\partial z} = K_T \left( \frac{\partial^2 T}{\partial x^2} + \frac{\partial^2 T}{\partial z^2} \right) + \mu \left( \frac{\partial w}{\partial x} \right)^2 \tag{12.129}$$

Moreover, with the specified boundary condition that $T = T_w$ at $x = \pm b$, the solution to (12.129) must be independent of vertical position $z$. Therefore $w \partial T / \partial z = 0$ and $\partial^2 T / \partial z^2 = 0$, so (12.129) reduces to

$$K_T \frac{\partial^2 T}{\partial x^2} = -\mu \left( \frac{\partial w}{\partial x} \right)^2 \tag{12.130}$$

The right side of this equation describes viscous generation of heat and therefore may be considered a "source" term; the left side describes dissipation of heat by conduction and therefore may be considered a "sink" term. Thus heat generated by friction is exactly balanced by heat lost through conduction.

Differentiating the velocity distribution $w(x)$ given by (12.112) with respect to $x$ gives $\partial w / \partial x = -\rho_0 g x / \mu$, where $\rho_0 = (\rho_c - \rho)$. Substituting this into (12.130) and rearranging then leads to

$$\frac{d^2 T}{dx^2} = -\frac{(\rho_0 g)^2}{\mu K_T} x^2 \tag{12.131}$$

where partial notation is no longer necessary. Integrating once,

$$\frac{dT}{dx} = -\frac{(\rho_0 g)^2}{3\mu K_T} x^3 + C_1 \tag{12.132}$$

The constant of integration $C_1$ is obtained from the symmetry of the problem; that is, $dT/dx = 0$ at $x = 0$, so $C_1 = 0$. Integrating (12.132) a second time (with $C_1 = 0$),

$$T = -\frac{(\rho_0 g)^2}{12\mu K_T} x^4 + C_2 \tag{12.133}$$

The constant of integration $C_2$ is evaluated using the boundary condition that $T = T_w$ at $x = \pm b$; thus $C_2 = (\rho_0 g)^2 b^4 / 12\mu K_T + T_w$, so (12.133) becomes

$$T(x) = -\frac{(\rho_0 g)^2}{12\mu K_T} (b^4 - x^4) + T_w \tag{12.134}$$

Thus the temperature distribution $T(x)$ is a fourth-degree parabola whose form arises from generation of heat by friction, and transverse conduction toward the dike walls (Figure 12.5). The maximum temperature $T_0$ at the center of the dike ($x = 0$) is then given by $T_0 = (\rho_0 g)^2 b^4 / 12 \mu K_T + T_w$. Observe that the temperature distribution $T(x)$ is relatively flat near the centerline. The thermal boundary layer—that portion of $T(x)$ that varies rapidly between $T_m$ and $T_w$—therefore is compressed relative to the flow boundary layer, which fully extends from the wall to the centerline.

According to Fourier's law, the heat flux density $q_T$ in the transverse direction is

$$q_T = -K_T \frac{dT}{dx} \tag{12.135}$$

Substituting (12.132) (with $C_1 = 0$) for the temperature gradient $\partial T / \partial x$ in (12.135), then evaluating this expression at $x = b$, the heat flux $q_w$ across the dike wall is

$$q_w = \frac{(\rho_0 g)^2}{3\mu} b^3 \tag{12.136}$$

Multiplying this quantity by 2 (two walls) gives the steady flux of heat out of the magma per unit length and breadth of wall. This flux therefore must represent the steady rate of heat generation within the magma due to friction. Note that $q_w$ does not depend on $K_T$; for given $K_T$, heat is generated such that the temperature gradient precisely necessary to export this heat is reached. Further, $q_w$ is proportional to $b^3$ and $\rho_0^2$; a wide dike and low-density magma (relative to the country rock) both lead to high velocities and therefore viscous shear. That $q_w$ is inversely proportional to $\mu$ may at first seem counterintuitive since viscosity characterizes the

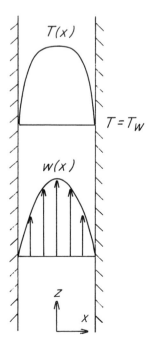

**Figure 12.5**   Vertical dike illustrating parabolic velocity profile $w(x)$, and fourth-order parabolic temperature profile $T(x)$ due to steady viscous heat generation.

frictional resistance to flow responsible for the generation of heat; more importantly, it governs heat generation via the rate of shear. Moreover, $q_w$ is independent of temperature (except indirectly through buoyancy). Thus the viscous generation of heat is independent of extant temperature conditions of the magma and wall, as it describes a conversion of mechanical energy (buoyancy) to heat that is superimposed on advected heat. Viewed from this perspective, assuming constant boundary conditions ($T = T_w$ at $x = \pm b$) is merely an artifice to obtain the rate of viscous generation of heat.

Let us now compare these results with upward advection of heat by hot magma, as in Example Problem 9.3.3. Here the magma temperature $T$ may vary with $z$. Let $T_m$ denote the temperature of the magma at the base of the dike, and let $T_s$ denote the eruption temperature at the surface. Consider a rectangular control volume within the dike with dimensions $2b\,dy\,dz$. Magma entering the control volume from below with average velocity $w_0$ and internal energy per unit mass $U$ advects into the control volume a quantity of heat per unit time equal to $\rho w_0 U 2b\,dy$. The net loss of heat out of the control volume per unit time equals to first order $\rho c w_0 (\partial T/\partial z)2b\,dy\,dz$, where we have made use of the relation $dU = c\,dT$. Neglecting viscous generation of heat, this quantity must represent the rate at which heat is lost through conduction into the dike walls. The vertically averaged value of $\partial T/\partial z$ is given by $(T_s - T_m)/h$. The average rate of heat loss per unit length and breadth of wall $q_w$ is therefore equal to

$$q_w = 2\rho c b w_0 \frac{T_s - T_m}{h} \qquad (12.137)$$

Let us choose for illustration the following values of physical quantities: $b = 0.5$ m, $h = 1$ km, $\rho = 2{,}660$ kg m$^{-3}$, $\rho_c = 2{,}720$ kg m$^{-3}$, $c = 730$ J kg$^{-1}$ °C$^{-1}$, $\mu = 100$ Pa s, $T_m = 1{,}200$ °C, and $T_s = 1{,}190$ °C. Using these values and (6.77), $w_0 = 0.50$ m s$^{-1}$. Based on (12.136) the rate of viscous heat generation ($2q_w$) equals about 288 W m$^{-2}$. (As a point of reference, this is about equal to the power expended by three ordinary light bulbs spread over an area of 1 m$^2$.) Based on (12.137) the average rate of heat loss, neglecting viscous generation, equals about 9,710 Wm$^{-2}$ for a temperature difference ($T_s - T_m$) of 10 °C. Viscous heat generation would therefore amount to only about 3 percent of the loss of heat advected from the magma chamber in this example. The temperature difference of 10 °C is used only for illustration. Late in an eruption after the country rock is heated and the transverse temperature gradient is reduced, it is conceivable that magma is hotter at the surface than at depth due to viscous generation of heat.

There are at least two other simple ways to compute the rate at which heat is generated by friction in this example, without using the energy equation. One way involves considering the rate at which the buoyant force performs work (against friction) in moving magma upward; another involves considering the average rate at which work is performed against the shear stress exerted by the dike walls on the magma. Both of these approaches assume that the average velocity $w_0$ is given.

### 12.5.4 Extending and compressing glacier flow

Large glaciers typically possess two zones: an accumulation zone where, on average, mass is added to the glacier as precipitation and condensation, and in some cases, by avalanching; and an ablation zone where, on average, mass is removed as meltwater, by sublimation and evaporation, and in some cases, by calving. This exchange of mass at the glacier surface generally requires a vertical component of ice velocity if the surface position does not change with time, in addition to a streamwise velocity. Moreover, usually the streamwise velocity $u$ on average increases with down valley distance $x$ over the accumulation zone ($\partial u/\partial x > 0$), and on average decreases over the ablation zone ($\partial u/\partial x < 0$). The former is referred to as *extending flow,* and the latter is referred to as *compressing flow.* Let us adopt the approach taken in Example Problem 8.4.1 to obtain a qualitative idea of the velocity field associated with extending and compressing flows. The development below follows the work of Nye (1957) and the treatment of these flows by Paterson (1981).

Envision a wide glacier with uniform thickness $b$ that is flowing steadily down an inclined bed (Figure 12.6). The $x$-axis is parallel to the bed and positive down valley. The $z$-axis is normal to the bed and positive upward, with origin at the bed. Assume the glacier slab is far from valley walls, so there is no influence of the walls on flow. Ice is added (or lost) at a constant rate $s$ uniformly over the surface. Thus, $s$ may be considered a time-averaged precipitation or ablation rate ($L\ t^{-1}$), in ice-equivalent units. For simplicity, assume that the ice density $\rho$ is constant.

Now consider a control volume with dimensions $b\,dx\,dy$ (Figure 12.6). Ice discharge $Q$ into the control volume from upstream is

$$Q = \left( \int_0^b u(x, z)dz \right) dy \tag{12.138}$$

where $u(x, z)$ is the unknown streamwise velocity that may in general vary with both $x$ and $z$. Between the upstream and downstream faces, mass is added at the steady rate $s\,dx\,dy$. Ice discharge out of the control volume is to first order $Q + (\partial Q/\partial x)\,dx$. By conservation of mass, $(\partial Q/\partial x)\,dx = s\,dx\,dy$. Thus

$$\frac{\partial}{\partial x} \int_0^b u(x, z)\,dz = s = \text{const} \tag{12.139}$$

A little reflection leads to the conclusion that this expression is satisfied only if $u(x, z)$ involves $x$ to the first power. This could include terms of the form $ax$, where $a$ is a constant, or more complicated terms involving products of $x$ and $z$.

With a wide ice slab, we may assume that $v = 0$, and that strain rates involving $y$ are also zero. Let us further assume that stress components are independent of $x$ for a long ice slab, as in Example Problem 12.5.2 above. This means that strain-rate components also are independent of $x$. Stated formally, derivatives of strain-rate

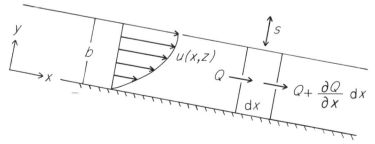

**Figure 12.6**   Definition diagram for glacier ice having uniform thickness $b$ flowing down an inclined bed, with unknown velocity profile $u(z)$, and control volume $b\,dx\,dy$.

components with respect to $x$ must equal zero. Thus, from (12.97) and (12.101),

$$\frac{\partial^2 u}{\partial x^2} = 0 \qquad (12.140)$$

$$\frac{\partial^2 u}{\partial x \partial z} + \frac{\partial^2 w}{\partial x^2} = 0 \qquad (12.141)$$

For reasons that will become clear below, it is useful to rewrite (12.141) as

$$\frac{\partial}{\partial z}\left(\frac{\partial u}{\partial x}\right) + \frac{\partial^2 w}{\partial x^2} = 0 \qquad (12.142)$$

In addition, the appropriate continuity equation is

$$\frac{\partial u}{\partial x} + \frac{\partial w}{\partial z} = 0 \qquad (12.143)$$

We now seek a solution $u(x, z)$ that satisfies (12.139), (12.140), (12.142), and (12.143). Two boundary conditions are useful. The shear stress $\tau_{zx}$ at the glacier surface equals zero, and the velocity component $w$ equals zero at the bed, assuming no basal melting. These conditions are written as

$$\frac{\partial u}{\partial z} + \frac{\partial w}{\partial x} = 0, \quad z = b; \qquad w = 0, \quad z = 0 \qquad (12.144)$$

To learn something about the form of $u(x, z)$, let us first integrate (12.140) once with respect to $x$. Note that, in performing this integration, we must allow for the possibility that the resulting antiderivative, denoted by $r$, is a function of $z$. (Why?) Thus,

$$\frac{\partial u}{\partial x} = r(z) \qquad (12.145)$$

Integrating again,

$$u(x, z) = r(z)x + u_z(z) \qquad (12.146)$$

Here $u_z(z)$ arises as the "constant" of integration, and denotes an unknown dependence of $u$ on $z$. Now, according to (12.143), $\partial u/\partial x = -\partial w/\partial z$; thus from (12.145),

$$\frac{\partial w}{\partial z} = -r(z) \qquad (12.147)$$

Integrating this gives

$$w = R(z) + C(x) \qquad (12.148)$$

where $R(z) = \int r(z)\,dz$, and the "constant" of integration $C(x)$ leaves open the possibility that this term is a function of $x$. Observe, however, that (12.148) can be made to satisfy the second boundary condition in (12.144) only if $C$ is, in fact, a constant or zero. (Why?) Thus, differentiating (12.148) with respect to $x$ equals zero, which means that $w$ is independent of $x$. This implies that the second term in (12.142) equals zero. Then, differentiating (12.145) with respect to $z$ suggests that (12.142) is satisfied only if $r(z)$ is, in fact, independent of $z$. It follows that $\partial u/\partial x = r$ is a constant. Returning to (12.146), this may now be written as

$$u = u_0 + rx + u_z(z) \qquad (12.149)$$

Here $u_z(z)$ must be specified such that $u_z(b) = 0$, so that $u_0$ is the surface stream-wise velocity at the origin ($x = 0$). It is left as an exercise to obtain for $w$:

$$w = -rz \qquad (12.150)$$

Thus the streamwise velocity component $u$ is a function of $x$ and $z$. In addition, a velocity component $w$ perpendicular to the surface exists, which is in contrast to the case of unidirectional ice flow presented in Example Problem 12.5.2 above.

The quantity $r = \partial u/\partial x$ serves as a measure of the extending or compressing conditions. Note that, since flow is incompressible, these expressions merely refer to a large-scale strain—an increase or reduction in the length—of the ice slab in a streamwise direction. This strain must be accompanied by a decrease or increase in the ice thickness or, as in this example, addition or loss of mass at the glacier surface. The velocity component $w < 0$ (downward) when $r > 0$, and $w > 0$ (upward) when $r < 0$. Since $r$ is independent of $z$, it can be measured at the surface. According to Paterson (1981), values of $r$ vary from about $10^{-5}$ yr$^{-1}$ for flow in central Antarctica to about $10^{-1}$ yr$^{-1}$ or more for temperate valley glaciers.

The analysis here is greatly simplified. Glacier thickness typically varies down valley along with surface and bed slopes. Valley walls influence the flow fields in valley glaciers. In addition, the result that $r$ is independent of $z$ implies that the streamwise velocity at the base of a glacier steadily increases or decreases down valley, which would require special stress conditions here. The assumption of a zero velocity component normal to the bed also is unrealistic for glaciers where basal melting (or freezing of water) occurs. In this regard, students are urged to complete the analysis and obtain an expression for $u_z(z)$ by following the work of Nye, or the presentation of Paterson. This involves evaluating the streamwise stress components making use of Glen's law and the equations of motion.

## 12.5.5 Very viscous (Hele–Shaw) flow

An important class of flows, referred to as *very viscous flows*, are those for which inertial terms in the dynamical equations are negligible relative to viscous terms. In effect we made this assumption in writing the equations of motion, (12.103), (12.104), and (12.105), for flow of glacier ice. Stokes's solution for the drag on a sphere also assumes very viscous flow, for which the condition $Re < 1$ must be satisfied. Another type of very viscous flow, which we will examine here, occurs in a *Hele–Shaw cell*.

Envision two flat, parallel plates separated by a small vertical distance equal to $2b$. Between the plates is a cylindrical object, with arbitrary cross section, whose axis is parallel to the $z$-axis (Figure 12.7). The object has a characteristic dimension $L$ such that the condition $b \ll L$ is satisfied. Assume that steady flow around the object is driven by pressure gradients. Consider, now, the conditions under which the inertial term, $\mathbf{u} \cdot \nabla \mathbf{u}$, can be neglected. Here we will loosely follow the discussion of Acheson (1990).

Observe first that, far from the object, the streamwise velocity at $z = 0$ is $U$; for convenience $U$ is parallel to the $x$-axis. Since the no-slip condition occurs at $z = \pm b$, the velocity profile has a parabolic (Poiseuille) form. Thus

$$u = U\left(1 - \frac{z^2}{b^2}\right), \quad v = w = 0; \qquad x = y = \infty \qquad (12.151)$$

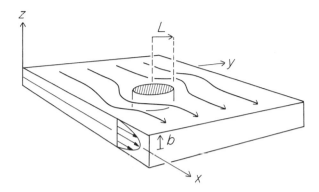

**Figure 12.7** Hele–Shaw flow in the vicinity of a cylinder with characteristic length $L$; adapted from Acheson (1990).

Turning now to flow in the vicinity of the object, let us estimate the magnitudes of the terms in the continuity equation for an incompressible fluid:

$$\frac{\partial u}{\partial x} + \frac{\partial v}{\partial y} + \frac{\partial w}{\partial z} = 0 \tag{12.152}$$

The terms $\partial u/\partial x$ and $\partial v/\partial y$ are each of order $U/L$. This implies that $\partial w/\partial z$ also is of order $U/L$, and therefore that $w$ is of order $Ub/L$. (Why?) Using this result, we can now estimate the magnitude of each term in the components of $\mathbf{u} \cdot \nabla\mathbf{u}$:

$$u\frac{\partial u}{\partial x} + v\frac{\partial u}{\partial y} + w\frac{\partial u}{\partial z}; \qquad u\frac{\partial v}{\partial x} + v\frac{\partial v}{\partial y} + w\frac{\partial v}{\partial z}; \qquad u\frac{\partial w}{\partial x} + v\frac{\partial w}{\partial y} + w\frac{\partial w}{\partial z}$$

$$\tag{12.153}$$

$$\frac{U^2}{L} \quad \frac{U^2}{L} \quad \frac{U^2}{L}; \qquad \frac{U^2}{L} \quad \frac{U^2}{L} \quad \frac{U^2}{L}; \qquad \frac{U^2 b}{L^2} \quad \frac{U^2 b}{L^2} \quad \frac{U^2 b}{L^2}$$

Observe that the magnitude of each term in the $z$-component is small relative to the magnitude of all other terms, which is $U^2/L$. We therefore may conclude that $\mathbf{u} \cdot \nabla\mathbf{u}$ is at least of order $U^2/L$. Now, estimating the magnitude of each term in the components of the viscous term $\nabla^2\mathbf{u}$:

$$\frac{\partial^2 u}{\partial x^2} + \frac{\partial^2 u}{\partial y^2} + \frac{\partial^2 u}{\partial z^2}; \qquad \frac{\partial^2 v}{\partial x^2} + \frac{\partial^2 v}{\partial y^2} + \frac{\partial^2 v}{\partial z^2}; \qquad \frac{\partial^2 w}{\partial x^2} + \frac{\partial^2 w}{\partial y^2} + \frac{\partial^2 w}{\partial z^2}$$

$$\frac{U}{L^2} \quad \frac{U}{L^2} \quad \frac{U}{b^2}; \qquad \frac{U}{L^2} \quad \frac{U}{L^2} \quad \frac{U}{b^2}; \qquad \frac{Ub}{L^3} \quad \frac{Ub}{L^3} \quad \frac{U}{Lb} \tag{12.154}$$

Observe that, since $b \ll L$, the magnitudes of terms involving $\partial^2/\partial z^2$ are much larger than those involving $\partial^2/\partial x^2$ and $\partial^2/\partial y^2$, and that the magnitude of the term $\partial^2 w/\partial z^2$ is small relative to that of $\partial^2 u/\partial z^2$ and $\partial^2 v/\partial z^2$, which is $U/b^2$. We may therefore conclude that $\nabla^2\mathbf{u}$ is at least of order $U/b^2$. These results suggest that we may neglect the inertial terms if $U^2/L \ll \nu U/b^2$, or

$$\frac{UL}{\nu}\left(\frac{b}{L}\right)^2 \ll 1 \tag{12.155}$$

This quantity is like a Reynolds number; indeed the left part, $UL/\nu$, is equivalent to a conventional Reynolds number. Thus, the condition (12.155) may be satisfied with large Reynolds number $UL/\nu$ if $b/L$ is sufficiently small.

Neglecting inertial terms and small viscous terms based on the analysis above, the equations of motion are greatly simplified:

$$\frac{\partial p}{\partial x} = \mu \frac{\partial^2 u}{\partial z^2} \tag{12.156}$$

$$\frac{\partial p}{\partial y} = \mu \frac{\partial^2 v}{\partial z^2} \tag{12.157}$$

$$\frac{\partial p}{\partial z} = \mu \frac{\partial^2 w}{\partial z^2} \tag{12.158}$$

Since, from (12.154), $\partial^2 w/\partial z^2$ is much smaller than $\partial^2 u/\partial z^2$ or $\partial^2 v/\partial z^2$, evidently $\partial p/\partial z$ is much smaller than $\partial p/\partial x$ or $\partial p/\partial y$, and we may assume that, to a good approximation, $p$ is a function of $x$ and $y$, but not $z$. We may therefore integrate (12.156) and (12.157) twice, then evaluate the constants of integration using the no-slip condition at $z = \pm b$ to obtain

$$u = -\frac{1}{2\mu}\frac{\partial p}{\partial x}(b^2 - z^2) \tag{12.159}$$

$$v = -\frac{1}{2\mu}\frac{\partial p}{\partial y}(b^2 - z^2) \tag{12.160}$$

Thus, whereas the direction of flow is independent of $z$, the magnitude of the flow varies with $z$. Writing this solution in vector form,

$$\mathbf{u} = -\frac{1}{2\mu}(b^2 - z^2)\nabla p \tag{12.161}$$

where $\mathbf{u} = \langle u, v \rangle$ and $\nabla p = (\partial p/\partial x)\mathbf{i} + (\partial p/\partial y)\mathbf{j}$.

Let us differentiate $u$ with respect to $y$, and $v$ with respect to $x$:

$$\frac{\partial u}{\partial y} = -\frac{1}{2\mu}\frac{\partial}{\partial y}\left(\frac{\partial p}{\partial x}\right)(b^2 - z^2) \tag{12.162}$$

$$\frac{\partial v}{\partial x} = -\frac{1}{2\mu}\frac{\partial}{\partial x}\left(\frac{\partial p}{\partial y}\right)(b^2 - z^2) \tag{12.163}$$

Noting that the order of differentiation of $p$ is immaterial, it follows that

$$\frac{\partial v}{\partial x} = \frac{\partial u}{\partial y} \tag{12.164}$$

Comparing this with (11.27), evidently (12.161) defines a two-dimensional irrotational flow field. Finally note that, whereas the no-slip condition exists at $z = \pm b$, this condition is not satisfied at the cylinder boundary in this analysis.

Geologic situations for which this Hele–Shaw configuration is a direct analogue are limited. Nonetheless, the similarity between this problem and viscous flow within a fracture (Example Problem 8.4.6) should be apparent. Indeed, in some respects the flow around an object between parallel plates resembles what might occur near a bump that protrudes into a fracture, although this description is still far from the variable-aperture conditions that exist in fractures. More importantly, the ideas of very viscous flow have analogues in several geological problems, including the spreading of lava and porous-media flows. We will also consider thermally driven flow in a Hele–Shaw configuration in Chapter 16.

### 12.5.6 Boundary-layer equations

Thus far we have emphasized boundary layers whose extents, normal to the boundary, are well defined. For example, we assumed that the boundary layer in a free-surface flow extends to the liquid surface; and we assumed with conduit flow that it extends to the centerline. In contrast, we have, in a couple of cases, implicitly allowed for the possibility that the boundary layer has no definite "edge" parallel to the boundary. For example, with flow around an object, as in the Hele–Shaw cell above, the edge of the boundary layer is described vaguely in terms of the velocity approaching a free-stream value "far" from the object. Likewise, the limit of the boundary layer associated with a settling sphere is not specified. An important problem, then, involves describing the case in which the limit of influence of the no-slip condition on nearby flow is not constrained by a free surface, or by symmetry arising from the existence of two parallel boundaries.

Envision a long, flat (thin) plate parallel to a streamwise flow, with a uniform one-dimensional velocity field approaching the plate. The $x$-axis is parallel to the plate, and the $y$-axis is normal to it. As flow reaches the plate, the no-slip condition begins to influence the flow (Figure 12.8). The streamwise velocity goes to zero at the plate, and a boundary layer begins to develop. Shear stresses next to the plate, where velocity gradients are nonzero, are transferred to fluid that is increasingly farther from the plate. The extent of the influence of the no-slip condition increases, the velocity profile evolves, and the boundary layer grows with distance along the plate. We may thus envision that flow approaching the plate, in absence of velocity gradients, has an inviscid quality. Once the plate is reached, however, two zones exist: a viscous boundary layer within which $\tau_{yx} = -\mu \partial u / \partial y$ is nonzero, and an exterior inviscid flow where $\partial u / \partial y = 0$.

This problem was first systematically treated by L. Prandtl in 1904. Let us start by defining the boundary layer thickness $\delta$ and a characteristic length $L$, which may be taken as the streamwise distance along the plate. Let $U$ denote the free-stream velocity parallel to $x$, and let $V$ denote a characteristic component of velocity parallel to $y$. Further, we may define a characteristic time interval $t_0 = L/U$, which is the time required for free-stream fluid to travel the distance $L$.

Because the exterior free-stream fluid is inviscid and one-dimensional, it can be described in terms of Bernoulli's equation (10.23). Differentiating this equation

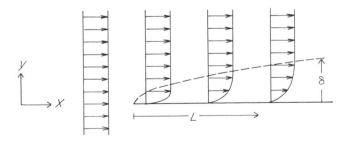

**Figure 12.8**   Definition diagram for flow in the vicinity of a flat plate showing development of the boundary-layer with thickness $\delta$.

with respect to $x$ we obtain

$$\frac{\partial p}{\partial x} = -\rho U \frac{\partial U}{\partial x} \qquad (12.165)$$

Thus $\partial p/\partial x$ in the exterior flow is of order $\rho U^2/L$. Turning to fluid within the boundary layer, with $w = 0$, the relevant equations of motion are the $x$-component and $y$-component of the Navier–Stokes equations (neglecting gravitational terms), and the two-dimensional equation of continuity. Let us now estimate the magnitude of each term in these equations, beginning with the continuity equation:

$$\frac{\partial u}{\partial x} + \frac{\partial v}{\partial y} = 0$$

$$(12.166)$$

$$\frac{U}{L} \qquad \frac{V}{\delta}$$

It follows that $V$ is of order $U\delta/L$. Making use of this result, the definition of $t_0$, and the estimate of the magnitude of $\partial p/\partial x$ obtained from (12.165),

$$u\frac{\partial u}{\partial x} + v\frac{\partial u}{\partial y} + \frac{\partial u}{\partial t} = -\frac{1}{\rho}\frac{\partial p}{\partial x} + \nu\left(\frac{\partial^2 u}{\partial x^2} + \frac{\partial^2 u}{\partial y^2}\right)$$

$$\frac{U^2}{L} \quad \frac{U^2}{L} \quad \frac{U^2}{L} \qquad \frac{U^2}{L} \qquad \nu\frac{U}{L^2} \quad \nu\frac{U}{\delta^2} \qquad (12.167)$$

$$u\frac{\partial v}{\partial x} + v\frac{\partial v}{\partial y} + \frac{\partial v}{\partial t} = -\frac{1}{\rho}\frac{\partial p}{\partial y} + \nu\left(\frac{\partial^2 v}{\partial x^2} + \frac{\partial^2 v}{\partial y^2}\right)$$

$$\frac{U^2\delta}{L^2} \quad \frac{U^2\delta}{L^2} \quad \frac{U^2\delta}{L^2} \qquad \frac{U^2\delta}{L^2} \qquad \nu\frac{U\delta}{L^3} \quad \nu\frac{U}{L\delta} \qquad (12.168)$$

Notice that the viscous term $\partial^2 u/\partial x^2$ is much smaller than the term $\partial^2 u/\partial y^2$ when $\delta \ll L$, in which case the former term may be neglected. In estimating the magnitude of the pressure term in (12.167), we have essentially assumed that the pressure field in the exterior flow is impressed throughout the boundary layer. Recall that we applied the same assumption regarding flow around a cylinder to describe the conditions leading to separation (Example Problem 11.4.2). Since $\partial p/\partial y = 0$ in the exterior flow, this suggests that this gradient is small within the boundary layer, and we may assume that the magnitude of this term is no greater than that of the largest term in (12.168), which is $U^2\delta/L^2$. Finally, notice that convective terms in (12.168) are small relative to those in (12.167), and viscous terms in (12.168) are small relative to the second viscous term in (12.167). Thus the $y$-component equation may be neglected, and the boundary-layer equations are simplified to

$$\frac{\partial u}{\partial x} + \frac{\partial v}{\partial y} = 0 \qquad (12.169)$$

$$u\frac{\partial u}{\partial x} + v\frac{\partial u}{\partial y} + \frac{\partial u}{\partial t} = -\frac{1}{\rho}\frac{\partial p}{\partial x} + \nu\frac{\partial^2 u}{\partial y^2} \qquad (12.170)$$

with boundary conditions

$$u = v = 0, \quad y = 0; \qquad u = U(x,t), \quad y = \infty \qquad (12.171)$$

Using (12.165), (12.170) may be written as

$$u\frac{\partial u}{\partial x} + v\frac{\partial u}{\partial y} + \frac{\partial u}{\partial t} = U\frac{\partial U}{\partial x} + v\frac{\partial^2 u}{\partial y^2} \qquad (12.172)$$

Let us now define a Reynolds number by

$$Re = \frac{UL}{\nu} \qquad (12.173)$$

Because the viscous term $\nu\partial^2 u/\partial y^2$, whose magnitude is $\nu U/\delta^2$, is of the same order as each convective term, whose magnitude is $U^2/L$, it follows that by using (12.173), the boundary-layer thickness $\delta$ is of order

$$\delta \sim \frac{L}{\sqrt{Re}} = L\sqrt{\frac{\nu}{UL}} \qquad (12.174)$$

The missing coefficient in (12.174) equals 5. Thus the boundary-layer thickness increases with $L$ and the square root of viscosity $\nu$, and inversely with the square root of $U$ and $L$. The essential condition leading to the boundary-layer equations (12.169) and (12.170), that $\delta \ll L$, is therefore most readily satisfied with high-velocity, low-viscosity flows (high $Re$).

These results can be generalized to cases in which the angle of incidence of the plate relative to the direction of the free-stream flow is nonzero. The boundary-layer equations therefore form the cornerstone of treatments of flow around objects immersed in a free-streaming fluid, including airfoil theory. In a geological context, the boundary-layer equations are applicable in computing forces that act on sediment particles, including drag and lift. In addition, the plate example above provides some insight regarding the parabolic velocity profile that we derived for conduit flow. Namely, viscous boundary layers on opposing conduit walls begin to develop near the conduit entrance, then eventually span the conduit only if its length is sufficiently long. The steady parabolic profile therefore pertains to a *fully developed boundary layer.* Similar remarks apply to steady flow down an inclined plane. We will use these ideas again in describing turbulent boundary layers (Chapter 15).

### 12.5.7 Flow around a vortex in a liquid with free surface

If you have carefully examined a vigorously swirling eddy that has reached the surface of a slowly flowing stream, you will know that the water surface is depressed within the core of the eddy, much like the water surface associated with a vortex or "whirlpool" that occurs when a bathtub drains. Let us examine the flow field around such a vortex, and relate this field to the shape of the water surface. We will start by considering a vertical cylinder having radius $r_0$ that is steadily rotating with angular velocity $\omega_0$ in an infinite, incompressible liquid with a free surface. Choosing a cylindrical $r\theta z$-coordinate system, the $z$-axis is vertical such that the origin of the radial $r$-axis ($r = 0$) coincides with the cylinder axis. Under these conditions, the effect of the no-slip condition at the outer cylinder surface is transmitted radially outward, such that flow occurs only in the tangential coordinate direction $\theta$. Then, a consideration of terms in the momentum equations for cylindrical coordinates (Appendix 17.2) suggests that these can be simplified to:

$$-\frac{v_\theta^2}{r} = -\frac{1}{\rho}\frac{\partial p}{\partial r} \tag{12.175}$$

$$0 = \frac{\partial^2 v_\theta}{\partial r^2} + \frac{1}{r}\frac{\partial v_\theta}{\partial r} - \frac{v_\theta}{r^2} \tag{12.176}$$

$$0 = -\frac{1}{\rho}\frac{\partial p}{\partial z} - g \tag{12.177}$$

The first of these indicates that the radially directed centrifugal force, proportional to $v_\theta^2$, is balanced by the force associated with a radial pressure gradient. The second indicates that velocity components contributing to the viscous force sum to zero. The third is equivalent to the equation of fluid statics.

Multiplying terms in (12.176) by $r^2$ and rearranging, and noting that partial notation is unnecessary,

$$r^2\frac{d^2 v_\theta}{dr^2} + r\frac{dv_\theta}{dr} - v_\theta = 0 \tag{12.178}$$

Those familiar with linear differential equations will recognize that (12.178) has the form of a *Cauchy–Euler equation*. This can be readily solved when it is transformed to an equation with constant coefficients by a change of independent variable, $r = e^t$, where $r > 0$ and $t = \ln r$ is a transformed variable. Without showing the steps, a general solution of (12.178) is

$$v_\theta = C_1 r + C_2\frac{1}{r} \tag{12.179}$$

The constants $C_1$ and $C_2$ can be evaluated with the boundary conditions:

$$v_\theta = r_0\omega_0 \qquad r = r_0 \tag{12.180}$$

$$v_\theta = 0 \qquad r \to \infty \tag{12.181}$$

The constant $C_1$ must equal zero to satisfy the second of these conditions. Using (12.180), the constant $C_2$ is then equal to $r_0^2\omega_0$, and the tangential velocity becomes

$$v_\theta = r_0^2\omega_0\frac{1}{r} \tag{12.182}$$

which we will return to below.

Now, rearranging (12.177) and integrating with respect to $z$ leads to an expression for pressure as a function of $z$, namely, $p = -\rho g z + C$. Evaluating the constant of integration $C$ using the boundary condition that $p = p_0$ at $z = \zeta$, where $\zeta$ is the vertical position of the liquid surface, then gives $p = p_0 + \rho g(\zeta - z)$. Differentiating this with respect to $r$,

$$\frac{\partial p}{\partial r} = \rho g\frac{\partial \zeta}{\partial r} \tag{12.183}$$

Substituting (12.182) and (12.183) into (12.175), separating variables, and then integrating,

$$\zeta = -\frac{r_0^4 \omega_0^2}{2g} \frac{1}{r^2} + C \qquad (12.184)$$

The constant of integration $C$ can be evaluated using the condition that $\zeta = \zeta_0$ as $r \to \infty$, whence

$$\zeta = \zeta_0 - \frac{r_0^4 \omega_0^2}{2g} \frac{1}{r^2} \qquad (12.185)$$

which indicates that the elevation of the water surface decreases from its value $\zeta_0$ far from the cylinder to a value equal to $\zeta_0 - r_0^2 \omega_0^2/2g$ at the cylinder.

Consider (12.182); this is equivalent to the expression we used in Chapter 11 to describe the tangential velocity around a free vortex, where $r_0^2 \omega_0$ was considered to be a measure of the intensity of the vortex. It follows from the discussion in Chapter 11 that (12.182) describes an irrotational flow field. Thus the rotating cylinder may be thought of as a vortex core surrounded by irrotational flow. It is interesting that (12.182) also can be obtained from Euler's equations for inviscid flow in cylindrical coordinates (which is left as an exercise). This result could be anticipated from the fact that (12.176) indicates that velocity components contributing to viscous forces sum to zero.

Turning to (12.185), this characterizes in terms of the water-surface elevation how the radially directed centrifugal force is balanced by the force associated with a radial pressure gradient. The pressure gradient is provided by the inward sloping liquid surface. The centrifugal force, according to (12.182), decreases away from the core; the liquid-surface gradient therefore also decreases away from the core. It is left as an exercise to infer the shape of the surface of the liquid core, assuming the core involves purely rotational motion (like that of a rigid body).

## 12.6 READING

Acheson, D. J. 1990. *Elementary fluid dynamics*. Oxford: Oxford University Press, 397 pp. Acheson presents a neat calculation (pp. 209–12) involving the stress on a boundary in simple shear flow to demonstrate the invariant property of the constitutive equations for incompressible Newtonian flow. This text also provides a good selection of problems involving very viscous flow, and it provides a nice historical, as well as technical, perspective of Prandtl's derivation of the boundary-layer equations.

Batchelor, G. K. 1970. *An introduction to fluid dynamics*. Cambridge: Cambridge University Press, 615 pp. This text contains a formal tensor analysis leading to the constitutive equations of a Newtonian fluid.

Bird, R. B., Stewart, W. E., and Lightfoot, E. N. 1960. *Transport phenomena*. New York: Wiley, 780 pp. This text provides a clear, alternative derivation (pp. 76–81) of the equation of motion in differential form, wherein normal and tangential surface stresses are described in terms of momentum fluxes.

Brown, R. A. 1991. *Fluid mechanics of the atmosphere*. San Diego: Academic, 486 pp. This text illustrates the structural homology between the tensor form of Hooke's law for an elastic substance and Stokes's law for a Newtonian fluid. A formal derivation of Hooke's law (tensor form) can be examined in Jaeger and Cook (1976).

de Groot, S. R. and Mazur, P. 1984. *Non-equilibrium thermodynamics*. New York: Dover, 510 pp. For those familiar with tensors, this text provides elegant derivations (pp.

304–10) of the Navier–Stokes equations and the energy equation for viscous flow, including treatments of the effects of bulk viscosity and rotational viscosity in the most general forms of these equations.

Huppert, H. E. 1986. The intrusion of fluid mechanics into geology. *Journal of Fluid Mechanics* 173:557–94. This paper provides a general discussion of the role of viscous-flow mechanics in treating geological processes (particularly magmatic and volcanic processes), including very viscous flows.

Jaeger, J. C. and Cook, N. G. W. 1976. *Fundamentals of rock mechanics,* 2nd ed. New York: Halsted Press, 585 pp. This text provides a comprehensive development of the tensor form of Hooke's law for an elastic substance.

Nye, J. F. 1957. The distribution of stress and velocity in glaciers and ice-sheets. *Proceedings of the Royal Society of London* Series A, 239:113–33. This paper introduces the ideas of effective shear stress and effective strain rate in the context of glaciological work. It is also the basis for Example Problem 12.5.4.

Paterson, W. S. B. 1981. *The physics of glaciers,* 2nd ed. Oxford: Pergamon, 380 pp. This classic text provides a development of the constitutive equations for glacier ice and systematically treats several cases of glacier flow, including flow over an adversely sloping bed, and flows that experience streamwise compression and extension. It also provides a development of the energy equation, including a discussion of the various boundary conditions required for its solution, and presents several important examples from the glaciological literature.

Prandtl, L. 1905. Über Flüssigkeitsbewegung bei sehr kleiner Reibung. Verhandlungen des III Internationalen Mathematiker-Kongresses, Heidelberg, 1904, Leipzig, 484–91. This is the published version of the paper that Prandtl presented in 1904 outlining the boundary-layer equations.

Prandtl, L. and Tietjens, O. G. 1957. *Fundamentals of hydro- and aeromechanics.* New York: Dover, 270 pp. This text provides a neat geometrical interpretation of stresses and strain rates, leading to the constitutive equations of a Newtonian fluid.

Schlichting, H. 1979. *Boundary-layer theory,* 7th ed. New York: McGraw-Hill, 817 pp. This text examines Stokes's hypothesis (pp. 60–61) in its development of the constitutive equations of a Newtonian fluid.

# CHAPTER 13

# Porous Media Flows

So far our treatment of fluid motions has not emphasized the behavior of fluids residing within porous geological materials. Let us now turn to this topic and, in doing so, make use of our insight regarding purely fluid flows. The general topic of fluid behavior within porous geological materials is an extensive one, forming the heart of such fields as groundwater hydrology, soils physics, and petroleum-reservoir dynamics. In addition, this topic is an essential ingredient in studies concerning the physical and chemical evolution of sedimentary basins, and the dynamics of accretionary prisms at convergent plate margins. In view of the breadth of these topics, the objective of this chapter is to introduce essential ingredients of fluid flow and transport within porous materials that are common to these topics.

Our first task is to examine the physical basis of Darcy's law, and to generalize this law to a form that can be used with an arbitrary orientation of the working coordinate system relative to the intrinsic coordinates of a geological unit that are associated with its anisotropic properties. We will likewise examine the basis of transport of solutes and heat in porous materials. We will then develop the equations of motion for the general case of saturated flow in a deformable medium. In this regard, several of the Example Problems highlight interactions between flow and strain of geological materials during loading, because this interaction bears on many geological processes. Examples include consolidation of sediments during loading, and responses of aquifers to loading by oceanic and Earth tides, and seismic stresses.

We will concentrate on the description of diffuse flows within the interstitial pores of granular materials, as opposed to flows within materials containing dual, or multiple, pore systems such as karstic media, or media containing both interstitial and fracture porosities. We will consider unsaturated, as well as saturated, conditions. For simplicity, the subscript $h$ is omitted from the notation of quantities such as specific discharge $q$ and hydraulic conductivity $K$.

## 13.1 HYDRAULIC CONDUCTIVITY AND DARCY'S LAW

### 13.1.1 Saturated media

Recall from Chapter 9 that Darcy's law,

$$\mathbf{q} = -K\nabla h \tag{13.1}$$

may be interpreted as a phenomenological law relating the specific discharge $\mathbf{q}$ to forces arising from a gravitational potential and a pressure gradient that are characterized in terms of the hydraulic head $h$, where the hydraulic conductivity $K$ is a phenomenological coefficient, or coefficient of proportionality. Since Henry Darcy first provided the experimental basis for this law in 1856, significant effort has been given to deriving Darcy's law from fluid mechanical principles, and to providing a physical explanation of the hydraulic conductivity $K$ in terms of its relation to properties of porous materials, including pore sizes and shapes. These treatments take several forms; the more sophisticated involve obtaining spatially averaged forms of the Navier–Stokes equations, where the averaging procedure is applied over a volume equal to the representative elementary volume (Chapter 2) or larger. (Students are urged to consider this topic in Bear [1972].) A particularly important part of such procedures is to take into account the shape, size, and interconnectedness of pores. This is a difficult problem given the infinity of possible pore arrangements in real geological media. In view of this, let us herein make use of dimensional analysis, appealing to the ideas of geometrical and dynamical similitude outlined in Chapter 5, to obtain a physical explanation of the hydraulic conductivity $K$.

Consider a set of geometrically similar porous media, each member of which is, in a statistical sense, an arbitrarily scaled version of all other members in the set. (See Example Problem 5.7.3 for the conditions necessary to satisfy this in the case of coarse, unconsolidated sediments.) Let us choose a characteristic length for each medium. Intuition suggests that this length ought to be characteristic of the pores, in which case we could choose a physical abstraction such as the mean pore radius $R$. For geometrically similar media, however, the mean pore radius $R$ is a constant proportion $k_0$ of the mean grain diameter $d$ (or of some other characteristic length based on grain size; Example Problem 5.7.3); that is $R = k_0 d$. For practical reasons that will become clear below, let us therefore choose $d$ as a characteristic dimension. Inasmuch as a condition of geometrical similitude is satisfied by the set of media, variations in $d$ completely specify any geometrical variation among them. Now, other relevant quantities include the fluid density $\rho$ and viscosity $\mu$, and a characteristic measure of fluid velocity, here taken to be represented by the specific discharge $q$.

As a fluid moves through a porous material it exerts a drag force on the grains of the material. The magnitude of this drag force must equal that of the driving force in the case of steady flow. Let us therefore define a force of drag per unit volume $F_V$, and assume that this drag force is a function of the characteristic quantities of the medium and fluid; thus $F_V = F_V(\mu, \rho, d, q)$. A dimensional analysis using the Buckingham Pi theorem (Example Problem 13.5.1) suggests that two dimensionless groups of quantities (Pi terms) can be formed such that

$$\frac{F_V d}{\rho q^2} = f\left(\frac{\rho d q}{\mu}\right) \tag{13.2}$$

The dimensionless quantity $F_V d/\rho q^2$ has the form of a coefficient of drag (compare this with [5.36]). The dimensionless quantity $\rho dq/\mu$ will be recognized as a Reynolds number. Thus, analogous to Equation (5.69) for a sphere, (13.2) suggests that the coefficient of drag $C_D = F_V d/\rho q^2$ is solely a function of the Reynolds number $Re = \rho dq/\mu$ for the case of geometrically similar porous media. Experiments indeed reveal that this is the case (Figure 13.1), and suggest that for $Re$ less than about 10, (13.2) has the form

$$C_D = \frac{F_V d}{\rho q^2} = \frac{C}{Re} \tag{13.3}$$

where $C$ is a constant that varies with pore shape, pore connectedness, and the distribution of pore sizes.

Recall from Hubbert's development of hydraulic head (Chapter 9) that the driving force per unit weight is equal to $-\nabla h$; therefore the driving force per unit volume is $-\rho g \nabla h$. With steady flow, this is equal in magnitude to $F_V$; substituting $-\rho g |\nabla h|$ for $F_V$ and $\rho dq/\mu$ for $Re$ in (13.3) then rearranging,

$$q = -Cd^2 \frac{\rho}{\mu} g |\nabla h| \tag{13.4}$$

Comparing this with Darcy's law, (13.1), evidently

$$K = Cd^2 \frac{\rho}{\mu} g = k \frac{\rho}{\mu} g \tag{13.5}$$

This suggests that the hydraulic conductivity $K$ varies with fluid properties, expressed as the ratio $\rho/\mu$, and medium properties, expressed by the product $Cd^2$.

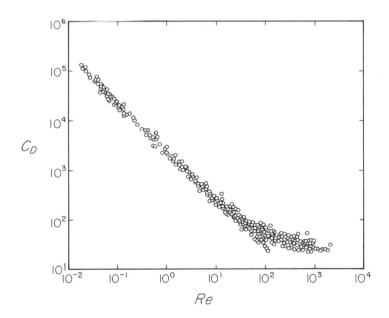

**Figure 13.1**  Relation between coefficient of drag $C_D$ and Reynolds number $Re$ for flow in granular porous media; adapted from Bear (1972), after Rose (1945).

The ratio $f = \rho/\mu$ is referred to as the *fluidity;* the product $k = Cd^2$ is referred to as the *intrinsic permeability* (or more simply, the *permeability*); this name arises because $k$ is solely a property of the medium, independent of fluid properties, whereas the hydraulic conductivity $K$ is a function of both medium and fluid properties.

More generally, dimensional analysis suggests that the coefficient of drag $C_D$ varies as a function of the Reynolds number $Re$ and a set of dimensionless groups involving variable quantities that include ratios formed from variable linear dimensions and the characteristic length used in defining the Reynolds number. Thus the constant $C$ in (13.3) can be described as a function of ratios such that

$$C = C\left(\frac{a}{d}, \frac{b}{d}, \frac{c}{d}, \cdots\right) \tag{13.6}$$

where $a, b, c, \ldots$ are auxiliary lengths that characterize variations in the size, shape, and arrangement of pores for geometrically dissimilar media. (Students should recognize the similarity between [13.6] and [5.74], the latter of which pertains to the description of drag on spinose foraminifera, as presented in Example Problem 5.7.2.) This suggests an experimental strategy for systematically examining effects of variations in pore geometry on $K$, although this topic is not examined here.

Darcy's law is valid for diffuse flows in which viscous forces dominate. It may not be valid for flows in certain media whose pores are exceedingly small, nor in media whose pores are sufficiently large that the hydraulic gradient is capable of inducing flow speeds where inertial effects become important. The upper limit of validity for Darcy's law was alluded to above; for $Re$ (as defined by $\rho dq/\mu$) greater than about 10, inertial effects can become sufficiently important that turbulence is induced and the rate of viscous dissipation increases for a given hydraulic gradient. This is manifest in the deviation of data from the curve given by (13.3) at large $Re$ (Figure 13.1). This topic is reexamined once ideas of turbulence are covered in Chapter 15. A lower limit on the validity of Darcy's law also may exist. This is related to flow within very small pores wherein molecular forces acting between the fluid and solid phases become sufficiently important that the rheological behavior of the fluid is different than its behavior in bulk, and when flow is within the Knudsen regime (Chapter 2). This topic is particularly important in the study of flow and solute transport within materials such as clays and shales; see Neuzil (1986). We shall assume throughout the remainder of this chapter that we are considering situations for which Darcy's law is valid.

The form of Darcy's law presented in (13.1) does not take into account the possibility that a porous material is more conductive in one direction than it is in another direction, a property referred to as *anisotropy*. For example, sediments often are more conductive in a direction parallel to bedding than in a direction normal to bedding, due in part to preferential alignment of platey grains parallel to bedding planes. This condition leads to the situation in which, at the pore scale, flow normal to bedding is necessarily more tortuous than flow parallel to bedding. Other possible factors contributing to anisotropy include the situation in which pore throats are on average larger, or more preferentially aligned, when viewed in one direction than when viewed in another. In such cases, for a given hydraulic gradient the rate of viscous dissipation of the mechanical energy of the flow is greater in one direction (for example, normal to bedding) than in another direction, manifest as a difference in the hydraulic conductivities associated with the two directions. In two dimensions, typically two *principal directions of anisotropy* can be defined. These coincide with the directions in which a medium exhibits its maximum and minimum hydraulic conductivities. More generally, in three dimensions there are three principal directions of anisotropy. A medium for which the hydraulic conductivities in two of the three principal directions are equal is said to be *transversely anisotropic* (or *transversely isotropic*). In the case where the hydraulic conductivities in all three principal directions are equal (or more correctly stated, where principal directions of anisotropy cannot be distinguished), the medium is *isotropic*. These points suggest that we need a more general mathematical description of the hydraulic conductivity than is embodied in (13.1).

In the case of a porous medium that is anisotropic with respect to hydraulic conductivity, normally a working coordinate system can be selected such that its axes coincide with the three principal directions that define the anisotropy. In situations where this is not possible, however, Darcy's law (13.1) must be generalized to tensor form. We will start by considering the two-dimensional case, then extend our results to three dimensions, and use this as an opportunity to examine tensor notation and several operations involving tensors.

Consider a porous sedimentary layer whose bedding planes are inclined at an angle $\phi$ from the $x$-axis of a global $xy$-coordinate system (Figure 13.2). Further define a local $x'y'$-coordinate system such that the $x'$-axis is parallel to the bedding planes. Suppose that the layer is anisotropic, with large hydraulic conductivity $K_{x'x'}$ parallel to bedding (and therefore parallel to the $x'$-axis), and relatively small hydraulic conductivity $K_{y'y'}$ normal to bedding (and therefore parallel to the $y'$-axis). The conductivities $K_{x'x'}$ and $K_{y'y'}$ are to be associated with the two principal directions of anisotropy of the layer, as would be measured in field or laboratory tests.

Now consider arbitrary equipotential lines that are not parallel to any of the global or local coordinate axes. In this situation, flow generally is not normal to the equipotential lines as it is under isotropic conditions. Rather, flow is deflected in a direction that is closer to the direction of the bedding—that is, toward the direction

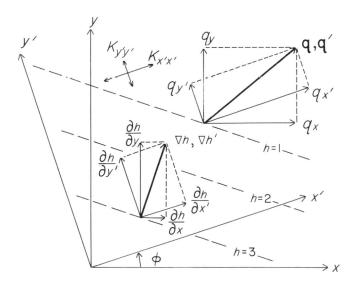

**Figure 13.2** Definition diagram relating specific discharge **q** (and **q**') and hydraulic gradient $\nabla h$ (and $\nabla h'$) in global $xy$-coordinate system and local $x'y'$-coordinate system.

of high conductivity. Indeed, in the case where $K_{y'y'}$ is essentially zero, flow can occur only in a direction parallel to the $x'$-axis. Alternatively, one may view the flow as being preferentially deflected toward a direction that exhibits a lower resistance to flow, as measured by the *hydraulic resistivity*, defined as the reciprocal, $1/K$, of the hydraulic conductivity.

The specific discharge $\mathbf{q}$ has components $q_x$ and $q_y$ in the global $xy$-coordinate system, and its counterpart $\mathbf{q}'$ has components $q_{x'}$ and $q_{y'}$ in the local $x'y'$-coordinate system (Figure 13.2). Likewise the vector $\nabla h$ has components $\partial h/\partial x$ and $\partial h/\partial y$ in the global coordinate system, and its counterpart $\nabla h'$ has components $\partial h/\partial x'$ and $\partial h/\partial y'$ in the local coordinate system. Consider a stream tube (which by definition is parallel to the direction of $\mathbf{q}$). Both components $\partial h/\partial x$ and $\partial h/\partial y$ are capable of contributing to the component of flow $q_x$ within the stream tube, as well as to the component $q_y$. To envision this, notice that in absence of the component gradient $\partial h/\partial x$, the component gradient $\partial h/\partial y$ would lead to a nonzero gradient along the flow tube, and therefore to a flow having component $q_x$. We thus write general forms for the components of Darcy's law:

$$q_x = -K_{xx}\frac{\partial h}{\partial x} - K_{xy}\frac{\partial h}{\partial y} \tag{13.7}$$

$$q_y = -K_{yx}\frac{\partial h}{\partial x} - K_{yy}\frac{\partial h}{\partial y} \tag{13.8}$$

These have the matrix form,

$$\mathbf{q} = \langle q_x, q_y \rangle \leftrightarrow \begin{bmatrix} q_x \\ q_y \end{bmatrix} = -\begin{bmatrix} K_{xx} & K_{xy} \\ K_{yx} & K_{yy} \end{bmatrix} \begin{bmatrix} \dfrac{\partial h}{\partial x} \\ \dfrac{\partial h}{\partial y} \end{bmatrix} \tag{13.9}$$

which may be written compactly using tensor notation:

$$\mathbf{q} = -\mathbf{K}\nabla h \tag{13.10}$$

Here $\mathbf{K}$ denotes a two-dimensional *hydraulic conductivity tensor*. We will see below that this tensor is symmetric, such that $K_{xy} = K_{yx}$.

The coefficients of hydraulic conductivity in these equations are to be interpreted as follows: $K_{xx}$ is the apparent conductivity necessary to provide the contribution to $q_x$ associated with the gradient component $\partial h/\partial x$, and $K_{xy}$ is the apparent conductivity necessary to provide the contribution to $q_x$ associated with the gradient component $\partial h/\partial y$. Similar remarks apply to $K_{yx}$ and $K_{yy}$. Note that $K_{xx}$ and $K_{yy}$ generally are not equal to the conductivities measured in the principal directions defining the anisotropy of the medium; that is, $K_{xx}$ generally does not equal $K_{x'x'}$ and $K_{yy}$ does not equal $K_{y'y'}$.

In a similar vein, we may write general forms for the components of Darcy's law with respect to the local $x'y'$-coordinate system:

$$q_{x'} = -K_{x'x'}\frac{\partial h}{\partial x'} - K_{x'y'}\frac{\partial h}{\partial y'} \tag{13.11}$$

$$q_{y'} = -K_{y'x'}\frac{\partial h}{\partial x'} - K_{y'y'}\frac{\partial h}{\partial y'} \tag{13.12}$$

Observe, however, that since components of the hydraulic gradient in the local co-ordinate system are parallel to the principal directions of anisotropy, $\partial h/\partial y'$ cannot contribute to the component $q_{x'}$, and $\partial h/\partial x'$ cannot contribute to the component $q_{y'}$. Therefore (13.11) and (13.12) reduce to

$$q_{x'} = -K_{x'x'} \frac{\partial h}{\partial x'} \tag{13.13}$$

$$q_{y'} = -K_{y'y'} \frac{\partial h}{\partial y'} \tag{13.14}$$

These have the matrix form,

$$\mathbf{q}' = \langle q_{x'}, q_{y'} \rangle \leftrightarrow \begin{bmatrix} q_{x'} \\ q_{y'} \end{bmatrix} = -\begin{bmatrix} K_{x'x'} & 0 \\ 0 & K_{y'y'} \end{bmatrix} \begin{bmatrix} \dfrac{\partial h}{\partial x'} \\ \dfrac{\partial h}{\partial y'} \end{bmatrix} \tag{13.15}$$

and the compact tensor notation,

$$\mathbf{q}' = -\mathbf{K}'\nabla h' \tag{13.16}$$

where $\mathbf{K}'$ denotes a two-dimensional hydraulic conductivity tensor whose nonzero diagonal elements coincide with the principal directions of anisotropy. The simpli-fied expressions for the components of Darcy's law, (13.13) and (13.14), are con-sistent with our earlier treatment of Darcy's law (Chapter 9), where we assumed that the axes of the coordinate system were selected to coincide with the principal directions of anisotropy. Our objective now is to obtain expressions for $K_{xx}$, $K_{zz}$, $K_{xy}$, and $K_{yx}$, in terms of $K_{x'x'}$ and $K_{y'y'}$.

Recall from analytical geometry that coordinate transformations between local and global systems involving a rotation through the angle $\phi$ are given by

$$x' = x \cos \phi + y \sin \phi; \qquad y' = -x \sin \phi + y \cos \phi \tag{13.17}$$

$$x = x' \cos \phi - y' \sin \phi; \qquad y = x' \sin \phi + y' \cos \phi \tag{13.18}$$

Consider $\mathbf{q}$ as a position vector (where its tail is at the origin); then the coordinates representing $\mathbf{q}$ are $q_x$ and $q_y$. It follows geometrically from (13.18) that $q_x$ and $q_y$ are

$$q_x = q_{x'} \cos \phi - q_{y'} \sin \phi \tag{13.19}$$

$$q_y = q_{x'} \sin \phi + q_{y'} \cos \phi \tag{13.20}$$

Likewise, from (13.17) the components $\partial h/\partial x'$ and $\partial h/\partial y'$ of the vector $\nabla h'$ are

$$\frac{\partial h}{\partial x'} = \frac{\partial h}{\partial x} \cos \phi + \frac{\partial h}{\partial y} \sin \phi \tag{13.21}$$

$$\frac{\partial h}{\partial y'} = -\frac{\partial h}{\partial x} \sin \phi + \frac{\partial h}{\partial y} \cos \phi \tag{13.22}$$

The two pairs of expressions above, (13.19) through (13.22), may be written as

$$\mathbf{q} = \mathbf{D}^{-1}\mathbf{q}'; \qquad \nabla h' = \mathbf{D}\nabla h \tag{13.23}$$

where $\mathbf{D}$ is a rotation matrix given by

$$\mathbf{D} = \begin{bmatrix} \cos\phi & \sin\phi \\ -\sin\phi & \cos\phi \end{bmatrix} \qquad (13.24)$$

and $\mathbf{D}^{-1}$ is equal to the transposed matrix $\mathbf{D}^{\mathrm{T}}$ formed by interchanging the rows and columns in $\mathbf{D}$:

$$\mathbf{D}^{-1} = \mathbf{D}^{\mathrm{T}} = \begin{bmatrix} \cos\phi & -\sin\phi \\ \sin\phi & \cos\phi \end{bmatrix} \qquad (13.25)$$

Substituting component gradients (13.21) and (13.22) into (13.13) and (13.14), substituting the resulting expressions for $q_{x'}$ and $q_{y'}$ into (13.19) and (13.20), then rearranging and collecting terms in $\partial h/\partial x$ and $\partial h/\partial y$,

$$q_x = -(K_{x'x'}\cos^2\phi + K_{y'y'}\sin^2\phi)\frac{\partial h}{\partial x} - (K_{x'x'} - K_{y'y'})\sin\phi\cos\phi\frac{\partial h}{\partial y} \qquad (13.26)$$

$$q_y = -(K_{x'x'} - K_{y'y'})\sin\phi\cos\phi\frac{\partial h}{\partial x} - (K_{x'x'}\sin^2\phi + K_{y'y'}\cos^2\phi)\frac{\partial h}{\partial y} \qquad (13.27)$$

Comparing these with (13.7) and (13.8), evidently

$$K_{xx} = K_{x'x'}\cos^2\phi + K_{y'y'}\sin^2\phi \qquad (13.28)$$

$$K_{yy} = K_{x'x'}\sin^2\phi + K_{y'y'}\cos^2\phi \qquad (13.29)$$

$$K_{xy} = K_{yx} = (K_{x'x'} - K_{y'y'})\sin\phi\cos\phi \qquad (13.30)$$

Symmetry of the tensor $\mathbf{K}$ is embodied in (13.30).

The algebraic steps leading to expressions for the elements of $\mathbf{K}$ above are straightforward since $\mathbf{K}$ (and $\mathbf{K}'$) are $2\times 2$ matrices. Such steps, however, are cumbersome with larger matrices. A solution can be obtained more directly from $\mathbf{D}$, $\mathbf{D}^{-1}$, and $\mathbf{K}'$. It is left as an exercise to demonstrate that

$$\mathbf{K} = \mathbf{D}^{-1}\mathbf{K}'\mathbf{D} \qquad (13.31)$$

Often (13.9) is a sufficient generalization of Darcy's law for field problems. Nonetheless it can be generalized to three dimensions such that

$$\mathbf{K} = \begin{bmatrix} K_{xx} & K_{xy} & K_{xz} \\ K_{yx} & K_{yy} & K_{yz} \\ K_{zx} & K_{zy} & K_{zz} \end{bmatrix} \qquad (13.32)$$

Like the conductivity tensor in (13.9), (13.32) is a symmetric tensor such that $K_{xy} = K_{yx}, K_{xz} = K_{zx}$, and $K_{yz} = K_{zy}$. If the global coordinate system coincides with the principal directions of anisotropy, these off-diagonal elements equal zero. In the transversely isotropic case, $K_{xx} = K_{yy}$. In the isotropic case, $K_{xx} = K_{yy} = K_{zz} = K$, a scalar quantity.

The set of expressions relating the elements of the conductivity tensor in (13.32) to the three values of conductivity that define the principle directions of anisotropy depend on the convention of rotation. A total of twelve conventions are possible, depending on the sequence of rotations used in the coordinate transformation. For example, the initial rotation can be about any of the three axes. Then, the only limitation in choosing the second and third rotations is that no two successive rotations can be about the same axis. Consider for example an $xyz$-convention. The first rotation is through the angle $\phi$ about the $z$-axis, the second rotation is through the angle $\theta$ about the intermediate $y$-axis, and the third rotation is through the angle $\psi$ about the final $x$-axis (Figure 13.3). The complete transformation matrix $\mathbf{A}$ can be obtained as the matrix product of the three matrices describing the separate rotations.

The first rotation about the $z$-axis involves the matrix

$$\mathbf{D} = \begin{bmatrix} \cos\phi & \sin\phi & 0 \\ -\sin\phi & \cos\phi & 0 \\ 0 & 0 & 1 \end{bmatrix} \tag{13.33}$$

which is a more general form of the rotation matrix in (13.24). The second rotation about the intermediate $y$-axis involves the matrix

$$\mathbf{C} = \begin{bmatrix} \cos\theta & 0 & -\sin\theta \\ 0 & 1 & 0 \\ \sin\theta & 0 & \cos\theta \end{bmatrix} \tag{13.34}$$

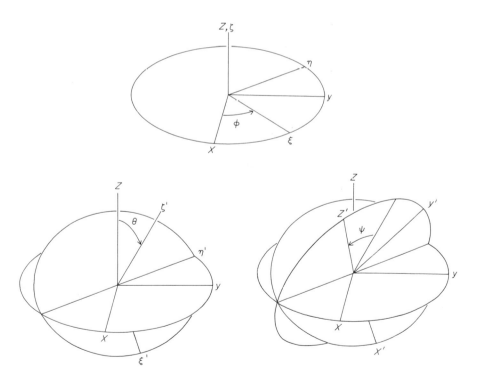

**Figure 13.3** Definition diagram for $xyz$-convention of rotation; $\xi$, $\eta$, and $\zeta$, and $\xi'$, $\eta'$, and $\zeta'$ denote intermediate axes positions between $x$, $y$, and $z$, and $x'$, $y'$, and $z'$.

and the third rotation about the final $x$-axis involves the matrix

$$\mathbf{B} = \begin{bmatrix} 1 & 0 & 0 \\ 0 & \cos\psi & \sin\psi \\ 0 & -\sin\psi & \cos\psi \end{bmatrix} \tag{13.35}$$

The complete transformation matrix $\mathbf{A}$, such that $\mathbf{q}' = \mathbf{Aq}$, is given by $\mathbf{A} = \mathbf{BCD}$:

$$\mathbf{A} = \begin{bmatrix} \cos\theta\cos\phi & \cos\theta\sin\phi & -\sin\theta \\ \sin\psi\sin\theta\cos\phi - \cos\psi\sin\phi & \sin\psi\sin\theta\sin\phi + \cos\psi\cos\phi & \cos\theta\sin\psi \\ \cos\psi\sin\theta\cos\phi + \sin\psi\sin\phi & \cos\psi\sin\theta\sin\phi - \sin\psi\cos\phi & \cos\theta\cos\psi \end{bmatrix} \tag{13.36}$$

The inverse transformation, $\mathbf{q} = \mathbf{A}^{-1}\mathbf{q}'$, is provided by the transposed matrix $\mathbf{A}^{\mathrm{T}}$:

$$\mathbf{A}^{-1} = \mathbf{A}^{\mathrm{T}}$$

$$= \begin{bmatrix} \cos\theta\cos\phi & \sin\psi\sin\theta\cos\phi - \cos\psi\sin\phi & \cos\psi\sin\theta\cos\phi + \sin\psi\sin\phi \\ \cos\theta\sin\phi & \sin\psi\sin\theta\sin\phi + \cos\psi\cos\phi & \cos\psi\sin\theta\sin\phi - \sin\psi\cos\phi \\ -\sin\theta & \cos\theta\sin\phi & \cos\theta\cos\psi \end{bmatrix} \tag{13.37}$$

It is left as an exercise to use the tensor product

$$\mathbf{K} = \mathbf{A}^{-1}\mathbf{K}'\mathbf{A} \tag{13.38}$$

where

$$\mathbf{K}' = \begin{bmatrix} K_{x'x'} & 0 & 0 \\ 0 & K_{y'y'} & 0 \\ 0 & 0 & K_{z'z'} \end{bmatrix} \tag{13.39}$$

to obtain expressions relating elements of the hydraulic conductivity tensor $\mathbf{K}$ to elements of the tensor $\mathbf{K}'$.

Anisotropy in hydraulic conductivity is just one way in which natural heterogeneities of geological materials are manifest in flow problems. In general, geological materials typically exhibit heterogeneity at many scales. Heterogeneities exist at the pore scale in relation to variations in the packing and orientation of grains; they exist at the outcrop scale in relation to variations in the extent, thickness, and composition of bedding; and heterogeneities exist at still larger scales in relation to changes in bedding composition associated with facies. Fractures also introduce a special form of heterogeneity in porosity and pore structure. All of these factors contribute to spatial variations in hydraulic conductivity. In addition, such heterogeneities lead to variations in conductivity that depend on scale. In general, the permeabilities of natural porous media increase with the scale of measurement or characteristic averaging volume (Chapter 2; also refer to Clauser [1992]). Herein arises an interesting question: How do local variations in hydraulic conductivity influence large-scale flow patterns that often are of interest to geologists; and what value of conductivity should be used to describe flow over large scales?

One conventional approach to this problem is to assume that an overall effective value of $K$ exists for a heterogeneous unit. In a formal analysis of flow within such a unit, the unit is conceptually "replaced" with an *equivalent homogeneous medium* that possesses this effective value of $K$. The equivalent homogeneous medium is therefore a physical abstraction having the property that it behaves hydraulically like the real medium it replaces, when viewed at a scale that is larger than the heterogeneities of the real medium. Certain techniques for estimating $K$, by their nature, provide an estimate of the effective value of $K$. For example, a pumping test (Example Problem 8.4.3) can involve flow within a significant volume of aquifer surrounding the pumping well; such a test therefore provides a volume-integrated estimate of $K$, which is in some sense an effective value for the aquifer. This value may or may not coincide with a value of $K$, say, for any individual bed within the aquifer; rather, it is a measure of how all beds (heterogeneities) collectively respond in transmitting flow during the pumping test.

Another approach involves collecting small samples of a medium, usually from cores taken during a well-drilling operation, then estimating $K$ for these samples using laboratory permeameter tests. This provides a statistical distribution of $K$ for the geological unit near the site of the well. The probability density function $f(K)$ of conductivity $K$ for geological units tends to possess a log-normal form. Since this is a skewed distribution (for arithmetic values of $K$), choices of an effective value of $K$ include the harmonic mean and the geometric mean, although use of these may be problematic (see references at the end of this chapter). Hereafter we will assume that an effective value of $K$ exists.

The existence of a distribution in $K$, even for a unit that is nominally uniform, suggests that the idea of a homogeneous unit ought to be defined in terms of $f(K)$. In particular, a unit may be considered (statistically) homogeneous with respect to $K$ if the mean and variance of $f(K)$ do not vary spatially. Then, use of an effective value of $K$ in an analysis of flow within a unit is justified inasmuch as the hydraulic head field, $h(x, y, z, t)$, is insensitive to local variations in $K$. More problematical are analyses of solute transport, because detailed flow patterns within a unit can be strongly influenced by local heterogeneities that lead to preferential paths of flow and transport. These cannot easily be reconciled with traditional views of flow in porous materials as consisting of uniformly diffuse, interstitial flow.

### 13.1.2 Unsaturated media

Consider a medium containing two fluids, for example, water and air. Here we are interested in the behavior of the wetting phase—the water. Further recall from Chapter 8 that the degree of saturation $\theta_s$ is in this case defined with respect to the water. Since $\theta_s < 1$, the water-filled pore volume involved in transmission of the water is less than the total pore volume, and the ability of the medium to transmit flow in response to a given hydraulic gradient decreases relative to its ability to transmit flow under saturated conditions.

It is conventionally assumed that flow under unsaturated conditions can be described by a form of Darcy's law. The saturated hydraulic conductivity $K_s$ in Darcy's law is replaced by the unsaturated hydraulic conductivity $K_u$, which varies as a function of $\theta_s$. Since our objective is to examine this relation, we will hereafter assume for simplicity that the medium is homogeneous and isotropic such that $K_u$ is a scalar quantity. Thus,

$$\mathbf{q} = -K_u \nabla h \qquad (13.40)$$

Numerous expressions for the functional relation between $K_u$ and $\theta_s$ have been proposed. A currently popular model is that of Brooks and Corey (1964); this model has a good theoretical basis, has a relatively simple form, and performs well empirically. Moreover, the Brooks–Corey model contains essential ingredients of most other available models, and therefore our examination of this model will serve to illustrate how the relation between $K_u$ and various physical factors has been typically conceptualized. Let us start with the characteristic curve—the relation between liquid content and capillary pressure. Here we shall neglect extremely small pores and the physico-chemical effects associated with them.

Recall from Chapter 3 that the pressure $p$ within a capillary wedge of liquid is influenced by the curvature of the liquid–gas interface such that $p$ is less than the pressure $p_0$ of the gas phase. Generally, as the liquid content of a porous material decreases, the liquid occupies progressively smaller pores and pore throats. Because the radii of pores containing liquid are decreasing, curvatures of the liquid–gas interfaces are, on average, increasing. Therefore, based on (3.58), the pressure drop across the interfaces, on average, increases. We defined this pressure drop as the capillary pressure, $p_c = p_0 - p$. This may be expressed in terms of (capillary) pressure head $\psi_c$, such that $\psi_c = p_c/\rho g$.

Using a pressure-plate apparatus (Chapter 3), one can obtain a relation between the liquid content and $\psi_c$ (Figure 13.4). This is referred to as a *characteristic moisture curve*, or *retention curve*, when the experiments are performed such that the porous material is initially saturated then subjected to increasing values of $\psi_c$ so that it progressively dewaters. For granular materials (Figure 13.5), Brooks and Corey (1964) found that this relation can be expressed by

$$\theta_e = \left(\frac{\psi_{cb}}{\psi_c}\right)^{\lambda} \qquad \psi_c \geq \psi_{cb} \qquad (13.41)$$

where $\theta_e$ is an effective saturation, $\psi_{cb}$ is the bubbling head, and $\lambda$ is a parameter that characterizes the distribution of pore sizes. The effective saturation $\theta_e$ is defined by

$$\theta_e = \frac{\theta - \theta_r}{n - \theta_r} \qquad (13.42)$$

Here $\theta$ is the moisture content, equal to the ratio of the volume of liquid to the total volume (with upper limit equal to the porosity $n$), and $\theta_r$ is a residual mois-

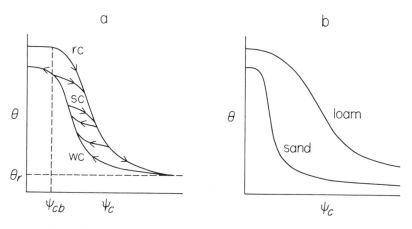

**Figure 13.4** Characteristic moisture curves relating moisture content $\theta$ to capillary pressure head $\psi_c$ (*a*), where arrows represent trajectoris along retention curve (rc), wetting curve (wc), and scanning curves (sc); and generalized characteristic moisture curves for different textures of media (*b*).

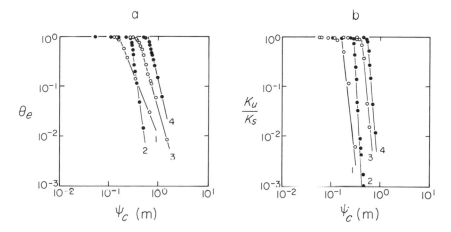

**Figure 13.5** Characteristic moisture curve relating effective saturation $\theta_e$ to capillary pressure head $\psi_c$ (a) and predicted relative unsaturated hydraulic conductivity $K_u/K_s$ (b) for granular media: volcanic sand, $\lambda = 2.29$, $\eta = 8.9$ (1); glass beads, $\lambda = 7.30$, $\eta = 23.9$ (2); fine sand $\lambda = 3.70$, $\eta = 13.1$ (3); hygiene sandstone, $\lambda = 4.17$, $\eta = 14.5$ (4); adapted from Brooks and Corey (1964).

ture content, a limiting value of $\theta$ as $\psi_c$ becomes large. (The effective saturation alternatively may be defined by $\theta_e = (\theta_s - \theta_{sr})/(1 - \theta_{sr})$, where $\theta_{sr}$ is the residual degree of saturation.) Recall from Chapter 3 that the bubbling head $\psi_{cb}$ is the value of $\psi_c$ at which dewatering first occurs. Note also that $\psi_c$ is a quantity obtained externally from a gauge on the pressure-plate apparatus; at the pore scale $\psi_c$ locally varies from pore to pore in the material, and thus $\psi_c$ is a quantity that is averaged over many pores and liquid–gas interfaces.

If the pressure-plate experiment is performed such that the porous material is initially dry, then progressively wetted (with decreasing $\psi_c$), a similar curve, the *wetting curve,* is obtained. The wetting curve, however, generally does not coincide with the retention curve (Figure 13.4). Thus, for given $\psi_c$, the moisture content is lower during wetting than during dewatering. This phenomena is referred to as *hysteresis.* Although a complete theory for hysteresis is not available, several mechanisms have been proposed that likely contribute to this phenomenon. First, air bubbles typically are trapped within pores and pore throats during wetting, thereby decreasing the moisture content for a given capillary head. In this regard the wetting curve typically does not reach the same value of moisture content that is associated with the retention curve at $\psi_c = 0$ (Figure 13.4). Second, the pore space occupied by liquid, and the associated curvatures of liquid–gas interfaces, may be different when a liquid "advances" on grain surfaces during wetting than when it "retreats" from grain surfaces during drying. Third, deformation of solid phases may occur with wetting, particularly if these phases include clay mineral grains that can incorporate water into their lattice structures.

An interesting part of the wetting–drying process are *scanning curves;* these are trajectories in the moisture-head state between the wetting and retention curves. For example, a sample that is initially dewatered from a saturated state to an arbitrary moisture content will, upon subsequent wetting, track along an intermediate path (scanning curve) to the wetting curve, and thereafter follow this limiting curve until it is again subjected to drying (Figure 13.4). With a second phase of drying, the soil will scan across to the retention curve. Hysteresis can be pronounced with fine-grained materials; coarse materials, for example, sands, exhibit less hysteresis such that the retention and wetting curves approximately coincide.

Let us now return to the retention curve and the parameter $\lambda$. Notice that for coarse material (for example, sorted sand) the retention curve declines sharply once the bubbling head $\psi_{cb}$ is exceeded. This reflects that the uniformly large pores typical of coarse, sorted materials rapidly dewater with increasing $\psi_c$. In contrast, loams (and finer-grained materials) dewater more gradually with increasing $\psi_c$, which reflects that these materials typically possess a more even distribution of pore sizes. The retention curves at large $\psi_c$, when liquid is present only in the smallest pores and in the form of thin films on grain surfaces, reflect effects of adhesion more than capillary forces. Therefore the form of this part of the retention curve for both coarse-grained and fine-grained materials is more related to the surface area and surface chemistry of the materials than to pore geometry.

The use of capillary head defined by $\psi_c = (p_0 - p)/\rho g$ is a matter of convenience; it provides a way to work with positive head quantities. However, when coupling descriptions of flow in unsaturated and saturated systems, it becomes important to choose a consistent measure of pressure head, and it is conventional to use the pressure head $\psi$ as normally defined for saturated media. Choosing $p_0$ as the pressure at datum state, $\psi = (p - p_0)/\rho g$ or $\psi = -\psi_c$, and the bubbling head $\psi_b = -\psi_{cb}$. Then the characteristic curve (Figure 13.4) is generally described by

$$\theta = \theta(\psi) \qquad \psi < \psi_b \qquad (13.43)$$

$$\theta = n \qquad \psi \geq \psi_b \qquad (13.44)$$

The local slope of $\theta(\psi)$ is denoted by $c$:

$$c = \frac{d\theta}{d\psi} \qquad (13.45)$$

This defines the *specific moisture capacity* of a porous material; it is a measure of the change in moisture content per unit change in pressure head. Notice that this definition is analogous to the definition (4.2) for the specific heat capacity of a substance. Judging from Figure 13.4, $c$ varies as a function of $\psi$; in general,

$$c = c(\psi) \qquad \psi < \psi_b \qquad (13.46)$$

$$c = 0 \qquad \psi \geq \psi_b \qquad (13.47)$$

Like $\theta(\psi)$, $c(\psi)$ may exhibit hysteresis.

With decreasing moisture content (and increasing $\psi_c$), the liquid-filled pore volume involved in transmission of the liquid is less than the total pore volume. In analogy with Hagen–Poiseuille flow in a capillary tube, the discharge through an individual pore is proportional to the fourth power of its radius. Since the liquid only occupies small pores, a given hydraulic gradient is not as effective in producing flow as when large pores are filled and involved in transmission. Flow occurs between interconnected small pores and along water films; thus the tortuosity of the flow increases with decreasing moisture content. For these reasons the ability of the medium to transmit flow in response to a given hydraulic gradient decreases relative to its ability to transmit flow under saturated conditions. This effect is embodied in the form of the function that relates $K_u$ to either moisture content or capillary head (or capillary pressure). Brooks and Corey (1964) proposed a function for $K_u$ having the form

$$K_u = K_s \left(\frac{\psi_{cb}}{\psi_c}\right)^\eta \qquad \psi_c \geq \psi_{cb} \qquad (13.48)$$

where $\eta = 2 + 3\lambda$ (Figure 13.5). As the moisture content approaches saturation, $\psi_c$ approaches $\psi_{cb}$ and $K_u$ approaches $K_s$. Formula (13.48) is based on a combination of dimensional analysis applied to geometrically similar pore arrangements, and consideration of Hagen–Poiseuille flow through pores with different cross-sectional areas. The underlying theory is due to Burdine (1953). The result $\eta = 2 + 3\lambda$ basically represents a compromise between the numerical values of coefficients that appear in the Hagen–Poiseuille equations for the discharge through a circular tube (Example Problem 13.5.1) and a conduit with parallel walls (Example Problem 3.7.1).

Numerous other expressions similar to (13.48) have been proposed for the functional relation between $K_u$ and either $\psi$ or $\theta$. Interested students should examine Hillel (1971) and the serial edited by Klute (1986). Letting $K_u(\psi)$ denote a general relation between $K_u$ and $\psi$, and recalling that $\psi_c = p_0 - \psi$, we may now write Darcy's law for unsaturated flow as

$$\mathbf{q} = -K_u(\psi)\nabla h \qquad (13.49)$$

## 13.2 EQUATIONS OF MOTION

### 13.2.1 Saturated media

Let us begin with the equation of continuity (8.45) developed in Chapter 8:

$$\nabla \cdot (\rho\mathbf{q}) + n\mathbf{u}_s \cdot \nabla\rho + \frac{\rho}{1-n}\mathbf{u}_s \cdot \nabla n = -n\frac{\partial\rho}{\partial t} - \frac{\rho}{1-n}\frac{\partial n}{\partial t} \qquad (13.50)$$

Recall that the specific discharge $\mathbf{q}$ describes relative motion between the fluid and matrix,

$$\mathbf{q} = n(\mathbf{u} - \mathbf{u}_s) \qquad (13.51)$$

where $\mathbf{u}$ is the average fluid velocity and $\mathbf{u}_s$ is the matrix velocity. Then, (13.50) is a general expression for conservation of total mass—fluid and solid matrix—under the condition that the matrix may be deforming. Also recall that (13.50) is based on the idea that any deformation of the matrix leading to a change in porosity $n$ involves a change in pore volume but not solid volume. In relation to matrix deformation, often it can be assumed that variations in porosity $n$ and fluid density $\rho$ with respect to space are small relative to local variations with respect to time; that is $\mathbf{u}_s \cdot \nabla n \ll \partial n/\partial t$ and $\mathbf{u}_s \cdot \nabla\rho \ll \partial\rho/\partial t$. Then (13.50) can be simplified to

$$\nabla \cdot (\rho\mathbf{q}) = -n\frac{\partial\rho}{\partial t} - \frac{\rho}{1-n}\frac{\partial n}{\partial t} \qquad (13.52)$$

If it is further assumed that the fluid is essentially homogeneous with respect to density, that is, $\mathbf{q} \cdot \nabla\rho \ll \partial\rho/\partial t$, then (13.52) is simplified to

$$\nabla \cdot \mathbf{q} = -\frac{n}{\rho}\frac{\partial\rho}{\partial t} - \frac{1}{1-n}\frac{\partial n}{\partial t} \qquad (13.53)$$

Let us now consider the right side of (13.53). Based on the definition (3.42) of fluid compressibility $\beta$, $\beta dp = d\rho/\rho$. Using this, the first term on the right of (13.53) becomes $-n\beta \, \partial p/\partial t$. The porosity $n = V_p/V_T$, where $V_p$ is the pore volume and $V_T$ is the total volume. Using the chain rule,

$$dn = \frac{1}{V_T}dV_p - \frac{V_p}{V_T^2}dV_T = \frac{1}{V_T}dV_p - \frac{n}{V_T}dV_T \tag{13.54}$$

If a change in total volume $dV_T$ involves only a change in pore volume, then $dV_T = dV_p$ and

$$\frac{1}{1-n}dn = \frac{1}{V_T}dV_p \tag{13.55}$$

Based on the definition (3.55) of matrix compressibility $\alpha$,

$$\alpha d\sigma_e = -\frac{1}{V_T}dV_p = -\frac{1}{1-n}dn \tag{13.56}$$

where $\sigma_e$ is the effective stress. From (3.54), $d\sigma_e = d\sigma - dp$, where $\sigma$ is the total stress; then

$$\alpha d\sigma - \alpha dp = -\frac{1}{1-n}dn \tag{13.57}$$

Using this, the last term on the right side of (13.53) is $-\alpha \, \partial p/\partial t + \alpha \, \partial\sigma/\partial t$, and (13.53) becomes

$$\nabla \cdot \mathbf{q} = -(n\beta + \alpha)\frac{\partial p}{\partial t} + \alpha\frac{\partial \sigma}{\partial t} \tag{13.58}$$

If the total stress $\sigma$ is constant, this reduces to

$$\nabla \cdot \mathbf{q} = -(n\beta + \alpha)\frac{\partial p}{\partial t} \tag{13.59}$$

Because $h = z + (p - p_0)/\rho g$, then $dp/dt = \rho g \, dh/dt$, and (13.59) may be written as

$$\nabla \cdot \mathbf{q} = -\rho g(n\beta + \alpha)\frac{\partial h}{\partial t} \tag{13.60}$$

Substituting Darcy's law (13.10) into (13.60),

$$\nabla \cdot (\mathbf{K}\nabla h) = \rho g(n\beta + \alpha)\frac{\partial h}{\partial t} \tag{13.61}$$

For anisotropic conditions, where the global coordinate system does not coincide with the principal directions of anisotropy, expansion of (13.61) will reveal that the resulting equation involves mixed partial differentials of the form $\partial^2 h/\partial x \, \partial y$. Problems involving such differentials generally are difficult to treat analytically except for simple cases, and normally are treated using numerical methods (for example, see Anderson et al. [1984]). When the global coordinate system can be selected to coincide with the principal directions of anisotropy, (13.61) becomes in the general case of heterogeneous and anisotropic conditions:

$$\frac{\partial}{\partial x}\left(K_x \frac{\partial h}{\partial x}\right) + \frac{\partial}{\partial y}\left(K_y \frac{\partial h}{\partial y}\right) + \frac{\partial}{\partial z}\left(K_z \frac{\partial h}{\partial z}\right) = \rho g(n\beta + \alpha)\frac{\partial h}{\partial t} \qquad (13.62)$$

Here $K_x$, $K_y$, and $K_z$ denote the conductivities in the three principal directions of anisotropy, where the notation used above involving double-letter subscripts has been discarded. For homogeneous, anisotropic conditions, (13.62) becomes

$$K_x \frac{\partial^2 h}{\partial x^2} + K_y \frac{\partial^2 h}{\partial y^2} + K_z \frac{\partial^2 h}{\partial z^2} = \rho g(n\beta + \alpha)\frac{\partial h}{\partial t} \qquad (13.63)$$

For isotropic conditions, where $K_x = K_y = K_z = K$, a scalar quantity,

$$\frac{\partial^2 h}{\partial x^2} + \frac{\partial^2 h}{\partial y^2} + \frac{\partial^2 h}{\partial z^2} = \frac{\rho g(n\beta + \alpha)}{K}\frac{\partial h}{\partial t} = \frac{1}{\kappa_h}\frac{\partial h}{\partial t} \qquad (13.64)$$

The ratio $\kappa_h = K/\rho g(n\beta + \alpha)$ is referred to as the *hydraulic diffusivity*. Under steady conditions this reduces to Laplace's equation,

$$\frac{\partial^2 h}{\partial x^2} + \frac{\partial^2 h}{\partial y^2} + \frac{\partial^2 h}{\partial z^2} = 0 \qquad (13.65)$$

Returning to (13.64), the quantity $\rho g(n\beta + \alpha)$ has an interesting physical interpretation. The part involving $\beta$ characterizes expansion or compression of fluid associated with a change in pressure head. The part involving $\alpha$ characterizes expansion or compression of the porous matrix associated with a change in pressure head via the effective stress $\sigma_e$. The quantity $\rho g(n\beta + \alpha)$ is referred to as the *specific storage* $S_s$. It is a measure of the volume of water that a unit volume of aquifer releases or incorporates with a unit change in pressure head. A release of water involves fluid expansion and matrix compression, whereas incorporation of water involves fluid compression and matrix expansion. In the case of flow within a confined aquifer with uniform thickness $b$, (13.64) reduces to the two-dimensional diffusion equation,

$$\frac{\partial^2 h}{\partial x^2} + \frac{\partial^2 h}{\partial y^2} = \frac{S}{T}\frac{\partial h}{\partial t} \qquad (13.66)$$

where $S = S_s b$ is referred to as the *storativity*, and $T = Kb$ is referred to as the *transmissivity*.

## 13.2.2 Unsaturated media: the Richards equation

Let us begin with the equation of continuity for unsaturated flow (8.34) developed in Chapter 8:

$$\nabla \cdot \mathbf{q} = -n\frac{\partial \theta_s}{\partial t} \qquad (13.67)$$

Here we are assuming for reasons outlined in Chapter 8 that the medium does not deform, and that the liquid is incompressible. Substituting the unsaturated form of Darcy's law (13.40) into (13.67), and noting that $d\theta = n\,d\theta_s$,

$$\frac{\partial}{\partial x}\left[K_u(\psi)\frac{\partial h}{\partial x}\right] + \frac{\partial}{\partial y}\left[K_u(\psi)\frac{\partial h}{\partial y}\right] + \frac{\partial}{\partial z}\left[K_u(\psi)\frac{\partial h}{\partial z}\right] = \frac{\partial \theta}{\partial t} \qquad (13.68)$$

According to the chain rule, $d\theta/dt = (d\theta/d\psi)(d\psi/dt)$. Thus, from (13.45), $d\theta/dt = c(\psi) \, d\psi/dt$. Because $h = z + \psi$, then for a coordinate system whose $z$-axis is vertical in the Earth reference frame, $\partial h/\partial x = \partial \psi/\partial x, \partial h/\partial y = \partial \psi/\partial y$ and $\partial h/\partial z = 1 + \partial \psi/\partial z$. With these substitutions, (13.68) becomes

$$\frac{\partial}{\partial x}\left[K_u(\psi)\frac{\partial \psi}{\partial x}\right] + \frac{\partial}{\partial y}\left[K_u(\psi)\frac{\partial \psi}{\partial y}\right] + \frac{\partial}{\partial z}\left[K_u(\psi) + K_u(\psi)\frac{\partial \psi}{\partial z}\right] = c(\psi)\frac{\partial \psi}{\partial t}$$
(13.69)

which is an equation for transient, unsaturated flow described in terms of the pressure head $\psi$. This is referred to as the *Richards equation*, named after the soil physicist who developed it in 1931.

## 13.3 ADVECTION–DISPERSION EQUATION

Many geological processes involve transport of mass in dissolved form through porous materials. The processes of dissolution and cementation of rock materials, for example, hinge on solute transport. Recall from Chapter 8 that the motion of a solute within a porous material involves diffusion at the molecular scale, advective transport associated with the motion of the fluid (solvent), and mechanical dispersion at a macroscopic scale that arises during advective transport. To clarify how the relative significance of these mechanisms of transport vary with scale, our development below will involve several views of the motion of a solute. For simplicity we will consider a *conservative* solute that does not react chemically with the solid phase.

Perhaps the easiest of these three processes to visualize is advective motion, since this involves wholesale transport of solute molecules associated with flow of the solvent. Superimposed on this mean motion or "drift" of solute molecules are effects of diffusion and dispersion; both contribute to a randomization of the motions of solute molecules. To envision this randomization of motions we will loosely follow the treatment of Chu and Sposito (1980) and consider characteristic time scales associated with diffusion in the absence of a solid phase, diffusion in the presence of a solid phase, and dispersion at the macroscopic scale associated with the porous medium.

Randomization of molecular motions at the smallest scale involves the molecular collisions that occur in all fluids whose molecules are thermally agitated; this is the mechanism underlying molecular diffusion of a solute molecule (in absence of fluid motion). A characteristic time scale $\tau_d$ for this mechanism can be defined by

$$\tau_d = \frac{m_H D_l}{k_B T}$$
(13.70)

where $m_H$ is the combined mass of the solute and solvent molecules, $k_B$ is Boltzmann's constant, $T$ is the absolute temperature, and $D_l$ is a diffusion coefficient for the solute–solvent system. Expression (13.70) is obtained from the theory of Brownian motion due to Albert Einstein. The time scale $\tau_d$ is a measure of the interval over which the velocity of a molecule measured at one instant is correlated with its velocity at some later instant. It is therefore a measure of the "memory" or persistence in molecular motion, manifest in the tendency for a molecule to maintain its velocity through successive collisions. If molecular velocity is on average correlated over the interval $\tau_d$, then randomization of molecular motions occurs over an interval longer than $\tau_d$. That is, the correlation in molecular velocity is lost (randomized) after this interval such that, when viewed at intervals longer than $\tau_d$, motions take on the appearance of a *random walk*. These random motions, which underly diffusion, become more important than motions over shorter intervals in characterizing molecular behavior. The time scale $\tau_d$ is on the order of tens of femtoseconds, depending on the solute and solvent.

A longer time scale, $\tau_f$, characterizes the motion of a solute molecule in traversing the distance across a pore. This time scale can be described by

$$\tau_f \simeq \frac{d}{v_m} \tag{13.71}$$

where $d$ is the pore diameter and $v_m$ is the mean molecular speed, which, according to kinetic theory, is given by

$$v_m = \left(\frac{8k_B T}{\pi m_H}\right)^{\frac{1}{2}} \tag{13.72}$$

Using (13.71) and (13.72), $\tau_f$ is on the order of a microsecond for typical pore sizes, depending on the solute and solvent. In general, $\tau_d \ll \tau_f$. The time scale $\tau_f$ is a measure of the interval over which the velocity of a solute molecule undergoes fluctuations due to interaction with the solid phase that bounds a pore. This mechanism provides a randomization of the motion of a solute molecule beyond that associated with molecular collisions between solute and solvent molecules; it is associated directly with the presence of the solid phase.

A third time scale, $\tau_c$, is associated with the porous medium. Consider a volume of medium that contains a large number of solute molecules, each of which has an instantaneous velocity. Let us envision taking an ensemble average of these individual velocities to define a volume-averaged molecular (solute) velocity. Then, $\tau_c$ is a measure of the interval over which the volume-averaged molecular velocity of a solute measured at one instant is correlated with its volume-averaged velocity at some later instant. Thus $\tau_c$ is analogous to $\tau_d$; $\tau_c$ is a measure of the persistence in the volume-averaged velocity over time, and reflects the effectiveness of the porous material in randomizing the motion of the solute at a macroscopic scale, just as $\tau_d$ reflects the effectiveness of molecular collisions in randomizing the motion of an individual solute molecule at the molecular scale within a single pore. If the volume-averaged molecular velocity is, on average, correlated over the interval $\tau_c$, then this mechanism of randomization occurs over an interval longer than $\tau_c$. The time scale $\tau_c$ is on the order of $10^2$ s to $10^3$ s or larger, depending on the solute, solvent, and porous material, and roughly coincides with the time required for molecules to traverse a few pore diameters or more during advective transport. Generally $\tau_f \ll \tau_c$. One significance of the conclusion that $\tau_d \ll \tau_f \ll \tau_c$ is that the mechanisms of randomization associated with these time scales can be readily distinguished in a formal discription of transport, to which we now turn.

In analogy with Hagen–Poiseuille flow, fluid at the center of a pore "far" from solid boundaries generally moves faster than fluid that is next to boundaries due to the influence of the no-slip condition. Fluid molecules that start at one instant infinitesimally close to each other thereafter are dispersed due to the individual tortuous paths that the molecules take during flow. Indeed, molecules that start in principle at the same mathematical point quickly separate due to their Brownian

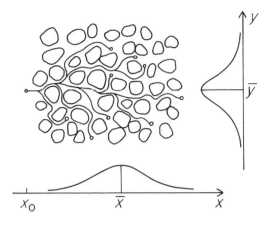

**Figure 13.6** Schematic diagram of longitudinal and transverse dispersion of solute molecules; variance of distribution of transverse positions $y$ about mean position $\bar{y}$ is less than variance of distribution of longitudinal positions $\bar{x}$ about mean position $\bar{x}$ after finite period.

motions, and then disperse along distinct paths. After an interval of time, some have traveled a significant distance whereas others have lagged behind; still others have become dispersed transversely with respect to the mean streamwise flow direction (Figure 13.6). This mechanism of dispersion operates only in the presence of advective motion of the solvent, and produces a diffusive-like behavior of solute when viewed at a macroscopic scale. Let us now examine diffusion and dispersion by turning to a Lagrangian description of this process wherein the significance of the time scales described above will become clear. Our development will loosely follow that provided by Phillips (1991).

Let $x$ denote the coordinate position of a solute molecule as it moves along a tortuous path through the pores of a medium. Then $x$ is a function of time; that is, $x = x(t)$, where $x(0) = x_0$ denotes the initial position of the molecule. The component velocity $u_s$ at time $t$ is by definition

$$u_s = \frac{d}{dt}x(t) = u_s(t) \tag{13.73}$$

which in general is a function of $t$. Since $dx = u_s\,dt$, the position $x$ of the molecule at time $t$ is

$$x = x_0 + \int_0^t u_s(t')\,dt' \tag{13.74}$$

where $t'$ denotes a variable of integration. The component velocity $u_s$ at any instant consists of the sum of the dispersive velocity $u_c$ associated with advective motion and the diffusive velocity $u_m$ associated with Brownian motion. These component velocities are independent; thus

$$u_s(t) = u_c(t) + u_m(t) \tag{13.75}$$

Substituting (13.75) into (13.74) and rearranging,

$$x - x_0 = \int_0^t [u_c(t') + u_m(t')]\,dt' \tag{13.76}$$

Let us now obtain an expression for the mean position $\bar{x}$ of a large number of solute molecules that start infinitesimally close to each other, but which then trace out many individual, tortuous pathlines through the porous material. This consists of taking the ensemble average of (13.76) for a specified interval $t$:

$$\overline{x - x_0} = \int_0^t \overline{[u_c(t') + u_m(t')]} \, dt' \tag{13.77}$$

Using (1.31), this reduces to

$$\overline{x} - x_0 = \int_0^t [\overline{u_c(t')} + \overline{u_m(t')}] \, dt' \tag{13.78}$$

Subtracting (13.78) from (13.76) then retrieves an expression for the displacement of an individual solute molecule from the mean position at time $t$:

$$x - \overline{x} = \int_0^t [u_c(t') + u_m(t')] \, dt' - \int_0^t [\overline{u_c(t')} + \overline{u_m(t')}] \, dt' \tag{13.79}$$

Combining integrands and rearranging,

$$x - \overline{x} = \int_0^t \left\{ [u_c(t') - \overline{u_c(t')}] + [u_m(t') - \overline{u_m(t')}] \right\} dt' \tag{13.80}$$

To simplify notation, here it is useful to note that the displacement $x$, and the velocity components $u_c$ and $u_m$, may be regarded as consisting of sums of mean values and deviations from mean values; thus

$$x(t) = \overline{x} + x'(t); \qquad u_c = \overline{u}_c + u'_c(t); \qquad u_m = \overline{u}_m + u'_m(t) \tag{13.81}$$

Because $x'(t) = x(t) - \overline{x}$, $u'_c(t) = u_c(t) - \overline{u}_c$ and $u'_m(t) = u_m(t) - \overline{u}_m$, (13.80) becomes

$$x' = \int_0^t [u'_c(t') + u'_m(t')] \, dt' \tag{13.82}$$

For reasons that will become clear below, let us take the derivative of (13.82) with respect to time:

$$\frac{dx'}{dt} = u'_c(t) + u'_m(t) \tag{13.83}$$

Multiplying (13.82) and (13.83), incorporating all functions of time within the integrand and expanding,

$$\frac{d}{dt} x'^2 = 2 \int_0^t [u'_c(t)u'_c(t') + u'_c(t)u'_m(t') + u'_m(t)u'_c(t') + u'_m(t)u'_m(t')] \, dt' \tag{13.84}$$

Taking averages of both sides of (13.84), and making the change of variable $t' = t + \tau$,

$$\frac{d}{dt} \overline{x'^2} = 2 \int_0^t [\overline{u'_c(t)u'_c(t + \tau)} + \overline{u'_c(t)u'_m(t + \tau)}$$

$$+ \overline{u'_m(t)u'_c(t + \tau)} + \overline{u'_m(t)u'_m(t + \tau)}] \, d\tau \tag{13.85}$$

The quantity $\overline{x'^2} = \overline{(x - \overline{x})^2}$ is the second moment of $x$ about the mean $\overline{x}$, or in the language of statistics, the variance of $x$; it is measure of the statistical dispersion of the distribution of $x$. In the present context, it also is a measure of the physical dispersion in the displacement $x$ of solute molecules about the mean displacement $\bar{x}$. The quantity $\tau$ may be regarded as a time lag separating values of $u'_c$ and $u'_m$. For example, $u'_c(t)u'_c(t + \tau)$ is the average of the product of values of $u'_c$ which occurred at time $t$ and at time $t + \tau$. We may therefore recognize that the quantity $U_c(\tau) = u'_c(t)u'_c(t + \tau)$ is the autocovariance of $u'_c$ expressed as a function of the lag $\tau$ (Chapter 1; Example Problem 1.4.3). Similarly, $U_{cm}(\tau) = u'_c(t)u'_m(t + \tau)$ is the covariance of $u'_c$ and $u'_m$. Recall that the component velocities, $u'_c = u_c - \bar{u}_c$ and $u'_m = u_m - \bar{u}_m$, are independent and therefore uncorrelated. This means that averaged products of $u'_c$ and $u'_m$, that is, $u'_c(t)u'_m(t + \tau)$ and $u'_m(t)u'_c(t + \tau)$, equal zero, so (13.85) simplifies to

$$\frac{\mathrm{d}}{\mathrm{d}t}\overline{x'^2} = 2\int_0^t U_c(\tau)\,\mathrm{d}\tau + 2\int_0^t U_m(\tau)\,\mathrm{d}\tau \tag{13.86}$$

This suggests that the statistical dispersion in the positions of solute molecules about the mean position $\bar{x}$ arises from a contribution due to (mechanical) dispersive velocities, as described by the first integral quantity, and from a contribution due to diffusive velocities, as described by the second integral quantity.

Momentarily neglecting velocities $u_m$ related to Brownian motion, $u'_c = 0$ when $\bar{u}_c = 0$, because the mechanism that produces the fluctuating part of $u_c$—advective motion—is not present. Thus, at any given instant (and local position within a pore), we may envision that $u'_c$ is proportional to the mean motion $\bar{u}_c$; that is, $u'_c = \gamma\bar{u}_c$. From (13.81),

$$u_c(t) = \bar{u}_c + \bar{u}_c\gamma(t) \tag{13.87}$$

We may envision, for example, that $\gamma$ is large when a solute molecule is at the center of a large pore, and $\gamma$ is small when a molecule is near a boundary. A positive $\gamma$ implies that the local velocity component $u_c$ is greater than $\bar{u}_c$, such that the molecule at that instant is speeding ahead of the mean motion. A negative $\gamma$ implies the molecule is at that instant lagging behind the mean motion; indeed one can envision the possibility that $\gamma < -1$, in which case a molecule possesses at some instant a negative velocity component $u_c$.

The quantity $\gamma$ is a function of the coordinate position of the molecule within a pore, and is independent of the fluid velocity. For a capillary tube with radius $R$, for example, assuming that the dispersive velocity equals the fluid velocity, one can use results of Example Problem (13.5.1) to illustrate that

$$\gamma = 1 - 2\left(\frac{r}{R}\right)^2 \tag{13.88}$$

where $r$ is the radial distance from the tube axis. For a complex arrangement of irregular pores, $\gamma$ must be a more complicated function. In this case the relative position of a molecule within a pore, analogous to $r/R$ in (13.88), varies with time. This is due to the sinuous path of a molecule, and to the changing pore geometry along this path. It also is due to randomization in the relative position of the molecule due to its diffusive (Brownian) motion. Indeed, even within a straight capillary tube involving one-dimensional flow, the relative position $r/R$ of an individual molecule must vary with time since the diffusive motion of the molecule transports it over varying $r$. For a large number of molecules distributed throughout many pores, $\gamma$ has at any instant a distribution of values. This suggests that we can treat $\gamma$ as a random variable characterized by the probability density function $f_\gamma(\gamma)$, with mean $\overline{\gamma}$ and variance $\sigma_\gamma^2$. The distribution $f_\gamma(\gamma)$ is a property of the porous material; it reflects the dispersive properties of the material arising from the geometry, arrangement, and interconnectedness of its pores, independently of the mean motion $u_c$. (We will return to this distribution below.)

Using (13.87), the definition of an autocorrelation function (Example Problem 1.4.3) and the averaging rules in (1.31), the first integral quantity in (13.86) becomes

$$\frac{d}{dt}\overline{x'^2} = 2\overline{u_c}^2\sigma_\gamma^2 \int_0^t \Gamma(\tau) \, d\tau \tag{13.89}$$

where $\Gamma(\tau)$ is the autocorrelation function of $\gamma$. When $\tau = 0$, the autocorrelation $\Gamma(0) = 1$. Otherwise $\Gamma(\tau)$ is a measure of the correlation between values of $\gamma(t)$ separated by the time lag $\tau$. Values of $\gamma$ separated by a small interval $\tau$ generally are strongly correlated. That is, $\gamma$ will tend to persist from moment to moment. With increasing $\tau$, $\gamma$ will exhibit less persistence (memory). With large $\tau$, values of $\gamma$ will become uncorrelated such that $\Gamma(\tau)$ approaches zero. In general, $\Gamma(\tau)$ converges to zero as $\tau \to \infty$. If the rate of this convergence is sufficiently rapid, then as $t \to \infty$, the integral in (13.89) approaches a finite quantity $\tau_c$:

$$\frac{d}{dt}\overline{x'^2} = 2\overline{u_c}^2\sigma_\gamma^2 \int_0^\infty \Gamma(\tau) \, d\tau = 2\overline{u_c}^2\sigma_\gamma^2\tau_c \tag{13.90}$$

The quantity $\tau_c$ is a characteristic time scale, often referred to as the *correlation interval*. It coincides with the characteristic time scale $\tau_c$ introduced above, and is therefore a measure of the persistence in the volume-averaged molecular velocity. Let $\delta^* = \overline{u_s}\tau_c$. The quantity $\delta^*$ is a *correlation length*; it is a measure of the distance over which the dispersive velocity is correlated. Then for $t \gg \tau_c = \delta^*/\overline{u_s}$,

$$d\overline{x'^2} = 2\overline{u_c}^2\sigma_\gamma^2\tau_c \, dt = 2\sigma_\gamma^2\delta^*\overline{u_c} \, dt \tag{13.91}$$

Integrating,

$$\overline{(x - \overline{x})^2} = 2\sigma_\gamma^2\delta^* u_c t = 2\alpha_d\overline{u_c}t \tag{13.92}$$

where $\alpha_d = \sigma_\gamma^2\delta^*$ is referred to as the *dispersivity*. The dispersivity is, in principle, a function of the porous material since it is obtained from the quantity $\gamma$, which depends only on pore structure according to our conceptualization of $\gamma$ above. Since the volume-averaged molecular velocity is, on average, correlated over the interval $\tau_c$, this mechanism of randomization of motions associated with pore structure occurs over an interval longer than $\tau_c$. Then, according to (13.92), the dispersion of $x$ increases steadily with time.

Diffusive molecular motions are superimposed on dispersive motions. This means that dispersive motions are irreversible. The pathline of a solute molecule obtained during advective motion in one direction is not retraced by the molecule when the direction of the mean advective motion is exactly reversed. Moreover, since these motions are independent and additive, the effect of diffusive motions can be readily incorporated into the formulation above. In particular, we can define an autocorrelation function associated with the autocovaiance $U_m$ of

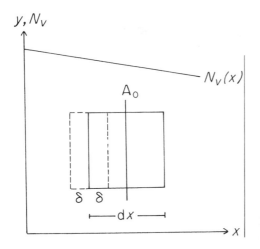

**Figure 13.7** Definition diagram for describing advective and dispersive solute fluxes.

diffusive velocities in (13.86), then integrate this function from zero to infinity to obtain a characteristic time interval $\tau_d$. This interval is a measure of the persistence in molecular motions that undergo randomization by collisions between solute and solvent molecules, and by interactions with the pore walls. Then (13.92) becomes

$$\overline{(x - \overline{x})^2} = 2(\alpha_d \overline{u_c} + D_c)t \tag{13.93}$$

where $D_c = \sigma_m 2\tau_d$ is the *coefficient of mass diffusion*, and $\sigma_m^2$ is the variance of the diffusive velocities $u_m$. If it is assumed that $\overline{u_c}$ equals the average fluid velocity $u$ for a conservative solute, then by convention the coefficients in (13.93) are combined to give $D^* = \alpha_d u + D_c = D + D_c$, where $D = \alpha_d u$ is referred to as the *coefficient of mechanical dispersion* and $D^*$ is referred to as the *coefficient of hydrodynamic dispersion*.

Let us now consider solute transport by advection and dispersion. Consider a small control volume with edge of length $dx$ that is moving with a velocity equal to the mean velocity $\overline{u_c}$ (Figure 13.7). The length $dx$ is chosen to equal the distance traveled by the volume during a small interval $2\,dt$ such that $dx = 2\overline{u_c}\,dt$. Our task is to determine the rate of flow of solute mass across a plane $A_0$ normal to the $x$-axis. This plane is fixed in space, and its coordinate position initially coincides with the center of the control volume at $x$. In general, solute molecules are dispersed into or out of the control volume during $dt$ such that the solute mass crossing $A_0$ during $dt$ may not equal the solute mass that is initially present in the control volume to the left of $A_0$.

A solute molecule that is initially to the left of the control volume must, on average, possess a velocity that is greater than $\overline{u_c}$ to move into the control volume then cross to the right of the plane $A_0$ during the small interval $d\,t$. The probability $P_0$ that a molecule has at any instant a velocity $u_c > \overline{u_c}$ is the same as the probability that $\gamma > 0$. Thus,

$$P_0 = \int_0^\infty f_\gamma(\gamma)\,d\gamma \tag{13.94}$$

and the probability that a molecule has a velocity $u_c < \bar{u}_c$ is

$$(1 - P_0) = 1 - \int_0^\infty f_\gamma(\gamma)\, d\gamma = \int_{-\infty}^0 f_\gamma(\gamma)\, d\gamma \tag{13.95}$$

Here we will make a heuristic assumption: that although $u_c$ varies with time for any individual molecule according to (13.87), the probability density function $f_\gamma(\gamma)$ is independent of time for a large number of molecules. (How might one justify this assumption?) Then the probability that a molecule will, on average, possess a velocity that is greater than $\bar{u}_c$ during $dt$ is equal to $P_0$, and the probability that a molecule will on average possess a velocity that is less than $\bar{u}_c$ during $dt$ is equal to $1 - P_0$.

Let $N_V$ denote the number of solute molecules per unit distance $x$ and area $dA$ normal to flow, at position $x$. Let $\delta$ denote a small distance within which molecules to the left of the control volume originate and move into the volume during $dt$. (We will consider the significance of $\delta$ further below.) The number of solute molecules within the small liquid volume $n\delta\, dA$ is to first order $[N_V - (1/2)(\partial N_V/\partial x)\, dx] n\delta\, dA$. The number of molecules within the exterior volume $n\delta\, dA$ with velocity greater than $\bar{u}_c$ (and which therefore move into the control volume during $dt$) is

$$P_0\left[N_V - \frac{1}{2}\frac{\partial N_V}{\partial x} dx\right] n\delta\, dA \tag{13.96}$$

The number of molecules within the interior volume $n\delta\, dA$ with velocity less than $\bar{u}_c$ (and which therefore move out of the control volume during $dt$) is

$$(1 - P_0)N_V n\delta\, dA \tag{13.97}$$

The number of molecules initially within the control volume to the left of $A_0$ is $\frac{1}{2}N_V n\, dx\, dA$. The total number of molecules $dN$ that cross the plane $A_0$ during $dt$ is therefore

$$dN = \frac{1}{2}N_V n\, dx\, dA + P_0\left[N_V - \frac{1}{2}\frac{\partial N_V}{\partial x} dx\right] n\delta\, dA - (1 - P_0)N_V n\delta\, dA \tag{13.98}$$

Here we will make an important assumption in keeping with conventional treatments of mechanical dispersion: that $f_\gamma(\gamma)$ is symmetrical, in which case $P_0 = 1 - P_0 = 1/2$. Using this, and recalling that $dx = 2\bar{u}_c\, dt$, the number of molecules that cross $A_0$ during $dt$ is

$$dN = n\bar{u}_c N_V\, dA\, dt - nP_0\delta\bar{u}_c\frac{\partial N_V}{\partial x} dA\, dt \tag{13.99}$$

Noting that the mass concentration $c = N_V M$, where $M$ is the molecular mass, and assuming that $\bar{u}_c \approx u$ (for a conservative solute), the rate of transport of solute mass per unit area, $q_{cx}$, is

$$q_{cx} = nuc - nD\frac{\partial c}{\partial x} \tag{13.100}$$

where $D = P_0\delta u$ is the coefficient of mechanical dispersion defined above. The first term on the right side of (13.100) describes an advective flux; the second term on the right side describes a dispersive flux. Comparing this expression for $D$ with that obtained above, $\delta$ in the formulation above is of order $\sigma_\gamma^2 \delta^* = \alpha_d$. It may be considered a rough measure of the distance that solute molecules must traverse before the randomization of motions due to pore geometry becomes sufficient to produce a dispersive flux.

A mass flux due to molecular diffusion must be added to the advective and dispersive fluxes. Since diffusion is governed by Fick's law, (13.100) becomes

$$q_{cx} = nuc - nD\frac{\partial c}{\partial x} - nD_c\frac{\partial c}{\partial x} \qquad (13.101)$$

It is important to note here that $D_c$ is an *effective* coefficient of diffusion that reflects the tortuosity of the pores. It is generally less than the coefficient of diffusion for a solvent–solute system in the absence of porous material. Combining terms in (13.101) that involve the concentration gradient,

$$q_{cx} = nuc - nD^*\frac{\partial c}{\partial x} \qquad (13.102)$$

Here $D^* = D + D_c$ is the coefficient of hydrodynamic dispersion described above, which characterizes effects of dispersion and diffusion.

Equation (13.102), when coupled with an expression of conservation of solute mass, leads to the advection–dispersion equation. This requires distinguishing between longitudinal and transverse coefficients of dispersion, $D_L^*$ and $D_T^*$. Recall that dispersion transverse to the mean (streamwise) motion occurs, despite zero transverse advective motion (Figure 13.6). This leads to a transverse dispersive flux in the presence of a transverse concentration gradient. The magnitude of transverse dispersion generally is less than longitudinal dispersion. We will start by considering a two-dimensional flow, then generalize the results to three dimensions. Assume for simplicity that the porous medium is homogeneous and isotropic.

Consider a global $xy$-coordinate system. In general, the average fluid velocity $\mathbf{u} = \langle u, v \rangle$ is a function $\mathbf{u} = \mathbf{u}(x, y)$ of coordinate position. Let us associate with each position $(x, y)$ a local $x'y'$-coordinate system such that the $x'$-axis is parallel to the vector $\mathbf{u}$. The $x'$-axis is locally inclined at an angle $\phi$ from the $x$-axis, where the rotation angle $\phi$ is at any position given by $\tan \phi = v/u$. In the local system the component $u' = \sqrt{(u^2 + v^2)} = |\mathbf{u}|$, and the component $v' = 0$. (The use of primes here to denote quantities associated with a local $x'y'$-coordinate system should not be confused with the use of primes above to denote fluctuating quantities.) The solute mass flux parallel to the $x'$-axis involves advective and dispersive parts whereas the solute mass flux parallel to the $y'$-axis involves only a dispersive part since $v' = 0$; thus

$$q_{cx'} = nu'c - nD_L^*\frac{\partial c}{\partial x'} \qquad (13.103)$$

$$q_{cy'} = -nD_T^*\frac{\partial c}{\partial y'} \qquad (13.104)$$

Here $D_L^* = \alpha_L u' + D_c$ and $D_T^* = \alpha_T u' + D_c$, where $\alpha_L$ and $\alpha_T$ denote longitudinal and transverse dispersivities. These dispersivities are, in principle, properties of the porous material, for reasons described above. Notice that the transverse flux $q_{cy'}$ is associated with the streamwise advective motion $u'$ via $D_T^*$. In matrix form (13.103) and (13.104) are

$$\langle q_{cx'}, q_{cy'} \rangle \leftrightarrow \begin{bmatrix} q_{cx'} \\ q_{cy'} \end{bmatrix} = n \begin{bmatrix} u'c \\ 0 \end{bmatrix} - n \begin{bmatrix} D_L^* & 0 \\ 0 & D_T^* \end{bmatrix} \begin{bmatrix} \dfrac{\partial c}{\partial x'} \\ \dfrac{\partial c}{\partial y'} \end{bmatrix} \qquad (13.105)$$

and using tensor notation,

$$\mathbf{q}_c' = n\mathbf{u}'c - n\mathbf{D}'\nabla c' \qquad (13.106)$$

where $\mathbf{D}'$ denotes a *local dispersion coefficient tensor.*

Notice that in the global coordinate system, a nonzero gradient $\partial c/\partial y$ contributes to a dispersive flux in the direction of $\mathbf{u}$, and therefore to the component flux $q_{cx}$. Likewise a nonzero gradient $\partial c/\partial x$ contributes to the flux $q_{cy}$. Thus in general we may write

$$q_{cx} = nuc - nD_{xx}^* \frac{\partial c}{\partial x} - nD_{xy}^* \frac{\partial c}{\partial y} \qquad (13.107)$$

$$q_{cy} = nvc - nD_{yx}^* \frac{\partial c}{\partial x} - nD_{yy}^* \frac{\partial c}{\partial y} \qquad (13.108)$$

In matrix form these are

$$\langle q_{cx}, q_{cy} \rangle \leftrightarrow \begin{bmatrix} q_{cx} \\ q_{cy} \end{bmatrix} = n \begin{bmatrix} uc \\ vc \end{bmatrix} - n \begin{bmatrix} D_{xx}^* & D_{xy}^* \\ D_{yx}^* & D_{yy}^* \end{bmatrix} \begin{bmatrix} \dfrac{\partial c}{\partial x} \\ \dfrac{\partial c}{\partial y} \end{bmatrix} \qquad (13.109)$$

and in tensor notation,

$$\mathbf{q}_c = n\mathbf{u}c - n\mathbf{D}^*\nabla c \qquad (13.110)$$

were $\mathbf{D}^*$ is a symmetric *dispersion coefficient tensor.* At this point we should recognize the similarity between this development and the development above regarding the generalization of Darcy's law to tensor form. Indeed, the elements of $\mathbf{D}^*$ are given by the rotation

$$\mathbf{D}^* = \mathbf{D}^{-1}\mathbf{D}'\mathbf{D} \qquad (13.111)$$

where $\mathbf{D}$ is the rotation matrix given by (13.24).

Following previous developments, conservation of mass requires that the divergence of the solute mass flux equals the negative of the rate of change in solute mass per unit volume of porous material; that is, $\operatorname{div}\mathbf{q}_c = \partial q_{cx}/\partial x + \partial q_{cy}/\partial y = -\partial(nc)/\partial t$. Using (13.107) and (13.108),

$$\frac{\partial}{\partial x}\left(nD_{xx}^* \frac{\partial c}{\partial x} + nD_{xy}^* \frac{\partial c}{\partial y}\right) + \frac{\partial}{\partial y}\left(nD_{yx}^* \frac{\partial c}{\partial x} + nD_{yy}^* \frac{\partial c}{\partial y}\right)$$

$$- \frac{\partial}{\partial x}(nuc) - \frac{\partial}{\partial y}(nvc) = \frac{\partial}{\partial t}(nc) \qquad (13.112)$$

More generally, in three dimensions:

$$\frac{\partial}{\partial x}\left[n\left(D_{xx}^*\frac{\partial c}{\partial x} + D_{xy}^*\frac{\partial c}{\partial y} + D_{xz}^*\frac{\partial c}{\partial z}\right)\right]$$

$$\frac{\partial}{\partial y}\left[n\left(D_{yx}^*\frac{\partial c}{\partial x} + D_{yy}^*\frac{\partial c}{\partial y} + D_{yz}^*\frac{\partial c}{\partial z}\right)\right]$$

$$\frac{\partial}{\partial z}\left[n\left(D_{zx}^*\frac{\partial c}{\partial x} + D_{zy}^*\frac{\partial c}{\partial y} + D_{zz}^*\frac{\partial c}{\partial z}\right)\right]$$

$$-\frac{\partial}{\partial x}(nuc) - \frac{\partial}{\partial y}(nvc) - \frac{\partial}{\partial z}(nwc) = \frac{\partial}{\partial t}(nc) \qquad (13.113)$$

Momentarily neglecting molecular diffusion, the local dispersion tensor for homogeneous and isotropic conditions is

$$\mathbf{D}' = \begin{bmatrix} D_L & 0 & 0 \\ 0 & D_T & 0 \\ 0 & 0 & D_T \end{bmatrix} = \begin{bmatrix} \alpha_L\bar{u} & 0 & 0 \\ 0 & \alpha_T\bar{u} & 0 \\ 0 & 0 & \alpha_T\bar{u} \end{bmatrix} \qquad (13.114)$$

which reflects that the transverse dispersivity $\alpha_T$ is the same in the local $y'$ and $z'$ directions (where the $x'$-axis is parallel to $\mathbf{u}$). Here $\bar{u} = |\mathbf{u}| = \sqrt{(u^2 + v^2 + w^2)}$. The relation between $\mathbf{D}^*$ and $\mathbf{D}'$ can be obtained from (13.111) above; a simpler version is given by (for example, see Bear [1972]):

$$D_{xx} = \frac{1}{\bar{u}}[\alpha_T(v^2 + w^2) + \alpha_L u^2] \qquad (13.115)$$

$$D_{yy} = \frac{1}{\bar{u}}[\alpha_T(u^2 + w^2) + \alpha_L v^2] \qquad (13.116)$$

$$D_{zz} = \frac{1}{\bar{u}}[\alpha_T(u^2 + v^2) + \alpha_L w^2] \qquad (13.117)$$

$$D_{xy} = D_{yx} = \frac{uv}{\bar{u}}(\alpha_L - \alpha_T) \qquad (13.118)$$

$$D_{xz} = D_{zx} = \frac{uw}{\bar{u}}(\alpha_L - \alpha_T) \qquad (13.119)$$

$$D_{yz} = D_{zy} = \frac{vw}{\bar{u}}(\alpha_L - \alpha_T) \qquad (13.120)$$

Writing (13.113) in tensor form,

$$\nabla \cdot (n\mathbf{D}^* \cdot \nabla c) - \nabla \cdot (n\mathbf{u}c) = \frac{\partial}{\partial t}(nc) \qquad (13.121)$$

With an incompressible liquid solvent and constant porosity $n$,

$$\nabla \cdot (\mathbf{D}^* \cdot \nabla c) - \mathbf{u} \cdot \nabla c = \frac{\partial c}{\partial t} \qquad (13.122)$$

Equation (13.121) (or [13.122]) is referred to as the advection–dispersion equation. Since the fluid velocity **u** appears in both terms on the left side of (13.122), direct use of (13.122) assumes that the velocity field $\mathbf{u}(x, y, z)$ is known.

Applications of (13.122) involving two-dimensional or three-dimensional flows normally require numerical simulations. For problems involving one-dimensional steady flow, analytical solutions are available for a variety of initial and boundary conditions. Analytical solutions also are available for simple two-dimensional problems, and for dispersion in radial flow (see Reading list). These have practical value inasmuch as they approximately mimic certain field situations, or can be adapted to situations in which boundary conditions vary as arbitrary functions of time. For one-dimensional transport of a conservative solute with constant $D^*$, (13.122) reduces to

$$D^* \frac{\partial^2 c}{\partial x^2} - u \frac{\partial c}{\partial x} = \frac{\partial c}{\partial t} \tag{13.123}$$

The relative importance of dispersion and diffusion is obtained from the ratio:

$$\frac{D}{D_c} \sim \frac{\delta^* u}{D_c} \sim \frac{du}{D_c} \tag{13.124}$$

where $d$ is the mean grain diameter. The last of these is referred to as the *Peclet number $Pe_c$* for solute transport. If $Pe_c$ is of order unity or smaller, diffusion will dominate; if it is significantly larger, dispersion will dominate. For heterogeneous materials that contain, say, bedding with varying grain and pore sizes, solutes may be differentially transported among units with varying permeability. This contributes to dispersion of solute at the bedding scale, referred to as *macrodispersion,* and suggests that the dispersivities $\alpha_L$ and $\alpha_T$ of such materials are scale dependent. Whereas, in practice, the dispersivity normally is assumed to be constant, significant evidence suggests that dispersivities, like permeabilities, generally increase with the scale of measurement or characteristic averaging volume. Students are urged to examine Gelhar et al. (1985) regarding this point.

The advection–dispersion equation must be generalized to treat the transport of solutes that react with the solid phase. For example, sorption and desorption of a solute represent processes that locally alter its concentration within the pore solution. In general, sorption and desorption may occur simultaneously and at different rates, such that the partitioning of solutes between the liquid and solid phases is in a transient (nonequilibrium) condition. These processes are governed by molecular diffusion rates and surface chemistry, and by the concentrations of the solute in the liquid phase and on surrounding pore surfaces. In certain situations, sorption and desorption processes may proceed sufficiently rapidly relative to local variations in concentration induced by advection and dispersion that the rates of sorption and desorption may be considered locally steady and equal. The effect of this equilibrium partitioning then can be readily incorporated into the advection–dispersion equation for cases where the partitioning—characterized as the ratio of the concentration of solute adsorbed on surfaces to that in solution—is independent of the solute concentration. For practical problems this conventionally involves defining a *retardation factor,* obtained semi-empirically, that characterizes how sorption leads to an average solute velocity that is less than the average fluid velocity. The advection–dispersion equation also can be modified to incorporate effects of chemical and biochemical reactions that lead to degradation or production of chemical species during transport, including radioactive decay. These topics are beyond the scope of this chapter; interested students are urged to examine the relevant texts and monographs in the references.

## 13.4 ENERGY EQUATION

The energy equation for flow in porous media is, in principle, like the advection–dispersion equation since heat energy is dispersed by the same mechanisms that disperse solutes. The difference is that heat is conducted through the solid phase as well as transported within the fluid phase. Let us start with the energy equation (9.33) obtained in Chapter 9:

$$\mathbf{q} \cdot \nabla T + \frac{c_e}{\rho c} \frac{\partial T}{\partial t} = \kappa \nabla^2 T \qquad (13.125)$$

Here $T$ denotes temperature, and $c_e$ is the *effective specific heat* defined by

$$c_e = n\rho c + (1 - n)\rho_s c_s \qquad (13.126)$$

where $c$ and $c_s$ are the specific heats of the liquid and solid phases, and $\rho$ and $\rho_s$ are the densities of the liquid and solid phases. The *effective thermal diffusivity* $\kappa$ is equal to

$$\kappa = \frac{K_{Te}}{\rho c} \qquad (13.127)$$

where $K_{Te}$ is the *effective thermal conductivity* of the solid and liquid phases acting together. Like hydraulic conductivity, the thermal conductivity takes the form of a second-order tensor for a thermally anisotropic medium. This anisotropy, however, is typically insignificant relative to anisotropy associated with hydraulic conductivity and is therefore normally neglected.

In principle a term must be added to (13.125) to take into account dispersive transport of heat. Analogous to (13.124), the relative importance of mechanical heat dispersion (through liquid-filled pores) and conduction (through solid plus liquid) is obtained from the ratio:

$$\frac{nD}{K_{Te}} \sim \frac{n\delta^* u}{K_{Te}} \qquad (13.128)$$

Conduction dominates over dispersion when this ratio is much less than unity. The thermal conductivities of saturated media normally are much larger than their coefficients of dispersion, so this condition is usually satisfied in geological problems, and heat dispersion may be neglected.

In addition, chemical reactions that occur during transport may provide an additional source (or sink) of heat energy. Then (13.125) becomes

$$\mathbf{q} \cdot \nabla T + \frac{c_e}{\rho c} \frac{\partial T}{\partial t} = \kappa \nabla^2 T + \frac{R_c}{\rho c} \qquad (13.129)$$

where $R_c$ denotes the rate at which heat is added (or extracted) per unit volume.

## 13.5 EXAMPLE PROBLEMS

### 13.5.1 Permeability and Hagen–Poiseuille flow

An interesting analogy exists between the definition of permeability $k$ and Hagen–Poiseuille flow within a capillary tube. Let us start with the dimensional analysis that leads to (13.2).

Consider the force of drag that a moving fluid exerts on the grains of a porous material. This force arises at the pore scale from viscous shear stresses acting on the pore walls, due to non-zero velocity gradients normal to the walls and the no-slip condition that exists at the walls. In principle the drag force could be computed by integrating this stress over the surface area of the pores. But what defines the total surface area that ought to be involved in the integration? Unlike computing the drag force on an isolated body with finite surface area, say a sphere, we need to specify a specific size of porous material to describe the drag force, since this force varies with the grain surface area, and thus with the volume of the medium involved. It therefore makes sense to consider instead the drag force per unit volume $F_V$, with units $(M L^{-2} t^{-2})$, then assume that $F_V = F_V(\mu, \rho, d, q)$.

Fundamental quantities in this problem include mass, length, and time. With $F_V$, there are a total of five physical quantities, so the dimensional analysis will involve two Pi terms. Choosing $\rho$, $d$, and $q$ as repeating quantities, and combining these first with $F_V$ and then with $\mu$,

$$\Pi_1 = \rho^E d^F q^G F_V \tag{13.130}$$

$$\Pi_2 = \rho^H d^I q^J \mu \tag{13.131}$$

It is left as an exercise to obtain the result that $E = -1$, $F = -2$, $G = 1$, $H = -1$, $I = -1$, and $J = -1$, from which it follows that $\Pi_1 = dF_V/\rho q^2$ and $\Pi_2 = \mu/\rho Dq$. This leads to the conclusion embodied in (13.3), that the coefficient of drag $C_D = dF_V/\rho q^2$ varies as a function of the Reynolds number $Re = \rho dq/\mu$.

Let us now turn to a similar analysis of Hagen–Poiseuille flow. Here we envision an individual pore as a segment of a capillary tube with constant radius $R$. Using results presented in Appendix 17.2 for a cylindrical coordinate system, the equation governing steady flow is

$$\frac{\partial^2 u}{\partial r^2} + \frac{1}{r}\frac{\partial u}{\partial r} = \frac{1}{\mu}\frac{\partial p}{\partial x} + \frac{\rho g}{\mu}\frac{\partial z_0}{\partial x} \tag{13.132}$$

where $x$ denotes the streamwise coordinate parallel to the tube axis, $r$ is a radial coordinate whose origin ($r = 0$) is at the tube axis, and $z_0$ denotes a vertical axis in the Earth reference frame. From the definition of pressure head $\psi$, $p = \rho g \psi + p_0$; and since $h = z_0 + \psi$, (13.132) may be written as

$$\frac{d^2 u}{dr^2} + \frac{1}{r}\frac{du}{dr} = \frac{\rho g}{\mu}\nabla h \tag{13.133}$$

where $\nabla h = dh/dx$, and partial notation has been omitted. Let us rewrite this as

$$\frac{d}{dr}\left(r\frac{du}{dr}\right) = \frac{\rho g \nabla h}{\mu}r \tag{13.134}$$

To see that (13.133) and (13.134) are indeed equivalent, apply the chain rule to the left side of (13.134), then rearrange the result. Further modifying (13.134),

$$d(rv) = \frac{\rho g \nabla h}{\mu} r \, dr \tag{13.135}$$

where $v = du/dr$. Integrating twice, replacing $v$ with $du/dr$ for the second integration,

$$u(r) = \frac{\rho g \nabla h}{4\mu} r^2 + C_1 \ln r + C_2 \tag{13.136}$$

where $C_1$ and $C_2$ are constants of integration. The constant $C_1$ must be zero if $u$ is not to become infinitely large at the tube axis ($r = 0$); $C_2$ is obtained from the no-slip condition at $r = R$. Then

$$u(r) = -\frac{\rho g \nabla h}{4\mu} (R^2 - r^2) \tag{13.137}$$

It is left as an exercise to integrate (13.137) over the tube area to obtain the result that the discharge $Q$ is given by

$$Q = -\frac{\pi \rho g \nabla h}{8\mu} R^4 \tag{13.138}$$

At this point we could divide by the tube area $\pi R^2$ to obtain an expression for the average velocity. Instead, envision that the area $\pi R^2$ is some proportion of $A_m$, the area of medium per capillary tube (normal to flow). Then $\pi R^2 = nA_m$, or $A_m = \pi R^2/n$, where $n$ is the porosity. Dividing (13.138) by $A_m$ then gives the specific discharge $q$ through the tube as defined for a porous medium:

$$q = -\frac{nR^2}{8} \frac{\rho}{\mu} g \nabla h \tag{13.139}$$

Comparing this with (13.4), the permeability $k = Cd^2$ is analogous to the quantity $nR^2/8$ in (13.139). This simple analogy between porous media flow and Hagen–Poiseuille flow again suggests that the permeability is an abstraction of pore size and shape, including porosity.

We can extend this analogy to anticipate the inverse relation between the coefficient of drag and the Reynolds number, as in (13.3). Let us start by computing the shear stress on the tube wall $\tau_R$. Taking the derivative of (13.137) with respect to $r$ and evaluating the result at $r = R$,

$$\left(\frac{du}{dr}\right)_R = \frac{\rho g \nabla h}{2\mu} R \tag{13.140}$$

where the subscript $R$ denotes the velocity gradient at the wall. By the definition of a Newtonian fluid, $\tau_R = -\mu(du/dr)_R$. The force of drag $F_D$ associated with $\tau_R$ over a tube of length $l$ is $2\pi R l \tau_R$; it follows that

$$F_D = -\rho g \nabla h \pi R^2 l \tag{13.141}$$

The force of drag per unit volume $F_V$ is obtained by dividing this by $A_m l = \pi R^2 l/n$:

$$F_V = -\rho g n \nabla h \tag{13.142}$$

Comparing this with (13.139), we may identify the quantity $-\rho gn\nabla h$ as the drag force per unit volme $F_V$, then rearrange this equation to

$$-\rho gn\nabla h = \frac{8\mu q}{R^2} \tag{13.143}$$

Multiplying by $R$ and dividing by $\rho q^2$,

$$C_D = \frac{F_V R}{\rho q^2} = 8\frac{\mu}{\rho Rq} \tag{13.144}$$

This indicates that the coefficient of drag $C_D = F_V R/\rho q^2$ is inversely proportional to the Reynolds number $Re = \rho Rq/\mu$, where the constant associated with this relation equals the numerical value 8. Inasmuch as flow in a capillary tube is an abstraction of the spatially complex flow in pores, (13.144) suggests that the simple inverse relation of (13.3) should be anticipated.

### 13.5.2 Propagation of pressure waves within a coastal aquifer

Consider a coastal aquifer that extends beneath land and ocean for a kilometer or more in both directions (Figure 13.8). For simplicity, assume that the aquifer is homogeneous and confined by virtually impermeable units above and beneath it. As the elevation of the ocean surface fluctuates with the tides, the load on the aquifer due to the water column above it varies. Recalling the concept of effective stress (Chapter 3), part of this fluctuating load is borne by the aquifer matrix, and part is borne by the fluid within its pores. The tidal fluctuations therefore alternately pressurize and depressurize the fluid within the aquifer. Since loading occurs over the oceanic extent of the aquifer but not over its landward extent, we may anticipate that these fluctuations in fluid pressure are dissipated in a landward direction. We will now see that, in fact, pressure waves are propagated inland, manifest as fluctuating water levels within wells that tap the aquifer near the coastline.

Tides typically sweep along a coastline, so tidal loading of a coastal aquifer normally involves a complex stress field, particularly if the coastline is irregular. For simplicity, however, let us examine pressure-wave propagation over a short distance inland, such that we may assume that the tide varies uniformly along the coastline, and seaward. We will thus envision the tide as merely involving a vertical rise and fall of the ocean surface at the coastline. Under these circumstances, loading of the aquifer occurs uniformly over its oceanic extent and pressure waves propagate in one dimension inland.

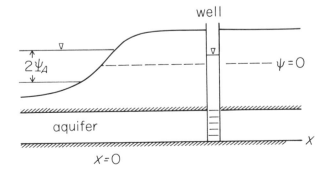

**Figure 13.8** Definition diagram for describing inland propagation of pressure waves within a coastal aquifer.

With reference to Figure 13.8, the governing equation is

$$\frac{\partial^2 h}{\partial x^2} = \frac{S}{T}\frac{\partial h}{\partial t} \tag{13.145}$$

Since the $z$-axis is vertical in the Earth reference frame, and since $h = z + \psi$, where $z$ also denotes elevation head and $\psi$ is pressure head, this may be written

$$\frac{\partial^2 \psi}{\partial x^2} = \frac{S}{T}\frac{\partial \psi}{\partial t} \tag{13.146}$$

Note for further reference below that this is structurally homologous to (9.56), which we used in Example Problem 9.3.2 to describe one-dimensional heat flow within glacier ice. Let us now examine two solutions of (13.146).

Suppose that pressure conditions within the aquifer are initially at an equilibrated (static) state. This defines a convenient datum state for which we set $\psi = 0$. Now envision the possibility that the tide increases as a step function by an amount equal to $\Psi$. Initial and boundary conditions are

$$\psi(x, t) = 0; \qquad t \le 0 \tag{13.147}$$

$$\psi(0, t) = \gamma\Psi \tag{13.148}$$

$$\psi(\infty, t) = 0 \tag{13.149}$$

The first of these states that $\psi = 0$ for all positions $x$ and times $t$ prior to the tidal increase $\Psi$. The second states that $\psi$ at $x = 0$ increases instantly to a value $\gamma\Psi$ and thereafter remains at this value for all $t > 0$. The third states that infinitely far from the coastline $\psi$ remains at the datum state for all $t > 0$. This is referred to as the *far-field* pressure head.

Returning to (13.148), notice that $\gamma = \psi(0, t)/\Psi$ is the ratio of the increase in pressure head to the increase in tidal level. This parameter is referred to as the *static-confined tidal efficiency* of the aquifer; it is a measure of the increase in aquifer fluid pressure (head) for a given increase in tidal load. The value of $\gamma$ generally is less than one due to the fact that the aquifer matrix bears part of the tidal load. From (3.54),

$$d\sigma = dp + d\sigma_e \tag{13.150}$$

Assuming vertical loading, the total load $\sigma$ relative to the datum state is due to the seawater. Because $\sigma = \rho_\Psi g\Psi$, where $\rho_\Psi$ is the density of seawater, and because $p = \rho g\psi$, substitution of these quantities into (13.150) leads to

$$d\psi = \frac{\rho_\Psi}{\rho}d\Psi - \frac{1}{\rho g}d\sigma_e \tag{13.151}$$

Dividing by $d\Psi$,

$$\gamma = \frac{d\psi}{d\Psi} = \frac{\rho_\Psi}{\rho} - \frac{1}{\rho g}\frac{d\sigma_e}{d\Psi} \approx 1 - \frac{1}{\rho g}\frac{d\sigma_e}{d\Psi} \tag{13.152}$$

which provides a definition of $\gamma$ in terms of the effective stress $\sigma_e$.

Without showing the steps, the solution to (13.146) using (13.147), (13.148), and (13.149) is

$$\psi(t; x) = \gamma\Psi(1 - \mathrm{erf}\,\eta) = \gamma\Psi\,\mathrm{erfc}\,\eta \tag{13.153}$$

where

$$\eta = \sqrt{\frac{Sx^2}{4Tt}} \tag{13.154}$$

The quantity $\eta$ becomes an argument of the *error function* erf, or the *complimentary error function* erfc, which are defined by

$$\text{erf}\,\eta = \frac{2}{\sqrt{\pi}} \int_0^{\eta} e^{-u^2} du \tag{13.155}$$

$$\text{erfc}\,\eta = \frac{2}{\sqrt{\pi}} \int_{\eta}^{\infty} e^{-u^2} du \tag{13.156}$$

where $u$ denotes a variable of integration. These are analytically nonintegrable functions; however, they can be readily integrated numerically. (These particular functions appear frequently in the sciences, and therefore their values for specified $\eta$ are available in tabled form in many handbooks of mathematical functions, as well as in groundwater hydrology texts.)

Notice that $x$ in (13.154) serves essentially as a parameter; for example, it might represent the distance that an observation well is from the coastline. For small time $t$, or very large distance $x$, erfc $\eta$ approaches zero. For small $x$, or large $t$, erfc $\eta$ approaches one. This suggests that if $\psi$ is monitored at the coastline ($x = 0$), its response to the tidal increase is immediate. At large $x$, however, $\psi$ fully responds, in the sense of reaching a value equal to $\gamma\Psi$, only if the step increase in the tide persists indefinitely such that the elevated fluid pressure at the coastline has sufficient time to diffuse inland. Implications for estimating $\gamma$ based on water-level fluctuations in observation wells are decribed below.

Now consider a more realistic boundary condition associated with a sinusoidal tidal fluctuation:

$$\psi(0, t) = \gamma A_{\Psi} \sin\left(\frac{2\pi}{\lambda}t\right) = \gamma A_{\Psi} \sin \omega t \tag{13.157}$$

Here, $A_{\Psi}$ is the amplitude, $\lambda$ is the period, and $\omega = 2\pi/\lambda$ is the angular frequency, of the tidal fluctuation. With this boundary condition our problem is structurally homologous to the heat-flow problem in Example Problem 9.3.2. Noting homologous parameters between (13.146) and (9.56), and between (13.157) and (9.59), we may immediately imitate (9.60) and write a steady (dynamical) solution:

$$\psi(x, t; \omega) = \gamma A_{\Psi} \exp\left(-x\sqrt{\frac{S\omega}{2T}}\right) \sin\left(\omega t - x\sqrt{\frac{S\omega}{2T}}\right) \tag{13.158}$$

This describes the propagation of tidally induced pressure waves, and suggests that the wave amplitude decreases with distance inland. This attenuation of waves increases with storativity $S$, and decreases with transmissivity $T$ (and therefore with hydraulic conductivity). Moreover, wave attenuation increases with angular frequency; short-term fluctuations are dissipated rapidly, whereas slower fluctuations propagate farther inland. The wave speed is equal to $\sqrt{2T\omega/S}$.

A finite wave speed means that the maximum (or minimum) water level associated with passage of a pressure wave, as monitored within an inland well, lags

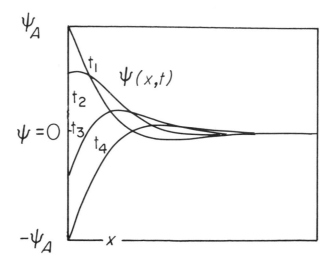

**Figure 13.9**  Instantaneous configurations of pressure wave forms
$\psi(x, t)$ propagating inalnd, where wave amplitude decreases with
distance from the coastline.

behind the tidal maximum (or minimum) (Figure 13.9). With wave attenuation, moreover, the ratio of the water-level amplitude in an inland well to the tidal amplitude provides an *apparent* tidal efficiency that is less than the static-confined tidal efficiency. Tidal efficiency is sometimes used to indirectly measure elastic behavior of an aquifer. The results above therefore suggest that both lag and attenuation must be taken into account to do this.

A technique related to this involves simultaneously measuring well-water level and tidal level over a period of time to estimate values of the aquifer properties $S$ and $T$. An observed water-level response to tidal loading is compared with its theoretically expected response in such a manner that values of $S$ and $T$ can be selected to "best" match the expected and observed responses. This is, in principle, similar to "curve matching" drawdown data associated with a pumping test (Example Problem 8.4.3), and is one of several techniques that make use of water-level responses to aquifer loading by natural phenomena, including Earth tides and atmospheric pressure fluctuations. These techniques are practically appealing when conventional aquifer tests, such as pumping tests, are not feasible. Use of these techniques also normally requires understanding how an aquifer responds to loading by several phenomena at once. For example, records of water-level fluctuations in coastal wells typically reflect superimposed effects of oceanic tides and atmospheric pressure fluctuations. Several papers on this topic are included in the reading list. Students who are familiar with time-series analysis, particularly spectral analysis, will appreciate how these mathematical techniques readily lend themselves in these papers to the analysis of aquifer responses to loading signals.

### 13.5.3  Excess-fluid pressure in thick sediments

Consider the situation in which sediments are continuously accumulating within water above an impermeable boundary (Figure 13.10). Fluid at the sediment-water interface is at a hydrostatic pressure state, as determined by the depth of the fluid

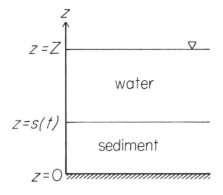

**Figure 13.10**  Definition diagram of sediment accumulating within water above an impermeable boundary.

column above it. As the level of the interface moves upward with sedimentation, fluid is incorporated within the sediment. Continued sedimentation then adds weight to the sediment and fluid beneath the interface. This vertical loading can elevate the pressure of pore fluid beneath the interface to a value above its initial hydrostatic pressure that existed when the fluid was initially incorporated within the sediment at the sediment-water interface. The magnitude of this pressurization is determined by the rate of loading (sedimentation) relative to the rate at which the elevated pressure can be dissipated upward. The rate of dissipation depends largely on the hydraulic conductivity of the sediment. If we assume that the areal extent of accumulating sediment is large relative to the thickness of the sediment, and that sedimentation is areally uniform, the vertical flow of water necessary to dissipate the elevated pressure may be treated as being essentially one-dimensional.

Significant attention has been given to this one-dimensional loading problem, and to extensions of it involving thermal effects and two-dimensional and three-dimensional loading. We will start by loosely following Bredehoeft and Hanshaw (1968), who first considered one-dimensional loading to assess whether sedimentation could account for anomalous pressures (in excess of hydrostatic pressures) observed during deep drilling in the Gulf Coast area of the United States.

With reference to Figure (13.10), let $z$ denote the vertical coordinate, where $z = 0$ is positioned at the impermeable lower boundary. The water-sediment interface is at $z = s$, and the surface of the water column is at $z = Z$, where the pressure $p$ is equal to atmospheric pressure $p_0$. Assuming isothermal conditions, the equation governing vertical flow is obtained from (13.58):

$$\frac{\partial}{\partial z}\left(K_z \frac{\partial h}{\partial z}\right) = \rho g(n\beta + \alpha)\frac{\partial h}{\partial t} - \alpha \frac{\partial \sigma}{\partial t} \tag{13.159}$$

For this problem, it is conventional to work with a quantity referred to as *excess-fluid pressure* $p'$ rather than the hydraulic head $h$. The hydraulic head at a position $z < s$ consists of a hydrostatic part and a part due to the excess-fluid pressure:

$$h = z + \frac{p - p_0 + p'}{\rho g} = z + \psi + \frac{p'}{\rho g} \tag{13.160}$$

Here $z$ is equal to the elevation head, $\psi = (p - p_0)/\rho g$ is the hydrostatic pressure head, and $p'/\rho g$ is referred to as *excess head*. Since $\psi = Z - z$, (13.160) becomes

$$h = Z + \frac{p'}{\rho g} \tag{13.161}$$

Assuming constant water level $Z$, taking derivatives of $h$ with respect to $z$ and $t$, then substituting these expressions into (13.159) and rearranging,

$$\frac{\partial}{\partial z}\left(K_z \frac{\partial p'}{\partial z}\right) = \rho g(n\beta + \alpha)\frac{\partial p'}{\partial t} - \rho g \alpha \frac{\partial \sigma}{\partial t} \tag{13.162}$$

This is referred to as an equation of one-dimensional consolidation, and characterizes how flow responds to excess-fluid pressure. The quantity $\rho g(n\beta + \alpha)$ is the specific storage; the quantity $\rho g \alpha$ is that part of the specific storage associated with pore compressibility.

Consider the term $\alpha \partial\sigma/\partial t$. The total load $\sigma$ at position $z$ within the sediment is due to the weight of the overlying water-filled sediment and the weight of the water above the sediment-water interface. Assume for simplicity that the water and sediments are essentially incompressible so that the density of the sediment $\rho_s$ and the density of the water $\rho$ are constants. Also assume that the porosity $n$ is uniform over $z$. (We will examine this simplification below.) It is left as an exercise to obtain, based on fluid statics, the result that

$$\sigma = \rho g n(s - z) + \rho_s g(1 - n)(s - z) + \rho g(Z - s) + p_0 \tag{13.163}$$

Taking the derivative of (13.163) with respect to time,

$$\frac{\partial \sigma}{\partial t} = g(\rho_s - \rho)(1 - n)\frac{\partial s}{\partial t} - g(\rho_s - \rho)(s - z)\frac{\partial n}{\partial t} \tag{13.164}$$

The quantity $\partial s/\partial t$ is the rate of sedimentation minus the rate of lowering of the sediment-water interface due to consolidation; $\partial n/\partial t$ describes the rate of consolidation. If it is assumed that consolidation is sufficiently slow that porosity does not change significantly—that is $\partial n/\partial t \ll \partial s/\partial t$—then $\partial s/\partial t$ is equal to the rate of sedimentation and (13.164) reduces to

$$\frac{\partial \sigma}{\partial t} = g(\rho_s - \rho)(1 - n)\frac{\partial s}{\partial t} = g(\rho_b - \rho)\frac{\partial s}{\partial t} \tag{13.165}$$

where $\rho_b$ is the wet bulk density of the sediment. Substituting this into (13.162), using the assumption of incompressible water ($\beta = 0$), and further assuming that $K_z$ is constant,

$$\kappa_h \frac{\partial^2 p'}{\partial z^2} - \frac{\partial p'}{\partial t} = -g(\rho_b - \rho)\frac{\partial s}{\partial t} \tag{13.166}$$

where $\kappa_h = K_z/\rho g \alpha$ may be considered a hydraulic diffusivity that is assumed to be constant. (This is not fully realistic since the hydraulic conductivity and specific storage can change as the sediment consolidates.)

Bredehoeft and Hanshaw, following the work of Gibson (1958), describe the solution of (13.166) for varying rates of sedimentation and values of hydraulic conductivity. In general, excess-fluid pressure increases with sedimentation rate and with decreasing hydraulic conductivity. Students are urged to examine the paper by Bethke and Corbet (1988), who examine implications of the assumptions leading to (13.166). Let us now turn to a different version of this problem.

For reasons that will become clear below, let us reconsider the general statement of conservation of mass (13.50):

$$\rho \nabla \cdot \mathbf{q} + \mathbf{q} \cdot \nabla \rho + n\mathbf{u}_s \cdot \nabla \rho + \frac{\rho}{1-n}\mathbf{u}_s \cdot \nabla n = -n\frac{\partial \rho}{\partial t} - \frac{\rho}{1-n}\frac{\partial n}{\partial t} \quad (13.167)$$

where $\mathbf{u}_s$ is the matrix velocity. Dividing all terms by $\rho$, then applying the relations $\beta dp = d\rho/\rho$ and (13.57) to both local and advective terms, (13.167) becomes

$$\nabla \cdot \mathbf{q} + \beta \mathbf{q} \cdot \nabla p + (n\beta + \alpha)\mathbf{u}_s \cdot \nabla p - \alpha \mathbf{u}_s \cdot \nabla \sigma = -(n\beta + \alpha)\frac{\partial p}{\partial t} + \alpha \frac{\partial \sigma}{\partial t}$$

$$(13.168)$$

This has the interesting interpretation that the total stress involves a local part, as in (13.58), and an advective part.

At this point, let us isolate the vertical one-dimensional part of (13.168). Substituting Darcy's law, assuming constant vertical hydraulic conductivity $K_z$, and dividing terms by $(n\beta + \alpha)$,

$$\frac{K_z}{n\beta + \alpha}\frac{\partial^2 h}{\partial z^2} + \frac{\beta K_z}{n\beta + \alpha}\frac{\partial h}{\partial z}\frac{\partial p}{\partial z} - w_s\frac{\partial p}{\partial z} + \frac{\alpha}{n\beta + \alpha}w_s\frac{\partial \sigma}{\partial z}$$

$$= \frac{\partial p}{\partial t} - \frac{\alpha}{n\beta + \alpha}\frac{\partial \sigma}{\partial t} \quad (13.169)$$

where $w_s$ is the vertical component of the matrix velocity. Now choose a $z$-axis that is positive downward, where $z = 0$ is positioned at the sediment-water interface. The free-water surface is at $z = Z$, where $Z$ is negative. To maintain $z = 0$ at the interface during accretion requires the Earth reference frame to be moving downward. This is the same as saying that the sediment, once accumulated, has a downward (positive) velocity $w_s$; this means that the free-water surface is moving downward with velocity $w_s$ also.

We now describe head $h$ and pressure $p$ in terms of excess-fluid pressure $p'$. First, since in general $p = \rho gh - \rho gz - p_0$,

$$\frac{\partial p}{\partial z} = \rho g\frac{\partial h}{\partial z} - \rho g; \qquad \frac{\partial p}{\partial t} - \rho g\frac{\partial h}{\partial t} \quad (13.170)$$

which, when substituted into (13.169) leads to a form of this equation expressed in terms of $h$. The head, in turn, must be defined in terms of the Earth reference frame. Define a datum that is fixed in the Earth reference frame, which moves downward with reference to the $z$-axis (Figure 13.11). The coordinate of this moving datum is $z = z_0$. The hydraulic head at position $z$ is then given by

$$h = (z_0 - z) + (z - Z) + \frac{p'}{\rho g} = z_0 - Z + \frac{p'}{\rho g} \quad (13.171)$$

Recalling that $Z$ is a negative quantity, the first parenthetical term is the elevation head and the second is the pressure head. Differentiating (13.171) with respect to $z$, and noting that $\partial z_0/\partial z = \partial Z/\partial z = 1$,

$$\frac{\partial h}{\partial z} = \frac{1}{\rho g}\frac{\partial p'}{\partial z} \quad (13.172)$$

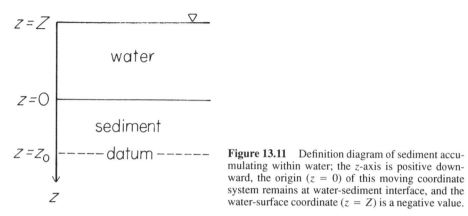

**Figure 13.11** Definition diagram of sediment accumulating within water; the $z$-axis is positive downward, the origin ($z = 0$) of this moving coordinate system remains at water-sediment interface, and the water-surface coordinate ($z = Z$) is a negative value.

Differentiating (13.171) with respect to $t$, and noting that $\partial z_0/\partial t = \partial Z/\partial t = w_s$,

$$\frac{\partial h}{\partial t} = \frac{1}{\rho g}\frac{\partial p'}{\partial t} \tag{13.173}$$

Turning to the total stress term in (13.169), again recalling that $Z$ is a negative quantity,

$$\sigma = \rho g n z + \rho_s g(1 - n)z - \rho g Z + p_0 \tag{13.174}$$

Differentiating this with respect to $z$ and $t$, again assuming a homogeneous fluid, incompressible sediment, and constant $n$,

$$\frac{\partial \sigma}{\partial z} = g(\rho_s - \rho)(1 - n) = g(\rho_b - \rho) \tag{13.175}$$

$$\frac{\partial \sigma}{\partial t} = -\rho g w_s \tag{13.176}$$

The second of these equations indicates that the rate at which the total stress changes is equal to the rate at which the depth of water over the sediment interface decreases. Using (13.170), then substituting (13.172), (13.173), and (13.175) into (13.176) and rearranging,

$$\kappa_h\frac{\partial^2 p'}{\partial z^2} + \beta\kappa_h\left(\frac{\partial p'}{\partial z}\right)^2 - \rho g\beta\kappa_h\frac{\partial p'}{\partial z} - w_s\frac{\partial p'}{\partial z} - \frac{\partial p'}{\partial t} = Aw_s \tag{13.177}$$

where $\kappa_h = K_z/\rho g(n\beta + \alpha)$ is the hydraulic diffusivity and

$$A = \frac{\rho g\alpha - \rho g(n\beta + \alpha) - (\rho_b - \rho)g\alpha}{n\beta + \alpha} \tag{13.178}$$

Using the assumption of incompressible water ($\beta = 0$), this reduces to $A = -(\rho_b - \rho)g$ and (13.178) becomes

$$\kappa_h\frac{\partial^2 p'}{\partial z^2} - w_s\frac{\partial p'}{\partial z} - \frac{\partial p'}{\partial t} = -(\rho_b - \rho)g w_s \tag{13.179}$$

where $\kappa_h$ in (13.179) is now equal to $K_z/\rho g \alpha$. This is the same as (13.166) except that it contains the advective term $-w_s \, \partial p'/\partial z$. This term occurs because excess-fluid pressure can be advected relative to the coordinate system with downward "motion" of the matrix.

Consider the following initial and boundary conditions:

$$p'(x,0) = 0; \qquad p'(0,t) = 0 \qquad (13.180)$$

The first of these indicates that hydrostatic conditions (zero excess-fluid pressure) initially exist over the thickness of the sediment, which is treated as an infinite half-space. We can envision, for example, a thick layer of sediment that is deposited sufficiently slowly that any excess-fluid pressure is fully dissipated during deposition, or which experiences a period of zero sedimentation during which excess pressure is dissipated. Then, (13.179) applies to sediments subsequently deposited on this basal layer for $t > 0$. The second condition indicates that, by definition, excess-fluid pressure must remain zero at the sediment–water interface. This problem has a counterpart in the theory of heat flow by conduction within a moving solid, from which the solution $p'(z,t)$ of (13.179) using (13.180) can be obtained using a *Laplace transformation* (for example, see Carslaw and Jaeger [1959]):

$$p'(z,t) = (\rho_b - \rho)gw_s t$$

$$\frac{1}{2}(\rho_b - \rho)g\left[(z + w_s t)e^{w_s z/\kappa_h} \text{ erfc } \frac{z + w_s t}{2\sqrt{\kappa_h t}} + (w_s - z) \text{ erfc } \frac{z - w_s t}{2\sqrt{\kappa_h t}}\right]$$

$$(13.181)$$

where erfc denotes the complementary error function defined in Example Problem 13.5.2.

As a point of reference, consider a plot of depth versus pressure head (Figure 13.12) computed using (13.181), based on the following values from Bredehoeft and Hanshaw: $w_s = 1.6 \times 10^{-11}$ m s$^{-1}$, $S_s = 0.003$ m$^{-1}$, $\rho_s = 2,300$ kg m$^{-3}$,

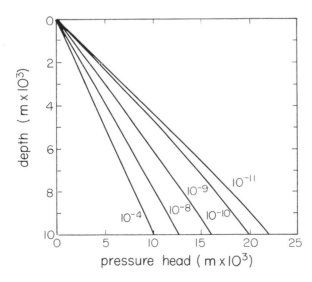

Figure 13.12 Plots of predicted pressure head versus depth within a thick sediment layer, for varying values of hydraulic conductivity (m s$^{-1}$); compare with Figure 6 in Bredehoeft and Hanshaw (1968).

and $t = 2 \times 10^7$ yr. With decreasing hydraulic conductivity $K_z$, dissipation of excess-fluid pressure during sedimentation is slow, and the pressure-head curve is displaced toward the lithostatic curve, with slope equal to $\rho_s g/\rho g$. With a value of $K_z = 1.0 \times 10^{-4}$ m s$^{-1}$, dissipation is sufficiently rapid during sedimentation that the pressure-head curve is essentially the same as the hydrostatic curve. Notice that the slopes of the curves approach the slope of the hydrostatic curve with increasing depth; this means that $\partial p'/\partial z$ approaches zero, so vertical dissipation of excess-fluid pressure becomes negligible at large depths.

The main difference between this formulation and that of Bredehoeft and Hanshaw resides in the different lower boundary conditions; in this formulation excess pressure can be dissipated downward, whereas it can only be dissipated upward with an impermeable lower boundary. It is also important to remember that, in both formulations, the assumption of constant hydraulic conductivity and specific storage is an oversimplification. For example, in the Barbados accretionary complex, measurements from drill cores indicate that porosity declines exponentially with depth. The rate of decline typically is greater for fine-grained sediments than for coarse sediments. Associated with this decline are changes in the hydraulic conductivity and specific storage. Incorporating these factors in the analysis requires reformulating the consolidation equation using explicit expressions for how porosity varies with the effective stress (see Bethke and Corbet [1988]), and distinguishing between the sediment accretion rate and the matrix velocity associated with consolidation.

The analyses above, despite the simplifications involved, clearly indicate that excess-fluid pressures due to loading by sedimentation can persist within thick sediments for geologically significant periods. In addition, excess-fluid pressures develop by other mechanisms. These include thermal expansion of interstitial fluids, and production of water associated with thermally driven, mineralogical phase changes. Two examples are the transformations of gypsum to anhydrite, and smectite to illite.

We will examine a related problem involving vertical flows within sediments in Example Problem 13.5.5.

### 13.5.4 Unsaturated flow as a diffusion phenomenon

Consider unsaturated flow in the horizontal direction. Since gravity does not influence such flow, from (13.69) the governing equation for flow parallel to a horizontal $x$-axis is

$$\frac{\partial}{\partial x}\left[K_u(\psi)\frac{\partial \psi}{\partial x}\right] = c(\psi)\frac{\partial \psi}{\partial t} \tag{13.182}$$

Let us now provide an interpretation of this flow that suggests it is a diffusion-like phenomenon.

Observe from the chain rule that

$$\frac{\partial \theta}{\partial x} = \frac{d\theta}{d\psi}\frac{\partial \psi}{\partial x} \tag{13.183}$$

Based on (13.45), the quantity $d\psi/d\theta$ is the reciprocal of the specific moisture capacity $c(\psi)$; thus

$$\frac{\partial \psi}{\partial x} = \frac{1}{c(\psi)} \frac{\partial \theta}{\partial x} \tag{13.184}$$

Substituting for $\partial \psi / \partial x$ in Darcy's law,

$$q_x = -K(\psi)\frac{\partial \psi}{\partial x} = -\frac{K(\psi)}{c(\psi)} \frac{\partial \theta}{\partial x} \tag{13.185}$$

The ratio $D(\psi) = K(\psi)/c(\psi)$ is referred to as the *soil–water diffusivity*, and we may thus write Darcy's law as

$$q_x = -D(\psi)\frac{\partial \theta}{\partial x} \tag{13.186}$$

Substituting this into the one-dimensional equation of continuity, $\partial q_x / \partial x = -\partial \theta / \partial t$,

$$\frac{\partial}{\partial x}\left[ D(\psi)\frac{\partial \theta}{\partial x} \right] = \frac{\partial \theta}{\partial t} \tag{13.187}$$

Let us assume that $D(\psi)$ is approximately constant with respect to position $x$. This is actually realistic only over a restricted range of $\psi$ (and $\theta$). Then

$$D\nabla^2 \theta = \frac{\partial \theta}{\partial t} \tag{13.188}$$

which has the form of a diffusion equation. In analogy with Fickian diffusion, (13.188) characterizes how moisture diffuses from positions in a porous material where the moisture content is high to positions where the moisture content is low.

This analogy also may be extended to the idea of diffusion of pressure. Dividing both sides of (13.182) by $c(\psi)$ then placing $1/c(\psi)$ within the differential,

$$\frac{\partial \psi}{\partial t} = \frac{\partial}{\partial x}\left[ D(\psi)\frac{\partial \psi}{\partial x} \right] \tag{13.189}$$

Again assuming that $D(\psi)$ is approximately constant,

$$\frac{\partial \psi}{\partial t} = D\nabla^2 \psi \tag{13.190}$$

This form of the diffusion equation suggests the intuitively appealing idea that horizontal unsaturated flow involves diffusion of pressure from positions in a porous material where the fluid pressure is high to positions where the pressure is low.

### 13.5.5 Use of temperature and solute-concentration curves to estimate vertical flow within sediments

Let us consider a problem that is closely related to the one treated in Example Problem 13.5.3.

Consider a general problem involving flows of water, heat, and solute within sediments whose upper boundary consists of an interface with water. This describes, for example, the sediment–seawater interface at the ocean floor. Describing such flows is an important part of understanding the rate of dewatering and chemical evolution of accretionary sediment prisms at convergent plate margins. Assume that the areal extent of the sediment is large relative to its thickness above a décollement. Then the flows of water, heat, and solutes may be treated as being essentially one-dimensional, at least locally.

Relevant one-dimensional forms of the energy equation and the advection–dispersion equation for a conservative solute are:

$$\kappa \frac{\partial^2 T}{\partial z^2} - q_z \frac{\partial T}{\partial z} = \frac{\partial T}{\partial t} \tag{13.191}$$

$$D^* \frac{\partial^2 c}{\partial z^2} - w \frac{\partial c}{\partial z} = \frac{\partial c}{\partial t} \tag{13.192}$$

where $q_z$ denotes the vertical specific discharge, and $w$ denotes the average vertical fluid velocity ($q_z = nw$). Here it is assumed that the effective thermal diffusivity $\kappa$ and the coefficient of hydrodynamic dispersion $D^*$ are constants. For simplicity, heat contributions due to chemical reactions are assumed to be negligible. Notice that (13.191) and (13.192) have identical forms.

Consider the possibility that the flows of water, heat, and solutes are steady. Then (13.191) and (13.192) become

$$\frac{\partial^2 T}{\partial z^2} - \frac{q_z}{\kappa} \frac{\partial T}{\partial z} = 0 \tag{13.193}$$

$$\frac{\partial^2 c}{\partial z^2} - \frac{w}{D^*} \frac{\partial c}{\partial z} = 0 \tag{13.194}$$

Let us position the $z$-axis so that it is positive downward with $z = 0$ at the sediment–seawater interface. Then $q_z$ and $w$ are positive downward. We now seek solutions $T(z)$ and $c(z)$, which arise from interactions between the advective and diffusive parts of (13.193) and (13.194). Consider the following boundary conditions:

$$T(0) = T_0; \qquad T(\zeta) = T_\zeta \tag{13.195}$$

$$c(0) = c_0; \qquad c(\zeta) = c_\zeta \tag{13.196}$$

In words, $T$ and $c$ are constants, $T_0$ and $c_0$, at the interface; these are equal to the temperature and solute concentration of seawater at the ocean floor. Moreover, $T$ and $c$ are fixed values, $T_\zeta$ and $c_\zeta$, at a specified distance $\zeta$ below the interface. General solutions of (13.193) and (13.194) are

$$T(z) = C_1 \frac{\kappa}{q_z} \exp\left(\frac{q_z}{\kappa} z\right) + C_2 \tag{13.197}$$

$$c(z) = C_1 \frac{D^*}{w} \exp\left(\frac{w}{D^*} z\right) + C_2 \tag{13.198}$$

Evaluating the constants $C_1$ and $C_2$ from the boundary conditions (13.195) and (13.196),

$$T(z) = T_0 + (T_\zeta - T_0)\frac{\exp\left(\dfrac{q_z}{\kappa}z\right) - 1}{\exp\left(\dfrac{q_z}{\kappa}\zeta\right) - 1} \tag{13.199}$$

$$c(z) = c_0 + (c_\zeta - c_0)\frac{\exp\left(\dfrac{w}{D^*}z\right) - 1}{\exp\left(\dfrac{w}{D^*}\zeta\right) - 1} \tag{13.200}$$

The ratio $q_z\zeta/\kappa = Pe_T$ in (13.199) is referred to as the *Peclet number* for heat transport. It may be regarded as the ratio of heat transported by advection and heat transported by conduction. The ratio $w\zeta/D^* = Pe_c^*$ in (13.200) is like a Peclet number $Pe_c$ for solute transport. Recall, however, that $Pe_c$ is defined in terms of the mean pore or particle diameter $d$ (instead of $\zeta$) and the coefficient of diffusion $D_c$ (instead of $D^*$). With these definitions,

$$\frac{T(z) - T_0}{T_\zeta - T_0} = \frac{\exp\left(Pe_T\dfrac{z}{\zeta}\right) - 1}{\exp(Pe_T) - 1} \tag{13.201}$$

$$\frac{c(z) - c_0}{c_\zeta - c_0} = \frac{\exp\left(Pe_c^*\dfrac{z}{\zeta}\right) - 1}{\exp(Pe_c^*) - 1} \tag{13.202}$$

Notice that the left sides of (13.201) and (13.202) are dimensionless ratios formed by deviations in temperature and concentration from values at $z = 0$, normalized by the total changes $(T_\zeta - T_0)$ and $(c_\zeta - c_0)$. These define sets of dimensionless temperature and concentration curves when the relative vertical position $z/\zeta$ is plotted against the right side of (13.201) or (13.202) (Figure 13.13). Each such curve is associated with a specific value of $Pe_T$ or $Pe_c^*$.

Because the relative contribution to heat transport due to advection increases with $Pe_T$, temperature curves are displaced upward with negative $Pe_T$ (associated with upward $q_z$), and downward with positive $Pe_T$ (positive $q_z$), assuming that $T_0 < T_\zeta$ (Figure 13.13). In the limit as $Pe_T$ approaches zero, transport by advection is zero; the straight curve reflects the steady temperature distribution that arises when the sediment acts like a static solid, transporting heat only by conduction. Similar remarks pertain to (13.202). In this case a straight curve reflects that solute transport occurs only by molecular diffusion.

This suggests an algorithm for estimating the discharge $q_z$ or velocity $w$ based on data for $T$ and $c$ obtained from temperature probes and core samples. An observed $T$ or $c$ curve is compared with its theoretically expected curve in such a manner that the value of the Peclet number can be selected to "best" match the expected and observed curves. In practice this involves plotting the data against $z/\zeta$, then matching the data with a curve to obtain a value of $Pe_T$ or $Pe_c^*$. The specific discharge $q_z$ or the fluid velocity $w$ then can be calculated, assuming $\kappa$ or $D^*$ is known. References that describe this technique in contexts of field and laboratory studies are listed below.

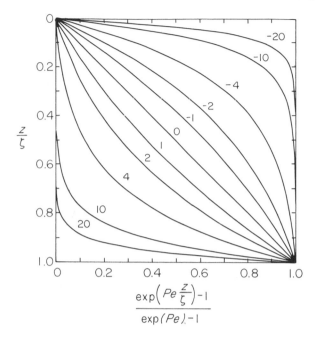

**Figure 13.13** Plots of predicted, dimensionless temperature and concentration values versus relative depth $z/\zeta$ in a sediment layer for varying Peclet number; adapted from Bredehoeft and Papadopulos (1965), copyright by the American Geophysical Union.

## 13.6 READING

Anderson, M. P. and Woessner, W. W. 1992. *Applied groundwater modeling: simulation of flow and advective transport*. San Diego: Academic Press, 381 pp. This text is a comprehensive treatment of how numerical methods are used in practice to treat groundwater flow and transport. The text's strength is that it provides clear guidance on how to conceptualize and incorporate essential parts of field problems into subsequent numerical analyses.

Bear, J. 1972. *Dynamics of fluids in porous media*. New York: Elsevier (Dover edition, 1988), 764 pp. This text provides a comprehensive coverage of the theory of classical groundwater dynamics. Of particular relevance to topics in this chapter are its treatment of the drag coefficient as related to the validity of Darcy's law, its coverage of mechanical models of Darcy's law and solute transport, and its discussion of the Peclet number in relation to the relative contribution to solute transport by diffusion versus dispersion.

Bethke, C. M. and Corbet, T. F. 1988. Linear and nonlinear solutions for one-dimensional compaction flow in sedimentary basins. *Water Resources Research* 24:461–67. The paper examines the assumptions of the one-dimensional consolidation model of Bredehoeft and Hanshaw, then reformulates the problem allowing for changes in hydraulic conductivity and specific storage associated with changes in porosity with depth.

Bredehoeft, J. D. and Hanshaw, B. B. 1968. On the maintenance of anomalous fluid pressures: I. Thick sedimentary sequences. *Geological Society of America Bulletin* 79:1097–106. This article forms the basis of Example Problem 13.5.3.

Bredehoeft, J. D., Roeloffs, E. A., and Riley, F. S. 1987. Dipping into the well to predict earthquakes. *Geotimes* 32:16–19. This article provides a general qualitative discussion of aquifer responses to natural loading phenomena, including the interesting idea that water-level fluctuations in a well can be amplified depending on the frequency of the loading signal, aquifer properties, and well dimensions.

Brooks, R. H. and Corey, A. T. 1964. Hydraulic properties of porous media. Hydrology Paper 3, Colorado State University, Fort Collins, 27 pp. This paper describes the basis of the Brooks–Corey model for unsaturated hydraulic conductivity, including the dimensional analyses regarding pore geometry, and the flow model due to Burdine (1953), that underlie the model.

Burdine, N. T. 1953. Relative permeability calculations from pore-size distribution data. *Transactions of the American Institute of Mineralogy, Metallurgy and Petroleum Engineering* 198:71–78. This paper provides the theory of unsaturated flow used by Brooks and Corey.

Carslaw, H. S. and Jaeger, J. C. 1959. *Conduction of heat in solids,* 2nd ed. Oxford: Oxford University Press, 510 pp. This classic text is a compendium of heat flow problems and their solutions. It describes the case of heat transport through a moving solid used in Example Problem 13.5.3, and has a nice outline of the use of Laplace transformations to solve the heat flow equation.

Chu, S. and Sposito, G. 1980. A derivation of the macroscopic solute transport equation for homogeneous, saturated, porous media. *Water Resources Research* 16:542–46. The introductory part of this article provides a clear description of the mechanisms, and their characteristic time scales, that contribute to randomization of molecular (solute) motions during transport.

Clauser, C. 1992. Permeability of crystalline rocks. *EOS, Transactions, American Geophysical Union* 73:233–238. This article provides a compilation of measurements of permeability for fractured rocks, showing how permeability varies with the scale of measurement. It is an update of a similar compilation provided by W. F. Brace.

Domenico, P. A. and Schwartz, F. W. 1990. *Physical and chemical hydrogeology.* New York: Wiley, 824 pp. This introductory text is an important, recent contribution to the field of geological hydrology. It has very good sections (among others) on aquifer responses to natural loading, forced convection, and coupled flow, and provides a nice derivation of the continuity equation for a deforming medium. It also has a discussion of the scale dependence in dispersivity following Gelhar et al. (1985), and it presents a nice introduction to the mathematics of solute transport involving reactions.

Freeze, R. A. and Cherry, J. A. 1979. *Groundwater.* Englewood Cliffs, N.J.: Prentice-Hall, 604 pp. This well-known text is a good introductory reference regarding flow and transport in porous media and groundwater hydrology.

Gelhar, L. W., Mantoglou, A., Welty, C., and Rehfeldt, K. R. 1985. A review of field-scale physical solute transport processes in saturated and unsaturated porous media. Electric Power Research Institute EPRI EA-4190 Project 2485-5, 116 pp. This report includes a compilation of laboratory and field measurements of dispersivity, showing how dispersivity varies with the scale of measurement.

Gibson, R. E. 1958. The progress of consolidation in a clay layer increasing in thickness with time. *Géotechnique* 8:171–82. This article provides the mathematical formulation of the one–dimensional loading problem used by Bredehoeft and Hanshaw in their analysis of the development of anomalous fluid pressures during sedimentation.

Goldstein, H. 1980. *Classical mechanics,* 2nd ed. Reading, Mass.: Addison-Wesley, 672 pp. This text contains a clear treatment of different conventions of coordinate transformations adopted by different fields in science.

Hillel, D. 1971. *Soil and water: physical principles and processes.* New York: Academic Press, 288 pp. This is a very readable, introductory text covering the field of soil physics, presented from the perspective of a soil scientist.

Hsieh, P. A., Bredehoeft, J. D., and Farr, J. M. 1987. Estimation of aquifer transmissivity from phase analysis of Earth-tide fluctuations of water levels in artesian wells. *Water Resources Research* 23:1824–32. This paper describes a method for estimating aquifer transmissivity based on the responses of water levels in wells to periodic loading by Earth tides.

Javandel, J. C., Doughty, C., and Tsang, C. F. 1984. Groundwater Transport: Handbook of Mathematical Models. Water Resources Monograph 10, American Geophysical Union, Washington, D. C., 228 pp. This monograph is a compendium of analytical and semi-analytical solutions of the advection–dispersion equation for a variety of initial and boundary conditions, including one-dimensional and simple two-dimensional problems. It provides a concise introduction to the advection–dispersion equation, but otherwise assumes that the reader is familiar with this topic.

Klute, A., ed. 1986. Methods of Soil Analysis, Part 1: Physical and Mineralogical Methods, 2nd ed. American Society of Agronomy and Soil Science Society of America, Agronomy Series No. 9, Madison, Wis., 1188 pp. This collection of chapters is a "handbook" regarding methods used to describe and analyze physical properties of soils in field and laboratory studies. These include methods to obtain characteristic moisture curves, and relations between unsaturated hydraulic conductivity and moisture content. It has a companion set of chapters edited by A. L. Page, R. H. Miller, and D. R. Keeney: Part 2, Chemical and Microbiological Properties.

Langseth, M. G. and Moore, J. C. 1990. Introduction to special section on the role of fluids in sediment accretion, deformation, diagenesis, and metamorphism in subduction zones. *Journal of Geophysical Research* 95:8737–41. This is the lead article of 31 papers in this special section of *JGR*. Several papers in the section examine the topics of Example Problems 13.5.3 and 13.5.5. This set of papers collectively provides a comprehensive review of flow and transport phenomena in accretionary prisms, and should be regarded as "essential" reading for students who are beginning to pursue these topics.

Neuzil, C. E. 1986. Groundwater flow in low-permeability environments. *Water Resources Research* 22:1163–95. This paper provides a comprehensive review of flow in low-permeability rocks, and examines factors that influence permeability and fluid behavior in small pores, and problems that arise in applying Darcy's law to describe flow in these rocks.

Phillips, O. M. 1991. *Flow and reactions in permeable rocks.* Cambridge: Cambridge University Press, 285 pp. This text is a significant, recent contribution that emphasizes the geological context of flow and transport. It has a very readable treatment of the processes of solute advection, diffusion and dispersion.

Roeloffs, E. A., Burford, S. S., Riley, F. S., and Records, A. W. 1989. Hydrologic effects on water level changes associated with episodic fault creep near Parkfield, California. *Journal of Geophysical Research* 94:12,3870–402. This paper examines the interrelation between subsurface flow and responses of aquifers in the vicinity of a fault zone to loading associated with strain and creep within the zone.

Rojstaczer, S. 1988. Determination of fluid flow properties from the response of water levels in wells to atmospheric loading. *Water Resources Research* 24:1927–38. This article provides a general treatment of how aquifers respond to atmospheric loading. It starts with the general case of a partially confined (leaky) aquifer, then examines the special case in which an aquifer responds as if it is confined. The article treats this problem using spectral analysis.

Rojstaczer, S. and Agnew, D. C. 1989. The influence of formation material properties on the response of water levels in wells to Earth tides and atmospheric loading. *Journal of Geophysical Research* 94:12,403–11. This paper examines the combined response of an aquifer to simultaneous loading by Earth tides and atmospheric pressure fluctuations.

van Genuchten, M. Th. and Alves, W. J. 1982. Analytical solutions of the one-dimensional convective–dispersive solute transport equation. U.S. Department of Agriculture Technical Bulletin 1661, Washington, D.C., 149 pp. This paper is a compendium of analytical solutions of the one-dimensional advection–dispersion equation, and is the main reference regarding one-dimensional problems compiled in the monograph by Javendel et al. (1984).

Wang, H. F. and Anderson, M. P. 1982. *Introduction to groundwater modeling.* San Francisco: W. H. Freeman, 237 pp. This text provides a clear introductory treatment of the theory and implementation of finite-difference and finite-element methods in solving the equations governing groundwater motion.

# CHAPTER 14

# Turbulent Flows

Many geological flows involve *turbulence,* wherein the velocity field involves complex, fluctuating motions superimposed on a mean motion. Flows in natural river channels are virtually always turbulent. Magma flow in dikes and sills, and lava flows, can be turbulent. Atmospheric flows involving eolian transport are turbulent. The complex, convective overturning of fluid in a magma chamber or geyser is a form of turbulence. Thus, a description of the basic qualities of these complex flows is essential for understanding many geological flow phenomena.

Turbulent flows generally are associated with large Reynolds numbers. Recall from Chapter 5 that the Reynolds number *Re* is a measure of the ratio of inertial to viscous forces acting on a fluid element,

$$Re = \frac{\rho U L}{\mu} \tag{14.1}$$

where the characteristic velocity $U$ and length $L$ are defined in terms of the particular flow system. Thus, turbulence is typically associated, for given fluid density $\rho$ and viscosity $\mu$, with high-speed flows (although we must be careful in applying this generality to thermally driven convective motions; see Chapter 16). A simple, visual illustration of this occurs when smoke rises from a cigar within otherwise calm, surrounding air. The smoke acts as a flow tracer. Smoke molecules at the cigar tip start from rest, since they are initially attached to the cigar. Upward fluid motion, as traced by the smoke, initially is of low speed, and viscous forces have a relatively important influence on its behavior. The flow is *laminar;* smoke streaklines are smooth and locally parallel. But as the flow accelerates upward, it typically reaches a point where viscous forces are no longer sufficient to damp out destabilizing effects of growing inertial forces, and the flow becomes turbulent, manifest as whirling, swirling fluid motions (see Tolkien [1937]).

Throughout this chapter we will consider only incompressible Newtonian fluids. Unfortunately, the complexity of turbulent fluid motions precludes directly using the Navier–Stokes equations to describe them. Instead, we will adopt a procedure whereby the Navier–Stokes equations are recast in terms of temporally averaged or spatially averaged values of velocity and pressure, and fluctuations about these averages. In this regard, there are many ways to describe and characterize the fluctuating velocities that occur in turbulent flows. It is noteworthy that a significant effort of current research on turbulence involves providing ways to characterize and visualize the three-dimensional motions of a fluid, including fluid eddies, as manifest in measurable, fluctuating velocities. Frisch and Orszag (1990) provide a good review of this topic, including how researchers in this field are making use of chaos theory and the mathematics of fractals. Historically, descriptions of turbulence have emphasized the statistics of fluctuating motions, based on the view that the fluctuating motions essentially represent a randomness and disorganization of the flow. Work on turbulence now suggests that this view is too simplistic, and points to the idea that turbulence exhibits a surprising structure and organization of eddying motions. Certain descriptions of turbulence now emphasize the generation, evolution, and decay of vorticity in the form of eddies and eddy-like structures, which are manifest in the complex, measurable velocity fluctuations. Classical statistical treatments nonetheless remain valid from a phenomenological point of view, and in their own way reveal much about the structure of turbulence. We will examine several statistical descriptions of turbulence in this chapter.

## 14.1 ONSET OF TURBULENCE

In 1883, Osborne Reynolds conducted a classic experiment that involved drawing water through tubes of varying radii, with incrementally increasing velocities. Each tube was flared at its intake, and immersed within a tank of water, to minimize disturbance of the water as it entered the tube. Dye was injected into the flared intake. With sufficiently low velocities, the dye followed a straight path, reflecting laminar flow. At a sufficiently high velocity, the dye would at some distance from the intake begin to mix with the surrounding clear water and, when viewed by the light of an electric spark, appear as distinct whirls of fluid motion, or eddies. Reynolds characterized this transition from laminar to turbulent fluid motion in terms of the ratio (14.1). In the context of the Reynolds experiment, $U$ is the mean water velocity and $L$ is the tube diameter. Under normal laboratory conditions, the value of $Re$ necessary for the onset of turbulence is about 2,000. However, no set value of $Re$ exists; transition to turbulence occurs over a range of $Re$. If disturbance of fluid at the intake is carefully avoided, for example, this value can be increased dramatically (to tens of thousands). Moreover, in such an experiment where a high value of $Re$ is reached while maintaining laminar flow, typically a small perturbation of the system—a vibration, for example—is sufficient to suddenly induce turbulent conditions.

Similar circumstances occur with flow past a cylinder (Figure 11.16). The transition to turbulence in the form of a wake of regular, whirling fluid motions occurs at a characteristic value of the Reynolds number. As $Re$ increases, a point is reached where separation and eddy growth occur. This is then followed by increasingly complex motions with further increases in $Re$. The reason for transition in this case, at least that stage involving separation and eddy growth, has the simple interpretation given in Example Problem 11.4.2: that viscous forces in the boundary layer consume energy so that the adverse pressure gradient on the downstream side of the cylinder initiates an upstream flow next to the cylinder. The reasons for transition in the Reynolds experiment, however, are not as clear.

The transition to turbulence is one of a general class of phenomena referred to as *flow instability*. Mechanisms of instability may take a variety of forms. Those of interest in geological applications normally involve flow next to boundaries, solid or fluid, where vorticity is present, as in the Reynolds experiment. In addition, certain turbulent flows arise from instabilities associated with thermally stratified fluids (Chapter 16). The instability mechanism reflected in the Reynolds number involves the idea that inertial effects become increasingly important relative to viscous effects. But what does this mean?

The convective terms in the equations of motion make these equations nonlinear. Qualitatively, the effects of these terms are such that, if a small perturbation occurs in an otherwise laminar flow, for example, involving a small variation in the local velocity field, internal feedbacks occur, which may amplify this perturbation. This is manifest in a variety of ways, including development of streamlines whose sinuosities grow, flow separation, and eddy growth. These, in turn, may destabilize neighboring parts of the flow, and thereby stimulate further eddy growth and development of an increasingly complex flow field. Viscous forces tend to dampen small perturbations such that the flow returns to its initial state. Thus if $Re$ is small, viscous forces exert a stabilizing influence that is sufficient to maintain what we see macroscopically as laminar flow. In fact, if the velocity of a turbulent flow is reduced sufficiently, it will revert to a laminar state. If $Re$ is large, these stabilizing forces are insufficient, and nonlinear inertial effects dominate the behavior of the flow. Note that we have previously used the phrases "very viscous flow" and "purely viscous flow." It should now be apparent that these refer to flows where viscous forces exert a strong, stabilizing influence on the flow behavior. The expression "viscous flow" is in many cases essentially the same as saying "laminar flow," but these expressions are not generally synonymous, as we will see.

Once a flow becomes turbulent, its overall motion consists of the complex, collective whirling motions of eddies and eddy-like structures. Here we must be careful to not envision an eddy exclusively as a distinct, whirling lump of fluid in the form of a vortex, although eddies may take this form. Rather, the term "eddy" merely provides a simple way to refer to any complex whirling motion. With this in mind, the notion of an intact, whirling lump of fluid nonetheless provides a useful abstraction in describing turbulent motions. For example, envision using a small meter to measure the streamwise velocity at a point within a turbulent flow. Suppose that an intact eddy moves with the mean flow past the measurement point (Figure 14.1). If the part of the eddy whose local motion is in the same direction as the mean motion traverses this point, the meter registers an instantaneous velocity whose value is greater than the mean velocity. Conversely, if the part of the eddy whose local motion is in the opposite direction traverses this point the meter registers a negative fluctuation about the mean. More generally, the complex whirling motions of turbulence involve varying speeds and directions relative to the mean

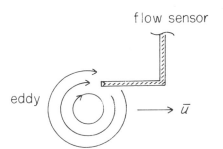

**Figure 14.1**   Schematic diagram of swirling eddy moving with mean speed $\bar{u}$ past stationary flow sensor.

motion, at least for brief intervals, and these motions contribute to fluctuations in the velocity measured at a point. At low to moderate $Re$, turbulence can exhibit a distinct structure with surprising regularity, as reflected by the von Kármán vortex street downstream of a cylinder or sphere (Figure 11.16). At high $Re$, turbulence typically exhibits less regularity, and increasing complexity. The generation (as opposed to onset) of turbulence in flows with high $Re$ occurs in different ways, again usually involving vorticity near boundaries, but also arising from locally unstable conditions in thermally driven flows.

In the remainder of this chapter, we will consider flows whose fluctuating velocity structure is maintained by a continuous generation of turbulence, and ignore, for the most part, the mechanisms that generate the turbulence. We will then turn to turbulent boundary layers in the next chapter, and briefly examine the generation of turbulence.

## 14.2 TIME-AVERAGED VELOCITIES AND PRESSURE

It is convenient in describing a turbulent flow to separate its motion into a mean part and a fluctuating part. Following the lead of Reynolds, it has become customary to describe the three components of velocity $u$, $v$, and $w$, and pressure $p$, in terms of mean values $\bar{u}$, $\bar{v}$, $\bar{w}$, and $\bar{p}$, and fluctuations $u'$, $v'$, $w'$, and $p'$ about the mean values. (Primes should not be confused with similar notation used to represent derivatives.) Thus,

$$u = \bar{u} + u'; \qquad v = \bar{v} + v'; \qquad w = \bar{w} + w'; \qquad p = \bar{p} + p' \quad (14.2)$$

Recalling our developments in Example Problem 1.4.3, the mean of a quantity is obtained as a time-average measured at a point over a suitable interval $T$. For example,

$$\bar{u} = \frac{1}{T} \int_{-T/2}^{T/2} u(t) dt \quad (14.3)$$

The limits of integration in this formula have the effect that the mean $\bar{u}$ is formally defined with respect to the center of the interval $T$. Operationally this is equivalent to

$$\bar{u} = \frac{1}{T} \int_0^T u(t) dt \quad (14.4)$$

if we refer to $\bar{u}$ as the mean over the measurement interval $T$. The interval $T$ must be sufficiently long that the average is independent of time. This means that $T$ must be much longer than the typical durations of fluctuating components. In addition, however, we must allow for the possibility of variations in mean quantities to describe unsteady conditions in the mean fluid motion, and thus $T$ cannot be "too" long. We will examine this further below.

By definition, means of fluctuating quantities must equal zero. For example, substituting $\bar{u} + u'$ for $u$ in (14.4),

$$\frac{1}{T} \int_0^T u(t) dt = \frac{1}{T} \int_0^T \bar{u} dt + \frac{1}{T} \int_0^T u' dt = \bar{u} \quad (14.5)$$

The integral involving $\bar{u}$ equals the "area" beneath the constant value $\bar{u}$, and the integral involving $u'$ equals the "area" between the curve of $u(t)$ and $\bar{u}$ (Figure 14.2). Since the magnitude and duration for which $u'$ is positive is, on average, equal to the magnitude and duration for which it is negative, the latter integral equals zero. Thus,

$$\bar{u'} = 0; \qquad \bar{v'} = 0; \qquad \bar{w'} = 0; \qquad \bar{p} = 0 \qquad (14.6)$$

It also is necessary to manipulate averages of means and fluctuating quantities, averages of their products, and averages of their derivatives with respect to space and time. For example, it will become essential to expand the time-averaged quantity $\overline{uv}$ into several terms. In addition, we will consider spatially averaged quantities obtained from measurements at different coordinate positions, at a single instant. For these reasons, let us list the rules for obtaining averages first introduced in Example Problem 1.4.3:

$$\bar{\bar{f}} = \bar{f}; \qquad \overline{f + g} = \bar{f} + \bar{g}$$

$$\overline{\bar{f} \cdot g} = \bar{f} \cdot \bar{g}; \qquad \overline{\bar{f} \cdot \bar{g}} = \bar{f} \cdot \bar{g}$$

$$\overline{\frac{\partial f}{\partial s}} = \frac{\partial \bar{f}}{\partial s} \qquad (14.7)$$

Here $f$ and $g$ denote any two dependent quantities, for example, $u$ and $v$, with which an average is to be formed, and $s$ denotes any of the independent coordinates $x$, $y$, $z$, or $t$. The first of these rules is equivalent to the first integral operation on the right side of (14.5).

Consider measuring two quantities $f$ and $g$ at regular intervals of time or space. Let us now recall from Example Problem 1.4.3 the definitions for the variances and covariance of $f$ and $g$,

$$s_f^2 = \overline{(f - \bar{f})^2} = \overline{f'^2}; \quad s_g^2 = \overline{(g - \bar{g})^2} = \overline{g'^2}; \quad c_{fg} = \overline{(f - \bar{f})(g - \bar{g})} = \overline{f'g'} \qquad (14.8)$$

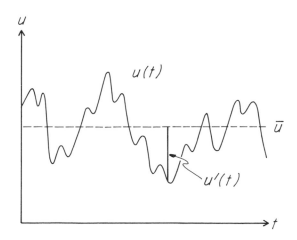

**Figure 14.2** Geometrical interpretation of the mean velocity component $\bar{u}$ and fluctuating values $u'(t)$.

and the correlation coefficient $r_{fg}$,

$$r_{fg} = \frac{\overline{f'g'}}{\sqrt{\overline{f'^2}\,\overline{g'^2}}} \tag{14.9}$$

The value of $r_{fg}$ may be positive or negative, and characterizes how strongly $f$ covaries with $g$ (or vice versa). Finally, let us recall the definition of autocorrelation. Suppose for illustration that $f$ is the velocity component $u(t)$ measured at a fixed position. Let $g$ denote the same velocity component $u(t + \tau)$ measured at an interval $\tau$ later, at the same fixed position. According to (14.8) the autocovariance $c(\tau)$ of values of $u$ separated by an interval $\tau$ is

$$c(\tau) = \overline{u'(t)u'(t + \tau)} \tag{14.10}$$

and according to (14.9) the autocorrelation function $r(\tau)$ is

$$r(\tau) = \frac{c(\tau)}{c(0)} = \frac{c(\tau)}{s_u^2} \tag{14.11}$$

Like the correlation coefficient (14.9), $r(\tau)$ can be positive or negative, and characterizes how $u(t)$ covaries with $u(t + \tau)$. Qualitatively, if one measures $u$ at any instant, then again a short interval later, the chance is good that $u$ has not changed much. That is, values of $u(t)$ separated by a small interval (small $\tau$) are, on average, strongly correlated. Indeed, when the interval goes to zero ($\tau = 0$), the correlation $r(0) = 1$ since, in the limit, each value of $u(t)$ is being compared with itself. As the interval between measurements increases, the chance is greater that $u$ will change significantly during the interval; that is, values of $u(t)$ are not as well correlated. Indeed, if the interval between measurements is sufficiently large, no correlation between measured values exists; observations of $u$ are independent and $r(\tau) \to 0$.

A characteristic time interval $T_0$ associated with the fluctuating part of $u(t)$ is obtained by

$$T_0 = \int_0^\infty r(\tau)d\tau \tag{14.12}$$

and is a measure of the interval beyond which observations of $u$ are statistically independent. Therefore, $T_0$ is a measure of the "memory" in the velocity signal $u$. (We will consider below a spatial counterpart to this.) This suggests that $T_0$ is longer than the typical durations of the fluctuating part $u'$ of $u$. This in turn suggests that the interval $T$ used in (14.4) to ensure that $u$ is a constant value must satisfy the condition: $T \gg T_0$.

Here it is useful to recall the idea of *stationarity* (Example Problem 1.4.3). A stationary time series is one whose statistical properties do not change with respect to a translation in the position of the origin of the time coordinate. Thus the mean, variance, and covariance for observations at any time $t$ are the same as those at time $t + \tau$, for arbitrary $\tau$. The developments above assume stationarity. Also of interest are *nonstationary* time series. Such series describe a condition that we wish to allow for: the possibility that mean quantities defining the mean motion vary. In this regard, $T$ must be defined such that stationarity is satisfied for any interval $T$ taken individually, but such that statistical properties (notably the mean) may change from one interval $T$ to the next. This suggests that $\partial \bar{u}/\partial t \approx 0$ during any one interval $T$. A turbulent flow satisfying these conditions is referred to as a *quasi-steady flow*. We will hereafter assume that a quasi-steady condition exists.

## 14.3 REYNOLDS STRESSES

The fluctuating velocity components $u'$, $v'$, and $w'$ interact with the mean motion in such a way that the apparent rheological behavior of the mean motion is altered from that of a Newtonian fluid. In particular, this interaction is manifest as an apparent increase in the viscosity associated with the mean motion; its resistance to deforming motions, as measured in terms of mean quantities, increases. This suggests that the fluctuating components give rise to stresses that resist such motions, and arc in addition to stresses associated with the Newtonian viscosity $\mu$.

Consider an elementary surface with area $dA$ embedded within a turbulent flow, and assume for illustration that the surface is normal to the $x$-axis. Fluid passing through the surface at some instant possesses a velocity $\mathbf{u} = \langle u, v, w \rangle$. The mass of fluid passing through the surface during a small interval $dt$ is, on geometrical grounds, equal to $\rho u \, dA \, dt$, so the small quantity of momentum transported through the surface in a direction parallel to the $x$-axis is $N_x = \rho u^2 \, dA \, dt$. Similarly, the quantities of momentum transported through the surface in directions parallel to the $y$-axis and $z$-axis are $N_y = \rho uv \, dA \, dt$ and $N_z = \rho uw \, dA \, dt$. With constant $\rho$, the time-averaged fluxes of momentum are

$$\frac{\overline{dN_x}}{dt} = \rho \overline{u^2} \, dA; \qquad \frac{\overline{dN_y}}{dt} = \rho \overline{uv} \, dA; \qquad \frac{\overline{dN_z}}{dt} = \rho \overline{uw} \, dA \qquad (14.13)$$

Substituting expressions in (14.2) for $u^2$, $uv$, and $uw$, then expanding,

$$u^2 = \overline{u}^2 + 2\overline{u}u' + u'^2 \qquad (14.14)$$

$$uv = \overline{u}\,\overline{v} + \overline{u}v' + \overline{v}u' + u'v' \qquad (14.15)$$

$$uw = \overline{u}\,\overline{w} + \overline{u}w' + \overline{w}u' + u'w' \qquad (14.16)$$

Applying (14.7) to obtain time averages, and using (14.6),

$$\overline{u^2} = \overline{u}^2 + \overline{u'^2}; \qquad \overline{uv} = \overline{u}\,\overline{v} + \overline{u'v'}; \qquad \overline{uw} = \overline{u}\,\overline{w} + \overline{u'w'} \qquad (14.17)$$

Substituting these into (14.13),

$$\frac{\overline{dN_x}}{dt} = \rho\left(\overline{u}^2 + \overline{u'^2}\right)dA; \qquad \frac{\overline{dN_y}}{dt} = \rho\left(\overline{u}\,\overline{v} + \overline{u'v'}\right)dA; \qquad \frac{\overline{dN_z}}{dt} = \rho\left(\overline{u}\,\overline{w} + \overline{u'w'}\right)dA$$
$$(14.18)$$

Notice that these quantities have the dimensions of force. When divided by the elementary area $dA$, the resulting components of the momentum flux density associated with fluid passing through $dA$ evidently are equivalent to stresses acting on $dA$. These stresses must be equal in magnitude, and act in the opposite direction, to the components of the momentum flux density. The first is a normal stress; the second and third are tangential stresses. Thus, for example, the surface $dA$ is subjected by surrounding fluid to a normal stress equal to $-\rho(\overline{u}^2 + \overline{u'^2})$. The term $\sigma_x = -\rho \overline{u}^2$ is the contribution of the mean motion; the term $\sigma'_x = -\rho \overline{u'^2}$ is the contribution of fluctuating motions. Thus the fluctuating motions give rise to stresses added to those associated with the mean motion:

$$\sigma'_x = -\rho\overline{u'^2}; \qquad \tau'_{xy} = -\rho\overline{u'v'}; \qquad \tau'_{xz} = -\rho\overline{u'w'} \qquad (14.19)$$

These are referred to as *Reynolds stresses,* and characterize the contribution of fluctuating motions in resisting deforming motions, which is in addition to the resistance arising from the Newtonian viscosity $\mu$. Similar expressions are obtained for elementary surfaces normal to the $y$-axis and $z$-axis. The nine expressions thus obtained form the elements of the *Reynolds stress tensor,* as we will see in the next section. Observe that the averaged quantities in (14.19) consist of the variance of $u$, and the covariances of $u$, $v$, and $w$.

## 14.4 TIME-AVERAGED CONTINUITY AND NAVIER–STOKES EQUATIONS

Let us now obtain expressions for the equation of continuity, and the equations of motion, which must be satisfied by time-averaged and fluctuating components of velocity and pressure. We begin with the equation of continuity for an incompressible fluid:

$$\frac{\partial u}{\partial x} + \frac{\partial v}{\partial y} + \frac{\partial w}{\partial z} = 0 \qquad (14.20)$$

Substituting the expressions from (14.2), then expanding and rearranging,

$$\frac{\partial \overline{u}}{\partial x} + \frac{\partial \overline{v}}{\partial y} + \frac{\partial \overline{w}}{\partial z} + \frac{\partial u'}{\partial x} + \frac{\partial v'}{\partial y} + \frac{\partial w'}{\partial z} = 0 \qquad (14.21)$$

Applying rules from (14.7) to obtain time averages, we see, for example, that $\overline{\partial u'/\partial x} = \partial \overline{u'}/\partial x$, and that $\partial \overline{u'}/\partial x = \partial \overline{u'}/\partial x = 0$ from (14.6). Thus

$$\frac{\partial \overline{u}}{\partial x} + \frac{\partial \overline{v}}{\partial y} + \frac{\partial \overline{w}}{\partial z} = 0 \qquad (14.22)$$

Comparing this with (14.21), it also follows that

$$\frac{\partial u'}{\partial x} + \frac{\partial v'}{\partial y} + \frac{\partial w'}{\partial z} = 0 \qquad (14.23)$$

Thus the time-averaged velocity components and the fluctuating components each satisfy an equation of continuity having the same form as (14.20).

Turning to the three components of the Navier–Stokes equations for incompressible fluid flow, (12.61), (12.62), and (12.63), the task of obtaining time averages will be simplified if these are rewritten, making use of (14.20). Consider, for example, the convective terms in (12.61). Because, according to the chain rule,

$$\frac{\partial}{\partial x}(u^2) = u\frac{\partial u}{\partial x} + u\frac{\partial u}{\partial x}; \qquad \frac{\partial}{\partial y}(uv) = v\frac{\partial u}{\partial y} + u\frac{\partial v}{\partial y}; \qquad \frac{\partial}{\partial z}(uw) = w\frac{\partial u}{\partial z} + u\frac{\partial w}{\partial z} \qquad (14.24)$$

the three convective terms in (12.61) may be written

$$\frac{\partial}{\partial x}(u^2) + \frac{\partial}{\partial y}(uv) + \frac{\partial}{\partial z}(uw) - u\left(\frac{\partial u}{\partial x} + \frac{\partial v}{\partial y} + \frac{\partial w}{\partial z}\right) \qquad (14.25)$$

According to (14.20) the parenthetical term equals zero. After similar applications of the chain rule to convective terms in (12.62) and (12.63), the Navier–Stokes equations for an incompressible fluid become:

$$\rho\left[\frac{\partial}{\partial x}(u^2) + \frac{\partial}{\partial y}(uv) + \frac{\partial}{\partial z}(uw) + \frac{\partial u}{\partial t}\right] = -\rho g\frac{\partial h}{\partial x} - \frac{\partial p}{\partial x} + \mu\nabla^2 u \quad (14.26)$$

$$\rho\left[\frac{\partial}{\partial x}(vu) + \frac{\partial}{\partial y}(v^2) + \frac{\partial}{\partial z}(vw) + \frac{\partial v}{\partial t}\right] = -\rho g\frac{\partial h}{\partial y} - \frac{\partial p}{\partial y} + \mu\nabla^2 v \quad (14.27)$$

$$\rho\left[\frac{\partial}{\partial x}(wu) + \frac{\partial}{\partial y}(wv) + \frac{\partial}{\partial z}(w^2) + \frac{\partial w}{\partial t}\right] = -\rho g\frac{\partial h}{\partial z} - \frac{\partial p}{\partial z} + \mu\nabla^2 w \quad (14.28)$$

Substituting the expressions from (14.2), then expanding and applying rules of (14.7) to obtain time averages, terms of the form $\partial u'/\partial t$, $\partial^2 u'/\partial x^2$, and $\overline{uu'}$ vanish according to (14.6). Moreover, terms of the form $\partial\overline{u}/\partial t$ equal zero to satisfy the quasi-steady assumption. Terms of the form $\overline{u'^2}$, $\overline{u'v'}$, etc., however, remain. Then, with simplifications arising from (14.22) and (14.23),

$$\rho\left(\overline{u}\frac{\partial\overline{u}}{\partial x} + \overline{v}\frac{\partial\overline{u}}{\partial y} + \overline{w}\frac{\partial\overline{u}}{\partial z}\right) = -\rho g\frac{\partial h}{\partial x} - \frac{\partial\overline{p}}{\partial x}$$

$$+ \mu\nabla^2\overline{u} - \rho\left[\frac{\partial}{\partial x}\overline{u'^2} + \frac{\partial}{\partial y}\overline{u'v'} + \frac{\partial}{\partial z}\overline{u'w'}\right] \quad (14.29)$$

$$\rho\left(\overline{u}\frac{\partial\overline{v}}{\partial x} + \overline{v}\frac{\partial\overline{v}}{\partial y} + \overline{w}\frac{\partial\overline{v}}{\partial z}\right) = -\rho g\frac{\partial h}{\partial y} - \frac{\partial\overline{p}}{\partial y}$$

$$+ \mu\nabla^2\overline{v} - \rho\left[\frac{\partial}{\partial x}\overline{u'v'} + \frac{\partial}{\partial y}\overline{v'^2} + \frac{\partial}{\partial z}\overline{v'w'}\right] \quad (14.30)$$

$$\rho\left(\overline{u}\frac{\partial\overline{w}}{\partial x} + \overline{v}\frac{\partial\overline{w}}{\partial y} + \overline{w}\frac{\partial\overline{w}}{\partial z}\right) = -\rho g\frac{\partial h}{\partial z} - \frac{\partial\overline{p}}{\partial z}$$

$$+ \mu\nabla^2\overline{w} - \rho\left[\frac{\partial}{\partial x}\overline{u'w'} + \frac{\partial}{\partial y}\overline{v'w'} + \frac{\partial}{\partial z}\overline{w'^2}\right] \quad (14.31)$$

These are the same as the steady Navier–Stokes equations with time-averaged quantities replacing $u$, $v$, $w$, and $p$, but with the addition of terms involving time-averaged products of fluctuating components. Comparing with (14.19), these time-averaged products form the elements of the Reynolds stress tensor

$$\boldsymbol{\tau}' = \begin{bmatrix} \sigma'_x & \tau'_{xy} & \tau'_{xz} \\ \tau'_{xy} & \sigma'_y & \tau'_{yz} \\ \tau'_{xz} & \tau'_{yz} & \sigma'_z \end{bmatrix} = -\rho\begin{bmatrix} \overline{u'^2} & \overline{u'v'} & \overline{u'w'} \\ \overline{u'v'} & \overline{v'^2} & \overline{v'w'} \\ \overline{u'w'} & \overline{v'w'} & \overline{w'^2} \end{bmatrix} \quad (14.32)$$

which is symmetric. The stresses described by this tensor are referred to as *apparent stresses,* and are added to those arising from ordinary viscous terms. Except near boundaries, the apparent stresses normally are much greater than viscous stresses, such that the latter can be ignored. This result has significant practical value, which we will use in the next chapter. The matrix in (14.32) involving time-averaged products may be referred to as the *velocity covariance tensor.*

The boundary conditions that must be satisfied by mean and fluctuating components are the same as for viscous flows. Mean components vanish at a solid boundary, and the no-slip condition applies. In addition, fluctuating components also must vanish at a solid boundary, and generally are small close to the boundary. Therefore, the apparent stresses must be small near a solid boundary, and vanish at the boundary. Indeed, only viscous stresses remain nonzero close to a solid boundary. In a turbulent flow, therefore, a layer of fluid next to a solid boundary exists where viscous stresses dominate. This is referred to as a *viscous sublayer.* Moreover, the stress and velocity fields must exhibit a smooth transition from the viscous sublayer to the adjacent, turbulent boundary layer. We will consider these topics further in the next chapter. Meanwhile, let us briefly examine the fluctuating velocity components near a solid boundary.

## 14.5 FLUCTUATING VELOCITY COMPONENTS

Hot-film, hot-wire, and laser anemometry can be used to obtain time series of $u$, $v$, and $w$ in a turbulent flow, from which mean quantities, fluctuating components, and their products can be derived. Such measurements reveal important information about the structure of a turbulent flow, including how apparent stresses vary in the vicinity of a solid boundary.

Consider a turbulent flow next to a smooth wall in a conduit. The quantities $\overline{u'^2}, \overline{v'^2}$, and $\overline{w'^2}$ typically increase away from the conduit centerline, reach maximum values near the boundary, then are constrained to approach zero at the boundary (Figure 14.3). These mean-squared fluctuations are measures of turbulence intensity, and reflect the local magnitude of the fluctuating part of the flow. We will see below that they are closely related to the kinetic energy of the fluctuating motions. The mean product $\overline{u'v'}$, which is a measure of the turbulent shearing stress $\tau'_{yx}$, increases steadily from a small value at the centerline to a maximum close to the wall, then is constrained to approach zero at the wall. Notice that $\overline{u'v'}$ varies nearly linearly, except near the wall, which reflects how the apparent stress systematically increases toward the wall. This suggests an analogy with the linear variation in the viscous shear stress $\tau_{yx}$ that exists with laminar flow in a conduit. We will, in fact, examine this analogy in the next chapter.

Recall from Chapter 9 that the kinetic energy per unit volume of fluid is

$$\frac{1}{2}\rho(u^2 + v^2 + w^2) \tag{14.33}$$

Substituting the relations (14.2), expanding and rearranging, then obtaining averages using the rules of (14.7),

$$\frac{1}{2}\rho(\overline{u}^2 + \overline{v}^2 + \overline{w}^2) + \frac{1}{2}\rho(\overline{u'^2} + \overline{v'^2} + \overline{w'^2}) = \frac{1}{2}\rho\Xi + \frac{1}{2}\rho\xi \tag{14.34}$$

The quantity $\frac{1}{2}\rho\Xi = \frac{1}{2}\rho(\overline{u}^2 + \overline{v}^2 + \overline{w}^2)$ is the kinetic energy per unit volume associated with the mean motion; the quantity $\frac{1}{2}\rho\xi = \frac{1}{2}\rho(\overline{u'^2} + \overline{v'^2} + \overline{w'^2})$ is the kinetic energy per unit volume associated with the fluctuating motions. The quantities

$$\sqrt{\overline{u'^2}}; \qquad \sqrt{\overline{v'^2}}; \qquad \sqrt{\overline{w'^2}} \tag{14.35}$$

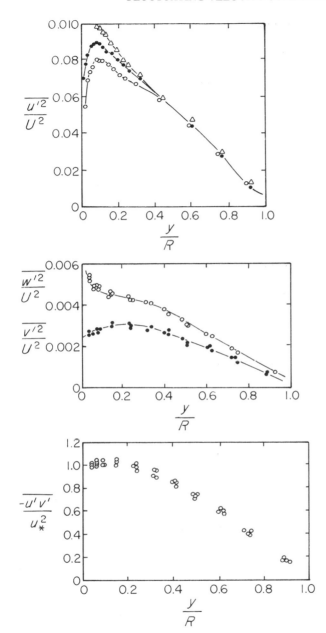

**Figure 14.3** Plot of Reynolds stresses within a unidirectional turbulent flow next to a solid boundary; adapted from Ligrani and Moffat (1986), copyright ©1986 Cambridge University Press; reprinted with the permission of Cambridge University Press.

are referred to as *turbulence component intensities,* and the quantity

$$\sqrt{\xi} \;=\; \sqrt{\overline{u'^2} + \overline{v'^2} + \overline{w'^2}} \qquad (14.36)$$

is referred to as the *turbulence intensity.* We will return to these in the next section.

Measurements of fluctuating velocity components also provide an indirect, statistical description of eddies. Suppose that $f$ and $g$ are two velocity components measured at different positions along a coordinate that is normal to the boundary. For example, suppose that $f$ is the component $u(s, t)$ measured at position $s$. Let $g$ denote the same velocity component $u(s + \sigma, t)$ simultaneously measured $\sigma$ intervals of space away. We may then define an autocorrelation function $r(\sigma)$ of $u$ with respect to space, analogous to (14.9). Then analogous to (14.12), the quantity

$$L_0 = \int_0^\infty r(\sigma)d\sigma \tag{14.37}$$

defines a characteristic length, referred to as the *scale of turbulence*, as first introduced by Taylor (1935). This is a measure of the distance over which fluctuations in a fluid velocity component are correlated, and reflects how the fluctuating motion, on average, spans a finite distance. This suggests, therefore, that $L_0$ is a measure of the average size of fluid lumps, or eddies.

## 14.6 PRODUCTION AND DISSIPATION OF TURBULENCE ENERGY

An intact eddy represents rotational kinetic energy. During its formation, this energy is extracted from the mean flow. More generally, the fluctuating motions in a turbulent flow, manifest as apparent stresses, continuously extract energy from the mean motion. This energy ultimately is dissipated as heat through viscous friction. This leads to the idea of a *turbulence energy cascade*, which has the following qualitative description.

Eddies are generated from concentrated vorticity, for example, in association with a boundary. These eddies then interact with surrounding flow and are disseminated throughout it as more or less intact lumps of fluid, but then mix with surrounding fluid, losing their individuality. This mixing process involves interactions among neighboring eddies and the mean motion that lead to *vortex stretching*, and generation of subsidiary vorticity in the form of smaller eddies and eddy-like structures. These also move for some distance as intact lumps, then mix with surrounding fluid. In this vein we may envision a turbulence field as being divided at any instant into all eddies larger than a given size and all eddies smaller than this size. The smaller eddies are subjected to the strain-rate field of the larger eddies. As a consequence the smaller eddies undergo straining and their vorticity increases. This increases their energy at the expense of the energy of the larger eddies. We may thus envision a cascade of energy from larger to smaller eddies, eventually down to the molecular scale where the mechanical energy is dissipated, via viscous friction, into the form of thermal agitation of molecules. Meanwhile, inasmuch as larger eddies are continuously generated within the flow, this cascade is replenished.

It is possible to characterize this dissipation of mechanical energy in the same way that we described dissipation of mechanical energy in a viscous flow. We begin with the dissipation function for an incompressible fluid (12.85). Substituting the expressions in (14.2), expanding, then applying the rules in (14.7) to obtain time averages, we obtain two groups of terms. The first group involves velocity gradients of the mean motion, and is referred to as the *direct dissipation function*. The second group involves gradients of fluctuating velocity components, and is referred to as the *turbulent dissipation function*. The direct dissipation function is

$$\Phi_d = 2\mu \left[ \overline{\left(\frac{\partial \overline{u}}{\partial x}\right)^2} + \overline{\left(\frac{\partial \overline{v}}{\partial y}\right)^2} + \overline{\left(\frac{\partial \overline{w}}{\partial z}\right)^2} \right]$$

$$+ \mu \left[ \overline{\left(\frac{\partial \overline{u}}{\partial y} + \frac{\partial \overline{v}}{\partial x}\right)^2} + \overline{\left(\frac{\partial \overline{u}}{\partial z} + \frac{\partial \overline{w}}{\partial x}\right)^2} + \overline{\left(\frac{\partial \overline{v}}{\partial z} + \frac{\partial \overline{w}}{\partial y}\right)^2} \right] \tag{14.38}$$

which has the same form as (12.85). This describes the irreversible transformation of energy of the mean motion to heat directly through work performed against viscous friction. The turbulent dissipation function has the same form:

$$\Phi_t = 2\mu \left[ \overline{\left(\frac{\partial u'}{\partial x}\right)^2} + \overline{\left(\frac{\partial v'}{\partial y}\right)^2} + \overline{\left(\frac{\partial w'}{\partial z}\right)^2} \right]$$

$$+ \mu \left[ \overline{\left(\frac{\partial u'}{\partial y} + \frac{\partial v'}{\partial x}\right)^2} + \overline{\left(\frac{\partial u'}{\partial z} + \frac{\partial w'}{\partial x}\right)^2} + \overline{\left(\frac{\partial v'}{\partial z} + \frac{\partial w'}{\partial y}\right)^2} \right] \tag{14.39}$$

This similarly describes the transformation of energy of the fluctuating motions to heat through work performed against viscous friction. The direct dissipation function is important only where gradients of the mean velocity components are strong—near boundaries—whereas the turbulent dissipation function may be significant throughout a flow.

Let us turn to the mechanical energy of the fluctuating motions. We start with the general energy equation (12.80) and observe that an expression for the mechanical energy is obtained by subtracting internal (thermal) energy terms, including the dissipation function, from (12.80). With incompressible flow, $\nabla \cdot \mathbf{u} = 0$, and (12.80) reduces to

$$\mathbf{u} \cdot \frac{D\mathbf{u}}{Dt} = -g\mathbf{u} \cdot \nabla h - \frac{1}{\rho}\mathbf{u} \cdot \nabla p + \nu\mathbf{u} \cdot \nabla^2 \mathbf{u} \tag{14.40}$$

Comparing this with (12.64), notice that (14.40) is equivalent to the inner product of $\mathbf{u}$ and the momentum equation, where each resulting term in (14.40) is a scalar quantity. Assuming steady flow ($\partial \mathbf{u}/\partial t = 0$) and substituting the relations (14.2) into (14.40),

$$(\overline{\mathbf{u}} + \mathbf{u}') \cdot [(\overline{\mathbf{u}} + \mathbf{u}') \cdot \nabla(\overline{\mathbf{u}} + \mathbf{u}')]$$

$$= -g(\overline{\mathbf{u}} + \mathbf{u}') \cdot \nabla h - \frac{1}{\rho}(\overline{\mathbf{u}} + \mathbf{u}') \cdot \nabla(\overline{p} + p') + \nu(\overline{\mathbf{u}} + \mathbf{u}') \cdot \nabla^2(\overline{\mathbf{u}} + \mathbf{u}') \tag{14.41}$$

where $\overline{\mathbf{u}} = \langle \overline{u}, \overline{v}, \overline{w} \rangle$ and $\mathbf{u}' = \langle u', v', w' \rangle$. This describes the kinetic energy of the total flow, both mean and fluctuating parts.

To obtain an expression for the kinetic energy of the fluctuating parts, we first need an expression for the kinetic energy associated with the mean motion; this is obtained by forming the inner product of $\overline{\mathbf{u}}$ and (14.29) through (14.31):

$$\overline{\mathbf{u}} \cdot (\overline{\mathbf{u}} \cdot \nabla \overline{\mathbf{u}}) = -g\overline{\mathbf{u}} \cdot \nabla h - \frac{1}{\rho}\overline{\mathbf{u}} \cdot \nabla \overline{p} + \nu \overline{\mathbf{u}} \cdot \nabla^2 \overline{\mathbf{u}} - \frac{1}{\rho}\overline{\mathbf{u}} \cdot (\nabla \cdot \boldsymbol{\tau}') \quad (14.42)$$

where $\boldsymbol{\tau}'$ denotes the Reynolds stress tensor. The inner product operations in (14.41) and (14.42) are equivalent to multiplying each component of the embedded momentum equation by the corresponding velocity component, then adding the results. It is left as an exercise to expand (14.41) and (14.42), then use the rules of (14.7) to obtain time averages. Subtracting results of the expansion of (14.42) involving mean quantities from results of the expansion of (14.41), making use of the continuity relations (14.22) and (14.23), then simplifying by grouping terms using $\xi = \overline{u'^2} + \overline{v'^2} + \overline{w'^2}$, we retrieve an expression for the kinetic energy associated with fluctuating motions (see Hinze [1959]):

$$\frac{1}{2}\rho\overline{u}\frac{\partial \xi}{\partial x} + \frac{1}{2}\rho\overline{v}\frac{\partial \xi}{\partial y} + \frac{1}{2}\rho\overline{w}\frac{\partial \xi}{\partial z}$$

$$= -\rho\overline{u'^2}\frac{\partial \overline{u}}{\partial x} - \rho\overline{u'v'}\frac{\partial \overline{u}}{\partial y} - \rho\overline{u'w'}\frac{\partial \overline{u}}{\partial z}$$

$$-\rho\overline{u'v'}\frac{\partial \overline{v}}{\partial x} - \rho\overline{v'^2}\frac{\partial \overline{v}}{\partial y} - \rho\overline{v'w'}\frac{\partial \overline{v}}{\partial z}$$

$$-\rho\overline{u'w'}\frac{\partial \overline{w}}{\partial x} - \rho\overline{v'w'}\frac{\partial \overline{w}}{\partial y} - \rho\overline{w'^2}\frac{\partial \overline{w}}{\partial z}$$

$$-\frac{1}{2}\rho\overline{u'\frac{\partial \xi}{\partial x}} - \frac{1}{2}\rho\overline{v'\frac{\partial \xi}{\partial y}} - \frac{1}{2}\rho\overline{w'\frac{\partial \xi}{\partial z}}$$

$$-\overline{u'\frac{\partial p'}{\partial x}} - \overline{v'\frac{\partial p'}{\partial y}} - \overline{w'\frac{\partial p'}{\partial z}}$$

$$+\mu\overline{u'\nabla^2 u'} + \mu\overline{v'\nabla^2 v'} + \mu\overline{w'\nabla^2 w'} \quad (14.43)$$

Recall from Section 14.5 that the kinetic energy per unit volume of fluid associated with fluctuating motions is $\frac{1}{2}\rho\xi$. Thus the first line in (14.43) describes advection of energy by the mean motion, and the fifth line describes advection of energy by fluctuating motions. These terms therefore represent transports of energy. Likewise, the sixth line represents a transport of energy. The last line represents both transport and viscous dissipation of energy. This can be illustrated by expanding and rearranging the terms in this line. Observe, for example, that

$$\overline{u'\frac{\partial^2 u'}{\partial x^2}} = \frac{1}{2}\frac{\partial^2}{\partial x^2}\overline{u'^2} - \overline{\left(\frac{\partial u'}{\partial x}\right)^2} \quad (14.44)$$

After writing similar expressions for the eight remaining terms in this line, using rules in (14.7) pertaining to sums and derivatives of averaged quantities, and again making use of the relation $\xi = \overline{u'^2} + \overline{v'^2} + \overline{w'^2}$:

$$\frac{1}{2}\mu\nabla^2\xi - \mu\left[\overline{\left(\frac{\partial u'}{\partial x}\right)^2} + \overline{\left(\frac{\partial u'}{\partial y}\right)^2} + \overline{\left(\frac{\partial u'}{\partial z}\right)^2} + \ldots + \overline{\left(\frac{\partial w'}{\partial z}\right)^2}\right] \quad (14.45)$$

Owing to its form, the first term in (14.45) represents diffusion of energy. The second term, containing the squares of all nine gradients formed by the fluctuating velocity components, represents dissipation of energy. Returning to (14.43), the second, third, and fourth lines represent the production of energy to compensate this dissipation. Each of the terms in these three lines is the product of a Reynolds stress and a mean velocity gradient. Thus the production of kinetic energy associated with fluctuating motions is greatest in those regions of a flow where both Reynolds stresses and mean velocity gradients are high. Such conditions typically occur near boundaries in geologically important flows.

The dissipation functions (14.38) and (14.39) are positive terms in the general energy equation. This reflects that these terms describe a source of internal (thermal) energy via viscous production of heat. The dissipation term in (14.45) is negative. This reflects that this term describes a viscous extraction of mechanical (kinetic) energy. We will examine the sign of those terms in (14.43) representing energy production in the next chapter.

## 14.7 EXAMPLE PROBLEMS

### 14.7.1 A statistical measure of the scale of turbulence

Let us examine one aspect of how fluctuating velocities in a turbulent flow are spatially correlated, and the significance of the *scale of turbulence* obtained from the integration (14.37) of the spatial correlation function $r(\sigma)$. For illustration, consider a unidirectional shear flow next to a planar boundary. The mean (streamwise) motion is parallel to the $x$-axis (and boundary), and the $y$-axis is normal to the boundary. In the case of a steady, fully developed flow (Chapter 15), the turbulence field does not vary in the streamwise direction nor in the $z$-direction, but it does vary in the $y$-direction.

Consider the streamwise velocity component $u$, which, in general, varies as a function $u(x, y, z, t)$ of coordinate position and time. Now, for constant $y$ and $z$, $u$ exhibits at any instant one realization of a spatial series, $u(x)$. Similarly, for constant $x$ and $z$, $u$ exhibits at any instant one realization of a spatial series, $u(y)$. In general, the statistical properties of $u(x)$ and $u(y)$—the mean, variance, and covariance—may be different. Moreover, for the particular flow considered here, a series $u(x)$ will be statistically homogeneous whereas a series $u(y)$ will be nonhomogeneous (Example Problem 1.4.3). Thus there are several possibilities for describing how $u$ is spatially correlated, depending on the orientation of the spatial series.

Let $s$ denote either $x$ or $y$. To obtain a spatial correlation function for $u(s)$, initially we might be tempted to measure directly at one instant a single realization $u(s)$, then follow procedures normally adopted for estimating the autocovariance of a time series (Example Problem 1.4.3). This strategy, however, would require placing many closely spaced velocity meters in the flow along the coordinate direction $s$, an experimentally difficult task. But there is a better way. Using two meters, the two streamwise components $u(s, t)$ and $u(s + \sigma, t)$ separated by the distance $\sigma$ are measured at the same time. These measurements must span a sufficient interval $T$ to obtain reliable estimates of statistical properties of the time series at the two positions. The time series thus obtained at the positions $s$ and $s + \sigma$ in effect provide an ensemble of realizations of the underlying process at $s$ and $s + \sigma$. The covariance of $u$ (obtained from the time series) is then expressed as a function of the separation distance $\sigma$, and the experiment is repeated, systematically varying $\sigma$.

Let us turn now to the specific case in which measurements are made along the $y$-axis ($s = y$) of a conduit. The spatial covariance is

$$c(\sigma) = \overline{u'(y)u'(y + \sigma)} \tag{14.46}$$

where $t$ has been omitted for simplicity. In the transverse $y$-direction, the variance of $u(y, t)$ and $u(y + \sigma, t)$ may be different, so the correlation function $r(\sigma)$ is defined by

$$r(\sigma) = \frac{\overline{u'(0)u'(\sigma)}}{\sqrt{\overline{[u'(0)]^2}\ \overline{[u'(\sigma)]^2}}} \tag{14.47}$$

where $u'(0)$ is the velocity fluctuation at the centerline of the conduit ($y = 0$) and $u'(\sigma)$ is the fluctuation at a distance $\sigma$ from the centerline. For such measurements, $r(\sigma)$ typically decreases rapidly with increasing $\sigma$, then exhibits small negative values over a range of $\sigma$ before approaching zero at larger $\sigma$ (Figure 14.4). Why does $r(\sigma)$ have this form?

When $\sigma = 0$, the two fluctuating velocity components are identical, so $r(0) = 1$. With large $\sigma$, the velocity fluctuations become independent and $r(\sigma)$ approaches zero. For small $\sigma$, motions are positively correlated. This suggests that, on average, fluid motions close to each other are coherent, and consist of the motions of parts of the same fluid lumps. With increasing $\sigma$, a point is reached where motions are negatively correlated such that a positive fluctuation $u'(0)$ is, on average, associated with a negative fluctuation $u'(\sigma)$, and visa versa. The reason for this arises from conservation of mass. The same amount of fluid mass must move through any plane normal to the $x$-axis per unit time. Thus, if at any instant a positive fluctuation occurs at one position, a negative fluctuation must occur somewhere else to compensate this.

These points lead to the idea that motion involves coherent fluid lumps. In particular, the distance defined by $0 \le \sigma \le \sigma_*$, where $\sigma_*$ denotes the distance where $r(\sigma)$ is a (negative) minimum, is a measure of the distance over which motion is coherent. The distance $\sigma_*$ therefore may be loosely interpreted as a measure of the average lump (eddy) size. More generally, a characteristic length $L_0$ of the turbulence structure is defined by (14.37). In the case of a conduit with diameter $2R$,

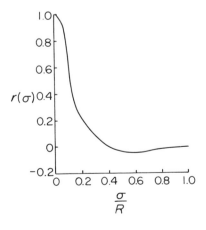

**Figure 14.4** Spatial correlation function $r(\sigma)$ of streamwise velocity component $u$ measured along $y$-axis normal to boundary.

(14.37) is specialized to

$$L_0 = \int_0^R r(\sigma)\mathrm{d}\sigma \qquad (14.48)$$

which is referred to as the scale of turbulence.

## 14.7.2 Turbulence kinetic energy in a boundary layer

In anticipation of several ideas presented in the next chapter regarding kinetic energy of turbulence, let us briefly consider experimental results of Ligrani and Moffat (1986) regarding the turbulence structure in a turbulent boundary layer.

Recall that the quantity

$$\frac{1}{2}\rho\xi = \frac{1}{2}\rho(\overline{u'^2} + \overline{v'^2} + \overline{w'^2}) \qquad (14.49)$$

is the kinetic energy per unit volume associated with the fluctuating motions of a turbulent flow. For constant density, $\xi$ is therefore a measure of this energy, and it is customary to refer to $\xi$ as the turbulence kinetic energy. With reference to Figure 14.3, the turbulence components $\overline{u'^2}$, $\overline{v'^2}$, and $\overline{w'^2}$ generally increase toward the boundary in a turbulent shear flow. Since $\xi$ consists of the sum of these components, $\xi$ likewise increases toward the boundary (Figure 14.5). In these experiments, $\overline{u'^2}$ is larger than the other components (which is typical for turbulent shear flows), and therefore provides the largest contribution to $\xi$.

In the next chapter, we will examine the production, transport, and dissipation of this kinetic energy. With regard to comments in Section 14.6, we can anticipate that production is greatest in regions where the Reynolds stresses and mean velocity gradients are large—that is, near the boundary—coincident with the position where the value of $\xi$ is large (Figure 14.5). We will also learn that, although kinetic energy is transported away from the boundary, most dissipation of kinetic energy occurs near the boundary. In addition, we will consider how kinetic energy is advected within the wake regions of dunes on a stream bed (Example Problem 15.8.2).

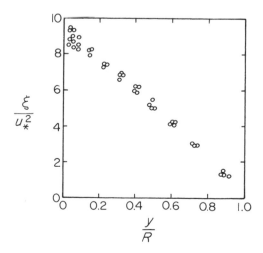

Figure 14.5 Plot of turbulence kinetic energy terms within a unidirectional turbulent flow next to a solid boundary; adapted from Ligrani and Moffat (1986), copyright ©1986 Cambridge University Press; reprinted with the permission of Cambridge University Press.

## 14.8 READING

Box, G. E. P. and Jenkins, G. M. 1976. *Time series analysis: forecasting and control.* San Francisco: Holden-Day, 575 pp. This is a classic treatise on discrete time series. Its treatment emphasizes the methods and strategies involved in estimating the statistical properties of time series. Although it is mainly concerned with describing data series in the temporal or spatial domain, it does include brief treatments of the relation between this style of analysis and spectral (Fourier) analysis. It is a wonderful text for learning the basics of time series as they may be applied to descriptions of turbulence.

Brown, R. A. 1991. *Fluid mechanics of the atmosphere.* San Diego: Academic Press, 486 pp. This text summarizes several views of turbulence that reveal how ideas on this subject have historically changed, including a poignant quotation attributed to Th. von Kármán: "To my mind, there are two great unexplained mysteries in our understanding of the universe. One is the nature of a unified generalized theory to explain both gravitation and electromagnetism. The other is an understanding of the nature of turbulence. After I die, I expect God to clarify general field theory for me. I have no such hope for turbulence."

Frisch, U. and Orszag, S. A. 1990. Turbulence: challenges for theory and experiment. *Physics Today* 43(1):24–32. This review article summarizes current work on turbulence research, and provides a nice historical perspective regarding the significance and lasting influence of Prandtl's mixing-length hypothesis (see Chapter 15).

Hinze, J. O. 1959. *Turbulence: an introduction to its mechanism and theory.* New York: McGraw-Hill, 586 pp. This text is a comprehensive treatment of traditional statistical descriptions of turbulence. The material is presented at an advanced level. For those familiar with Einstein summation notation, Hinze provides a compact derivation of the kinetic energy equation. If you expand equations (14.41) and (14.42) to obtain (14.43), you will appreciate the compact form provided by this notation.

Jenkins, G. M. and Watts, D. G. 1968. *Spectral analysis and its applications.* San Francisco: Holden-Day, 525 pp. This is a classic treatise on spectral analysis. Like Box and Jenkins (1976), its treatment emphasizes the methods and strategies involved in estimating time-series spectra from the point of view of a statistician. Its level of presentation assumes familiarity with Fourier series, complex notation involving Euler's representation of sinusoids, and Fourier transforms.

Ligrani, P. M. and Moffat, R. J. 1986. Structure of transitionally rough and fully rough turbulent boundary layers. *Journal of Fluid Mechanics* 162:69–98. This article presents the results of hot-wire measurements of velocity components in a turbulent boundary-layer shear flow, and is the source of Figures 14.3 and 14.5. The article also presents an interesting evaluation of how variations in the length of the hot-wire sensor affect measurements of small-scale turbulence.

Orszag, S. A. and Zabusky, N. J. 1993. High-performance computing and physics. *Physics Today* 46(3):22–23. This is the introductory article for several pertaining to this topic in this issue of *PT.* Several of the papers in this issue are concerned with computational treatments of flow, including turbulence, and visualization of turbulence: Zabusky et al. (pp. 24–31); Karniadakis and Orszag (pp. 34–42); and Dritschel and Legras (pp. 44–51). Such computational treatments are becoming increasingly important in studies of the complex motions and structures of turbulent flows.

Priestley, M. B. 1981. *Spectral analysis and time series.* London: Academic Press, 940 pp. This is a comprehensive treatise on spectral analysis and time series from the point of view of a mathematician. The material is presented at an advanced level.

Reynolds, O. 1883. An experimental investigation of the circumstances that determine whether the motion of water shall be direct or sinuous, and of the law of resistance in parallel channels. *Philosophical Transactions of the Royal Society of London* 174:935–82. Reynolds describes his experiment involving injection of dye into flows within tubes.

Schlichting, H. 1979. *Boundary-layer theory,* 7th ed. New York: McGraw-Hill, 817 pp. This text contains two chapters (16 and 17) that treat the origin and generation of turbulence, three chapters (18, 19, and 20) that treat general properties of turbulent boundary layers, and several other chapters that treat specialized topics of turbulent boundary layers.

Taylor, G. I. 1935. Statistical theory of turbulence: Parts 1–4. *Proceedings of the Royal Society* A 151:421–78. Taylor introduces the characteristic length defined by (14.37) in Part 1 of this series of papers. These papers provide an interesting historical and technical perspective of statistical treatments of turbulence. Students should also refer to the 1971 volume "The Scientific Papers of Sir Geoffrey Ingram Taylor, Volume IV, Mechanics of Fluids; Miscellaneous Papers" (Cambridge University Press, Cambridge, 579 pp.) to gain a sense of the breadth of Taylor 's work. Included are such titles as "On the decay of vortices in a viscous fluid" and "Analysis of the swimming of long and narrow animals."

Tennekes, H. and Lumley, J. L. 1972. *A first course in turbulence.* Cambridge, Mass.: MIT Press, 300 pp. This text provides a clear description of the turbulence energy cascade, and how this cascade is characterized in terms of spectral analysis.

Tolkien, J. R. R. 1937. *The hobbit;* 1954. *The lord of the rings* (trilogy). London: Allen & Unwin. If you have read this delightful epic fantasy, you are aware that blowing smoke rings, and making them do neat tricks, is one of the chief amusements of Bilbo, Frodo, and other Hobbiton characters when they have an opportunity to relax and smoke their pipes. A smoke ring is a special turbulence structure, or eddy, referred to as a *ring vortex.* The fluid motion associated with a ring vortex is such that the vortex can travel in a coherent form for a significant distance within an otherwise calm fluid. You will also recall that Gandalf the Grey had a knack for making smoke rings interact with each other, and with solid surfaces, in complex ways, although portrayed with ample artistic license, Tolkien 's descriptions of these interactions nonetheless have a grounding to some extent in fluid dynamics. Interested students may wish to consult any of a number of texts, for example, that by G. K. Batchelor, that treat ring vortices.

Tritton, D. J. 1988. *Physical fluid dynamics.* Oxford: Oxford University Press, 519 pp. Chapter 24 in this text provides a nice introduction to chaos theory as applied to descriptions of turbulence. Basically, a system can exhibit chaotic behavior in a mathematical sense when the equations governing the system are nonlinear, such that complex interactions among system components are possible. The Navier–Stokes equations fall in this class of equations, and therefore describe a system whose behavior is of interest in the purely mathematical language of chaos, particularly in view of the fact that this behavior in many ways resembles the chaotic behaviors of other nonlinear systems that have little to do with fluids. Thus, turbulence, a fluid-flow thing, is one example of a more general class of phenomena, chaos. This text also has several figures and photographs that clarify aspects of the complex eddy structures in turbulent boundary flows.

# CHAPTER 15

# Turbulent Boundary-Layer Shear Flows

Turbulent shear flows next to solid boundaries are one of the most important types of flow in geology. In such flows, turbulence is generated primarily by boundary effects; vorticity originates near a boundary in association with the velocity gradients that arise from the no-slip condition at the boundary. Such gradients provide a ready source of vorticity for eddies and eddy-like structures to develop in response to the destabilizing effects of inertial forces, and then move outward into the adjacent flow. Eddies are also generated within the wakes of bumps that comprise boundary roughness, for example, sediment particles on the bed of a stream channel (Example Problem 11.4.2). As we have seen in Chapter 14, the fluctuating motions of turbulence involve, over any elementary area, fluxes of fluid momentum that are manifest as apparent (Reynolds) stresses. In addition, the complex motions of eddies and eddy-like structures efficiently advect heat and solutes from one place to another within a turbulent flow, and thereby facilitate mixing of heat and solutes throughout the fluid. For similar reasons, turbulent motions are responsible for lofting fine sediment into the fluid column of a stream channel and in the atmosphere.

We will concentrate in this chapter on steady unidirectional flows where the mean streamwise velocity varies only in the coordinate direction normal to a boundary and the mean velocity normal to the boundary is zero. We also will adopt a classic treatment of turbulent boundary flow in developing the idea of L. Prandtl's *mixing-length hypothesis,* from which we will obtain the *logarithmic velocity law,* a function that describes how the mean streamwise velocity varies in the coordinate direction normal to a boundary. In developing Prandtl's hypothesis, we will see that the presence of apparent stresses associated with fluctuating motions leads to the idea of an *eddy viscosity* or *apparent viscosity.* Unlike the Newtonian viscosity, the eddy viscosity is a function of the mean velocity, and therefore coordinate position. This means that the eddy viscosity cannot, in general, be removed from stress terms involving spatial derivatives, as we previously did with the Newtonian viscosity in simplifying the Navier–Stokes equations. But there is another side to this: we will see that apparent stresses in fully turbulent flows are much more important than viscous stresses, such that the Newtonian viscosity can be neglected except within the *viscous sublayer* very close to a boundary. We will also see that Prandtl's hypothesis provides an interesting heuristic description of eddy motions and momentum fluxes that bears a clear resemblance to the kinetic theory of viscosity for gases (Section 3.4.1).

Most boundaries in geological problems are, qualitatively speaking, rough. The idea of "rough" needs to be qualified, however, in the context of turbulent boundary flows. The reason for this resides in how smooth versus rough boundaries influence the velocity field next to them via generation of turbulence, and how this in turn influences the form of the logarithmic velocity law close to the boundary. We will therefore distinguish between boundaries that are *hydrodynamically smooth* and boundaries that are *hydrodynamically rough*. In this regard, one of the main applications of turbulent boundary-layer theory in geology is in relation to turbulent shear flows over boundaries composed of sediment. Under this heading fall studies of the mechanics of sediment motions, the dynamics of sediment bedforms, and the flow resistance produced by the roughness of sediment grains and bedforms. One objective of such studies is to describe how the complex geometries of sediment surfaces co-evolve with the turbulence fields above them. Although we will only touch on these topics herein, it is important to recognize that our introductory description of turbulent boundary flows, following the classic work of Prandtl, von Kármán, Nikuradse, Taylor, and others, forms a cornerstone for studies of sediment dynamics.

Finally, we also will take this opportunity to consider turbulence at the pore scale in relation to Darcy's law (Chapter 13).

## 15.1 TURBULENT BOUNDARY-LAYER DEVELOPMENT

Consider incompressible fluid flow next to a planar boundary. The $x$-axis is parallel to the boundary and coincides with the direction of the mean fluid (streamwise) motion; the $z$-axis is normal to the boundary. We shall leave open the possibility that the boundary is inclined with respect to the Earth reference frame. In this situation the relevant equations of motion are the two-dimensional boundary-layer equations obtained in Example Problem 12.5.6, to which we add a gravitational term:

$$\frac{\partial u}{\partial x} + \frac{\partial w}{\partial z} = 0 \tag{15.1}$$

$$u\frac{\partial u}{\partial x} + w\frac{\partial u}{\partial z} + \frac{\partial u}{\partial t} = -g\frac{\partial h}{\partial x} - \frac{1}{\rho}\frac{\partial p}{\partial x} + \nu\frac{\partial^2 u}{\partial z^2} \tag{15.2}$$

where $h$ denotes the height relative to a horizontal datum. It is left as an exercise to go through the necessary time-averaging procedures described in Chapter 14 to obtain the following two-dimensional equations for turbulent flow:

$$\frac{\partial \overline{u}}{\partial x} + \frac{\partial \overline{w}}{\partial z} = 0 \tag{15.3}$$

$$\overline{u}\frac{\partial \overline{u}}{\partial x} + \overline{w}\frac{\partial \overline{u}}{\partial z} = -g\frac{\partial h}{\partial x} - \frac{1}{\rho}\frac{\partial \overline{p}}{\partial x} + \frac{\partial}{\partial z}\left(\nu\frac{\partial \overline{u}}{\partial z} - \overline{u'w'}\right) \tag{15.4}$$

The reason for including viscous and fluctuating parts within the parentheses in (15.4) will become clear below.

Let us write the viscous stress associated with the mean flow as

$$\tau_{zx} = \mu \frac{\partial \bar{u}}{\partial z} = \rho \nu \frac{\partial \bar{u}}{\partial z} \tag{15.5}$$

By analogy with (15.5) let us write the Reynolds stress as

$$\tau_{zx}' = -\rho \overline{u'w'} = A \frac{\partial \bar{u}}{\partial z} = \rho \varepsilon \frac{\partial \bar{u}}{\partial z} \tag{15.6}$$

The factor $A$ is referred to as an *eddy viscosity* or *apparent viscosity* in analogy to the Newtonian viscosity $\mu$; the *kinematic eddy viscosity* $\varepsilon = A/\rho$ is in analogy to the kinematic viscosity $\nu$. Unlike $\mu$, however, $A$ is a function of the mean velocity $\bar{u}$ rather than being a property of the fluid. Equation (15.6) is referred to as the *Boussinesq hypothesis*. (This should not be confused with what is referred to as the Boussinesq approximation, which pertains to buoyancy driven convection; see Chapter 16.)

The parenthetical term in (15.4) is equivalent to $(1/\rho)\partial(\tau_{zx} + \tau_{zx}')/\partial z$. Comparing this with (15.5) and (15.6), the boundary-layer equation (15.4) becomes

$$\bar{u} \frac{\partial \bar{u}}{\partial x} + \bar{w} \frac{\partial \bar{u}}{\partial z} = -g \frac{\partial h}{\partial x} - \frac{1}{\rho} \frac{\partial \bar{p}}{\partial x} + \frac{\partial}{\partial z} \left[ (\nu + \varepsilon) \frac{\partial \bar{u}}{\partial z} \right] \tag{15.7}$$

To use this requires knowing how $\varepsilon$ varies with the mean velocity $\bar{u}$. We will return to this momentarily; meanwhile let us simplify (15.7) further by considering the idea of a fully developed, turbulent boundary layer. We again envision the problem of flow along a flat plate, as Prandtl did in developing the viscous boundary-layer equations (Example Problem 12.5.6). A flow with uniform velocity approaches the leading edge of the plate (Figure 15.1). As with viscous flow, the no-slip condition begins to influence flow reaching the plate, and a viscous boundary layer of thickness $\delta$ begins to develop. At some finite distance along the plate, the destabilizing influence of inertial forces becomes significant, the flow becomes unstable, and turbulence is induced. A region of turbulence develops above the viscous boundary layer, and extends to a distance $\delta'$ from the boundary; this represents the onset of a *turbulent boundary layer*. This region of turbulence also extends toward the boundary, which suppresses the *viscous sublayer*. With increasing distance downstream, the thickness of the turbulent boundary layer further increases, and the viscous sublayer is suppressed to a thin region close to the boundary. In absence

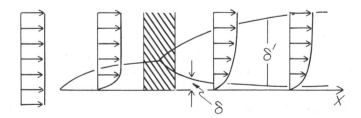

**Figure 15.1** Schematic diagram of development of turbulent boundary layer of thickness $\delta'$ and suppression of viscous sublayer of thickness $\delta$ along flat plate; shaded area represents transition zone.

of a free surface or opposing boundary, the turbulent boundary layer thickness increases indefinitely, albeit slowly. In contrast, the presence of either a free surface or opposing boundary limits the thickness of the turbulent boundary layer. For example, the turbulent boundary layer extends to the free surface in an open stream channel; in a conduit, turbulent boundary layers extend from each opposing boundary to the conduit axis. We will take a closer look below at different regions within a turbulent boundary layer.

At some distance downstream from the leading edge of the plate, the thicknesses of the turbulent boundary layer and viscous sublayer reach a steady condition in the presence of a free surface or opposing boundary. Note that the viscous sublayer thickness is a time-averaged quantity. Eddies continually impinge on this layer, momentarily depressing it. That is, fluctuating motions intermittently exist within the nominal thickness $\delta$. The viscous sublayer is sometimes referred to as a "laminar sublayer"; the term "viscous sublayer" is preferred, however, to emphasise the idea that it is a region where viscous stresses, on average, dominate over apparent stresses. (The viscous sublayer is sometimes incorrectly referred to as the "boundary layer.") In addition, the turbulence structure (a complex idea) is steady. This means that time derivatives of mean quantities are zero, the rate of turbulence production is equal to the rate of dissipation (Section 15.6), and the intensity of the turbulence at any given position $z$ has reached a steady value. Furthermore, $\overline{v} = \overline{w} = 0$ and the derivative $\partial \overline{u}/\partial x = 0$ such that $\overline{u}$ is only a function $\overline{u}(z)$ of the coordinate $z$. This describes a steady, fully developed boundary layer. (We will modify this idea slightly for rough boundaries.) Under these circumstances, (15.7) reduces to

$$\frac{\partial}{\partial z}\left[(\nu + \varepsilon)\frac{\partial \overline{u}}{\partial z}\right] = g\frac{\partial h}{\partial x} + \frac{1}{\rho}\frac{\partial \overline{p}}{\partial x} \qquad (15.8)$$

In a fully developed turbulent boundary layer, Reynolds stresses completely dominate over viscous stresses (Example Problem 15.8.1). This means that within the turbulent region of a boundary layer flow, the viscous part of (15.8) may be neglected so that it reduces to

$$\frac{\partial}{\partial z}\left(\varepsilon\frac{\partial \overline{u}}{\partial z}\right) = g\frac{\partial h}{\partial x} + \frac{1}{\rho}\frac{\partial \overline{p}}{\partial x} \qquad (15.9)$$

Let us now turn to the relation between the kinematic eddy viscosity $\varepsilon$ and the mean velocity $\overline{u}$.

## 15.2 PRANDTL'S MIXING-LENGTH HYPOTHESIS

Consider a steady, fully developed boundary-layer flow in which, as above, the main streamwise motion is parallel to the $x$-axis. Under these conditions,

$$\overline{u} = \overline{u}(z); \qquad \overline{v} = \overline{w} = 0 \qquad (15.10)$$

which describes a unidirectional turbulent shear flow. The complex fluid motions within this flow in part involve motions of fluid lumps that continuously move transversely to the $x$-axis—this is, with some component of motion parallel to the $z$-axis (Figure 15.2). These fluid lumps may be envisioned as more or less intact parcels of fluid that retain their individuality, at least for brief periods. An

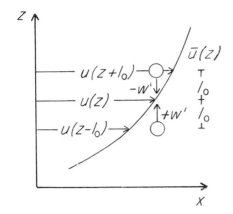

**Figure 15.2**  Definition diagram depicting motions of fluid lumps transverse to mean streamwise motion in presence of a mean velocity gradient.

individual lump therefore (momentarily) retains its streamwise momentum during its transverse motion. We may think of a fluid lump as an eddy.

Consider a position $z$ in this flow. Fluid lumps reaching $z$ from initial positions below $z$, on average, start with a streamwise velocity equal to

$$u(z - l_0) \approx \bar{u}(z) - l_0 \left( \frac{\mathrm{d}\bar{u}}{\mathrm{d}z} \right)_z \tag{15.11}$$

where $l_0$ is the distance traversed by the lumps, and the subscript $z$ denotes that this is the gradient of the average velocity $\bar{u}$ at $z$. Because a lump is assumed to retain its momentum in traversing the distance $l_0$, it reaches position $z$ with a smaller streamwise velocity than that prevailing at $z$, and therefore, on average, induces a negative fluctuation $u'$ at $z$. Because $u' = u - \bar{u}$, this fluctuation $u' \approx -l_0 \mathrm{d}\bar{u}/\mathrm{d}z$. Notice that (15.11) is like a first-order Taylor expansion, although $l_0$ is envisioned here as being a finite, rather than infinitesimal, quantity. Similarly, fluid lumps reaching $z$ from initial positions above $z$, on average, start with a streamwise velocity equal to

$$u(z + l_0) \approx \bar{u}(z) + l_0 \left( \frac{\mathrm{d}\bar{u}}{\mathrm{d}z} \right)_z \tag{15.12}$$

Such lumps reach position $z$ with a larger streamwise velocity than that prevailing at $z$, and therefore, on average, induce a positive fluctuation $u' \approx l_0 \mathrm{d}u/\mathrm{d}z$ at $z$. In general the fluctuation $u'$ at $z$ may be written as

$$u' \approx l_0 \frac{\mathrm{d}\bar{u}}{\mathrm{d}z} \tag{15.13}$$

where now $l_0$ may be positive (for a lump arriving from above $z$) or negative (for a lump arriving from below $z$).

A fluid lump arriving at $z$ from below with a positive component of velocity is by definition associated with a positive fluctuation $w'$, whereas a fluid lump arriving at $z$ from above is associated with a negative fluctuation $w'$. Therefore, a positive fluctuation $w'$ is, on average, associated with a negative fluctuation $u'$ at $z$, and a negative fluctuation $w'$ is, on average, associated with a positive fluctuation $u'$. Fluid arriving at $z$ from a transverse position must displace (or replace) fluid at $z$. By conservation of mass we may therefore assume that $w'$ is of the same order as

$u'$; thus

$$w' = -cu' \approx -cl_0 \frac{d\bar{u}}{dz} \qquad (15.14)$$

where $c$ is an unknown coefficient, and the last part of (15.14) is obtained from (15.13). The product $u'w'$ is then given by the product of (15.13) and (15.14):

$$u'w' = -cl_0 \left(\frac{d\bar{u}}{dz}\right)^2 \qquad (15.15)$$

Applying rules of (14.7) to obtain the time average of (15.15), and combining $c$ and $l_0$ into a single quantity $l$,

$$\overline{u'w'} = -l^2 \left(\frac{d\bar{u}}{dz}\right)^2 \qquad (15.16)$$

The quantity $l$ is referred to as *Prandtl's mixing length,* and has a simple heuristic interpretation: With reference to Figure 15.2, where $l \sim l_0$, the mixing length $l$ is the distance parallel to the $z$-axis that a fluid lump must, on average, traverse to provide the mean (positive or negative) fluctuation $u'$ at $z$. Comparing (15.16) with (15.6), we may write

$$\tau'_{zx} = -\rho \overline{u'w'} = \rho l^2 \left(\frac{d\bar{u}}{dz}\right)^2 \qquad (15.17)$$

Because the sign of $\tau'_{zx}$ must change with the sign of $d\bar{u}/dz$, this should actually be written as

$$\tau'_{zx} = \rho l^2 \left|\frac{d\bar{u}}{dz}\right| \frac{d\bar{u}}{dz} \qquad (15.18)$$

Comparing this further with (15.6), evidently

$$A = \rho l^2 \left|\frac{d\bar{u}}{dz}\right|; \qquad \varepsilon = l^2 \left|\frac{d\bar{u}}{dz}\right| \qquad (15.19)$$

Equation (15.18) is referred to as *Prandtl's mixing-length hypothesis.* The mixing length $l$ is still unknown. Like the eddy viscosity $\varepsilon$, it is not a property of the fluid, but rather is a function of coordinate position as we will see below.

   Notice that, according to (15.19), the eddy viscosity vanishes when the mean velocity gradient $d\bar{u}/dz$ goes to zero, for example, at the axis of a conduit. Prandtl recognized this as problematic since the fluctuating velocity components do not vanish at such positions in a turbulent flow. Readers are urged to examine Schlichting (1979, pp. 583–85) regarding this point.

## 15.3 MIXING-LENGTH AND EDDY VISCOSITY DISTRIBUTIONS

Consider a steady, fully turbulent flow within a conduit of uniform half-width $R$, where the boundaries are parallel to the $x$-axis, and the origin $z = 0$ is placed at one boundary. For simplicity assume that the flow is induced by a constant pressure gradient $\partial \overline{p}/\partial x = C$, neglecting gravity. Let us now rewrite (15.8) as

$$\frac{\partial T}{\partial z} = \frac{\partial \overline{p}}{\partial x} = C \tag{15.20}$$

where $T = \tau_{zx} + \tau'_{zx}$ denotes the total stress. Integrating this,

$$T(z) = Cz + C_0 \tag{15.21}$$

Symmetry of the flow suggests that the stress at the conduit axis $T(R) = 0$. Let us also specify that $T(0) = T(2R) = \tau_0$; thus $\tau_0$ denotes the stress at each of the two boundaries. From these boundary conditions, $C_0 = \tau_0$ and $C = -\tau_0/R$. This suggests that the total stress $T$ is a linear function over the half-width $R$:

$$T = \tau_0 \left( 1 - \frac{z}{R} \right) \tag{15.22}$$

Before considering this further, let us momentarily examine the boundary stress $\tau_0$. Since the fluctuating components $u'$ and $w'$ must vanish at the boundary, the Reynolds stress $\tau'_{zx} = -\overline{\rho u' w'}$ must also. Thus the only stress acting on the boundary ($z = 0$) is the viscous stress $\tau_{zx} = \mu(d\overline{u}/dz)_0$; and very close to the boundary, well within the viscous sublayer, the total stress $T = \tau_{zx}$. Away from the boundary where viscous stresses are negligible, the total stress $T$ becomes equal to the Reynolds stress $\tau'_{zx}$. In approaching the boundary, therefore, the viscous stress must increase to compensate the decreasing Reynolds stress. This transition must be smooth. We may therefore think of $\tau_0$ as a Reynolds stress that would result in the limit as the linear stress distribution (15.22) is projected to the boundary (more on this below). For distances sufficiently far from the boundary, then, (15.22) may be written as

$$\tau'_{zx} = \tau_0 \left( 1 - \frac{z}{R} \right) \tag{15.23}$$

Equating (15.17) and (15.23) and rearranging,

$$l = u_* \sqrt{1 - \frac{z}{R}} \left( \frac{d\overline{u}}{dz} \right)^{-1} \tag{15.24}$$

Here $u_* = \sqrt{\tau_0/\rho}$ is referred to as the *shear velocity* or *friction velocity*. The shear velocity is characteristic of the overall flow since it is defined with respect to the boundary stress $\tau_0$, which must be equal in magnitude to the total force, per unit area of boundary, that is driving the flow (Example Problem 15.8.1).

Consider experiments in which the velocity distribution $\overline{u}(z)$ is obtained from measurements in a conduit. The distribution $\overline{u}(z)$ can then be used together with measured value of $u_*$ to obtain a distribution of $l$ based on (15.24). Such measurements were obtained by J. Nikuradse for smooth boundaries, and for boundaries roughened with a uniform layer of sand (Figure 15.3). On empirical grounds, the distribution of $l$ is given by

$$\frac{l}{R} = 0.14 - 0.08 \left( 1 - \frac{z}{R} \right)^2 - 0.06 \left( 1 - \frac{z}{R} \right)^4 \tag{15.25}$$

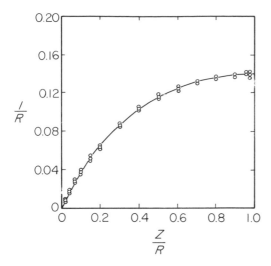

**Figure 15.3** Distribution of Prandtl's mixing length $l$ next to a solid boundary; adapted from Schlichting (1979).

It is left as an exercise to demonstrate that for small $z$ (close to the boundary) this can be simplified to

$$l \approx 0.4z \qquad (15.26)$$

The factor 0.4 in (15.26), which we will denote by $\kappa$, is referred to as the *von Kármán constant*. (It is now generally accepted that $\kappa = 0.41$.) We will return to this constant below.

Recalling the heuristic interpretation of $l$ above, the distribution of $l$ (Figure 15.3) suggests that the distance a fluid lump must, on average, traverse to provide the mean (positive or negative) fluctuation $u'$ at any specified $z$ increases with distance from the boundary. Since a fluid lump is envisioned as maintaining its initial streamwise velocity while traversing $l$, this suggests that, based on (15.17) and (15.23), either the intensity of the fluctuations or the mean velocity gradient, or both, decrease with distance from the boundary. (Why?)

By a similar approach, we can substitute (15.24) into (15.19) to obtain for the kinematic eddy viscosity:

$$\varepsilon = u_*^2 \left(1 - \frac{z}{R}\right)\left(\frac{d\bar{u}}{dz}\right)^{-1} \qquad (15.27)$$

Let us divide this by the product $u_* R$:

$$\varepsilon_R = \frac{\varepsilon}{u_* R} = \frac{u_*}{R}\left(1 - \frac{z}{R}\right)\left(\frac{d\bar{u}}{dz}\right)^{-1} \qquad (15.28)$$

Here $\varepsilon_R = \varepsilon/u_* R$ is referred to as the *dimensionless kinematic eddy viscosity*. A plot of $\varepsilon_R$ against the dimensionless distance $z/R$ (Figure 15.4) reveals two important points. First, the eddy viscosity exhibits a maximum midway between the boundary and the conduit axis. As we will see below, the mean velocity gradient $d\bar{u}/dz$ at this position is less than the gradient near the boundary but greater than that near the conduit axis. This suggests that, according to (15.19), the mixing length must be relatively large at this intermediate position, which, in turn, suggests that

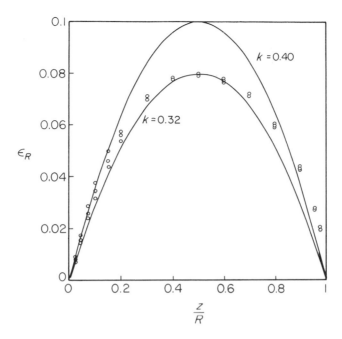

**Figure 15.4**  Distribution of dimensionless kinematic eddy viscosity $\varepsilon_R$ next to a solid boundary; adapted from Schlichting (1979).

strong turbulent mixing occurs here. Second, the eddy viscosity does not return to zero at the conduit axis (where $d\bar{u}/dz = 0$) as is implied by (15.28).

Knowing how either $l$ or $\varepsilon_R$ varies with coordinate position $z$ opens the possibility of obtaining an expression for the mean velocity distribution $\bar{u}(z)$ using (15.24) or (15.28). Suppose, for example, that we approximate the function $\varepsilon_R$ by

$$\varepsilon_R = k\frac{z}{R}\left(1 - \frac{z}{R}\right) \tag{15.29}$$

where $k$ is a fitting coefficient. With $k = 0.32$, the fit is faithful to the data near $z/R = 0.5$, but otherwise underestimates measured values of $\varepsilon_R$. With $k = 0.40$, the fit overestimates data near $z/R = 0.5$, but is faithful to data for $z/R < 0.3$, and only slightly underestimates measured values for $z/R > 0.7$. Based on this we might choose a happy medium of, say, $k = 0.36$. In any case, equating (15.28) and (15.29) and rearranging,

$$\frac{d\bar{u}}{dz} = \frac{u_*}{k}\frac{1}{z} \tag{15.30}$$

This suggests that the mean velocity gradient is a function of the shear velocity $u_*$ and position $z$, and decreases with increasing distance from the boundary. (This conclusion can be obtained directly from dimensional analysis; see Tritton [1988], pp. 341–42.) The next step would involve separating (15.30) and integrating this with respect to $z$ to obtain an expression for the velocity profile $\bar{u}(z)$, assuming $k$ is known. We will see below using a different approach that $k$ is usually taken to equal the von Kármán constant; that is, $k = \kappa = 0.4$.

We started here by assuming that the mean velocity profile $\bar{u}(z)$ was obtained empirically, then found expressions for $l$ and $\varepsilon_R$, which provided insight regarding the concepts of mixing-length and eddy viscosity introduced by Prandtl and others. Of more general value would be to obtain one or both of these expressions independently of empirical measurements. This has been attempted using several approaches, notably early on by Prandtl and von Kármán; we will turn to Prandtl's treatment next. Students are also urged to examine this problem as treated by Schlichting (1979, pp. 606–7) regarding use of (15.25) to obtain a general expression for $\bar{u}(z)$, and as treated by by Smith and McLean (1984) in the context of turbulent mixing within a natural stream whose bed possesses roughness due to ripples and dunes.

## 15.4 LOGARITHMIC VELOCITY LAW

Let us reconsider (15.17) and, following Prandtl, assume that the mixing length $l$ is given by

$$l = \kappa z \tag{15.31}$$

This simple proportionality is intuitively appealing. Inasmuch as turbulent fluctuations must vanish at the boundary ($z = 0$), $l$ must also; and it is reasonable to suspect that $l$ ought to steadily increase with distance $z$ from the boundary. Notice that this is essentially equivalent to the result embodied in (15.26) that Nikuradse obtained experimentally. We may therefore anticipate that (15.31) is reasonable only near the boundary. Substituting (15.31) into (15.17),

$$\tau'_{zx} = \rho \kappa^2 z^2 \left(\frac{d\bar{u}}{dz}\right)^2 \tag{15.32}$$

Prandtl then made an interesting, and perhaps surprising, assumption regarding (15.32): that the stress $\tau'_{zx}$ is constant throughout the flow and equal to the stress at the boundary $\tau_0$. According to the definition of the shear velocity, $\tau_0 = \rho u_*^2$. Equating this with (15.32) and rearranging,

$$\frac{d\bar{u}}{dz} = \frac{u_*}{\kappa} \frac{1}{z} \tag{15.33}$$

Notice that this is equivalent to the result (15.30) obtained by assuming a distribution of eddy viscosity given by (15.29), where the constant $k$ in (15.30) is equal to $\kappa$. Rearranging (15.33) and integrating,

$$\bar{u}(z) = \frac{u_*}{\kappa} \ln z + C \tag{15.34}$$

The constant of integration $C$ is determined by assuming that the velocity $\bar{u} = 0$ at a position $z_0$ near the boundary. Thus,

$$\bar{u}(z) = \frac{u_*}{\kappa}(\ln z - \ln z_0) \tag{15.35}$$

Consider a smooth boundary. The position $z_0$ is on the order of the thickness $\delta$ of the viscous sublayer. Dimensional analysis suggests that $z_0$ is proportional to the ratio $\nu/u_*$, which has the dimensions of length; thus

$$z_0 = k_0 \frac{\nu}{u_*} \tag{15.36}$$

where $k_0$ is a dimensionless constant that must be determined experimentally. Substituting this into (15.35), then making use of rules of logarithms,

$$\bar{u}(z) = \frac{u_*}{\kappa} \left( \ln \frac{u_* z}{\nu} - \ln k_0 \right) \tag{15.37}$$

This is referred to as the *logarithmic velocity law*. Notice that the ratio $u_* z/\nu$ has the form of a Reynolds number.

The velocity law (15.37) is valid for that part of a turbulent boundary flow where viscous forces can be neglected. It cannot be expected to pertain to the viscous sublayer. In this regard, let us rewrite (15.37) as

$$\frac{\bar{u}}{u_*} = A \ln \frac{u_* z}{\nu} + B \tag{15.38}$$

where

$$A = \frac{1}{\kappa}; \qquad B = -\frac{1}{\kappa} \ln k_0 \tag{15.39}$$

Consider a plot of $\bar{u}/u_*$ versus $\ln(u_* z/\nu)$ (Figure 15.5). Based on the straight-line plot of (15.38), the slope $A = 2.5$, giving $\kappa = 0.4$. We also can obtain $B = 5.5$ from the intercept. Assuming that $\kappa = 0.4$, this gives $k_0 = 0.11$. This plot also reveals three regions in the flow. In the region where $u_* z/\nu > 30$, flow is fully turbulent and apparent stresses dominate over viscous stresses. In the transitional

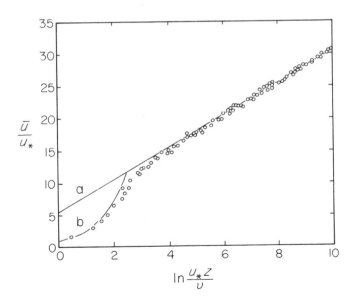

**Figure 15.5**  Plot of mean velocity versus distance from boundary showing three regions in boundary layer: lines represent logarithmic velocity law (a) and linear velocity profile within viscous sublayer (b); adapted from Schlichting (1979).

region where $5 < u_* z/\nu < 30$, both viscous and apparent stresses are important. In the region where $u_* z/\nu < 5$, viscous forces dominate; this is the viscous sublayer.

Within the viscous sublayer the shear stress at the boundary is $\tau_0 = \mu(d\bar{u}/dz)_0$. From above, $\tau_0 = \rho u_*^2$. Assuming $\tau_0$ is constant throughout the viscous sublayer in analogy with Couette flow, we may equate these expressions, then rearrange them to obtain the linear velocity distribution within the viscous sublayer:

$$\bar{u} = \frac{u_*^2}{\nu} z \qquad (15.40)$$

Notice that, although the viscous sublayer is thin, a significant proportion of the total change in the mean velocity over the entire profile occurs within the sublayer (Figure 15.5).

Let us evaluate the constant $C$ in (15.34) in a way that does not require specifying anything about the position $z_0$. To do this we use the boundary condition that $\bar{u} = \bar{u}_m$, where $\bar{u}_m$ denotes the maximum mean velocity at, say, the conduit axis ($z = R$). Then (15.34) becomes

$$\frac{\bar{u}_m - \bar{u}}{u_*} = \frac{1}{\kappa} \ln \frac{z}{R} \qquad (15.41)$$

Consider a plot of $(\bar{u}_m - \bar{u})/u_*$ versus $z/R$ (Figure 15.6). The data in this figure are for both smooth and rough boundaries. Evidently the logarithmic law for flow outside the viscous sublayer remains valid for both. When $\bar{u}/\bar{u}_m$ is plotted against $z/R$, however, the velocity profiles are different; the dimensionless velocity gradient $d(\bar{u}/\bar{u}_m)/d(z/R)$ associated with a rough boundary is, on average, smaller than that associated with a smooth boundary. This difference resides in the effect of the boundary roughness. Individual bumps composing the roughness extend sufficiently far into the flow that they induce drag on the flow (in the same sense that a drag force exists with flow around a sphere), and thereby contribute to the frictional resistance of the boundary to flow. To take this into account requires modifying the description of the position $z_0$.

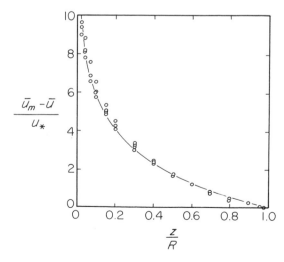

**Figure 15.6** Plot of logarithmic velocity profiles with data for smooth and rough boundaries; adapted from Schlichting (1979).

## 15.5 TURBULENT FLOW OVER ROUGH BOUNDARIES

### 15.5.1 Nikuradse sand roughness

Recall from Chapter 5 that we introduced the idea of *Nikuradse sand roughness* in the context of defining geometrical similitude. We envisioned an experiment where well-sorted, equidimensional sand grains are closely, but randomly, packed in a single-grain layer and glued to an underlying smooth surface. The roughness thus produced is (statistically) geometrically similar such that a single linear dimension equal to one grain diameter, the roughness height $k_s$, is sufficient to distinguish one surface prepared in this manner from another. We will initially assume in the developments below that the solid boundary possesses a Nikuradse sand roughness.

Reconsider (15.35). For a rough boundary, $z_0$ is on the same order as the roughness height $k_s$, and we may therefore assume that

$$z_0 = k_1 k_s \tag{15.42}$$

where $k_1$ is a dimensionless coefficient. Substituting this into (15.35) and rearranging,

$$\frac{\bar{u}}{u_*} = \frac{1}{\kappa}\left(\ln\frac{z}{k_s} - \ln k_1\right) \tag{15.43}$$

where $k_1$ must depend on the geometrical nature of the roughness. This is the *logarithmic velocity law for rough boundaries*.

Let us rewrite (15.43) as

$$\frac{\bar{u}}{u_*} = \frac{1}{\kappa}\ln\frac{z}{k_s} + B_s \tag{15.44}$$

where

$$B_s = -\frac{1}{\kappa}\ln k_1 \tag{15.45}$$

Before evaluating $B_s$ (and $k_1$) in relation to rough boundaries, it is reasonable to first insist that (15.44) ought to reduce to (15.38) for the case of a smooth boundary. Equating (15.38) and (15.44), solving for $B_s$, then using rules of logarithms,

$$B_s = \frac{1}{\kappa}\ln\frac{u_* k_s}{\nu} + B \tag{15.46}$$

which gives the value of $B_s$ for a smooth boundary. Notice that the ratio $u_* k_s/\nu$ is a Reynolds number for sand roughness defined in terms of the shear velocity $u_*$. This suggests that $B_s$ is, in general, a function of the Reynolds number $u_* k_s/\nu$.

Consider a plot of $B_s$ versus $\ln(u_* k_s/\nu)$ (Figure 15.7). This figure reveals three regimes. For $u_* k_s/\nu < 5$, the roughness is sufficiently small that it is contained within the viscous sublayer; this is a *hydrodynamically smooth regime* for which the value of $B_s$ is given by (15.46). For $5 < u_* k_s/\nu < 70$, the roughness partially protrudes outside the viscous sublayer, and resistance to flow increases due to form drag on individual bumps composing the roughness; this is a *transitional regime*. For $u_* k_s/\nu > 70$, the roughness extends beyond the viscous sublayer, and resistance to flow is virtually entirely due to form drag on the roughness bumps; this is a *hydrodynamical rough regime*. In this regime, $B_s$ is essentially constant,

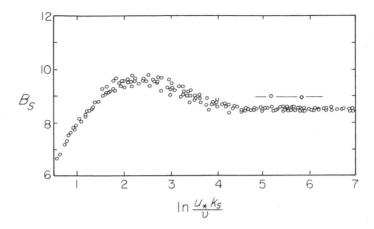

**Figure 15.7**   Variation in coefficient $B_s$ with Reynolds number $u_* k_s / \nu$; adapted from Schlichting (1979).

independent of the Reynolds number. The numerical value of $B_s = 8.5$, according to (15.45), gives $k_1 = 0.033$, assuming $\kappa = 0.4$.

Here it is worthwhile to emphasize an important point: the numerical values for $u_* k_s / \nu$, $B_s$, and $k_1$ presented above are correct only for Nikuradse sand roughness. To clarify this, let us recall a caveat mentioned in Chapter 5: that Reynolds numbers defined for geometrically dissimilar systems are not directly comparable. A specific value of $u_* k_s / \nu$ associated with a Nikuradse roughness with height $k_s$ does not represent the same hydrodynamical conditions that are associated with an identical value of $u_* k_s / \nu$ obtained for a boundary whose roughness does not resemble a Nikuradse roughness. In particular, the roughness height $k_s$ as used above represents a geometrical abstraction that pertains only to sand grains closely packed in a single layer. To illustrate this, consider artificial roughness consisting of equal-diameter spheres glued in regular arrangements on a flat surface, where the spacing of the spheres systematically varies (Figure 15.8). As defined above,

| DIMENSIONS | $D$ (cm) | $d$ (cm) | $k$ (cm) | $k_s$ (cm) | $B_s$ | $k_1$ |
|---|---|---|---|---|---|---|
| 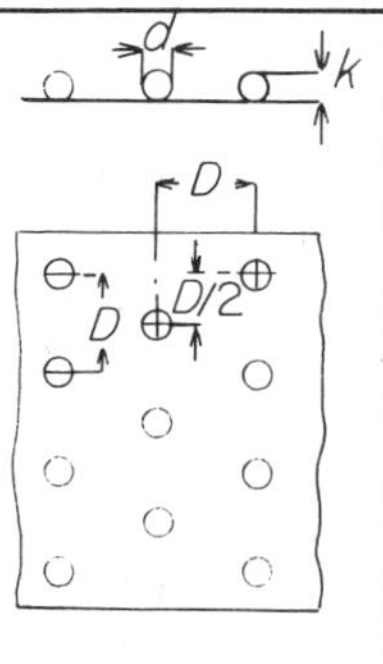 | 4 | 0.41 | 0.41 | 0.093 | 12.2 | 0.0076 |
| | 2 | 0.41 | 0.41 | 0.344 | 8.9 | 0.028 |
| | 1 | 0.41 | 0.41 | 1.26 | 5.7 | 0.10 |
| | 0.6 | 0.41 | 0.41 | 1.56 | 5.2 | 0.12 |
| | densest arrgt. | 0.41 | 0.41 | 0.257 | 9.7 | 0.021 |
| | 1 | 0.21 | 0.21 | 0.172 | 9.0 | 0.027 |
| | 0.5 | 0.21 | 0.21 | 0.759 | 5.3 | 0.12 |

**Figure 15.8**   Artificial roughness arrangements with associated values of $B_s$ and $k_1$; adapted from Schlichting (1979).

$k_s$ is the same for each arrangement. The coefficients $B_s$ and $k_1$, however, vary among the arrangements based on measurements of flow over them. Since these are geometrically dissimilar roughnesses, a given value of the Reynolds number $u_* k_s / \nu$ therefore does not ensure that a condition of dynamical similitude exists among the arrangements. This suggests, conversely, that similar values of coefficients ($B_s$ and $k_1$) ought to be obtained for artificial roughnesses that are geometrically similar. Students are urged to examine Example Problem 15.8.4 regarding this point.

## 15.5.2 Natural roughness

The geometry of natural-surface roughness generally is more complex than Nikuradse sand roughness. For example, the roughness on a rock-fracture surface involves small protrusions created by individual mineral grains, and larger bumps, cavities, and notches related to cleavage planes and fracture textures. Desert pavements consist of mixtures of gravel and pebbles with interdispersed sand, which provide a range of bump sizes. Beds of alluvial streams likewise consist of mixtures of sediment sizes, and in addition possess bumpiness created by ripples, dunes, and particle clusters. If a generalization is possible, it is that such surfaces typically possess roughness at multiple scales.

The immediate implication is that a single linear dimension analogous to the roughness height $k_s$ generally is not adequate to characterize the roughness of such surfaces for use in, say, (15.44). The roughness of a smooth, sand stream bed is, perhaps, close to a Nikuradse roughness. But a natural, sand stream bed rarely is smooth; ripples and dunes normally occur and significantly influence the turbulence field above them (Example Problem 15.8.2). Likewise, poorly sorted sediment mixtures on a stream bed do not mimic a Nikuradse roughness.

The logarithmic velocity law nonetheless provides a good approximation of the mean velocity profile for unidirectional flows next to natural surfaces when certain conditions are satisfied. One condition is that the *relative smoothness* of the flow must be large. The relative smoothness is defined as the ratio of the turbulent boundary-layer thickness to a characteristic bump size composing the roughness. In the case of channel flow this corresponds to the ratio of the flow depth to the bump size. Thus flows in which the roughness bumps are of the same order as the flow depth are excluded. (The reciprocal of the relative smoothness is referred to as the *relative roughness*.) A second condition is that the local velocity $u(z)$ must be treated as an average over the $xy$-field of a *homogeneous roughness*. This idea will be clarified below. Let us meanwhile revisit the idea of a fully developed boundary layer. For simplicity we will consider the flow and roughness within a stream channel, although the developments equally pertain to other flows and rough surfaces. Here it will be useful to distinguish between roughness that is random in character (for example, the roughness produced by coarse gravel) and roughness whose bumps possess a sense of geometrical regularity (for example, ripples on a sand bed).

Envision an idealized situation in which two geometrically similar rough surfaces with different characteristic linear dimensions (for example, $k_s$) are juxtaposed in a streamwise direction. A steady, fully developed turbulent boundary layer exists over the upstream surface, as manifest by a steady velocity profile $\bar{u}(z)$ and a steady turbulence field that does not vary in the streamwise direction. As flow reaches the downstream surface, the "new" roughness begins to influence the turbulence and velocity fields above it, the boundary stress $\tau_0$ changes, and the overall flow accelerates (or decelerates) in response. After some distance downstream the turbulence and velocity fields adjust to the downstream roughness in the sense of reaching a new steady condition; derivatives of mean and fluctuating quantities with respect to the streamwise direction have vanished. (In general, such an adjustment is more rapid with increasing roughness.) In this context, then, a fully developed boundary layer refers to the idea that steady conditions, compatible with the local boundary conditions, have been reached. But inasmuch as the geometrical properties of roughness vary over space in real channels, we may envision that the turbulence structure is in a state of continuous adjustment from one position to another.

Local variations in turbulence structure associated with individual bumps— individual particles, particle clusters, ripples, dunes—are related to local flow accelerations (and decelerations) in the vicinity of the bumps, flow separation in the lees of the bumps, and turbulent wakes behind the bumps (Example Problem 11.4.2). Zones of separation, in particular, are sites where turbulence can be unusually complex, as characterized by large turbulence intensities, strong vertical fluctuating motions, and vortex generation (Example Problem 15.8.2). For large relative smoothness, on the order of 10 or more, the turbulence structure close to the bed is strongly influenced by turbulence generated by bumps on the bed, whereas the turbulence structure far above the bed is determined more by dynamical effects that generate eddies within the upper part of the turbulent layer. (Nonetheless, eddies generated at the bed can make their way upward and affect the turbulence structure high in the flow.) As a consequence the velocity profile $\bar{u}(z)$ locally can deviate from the logarithmic form (15.44) near the bed, while conforming reasonably well to a logarithmic form well above the bed (Figure 15.9). As the relative smoothness decreases to a value on the order of 10 or less, the influence of bed roughness becomes more important throughout the flow column, and the mean velocity profile $\bar{u}(z)$ locally can deviate significantly from a logarithmic form (Figure 15.9). But let us return to the case of large relative smoothness.

At the grain scale, the small-scale velocity and turbulence structures in the vicinity of an individual bump are different from those in the vicinity of a neighboring bump. Enlarging our scale of view, however, we may envision the turbulence structure as being homogeneous when suitably averaged over time *and* space, at a scale much larger than individual grains. In particular, we may envision the profile $\bar{u}(z)$ near the bed as being an ensemble average obtained from the infinite set of such profiles located over the $xy$-field, in which case $\bar{u}(z)$ may possess a

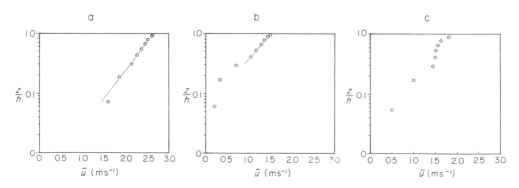

**Figure 15.9**   Plot of velocity data for flow over a very rough bed illustrating logarithmic form near surface with poor fit near bed, (*a*) and (*b*), and nonlogarithmic form over entire fluid column (*c*).

logarithmic form with which we associate a uniform boundary stress $\tau_0$, as with a Nikuradse sand roughness. Thus, whereas the stress $\tau_0$ varies from point to point at the grain scale, it becomes a macroscopically uniform quantity at the scale of a homogeneous Nikuradse-like texture.

Consider in this context a unidirectional flow over a rough channel bed composed of a coarse gravel mixture, where bedforms like ripples and dunes are absent. Assume that the relative smoothness of the flow is large. Unlike grains composing a Nikuradse roughness, the bed-surface particles do not rest on a common plane. With a mixture of particle sizes, moreover, no individual particle diameter characterizes the bump size on the bed in the sense that $k_s$ does for a Nikuradse sand roughness. It has therefore become conventional to choose a single grain diameter, based on the statistical distribution of sizes present, to serve as a linear dimension to characterize the roughness. Normally this is selected to be the diameter $D$ that represents one (positive) standard deviation from the mean diameter, or $D_{84}$. The subscript denotes that 84 percent of the particles are smaller than $D_{84}$. This choice is heuristically based (although Wiberg and Smith [1991] present some physical basis for this choice). Large bumps, presumably associated with large particles, account for most of the boundary drag on flow. But the largest grains on a bed normally represent a small proportion of the distribution of grain sizes. Conversely, the proportion represented by small grains is large, but these individually do not account for much drag. Thus $D_{84}$ is a compromise, intended to represent the particles that are both sufficiently large and numerous to contribute most to flow resistance. Then $D_{84}$ is used in place of $k_s$ in (15.44), where the coefficient $B_s$ must be determined empirically since a value of $B_s = 8.5$ applies only to Nikuradse roughness. Two noteworthy assumptions are implicit in this approach. First, the choice of a single linear dimension based on grain size is equivalent to assuming that sediment mixtures for different stream sites are geometrically similar, albeit possessing different mean diameters (Example Problem 5.7.3). Second, it is assumed that geometrically similar mixtures produce geometrically similar bed surface textures. Although this approach is useful in certain practical situations, these conditions are rarely satisfied. In this regard, various authors have recommended using $D_{50}$, $D_{84}$, $D_{90}$, and linear combinations of these, but with inconsistent results.

Consider now the roughness of two-dimensional ripples on a sand bed. Viewed at the grain scale, individual bumps produced by sand particles vary in size and shape. Moreover, no small cluster of five or six grains produces a bump texture that is geometrically similar to that of any other cluster. But when viewed at a scale much larger than this (but smaller than a ripple), the sand produces a roughness texture that is statistically geometrically similar, or homogeneous, in the sense of a Nikuradse sand roughness. A first characteristic dimension, for example, a roughness height analogous to the Nikuradse height $k_s$, therefore is associated with the roughness of the sand grains. At the scale of an individual ripple, however, roughness is characterized in terms of the dimensions of the ripple. But no two ripples are geometrically similar. Nonetheless, by again increasing our scale of view, a field of ripples might provide a roughness texture that is statistically geometrically similar, or homogeneous, from one location to another. Characteristic dimensions are, say, the average ripple height and spacing; or if spacing is reasonably consistent, then the average height might suffice.

For reasons described above, we would expect that flow over a sand texture, in absence of ripples, would possess a near-bed turbulence field like that over Nikuradse roughness. The time-averaged turbulence structure, however, varies from stoss to lee of a ripple, despite the fact that the grain-scale roughness is homogeneous from stoss to lee. At the scale of a ripple field, however, we may envision the turbulence structure as being homogeneous when averaged over time and space, at a scale much larger than individual ripples, and we may therefore associate a macroscopic value $\tau_0$ with the ripple field. Thus, whereas the stress $\tau_0$ varies from point to point at the ripple scale, it becomes a macroscopically uniform quantity at the scale of a homogeneous ripple field, such that the logarithmic law can be used to characterize the mean velocity profile for large relative smoothness. Conversely, at a larger scale, the average ripple spacing may change in a streamwise direction, implying heterogeneous roughness at this larger scale.

Whereas the logarithmic law may in practice adequately characterize the mean velocity profile for large relative smoothness, this condition imposes a severe restriction since many flows of geological interest involve relative smoothnesses on the order of 10 or smaller. The logarithmic law therefore must be viewed as a special case (in the limit of large relative smoothness) of some more general description of the mean velocity field above a rough bed. In addition, local near-bed velocity, turbulence, and stress conditions are particularly significant in terms of describing the forces involved in sediment transport. The logarithmic law based on an ensemble average over space reveals little about these conditions.

For these reasons, attempts have been made to distinguish contributions to the bed stress $\tau_0$ due to the various morphological features on a channel bed—for example, the contribution of form drag associated with dunes and the "skin friction" associated with the Nikuradse-like roughness of dune surfaces. Students are urged to examine relevant papers in the Reading list regarding this point. In the case of the random roughness of coarse sediment, it is, in principle, possible to integrate the drag over all bump sizes composing a rough surface. Recall that the coefficient of drag $C_D$ on a sphere becomes approximately constant ($C_D \approx 0.4$) at high Reynolds numbers (Figure 6.9). This means that the force of drag is proportional to the square of the flow velocity. (Why?) Reynolds numbers associated with flow around bumps on a stream bed typically are sufficiently high that a similar relation can be expected to hold for these bumps. The drag associated with an individual bump extending from a rough bed into the flow above it, however, is not the same as the drag on an isolated bump (without surrounding bumps) in the sense of an isolated sphere. In addition, the drag on a bump varies with the velocity fluctuations associated with turbulence. Thus, although there is uncertainty in the precise values of the coefficient of drag that ought to be associated with bumps on a channel bed, the problem of obtaining a spatially averaged form of the velocity profile $\bar{u}(z)$ by taking into account the drag on many bumps is nonetheless tractable. Students are urged to examine Wiberg and Smith (1991) regarding this point. We shall conclude this section by noting that describing the drag on individual particles as outlined above is one of the early steps in treating the mechanics of sediment motions on a stream bed.

## 15.6 PRODUCTION AND DISSIPATION OF TURBULENCE ENERGY

Recall from Chapter 14 that the fluctuating motions of turbulence continuously extract energy from the mean motion, ultimately dissipating it as heat via viscous friction. Our description of the kinetic energy of the fluctuating motions as embodied in (14.43) revealed that this dissipation, represented by terms involving squared gradients of fluctuating velocity components, is compensated by energy production, represented by terms involving products of Reynolds stresses and gradients of mean velocity components. The remaining terms in (14.43) represent transfers of energy within a flow by advective and diffusive processes. Let us now consider these mechanisms in relation to a steady, unidirectional shear flow.

Suppose as above that the mean streamwise motion of such a flow is parallel to the $x$-axis. With steady conditions, time-averaged fluctuating quantities may vary in a direction normal to the boundary, but terms in (14.43) involving gradients with respect to the $x$-axis and $y$-axis vanish. Moreover, since $\bar{v} = \bar{w} = 0$, and using (14.45), (14.43) reduces to

$$\frac{1}{2}\overline{w'\frac{\partial \xi}{\partial z}} + \frac{1}{\rho}\overline{w'\frac{\partial p'}{\partial z}} - \frac{1}{2}\nu\overline{\frac{\partial^2 \xi}{\partial z^2}} = -\overline{u'w'}\frac{\partial \bar{u}}{\partial z} - \nu\left[\overline{\left(\frac{\partial u'}{\partial z}\right)^2} + \overline{\left(\frac{\partial v'}{\partial z}\right)^2} + \overline{\left(\frac{\partial w'}{\partial z}\right)^2}\right]$$

(15.47)

where $\xi = \overline{u'^2} + \overline{v'^2} + \overline{w'^2}$ and the kinetic energy per unit volume of fluid associated with fluctuating motions is $\frac{1}{2}\rho\xi$. Terms on the left side of (15.47) represent transports of energy; the first term on the right side represents energy production; and the last set of terms represents dissipation.

Consider the production term in (15.47). With reference to Figures 14.3 and 15.6, the covariance $\overline{u'w'}$ is, in general, a negative quantity when the mean velocity gradient $\partial\bar{u}/\partial z$ is positive; thus the production term is positive. Moreover, the magnitudes of both the covariance $\overline{u'w'}$ and the gradient $\partial\bar{u}/\partial z$ increase toward the boundary, so most of the energy associated with fluctuating motions is generated close to the boundary. Since fluctuating motions vanish at the boundary, however, generation of turbulence energy must decrease close to (and within) the viscous sublayer then vanish at the boundary also. Conversely, the magnitude of the gradient $\partial\bar{u}/\partial z$ decreases away from the boundary, so energy production decreases away from the boundary also. From Chapter 11, the vorticity associated with the mean motion in this shear flow is equal to $-\partial\bar{u}/\partial z$. For this reason it is sometimes stated that turbulence energy is generated from near-boundary vorticity, which provides a context for the idea stated in the introductory paragraph of this chapter: that near-boundary velocity gradients provide a ready source of vorticity for eddies and eddy-like structures to develop in response to the destabilizing effects of inertial forces.

If energy production is greatest near the boundary, we may anticipate that any transports of this energy are mainly away from the boundary. Judging from Figure 14.3, the average quantities $\overline{u'^2}$, $\overline{v'^2}$, and $\overline{w'^2}$ have maxima very close to the boundary. Therefore the sum $\xi = \overline{u'^2} + \overline{v'^2} + \overline{w'^2}$ has a maximum near the boundary (Figure 14.5). Energy is transported outward from this position of maximum energy by turbulence, as represented by the terms on the left of (15.47).

Turning to the term in (15.47) representing dissipation of energy, observe that, regardless of the signs of the gradients in this term, the full term remains negative, which reflects that this term represents an extraction of energy from the kinetic energy balance. Moreover, since dissipation is the result of work performed against viscous friction, this extraction ought to be independent of the sense of shear (and therefore independent of the signs of the gradients of the fluctuating velocity components). Dissipation of energy, like production, is greatest near the boundary.

Thus, turbulence energy production is greatest near the boundary and decreases away from it. Dissipation of energy mainly occurs near the boundary and decreases away from it also. Whereas energy production near the boundary is nearly balanced by dissipation near the boundary, dissipation is greater than production in the outer part of the boundary layer. The deficit in energy in the outer part of the boundary layer is supplied by outward transport of excess energy produced near the boundary. Turbulence in the outer region therefore is supplied in part by turbulence generated in the inner region.

With a two-dimensional boundary layer, for example, a boundary layer that is developing downstream from the leading edge of plate, or the turbulent wake behind a blunt object, advective terms in (14.43) involving the mean motion also become important in the energy balance. Students are urged to consult Tritton (1988) and Ligrani and Moffat (1986) regarding this point.

## 15.7  TURBULENT FLOW AND DARCY'S LAW

Flow within a porous medium is, at the pore scale, a complex boundary-layer flow. We have previously seen, for example, that pore-scale velocity gradients associated with the no-slip condition at pore walls contribute to mechanical dispersion of solutes (Chapter 13). These velocity gradients likewise induce a drag between the fluid and solid phases (Example Problem 13.5.1). Our treatment of porous media flows in Chapter 13 centered on conditions for which we could assume validity of Darcy's law. Recall, however, that for sufficiently large Reynolds numbers defined by $\rho d q/\mu$ (where $d$ is average particle diameter and $q$ is specific discharge), inertial effects can become significant enough to induce turbulence, in which case Darcy's law is not valid.

Darcy's law, (13.1), is based on the assumption that a balance exists between the pressure and body forces inducing fluid motion, and the drag force that derives from viscous stresses integrated over a finite volume of porous material. It neglects any pore-scale variations in pressure associated with inertial effects. Suppose that the trajectory of a fluid particle moving along a tortuous path within a pore has at some point a curvature $C = 1/r$, where $r$ is the local radius of curvature of the trajectory. Associated with this curved trajectory is an inertial (centrifugal) force per unit volume equal to $\rho C u_\theta^2 = \rho u_\theta^2/r$, where $u_\theta$ is the tangential velocity component. This centrifugal force must be balanced by a centripetal force; this force is supplied by a pressure gradient directed toward the center of curvature. We may therefore envision that the fluid pressure $p$ varies in a complex manner at the pore scale; the pressure gradient in the average direction of flow consists of a macroscopic part as embodied in Darcy's law, and a complex part due to inertial effects associated with curved flow trajectories. The viscous force per unit volume is $\mu \partial^2 u_\theta/\partial r^2$. With $r$ on the order of $d$, and $u_\theta$ on the order $q/n$ (where $n$ is the porosity), the viscous force per unit volume is on the order of $\mu q/nd^2$, and the centripetal force supplied by the local pressure gradient is of order $\rho q^2/n^2 d$. We may therefore anticipate that inertial forces are of the same order as viscous forces when

$$\frac{\rho q^2}{n^2 d} \sim \frac{\mu q}{nd^2} \tag{15.48}$$

This in turn suggests that viscous forces dominate over inertial forces, such that Darcy 's law is valid, when

$$\frac{\rho q d}{\mu} < n \tag{15.49}$$

where the ratio on the left side is the Reynolds number $Re$ as defined in Chapter 13 and above.

With reference to Figure 13.1, a transitional regime exists where viscous and inertial forces are of the same order. The lower limit of this regime occurs at a value of $Re$ between one and 10, and the upper limit occurs at about $Re = 100$. For $Re$ less than the lower limit, viscous forces dominate, flow is laminar, and Darcy's law is valid. For $Re$ at the upper limit, inertial forces dominate and a transition to turbulence occurs. For $Re$ greater than this limit, flow is turbulent. Since turbulence extracts energy from the mean fluid motion, there is an apparent increase in the resistance to flow as measured by the mean motion $q$. This means that for a given hydraulic-head gradient the average fluid velocity will be less with turbulent flow than with viscous flow. It is left as an exercise to use typical values of $\mu$, $q$, and $d$ to illustrate that (15.49) is normally satisfied for groundwater flows within sedimentary materials, except for very coarse materials such as gravel. Students are urged to examine Bear (1972) regarding attempts to generalize Darcy's law to a form that incorporates inertial forces.

## 15.8 EXAMPLE PROBLEMS

### 15.8.1 Turbulent magma flow in a dike

Let us consider turbulent magma flow within a dike for comparison with the results of Example Problems 6.7.2 and 12.5.1 regarding viscous magma flow. For vertical $z$-axis, uniform dike half-width $b$, and steady flow in the mean, the appropriate momentum equation has the form of (15.9):

$$\frac{\partial}{\partial x}\left(\varepsilon \frac{\partial \overline{w}}{\partial x}\right) = g + \frac{1}{\rho}\frac{\partial \overline{p}}{\partial z} \tag{15.50}$$

Consider terms on the right side of (15.50). For reasons that will become clear below, let us describe these terms with respect to the boundary shear stress $\tau_0$. With reference to Figure 15.10, it is left as an exercise to obtain

$$g + \frac{1}{\rho}\frac{\partial \overline{p}}{\partial z} = -\frac{\tau_0}{\rho b} = -\frac{u_*^2}{b} \tag{15.51}$$

where $u_* = \sqrt{\tau_0/\rho}$ denotes the shear velocity. Substituting this into (15.50) and expanding the left side,

$$\varepsilon\frac{\partial^2 \overline{w}}{\partial x^2} + \frac{\partial \varepsilon}{\partial x}\frac{\partial \overline{w}}{\partial x} = -\frac{u_*^2}{b} \tag{15.52}$$

which requires that expressions for $\varepsilon$ and $\partial\varepsilon/\partial x$ be specified. Let us assume that the kinematic eddy viscosity $\varepsilon$ has the form of (15.29); then

$$\varepsilon = \kappa u_* x\left(1 - \frac{x}{b}\right); \qquad \frac{d\varepsilon}{dx} = \kappa u_*\left(1 - 2\frac{x}{b}\right) \tag{15.53}$$

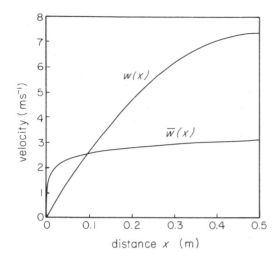

**Figure 15.10** Dike with logarithmic velocity profile $\overline{w}(x)$ and parabolic velocity profile $w(x)$ for the same boundary stress $\tau_0$.

Since the first expression in (15.53) is a symmetric function, we may place the origin $x = 0$ at either the dike wall or the dike axis. For convenience we will choose the dike wall. The expressions (15.53) may be substituted into (15.52) to obtain a differential equation that, in principle, can be solved for $\overline{w}$. Alternatively, analogous to the relation between (15.29) and (15.33), let us anticipate that (15.52) is satisfied by

$$\frac{d\overline{w}}{dx} = \frac{u_*}{\kappa}\frac{1}{x}; \qquad \frac{d^2\overline{w}}{dx^2} = -\frac{u_*}{\kappa}\frac{1}{x^2} \qquad (15.54)$$

where it is left as an exercise to demonstrate that this indeed is the case. For simplicity, let us assume that the dike walls are smooth. Following developments presented in Section 15.4, the velocity distribution $\overline{w}(x)$ is then given by

$$\overline{w}(x) = u_*\left(\frac{1}{\kappa}\ln\frac{u_*x}{\nu} + B\right) = \sqrt{\frac{\tau_0}{\rho}}\left(\frac{1}{\kappa}\ln\sqrt{\frac{\tau_0}{\rho}}\frac{x}{\nu} + B\right) \qquad (15.55)$$

where we will assume that $B = 5.5$.

Turning now to the same problem assuming viscous flow, the momentum equation becomes

$$\nu\frac{\partial^2 w}{\partial x^2} = -\frac{\tau_0}{\rho b} = -\frac{u_*^2}{b} \qquad (15.56)$$

from which it is left as an exercise to demonstrate that the velocity distribution $w(x)$ is given by

$$w(x) = \frac{\tau_0}{\rho\nu}\left(x - \frac{x^2}{2b}\right) \qquad (15.57)$$

A plot of $\overline{w}(x)$ and $w(x)$ for the same value of $\tau_0$ indicates that $\overline{w}(x)$ is significantly less than $w(x)$ over most of the dike width (Figure 15.10). Since $\tau_0$ is the same, this means that the pressure gradient inducing upward flow is less effective in the case of turbulent flow. Turbulence extracts energy from the flow; thus the pressure potential is consumed not only by viscous friction, but also by transformation to the kinetic energy of the fluctuating motions. With increasing Reynolds number, the difference between $\overline{w}(x)$ and $w(x)$ becomes more significant.

### 15.8.2  Turbulence field over two-dimensional dunes

Beds of alluvial channels typically consist of bedforms—ripples, dunes, and bars—whose sizes vary from small to very large. One objective of studies concerned with sediment motion and bedform dynamics is to understand how these bedforms interact with the mean flow and turbulence field above them. Let us examine a part of this problem, going beyond unidirectional flows, by turning to experiments performed by Nelson, McLean, and Wolfe (1993), who used laser-Doppler techniques to measure fluctuating velocity components in the vicinity of static two-dimensional dunes (Figure 15.11).

Consider a straight channel with uniform width. Let $x$ denote a streamwise coordinate that is positive downstream; let $y$ denote a transverse coordinate that is positive toward the left bank with origin at the centerline; let $z$ denote a nearly vertical coordinate that is positive upward and normal to the average plane of the bed; and let $z_0$ denote a vertical coordinate in the Earth reference frame. Neglecting flow

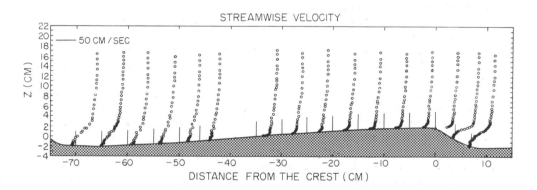

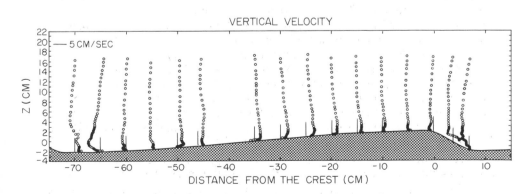

**Figure 15.11**    Plot of measured, mean velocity components $u$ and $w$ over two-dimensional dunes; from Nelson, McLean Wolfe (1993), copyright by the American Geophysical Union.

near the banks, where the no-slip condition induces a transverse boundary layer, the appropriate equations of motion are

$$u\frac{\partial u}{\partial x} + w\frac{\partial u}{\partial z} = -\frac{1}{\rho}\frac{\partial p}{\partial x} - g\frac{\partial z_0}{\partial x} + \frac{1}{\rho}\left(\frac{\partial \tau_{xx}}{\partial x} + \frac{\partial \tau_{zx}}{\partial z}\right) \qquad (15.58)$$

$$u\frac{\partial w}{\partial x} + w\frac{\partial w}{\partial z} = -\frac{1}{\rho}\frac{\partial p}{\partial z} - g\frac{\partial z_0}{\partial z} + \frac{1}{\rho}\left(\frac{\partial \tau_{xz}}{\partial x} + \frac{\partial \tau_{zz}}{\partial z}\right) \qquad (15.59)$$

$$\frac{\partial u}{\partial x} + \frac{\partial w}{\partial z} = 0 \qquad (15.60)$$

The overbar notation for time-averaged values of $u$, $w$, and $p$ has been omitted here for simplicity. With negligible viscous stresses, components of the deviatoric stress terms consist of Reynolds stresses. Written in terms of fluctuating velocity components $u'$ and $w'$,

$$\tau_{zx} = \tau_{xz} = -\rho\overline{u'w'}; \qquad \tau_{xx} = -\rho\overline{u'^2}; \qquad \tau_{zz} = -\rho\overline{w'^2} \qquad (15.61)$$

All of these fluctuating components may be important in the momentum balance of two-dimensional flows over dunes. Let us therefore consider the results of the laser-Doppler measurements of these components to obtain insight regarding the turbulence structure over dunes.

Flow and bedform conditions in the experiments were as follows: The mean flow depth was 0.195 m, the vertically averaged velocity at the mean depth was 0.51 m s$^{-1}$, the dune height was 0.04 m, and the dune wavelength was 0.80 m. Measurements were obtained in the vicinity of the fourth dune in a series of dunes on the bed of a 0.70 m wide racetrack flume.

Plots of mean streamwise velocity $u$ and mean vertical velocity $w$ (Figure 15.11) reveal several regions within the flow. A separation zone occurs in the lee of the dune, with reattachment at a downstream distance of about four dune heights. Velocity profiles are irregular within this zone. Streamwise velocities locally are negative near the bed, reflecting an upstream counterflow in the mean. Vertical velocities over the crest of the dune and immediately downstream are negative. In addition, the separation zone contains intense turbulence and strong vertical motions (described below). A wake region associated with the separation zone extends downstream of the lee of a dune; the wake is a zone of high turbulence intensity and therefore strongly influences the development of the overall turbulence structure over a series of bedforms. Beneath the wake and downstream of the reattachment point is an internal boundary layer (which should not to be confused with the viscous sublayer). This internal boundary layer is apparent as a streamwise strengthening of the near-bed velocity gradient $\partial u/\partial z$. Above the wake is a region where the velocity profiles are smooth and nearly vertical; this region therefore has an inviscid quality in the sense that the vertical shear ($\partial u/\partial z$) is negligible. This part of the flow uniformly accelerates and decelerates in response to the dune topography, and may be considered a *potential flow* whose response is approximately like that of a Bernoulli flow (Example Problem 10.3.1).

Plots of the Reynolds stresses $-\rho\overline{u'^2}$ and $-\rho\overline{w'^2}$ (Figure 15.12) reveal that strong turbulence intensities occur in the separation zone. The largest values of $-\rho\overline{u'^2}$ and $-\rho\overline{w'^2}$ occur at the upper part of the separation zone. This is related to the interaction of fluid within the lee of the dune and the shearing motion of fluid above. Here, eddies develop within the separation zone and are shed into the wake

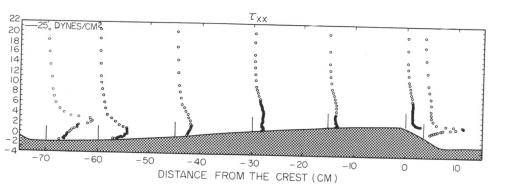

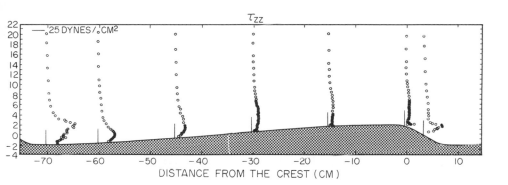

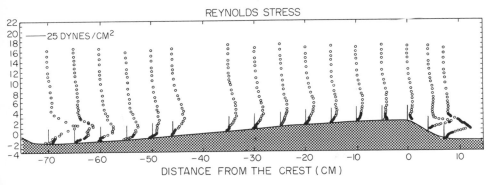

**Figure 15.12** Plot of measured Reynolds stresses over two-dimensional dunes; from Nelson, McLean, and Wolfe (1993), copyright by the American Geophysical Union.

region. This also coincides with a position of strong vertical motions in the mean (Figure 15.11) associated with an overall circulation of fluid related to the streamwise counterflow. The turbulence intensities $-\overline{u'^2}$ and $-\overline{w'^2}$ then weaken downstream within the wake; the vertical extent of the wake is well defined by the elevated turbulence intensities, relative to those in the overlying potential region. The Reynolds stress $-\rho\overline{u'w'}$ also is large in the separation zone; maximum values again occur in the vicinity of the upper part of the separation, and reflect interactions with the shearing motion above, and the strong vorticity at this site. Notice that small negative values of $-\rho\overline{u'w'}$ occur close to the bed in the separation zone. The Reynolds stress $-\rho\overline{u'w'}$ also weakens downstream and vertically, and approaches zero within the region of potential flow. This should be compared with the distribution of $-\rho\overline{u'w'}$ in a unidirectional shear flow (Figure 14.3).

Based on (15.6) the eddy viscosity is approximated by

$$-\frac{\overline{u'w'}}{\partial u/\partial z} \tag{15.62}$$

This expression is strictly correct for a mean flow streamline that is parallel to the $x$-axis. Nonetheless, the horizontal velocity gradient is much smaller than the vertical gradient over much of the flow field, so use of (15.62) is reasonable. The eddy viscosity increases from zero at the bed, reaches its largest value in the wake, which reflects the strong turbulent mixing in this zone, then decreases vertically within the region of potential flow (Figure 15.13). This distribution should be compared with that occurring in a unidirectional shear flow (Figure 15.4).

With the possible exception of the site just downstream of the point of separation at the dune crest, production of turbulence energy is largely associated with vertical shear in the mean flow. Thus, from (15.47), the production of energy can be approximated by

$$-\overline{u'w'}\frac{\partial u}{\partial z} \tag{15.63}$$

Production of energy is strong in the separation zone (Figure 15.13), and is due more to large values of $-\overline{u'w'}$ than to a strong velocity gradient $\partial u/\partial z$ in this zone (Figures 15.11 and 15.12). Production then rapidly decreases downstream; energy produced within the separation zone is advected downstream within the wake. With strengthening vorticity near the bed in the inner boundary layer, energy production in this layer strengthens downstream. Since the Reynolds stress $-\rho\overline{u'w'}$ decreases near the bed (Figure 15.12), this production of energy in the inner boundary layer is due more to the strong velocity gradient within this layer (Figure 15.11) than to the Reynolds stress. Thus, over the stoss of the downstream dune, production of turbulence is limited to a region very close to the bed. Moreover, this production associated with the internal boundary layer constitutes only a fraction of the total production of turbulence energy within the flow.

The measurements described here reveal a rich turbulence structure within two-dimensional flows over bedforms. Students are urged to examine Nelson, McLean, and Wolfe (1993) (and references in this paper to related work) regarding how this turbulence structure bears on computations of bed stress, bedform stability, and sediment transport. Also refer to Nowell and Church (1979), and Best (1993), regarding turbulence over bed bumps.

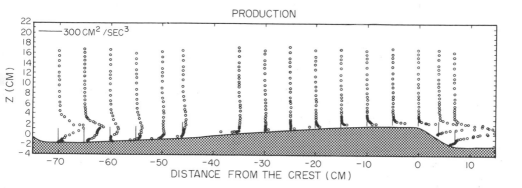

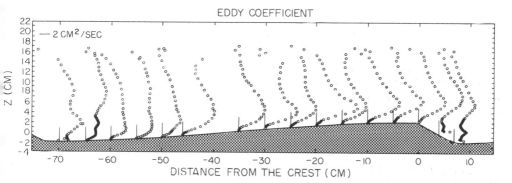

**Figure 15.13** Plot of eddy viscosity and turbulence energy production over two-dimensional dunes; from Nelson, McLean, and Wolfe (1993), copyright by the American Geophysical Union.

## 15.8.3 Flow over a two-dimensional wavy bed

Let us consider a simplified description of the flow in the problem above, neglecting the detailed turbulence structure. Here we will assume that the bed has a wavy sinusoidal form and thus neglect the asymmetry of the dunes. In addition, we will assume that the bed waveform has infinitesimal amplitude. With this restriction we may apply an important technique referred to as *linearization* to greatly simplify the mathematics of the problem. We will then relax the assumption of an infinitesimal amplitude to obtain an approximate description of essential features of the flow above the bed.

Consider a straight channel with uniform width, where the coordinate system is the same as that in the Example Problem above. Again neglecting flow near the banks, the appropriate equations of motion are given by (15.58), (15.59), and (15.60). Here we wish to use depth-integrated versions of these. We start by rewriting the convective terms in (15.58) and (15.59) following the procedure that leads to (14.25), making use of the statement of continuity (15.60):

$$\frac{\partial}{\partial x}(u^2) + \frac{\partial}{\partial z}(uw) = -\frac{1}{\rho}\frac{\partial p}{\partial x} - g\frac{\partial z_0}{\partial x} + \frac{1}{\rho}\left(\frac{\partial \tau_{xx}}{\partial x} + \frac{\partial \tau_{zx}}{\partial z}\right) \qquad (15.64)$$

$$\frac{\partial}{\partial x}(uw) + \frac{\partial}{\partial z}(w^2) = -\frac{1}{\rho}\frac{\partial p}{\partial z} - g\frac{\partial z_0}{\partial z} + \frac{1}{\rho}\left(\frac{\partial \tau_{xz}}{\partial x} + \frac{\partial \tau_{zz}}{\partial z}\right) \qquad (15.65)$$

Now, let $U_*$ denote a characteristic streamwise velocity, and let $W_*$ denote a characteristic velocity parallel to the $z$-axis. The velocity $W_*$ is of order $4U_*\eta_0/\lambda$ where $\eta_0$ is the amplitude of the bedform and $\lambda$ is its wavelength. (How is this obtained?) We may therefore conclude that, for infinitesimal bedform amplitude ($\eta_0 \ll \lambda$), the term $\partial(uw)/\partial x \ll \partial(u^2)/\partial x$ and, further, the term $\partial(w^2)/\partial z \ll \partial(u^2)/\partial x$. Turning to stress terms in (15.64) and (15.65), experience suggests that at a given position $z$, $\tau_{xx} = -\rho\overline{u'^2}$ is of the same order as $\tau_{zx} = -\rho\overline{u'w'}$. With $\eta_0 \ll \lambda$, however, both $\partial\tau_{xx}/\partial x$ and $\partial\tau_{xz}/\partial x$ are negligibly small. Then $\tau_{zx}$ and $\tau_{zz}$ are the only important stresses in (15.64) and (15.65). Introducing these simplifications, (15.64) and (15.65) become

$$\frac{\partial}{\partial x}(u^2) + \frac{\partial}{\partial z}(uw) = -\frac{1}{\rho}\frac{\partial p}{\partial x} - g\frac{\partial z_0}{\partial x} + \frac{1}{\rho}\frac{\partial \tau_{zx}}{\partial z} \qquad (15.66)$$

$$0 = -\frac{1}{\rho}\frac{\partial p}{\partial z} - g\frac{\partial z_0}{\partial z} + \frac{1}{\rho}\frac{\partial \tau_{zz}}{\partial z} \qquad (15.67)$$

Consider the procedure of integrating these equations, together with the equation of continuity (15.60), over the local depth $h$ from the bed ($z = \eta$) to the water surface ($z = \zeta$). Here it is useful to start with a result from Example Problem 8.4.2; namely, with $v = 0$, (8.88) becomes

$$w_\zeta = u_\zeta \frac{\partial \zeta}{\partial x} \qquad (15.68)$$

where the subscript $\zeta$ indicates that this is the component of the fluid velocity parallel to the $z$-axis at the water surface. Based on Example Problem 8.4.2, the appropriate depth-integrated version of (15.60) is

$$\frac{\partial}{\partial x}(hu) = 0 \qquad (15.69)$$

where $u$ in (15.69) now denotes the depth-averaged streamwise velocity. We will use (15.68) and (15.69) below.

Integrating the second term in (15.66), we obtain

$$\int_\eta^\zeta \frac{\partial}{\partial z}(uw)dz = uw\Big|_\eta^\zeta = u_\zeta w_\zeta - u_\eta w_\eta \qquad (15.70)$$

Applying Leibnitz's rule (Example Problem 8.4.2) to the first term in (15.66),

$$\int_\eta^\zeta \frac{\partial}{\partial x}(u^2)dz = \frac{\partial}{\partial x}\int_\eta^\zeta u^2 dz - u_\zeta^2\frac{\partial \zeta}{\partial x} + u_\eta^2\frac{\partial \eta}{\partial x} \qquad (15.71)$$

then combining this with (15.70), the nonlinear terms in (15.66) become

$$\frac{\partial}{\partial x}\int_\eta^\zeta u^2 dz - u_\zeta^2\frac{\partial \zeta}{\partial x} + u_\eta^2\frac{\partial \eta}{\partial x} + u_\zeta w_\zeta - u_\eta w_\eta \qquad (15.72)$$

The no-slip condition leads to the result that $u_\eta = 0$; using this and rearranging (15.72),

$$\frac{\partial}{\partial x}\int_\eta^\zeta u^2 \, dz - u_\zeta\left(u_\zeta\frac{\partial \zeta}{\partial x} - w_\zeta\right) \tag{15.73}$$

Using (15.68) and the mean value theorem, this reduces to

$$\frac{\partial}{\partial x}\int_\eta^\zeta u^2 \, dz = \frac{\partial}{\partial x}\left(\langle u^2\rangle h\right) \tag{15.74}$$

where $h = \zeta - \eta$, and $\langle u^2\rangle$ denotes the depth-average of the squared streamwise component $u$. Here we will assume that $\langle u^2\rangle \approx \langle u\rangle^2$; then we may use $\partial(\langle u^2\rangle h)/\partial x \approx \partial(\langle u\rangle^2 h)/\partial x$, or more simply $\partial(u^2 h)/\partial x$, where now $u$ represents the depth-averaged streamwise velocity (with braces omitted). Using the chain rule, the term $\partial(u^2 h)/\partial x$ may be written as $hu\partial u/\partial x + u\partial(hu)/\partial x$. According to (15.69), the second term in this expression equals zero; thus the depth-integrated form of the nonlinear terms in (15.66) is approximated by

$$hu\frac{\partial u}{\partial x} \tag{15.75}$$

Turning to the stress terms in (15.66) and (15.67), that in (15.66) is integrated as

$$\frac{1}{\rho}\int_\eta^b \frac{\partial \tau_{zx}}{\partial z}\, dz = \frac{1}{\rho}\tau_{zx}\Big|_\eta^\zeta = \frac{1}{\rho}(\tau_{\zeta x} - \tau_{\eta x}) \tag{15.76}$$

Treating the water surface as a zero-stress boundary, the stress at the water surface $\tau_{\zeta x}$ equals zero, and (15.76) reduces to $-\tau_{\eta x}/\rho$, where $\tau_{\eta x}$ denotes the stress at the bed. It is left as an exercise to obtain the result that the stress term in (15.67) integrates to zero given that $\overline{w'^2}$, and therefore $\tau_{zz}$, must vanish at the bed and water surface.

Noting that $\partial z_0/\partial x = \sin\theta$, where $\theta$ is the angle at which the $x$-axis is inclined from the horizontal, the gravitational term in (15.66) is integrated as

$$-g\frac{\partial z_0}{\partial x}\int_\eta^\zeta dz = -g\frac{\partial z_0}{\partial x}z\Big|_\eta^\zeta = -gh\frac{\partial z_0}{\partial x} = -gh\sin\theta \tag{15.77}$$

Turning to the pressure term in (15.66), notice that (15.67), neglecting the stress term, is equivalent to the equation of hydrostatics, modified by the factor $\partial z_0/\partial z = \cos\theta$, where $\theta$ is a negative quantity. Integrating this with respect to $z$, evaluating the constant of integration using the boundary condition that $p$ equals atmospheric pressure $p_0$ at the water surface $z = \zeta$, then differentiating the resulting expression with respect to $x$,

$$\frac{\partial p}{\partial x} = \rho\cos\theta\frac{\partial \zeta}{\partial x} \tag{15.78}$$

Our use of this expression below is equivalent to assuming that hydrostatic conditions exist throughout the flow column. Substituting (15.78) for $\partial p/\partial x$ in (15.66), then integrating the resulting term over the water depth, noting that $\partial\zeta/\partial x$ is independent of $z$,

$$-\frac{1}{\rho}\int_\eta^\zeta \frac{\partial p}{\partial x}dz = -g\cos\theta\frac{\partial\zeta}{\partial x}\int_\eta^\zeta dz = -g\cos\theta h\frac{\partial\zeta}{\partial x} \qquad (15.79)$$

Now, using the results above, the depth-averaged version of (15.66) is

$$u\frac{\partial u}{\partial x} = -g\cos\theta\frac{\partial\zeta}{\partial x} - g\sin\theta - \frac{\tau_{\eta x}}{\rho h} \qquad (15.80)$$

where all terms have been divided by $h$.

Turning to the bed stress $\tau_{\eta x}$, it is conventional, particularly in the engineering literature, to describe $\tau_{\eta x}$ in terms of the velocity $u$ as

$$\tau_{\eta x} = \rho\Gamma u^2 \qquad (15.81)$$

Here $\Gamma$ is a friction coefficient; it is like a *Darcy–Weisbach* coefficient, originally derived from dimensional analysis of turbulent flow in a pipe. Substituting (15.81) into (15.80),

$$u\frac{\partial u}{\partial x} = -g\cos\theta\frac{\partial\zeta}{\partial x} - g\sin\theta - \frac{\Gamma u^2}{h} \qquad (15.82)$$

This and (15.69) form the basis of the remainder of the analysis. Notice that these equations are nonlinear; the next task is to obtain linearized forms.

Let $U$, $H$, $Z$, and H denote reach-averaged values of $u$, $h$, $\zeta$, and $\eta$, where averaging occurs over a long streamwise distance (over many bedform wavelengths $\lambda$). Then let $u'$, $h'$, $\zeta'$, and $\eta'$ denote perturbed values (small deviations) about the reach-averaged values (Figure 15.14) such that

$$u = U + u'; \qquad h = H + h'; \qquad \zeta = Z + \zeta'; \qquad \eta = H + \eta' \qquad (15.83)$$

The reach-averaged values are referred to as *zeroth-order* quantities, and collectively define a "zeroth-order flow state." Perturbed values are referred to as *first-order* quantities. Notice that

$$h' = \zeta' - \eta' \qquad (15.84)$$

Substituting expressions in (15.83) for $u$, $\zeta$, $h$, and $\eta$ in (15.82), and making use of (15.84),

$$(U + u')\frac{\partial}{\partial x}(U + u') = -g\cos\theta\frac{\partial}{\partial x}(Z + \zeta') - g\sin\theta - \frac{\Gamma(U + u')^2}{H + \zeta' - \eta'} \qquad (15.85)$$

Rearranging and expanding, and noticing that derivatives of zeroth-order quantities, for example, $\partial H/\partial x$, equal zero, then leads to

$$HU\frac{\partial u'}{\partial x} + Hu'\frac{\partial u'}{\partial x} + U\zeta'\frac{\partial u'}{\partial x} + \zeta'u'\frac{\partial u'}{\partial x} - U\eta'\frac{\partial u'}{\partial x} - \eta'u'\frac{\partial u'}{\partial x}$$

$$+ g\cos\theta H\frac{\partial\zeta'}{\partial x} + g\cos\theta\zeta'\frac{\partial\zeta'}{\partial x} - g\cos\theta\eta'\frac{\partial\zeta'}{\partial x}$$

$$+ g\sin\theta H + g\sin\theta\zeta' - g\sin\theta\eta' + \Gamma U^2 + 2\Gamma Uu' + \Gamma u'^2 = 0 \quad (15.86)$$

Assume for illustration that $H$ and $U$ are of order unity. Then, since perturbations are much smaller than unity, say 0.1 or less, this means that the contribution to the momentum balance of terms involving perturbations are on the order of 0.1 of the zeroth-order terms, or less. For example, since $\zeta' \ll H$, $g\sin\theta\zeta' \ll g\sin\theta H$.

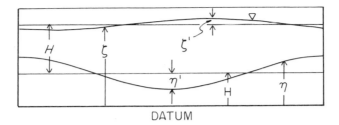

**Figure 15.14** Definition diagram for reach-averaged values of velocity $U$, depth $H$, water-surface Z, and bed elevation H, and perturbations $u'$, $h'$, $\zeta'$, and $\eta'$ about average values.

Thus, to describe the zeroth-order (average) momentum balance neglecting terms involving $g \sin \theta \zeta'$ and $g \sin \theta \eta'$, while retaining the term $g \sin \theta H$, would involve an error of 10 percent or less. Retaining only zeroth-order terms then gives a zeroth-order expression of the momentum balance:

$$g \sin \theta H + \Gamma U^2 = 0 \quad (15.87)$$

It is left as an exercise to show that a similar substitution and expansion of the continuity equation (15.69) provides no zeroth-order expression.

Now consider first-order terms. Because $u'$, $\zeta'$, and $\eta'$ are of order 0.1 or less, it follows that terms in (15.86) involving products of these will be of order 0.01 or less. Thus, retaining first-order terms in (15.86) that do not involve such products provides a first-order expression of the momentum balance:

$$HU\frac{\partial u'}{\partial x} + g \cos \theta H\frac{\partial \zeta'}{\partial x} + g \sin \theta \zeta' - g \sin \theta \eta' + 2\Gamma U u' = 0 \quad (15.88)$$

It is left as an exercise to show that the continuity equation provides the first-order expression

$$H\frac{\partial u'}{\partial x} + U\frac{\partial \zeta'}{\partial x} - U\frac{\partial \eta'}{\partial x} = 0 \quad (15.89)$$

Notice that (15.88) and (15.89) are linear equations. The task now consists of solving for $u'$ and $\zeta'$ in these first-order equations, knowing how $\eta'$ varies with $x$ along the channel.

Equations (15.88) and (15.89) are conditions that must be simultaneously satisfied at each coordinate position $x$. Note that the physical units of each term in both equations must be the same to solve them simultaneously; for this and other reasons that will become clear below, let us divide (15.88) by the product $HU^2$, and (15.89) by the product $HU$. Then, rearranging (15.87) to obtain $g \sin \theta = -\Gamma U^2/H$, substituting for $g \sin \theta$ in (15.88), and noting that with small $\theta$, $\cos \theta \approx 1$,

$$\frac{1}{U}\frac{\partial u'}{\partial x} + \frac{2\Gamma}{HU}u' + \frac{g}{U^2}\frac{\partial \zeta'}{\partial x} - \frac{\Gamma}{H^2}\zeta' = -\frac{\Gamma}{H^2}\eta' \quad (15.90)$$

$$\frac{1}{U}\frac{\partial u'}{\partial x} + \frac{1}{H}\frac{\partial \zeta'}{\partial x} = \frac{1}{H}\frac{\partial \eta'}{\partial x} \quad (15.91)$$

There are basically two ways to solve these simultaneously. A relatively quick way is to use complex numbers involving the imaginary number $i$, defined by $i^2 = -1$, and the Euler identity $e^{ia} = \cos a + i \sin a$, where $a$ is an argument. We will return to this method below for those familiar with this notation. Here we will start with a longer procedure that involves only ordinary trigonometric functions.

Recall that the bed is described as a sinusoid; namely

$$\eta' = \eta_0 \cos \omega x \tag{15.92}$$

Here $\eta_0$ is the amplitude of the bedform, and $\omega = 2\pi/\lambda$ is referred to as the *wavenumber*, where $\lambda$ is the wavelength of the bedform. Experience with linear differential equations suggests that the responses, $u'$ and $\zeta'$, also will be sinusoids that possess the same wavelength but possibly different amplitudes, and may be shifted upstream or downstream a finite distance. Thus, we assume solutions of the form

$$u' = u_0 \cos(\omega x + \phi_u); \qquad \zeta' = \zeta_0 \cos(\omega x + \phi_\zeta) \tag{15.93}$$

Here $u_0$ and $\zeta_0$ denote unknown amplitudes, and $\phi_u$ and $\phi_\zeta$ denote unknown phase shifts, of the sinusoidal responses $u'$ and $\zeta'$.

Taking derivatives of (15.92) and (15.93), substituting into (15.90) and (15.91), making use of the identities $\sin(a_1 + a_2) = \sin a_1 \cos a_2 + \cos a_1 \sin a_2$ and $\cos(a_1 + a_2) = \cos a_1 \cos a_2 - \sin a_1 \sin a_2$, where $a_1$ and $a_2$ are any two arguments, then expanding and regrouping, we obtain

$$(U_0\omega \cos \phi_u + 2\alpha U_0 \sin \phi_u + \beta Z_0\omega \cos \phi_\zeta - \alpha Z_0 \sin \phi_\zeta) \sin \omega x$$
$$+ (U_0\omega \sin \phi_u - 2\alpha U_0 \cos \phi_u + \beta Z_0\omega \sin \phi_\zeta + \alpha Z_0 \cos \phi_\zeta) \cos \omega x = \alpha H_0 \cos \omega x$$
$$\tag{15.94}$$

$$(U_0\omega \cos \phi_u + Z_0\omega \cos \phi_\zeta) \sin \omega x + (U_0\omega \sin \phi_u + Z_0\omega \sin \phi_\zeta) \cos \omega x$$
$$= H_0\omega \sin \omega x$$
$$\tag{15.95}$$

where

$$U_0 = \frac{u_0}{U}; \qquad Z_0 = \frac{\zeta_0}{H}; \qquad H_0 = \frac{\eta_0}{H}; \qquad \alpha = \frac{\Gamma}{H}; \qquad \beta = \frac{gH}{U^2} \tag{15.96}$$

Notice that $\beta = Fr^{-2}$, where $Fr$ is a Froude number (Chapter 5) defined in terms of the linear dimension $H$. The quantity $\omega$ is not divided out of (15.95) to retain the same units between (15.94) and (15.95). Observe that these two equations are satisfied simultaneously if

$$U_0\omega \cos \phi_u + 2\alpha U_0 \sin \phi_u + \beta Z_0\omega \cos \phi_\zeta - \alpha Z_0 \sin \phi_\zeta = 0 \tag{15.97}$$

$$U_0\omega \sin \phi_u - 2\alpha U_0 \cos \phi_u + \beta Z_0\omega \sin \phi_\zeta + \alpha Z_0 \cos \phi_\zeta = \alpha H_0 \tag{15.98}$$

$$U_0\omega \cos \phi_u + Z_0\omega \cos \phi_\zeta = H_0\omega \tag{15.99}$$

$$U_0\omega \sin \phi_u + Z_0\omega \sin \phi_\zeta = 0 \tag{15.100}$$

(Why?) Our notation will be simplified at this point by letting

$$A = U_0 \cos \phi_u; \qquad B = U_0 \sin \phi_u; \qquad C = Z_0 \cos \phi_\zeta; \qquad D = Z_0 \sin \phi_\zeta \tag{15.101}$$

Substituting (15.101) into (15.97) through (15.100), then reversing the order of these equations (the reason for this will become clear below),

$$0A + \omega B + 0C + \omega D = 0 \tag{15.102}$$

$$\omega A + 0B + \omega C + 0D = H_0\omega \tag{15.103}$$

$$-2\alpha A + \omega B + \alpha C + \beta\omega D = \alpha H_0 \tag{15.104}$$

$$\omega A + 2\alpha B + \beta\omega C - \alpha D = 0 \tag{15.105}$$

These may be written in matrix form:

$$\begin{bmatrix} 0 & \omega & 0 & \omega \\ \omega & 0 & \omega & 0 \\ -2\alpha & \omega & \alpha & \beta\omega \\ \omega & 2\alpha & \beta\omega & -\alpha \end{bmatrix} \cdot \begin{bmatrix} A \\ B \\ C \\ D \end{bmatrix} = \begin{bmatrix} 0 \\ H_0\omega \\ \alpha H_0 \\ 0 \end{bmatrix} \tag{15.106}$$

There arc several ways to solve for the unknowns, $A$, $B$, $C$, and $D$. Perhaps the most straightforward way, although not necessarily the most efficient, is to use *Cramer's rule*. Those familiar with this method will appreciate the value of reversing the order of the equations above; this places three zeros in the top row, which greatly simplifies calculation of determinants. (Students not familiar with matrix algebra are urged to consult the references for introductory treatments of Cramer's rule and other methods for solving systems of equations having the form of [15.106].) A bit of algebra then gives

$$A = U_0 \cos \phi_u = \frac{(\beta - 1)\beta\omega^2}{9\alpha^2 + (\beta^2 - 2\beta + 1)\omega^2} H_0 \tag{15.107}$$

$$B = U_0 \sin \phi_u = \frac{-3\alpha\beta\omega}{9\alpha^2 + (\beta^2 - 2\beta + 1)\omega^2} H_0 \tag{15.108}$$

$$C = Z_0 \cos \phi_\zeta = \frac{9\alpha^2 + (1 - \beta)\omega^2}{9\alpha^2 + (\beta^2 - 2\beta + 1)\omega^2} H_0 \tag{15.109}$$

$$D = Z_0 \sin \phi_\zeta = \frac{3\alpha\beta\omega}{9\alpha^2 + (\beta^2 - 2\beta + 1)\omega^2} H_0 \tag{15.110}$$

From these we solve for the phase angles and amplitudes:

$$\phi_u = \tan^{-1} \frac{3\alpha}{(1 - \beta)\omega} \tag{15.111}$$

$$\phi_\zeta = \tan^{-1} \frac{3\alpha\beta\omega}{9\alpha^2 + (1 - \beta)\omega^2} \tag{15.112}$$

$$u_0 = \frac{U}{H} \frac{(\beta - 1)\beta\omega^2}{[9\alpha^2 + (\beta^2 - 2\beta + 1)\omega^2] \cos \phi_u} \eta_0 \tag{15.113}$$

$$\zeta_0 = \frac{9\alpha^2 + (1 - \beta)\omega^2}{[9\alpha^2 + (\beta^2 - 2\beta + 1)\omega^2] \cos \phi_\zeta} \eta_0 \tag{15.114}$$

Thus, for sinusoidal bedforms described by $\eta' = \eta_0 \cos \omega x$, the depth-averaged velocity and water surface perturbations, $u'$ and $\zeta'$, are given by substituting (15.111) through (15.114) into equations (15.93).

It is important to verify that the solutions for $u'$ and $\zeta'$ are correct. This might involve, for example, taking derivatives of $u'$, $\zeta'$, and $\eta'$ with respect to $x$, then substituting these expressions into the equation of continuity (15.91) to determine whether the three terms of this equation sum to zero for arbitrary $\omega$ and $x$. It is particularly important to do this since the arctan function used in (15.111) and (15.112) may be satisfied by integer multiples of $\pi$.

In a strict sense, our analysis pertains to infinitesimal perturbations. Let us relax this condition, nonetheless, to gain an idea of how $u'$ and $\zeta'$ vary over a bedform with finite (but small) amplitude $\eta_0$. In this regard, we should pay more attention to the phase angles than to the amplitudes of the waveforms of $u'$ and $\zeta'$. With decreasing wavenumber $\omega$ (increasing wavelength $\lambda$), $\phi_u$ approaches $-\pi/2$. The waveform of $u'$ is therefore one-quarter wavelength out of phase with the bedform; highest velocities occur over lees of bedforms, and lowest velocities occur over stosses of bedforms. With increasing $\omega$, $\phi_u$ monotonically increases (becomes less negative); highest velocities shift to positions closer to bedform crests. With small $\omega$, the waveform of $\zeta'$ is nearly in phase with the bedform. With increasing $\omega$, crests of the water-surface waveform shift to positions over bedform stosses. Similar effects are manifest with increasing Froude number. It is left as an exercise to consider effects of varying friction $\Gamma$ and flow depth $H$.

These results suggest a simple physical interpretation. At low wavenumbers $\omega$, effects of friction dominate. Flow decelerates as it shoals onto a bedform stoss (Figure 15.15a). With increasing wavenumber, or Froude number, inertial effects become more important. Flow accelerates up the bedform stoss, driven by an increasing pressure gradient (Figure 15.15b). Recall from the Example Problem above that this flow acceleration is apparent over the stoss of two-dimensional dunes, although we must be careful to not extend our comparison of these two problems much beyond this qualitative result. We also should not extend our results to high $\omega$, since vertical flows, and perhaps separation, become an important part of the momentum balance. In addition, our analysis applies only to the case of low to moderate Froude numbers (subcritical flow).

Let us return briefly to the method of solving (15.90) and (15.91) that involves complex numbers. Assume that $u'$, $\zeta'$, and $\eta'$ have the forms

$$u' = u_0 e^{i\omega x}; \qquad \zeta' = \zeta_0 e^{i\omega x}; \qquad \eta' = \eta_0 e^{i\omega x} \qquad (15.115)$$

where $u_0$ and $\zeta_0$ are complex amplitudes and $\eta_0$ is a real amplitude. Differentiating these with respect to $x$, substituting the resulting expressions together with those of (15.115) into (15.90) and (15.91), dividing through by $e^{i\omega x}$, and using the relations of (15.96),

$$\begin{bmatrix} 2\alpha + i\omega & -\alpha + i\beta\omega \\ \omega & \omega \end{bmatrix} \cdot \begin{bmatrix} U_0 \\ Z_0 \end{bmatrix} = \begin{bmatrix} -\alpha H_0 \\ \omega H_0 \end{bmatrix} \qquad (15.116)$$

Solving for $U_0$ and $Z_0$ (using Cramer's rule, for example), grouping real and imaginary terms, and dividing by $H_0$,

$$\frac{U_0}{H_0} = \frac{-i\beta\omega}{3\alpha + i(1 - \beta)\omega} \qquad (15.117)$$

$$\frac{Z_0}{H_0} = \frac{3\alpha + i\omega}{3\alpha + i(1 - \beta)\omega} \qquad (15.118)$$

The amplitudes $u_0$ and $\zeta_0$ are obtained by computing the *magnitudes* of these relations. For complex numbers $A = A_1 + iA_2$ and $B = B_1 + iB_2$, the magnitude of the ratio $A/B$ is denoted by mag $(A/B) = |A/B| = |A|/|B|$, and is given by

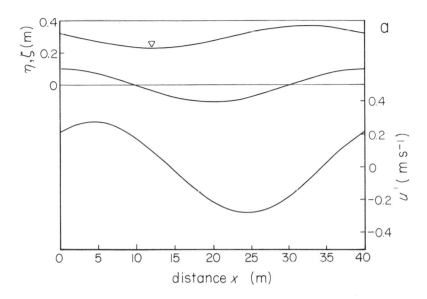

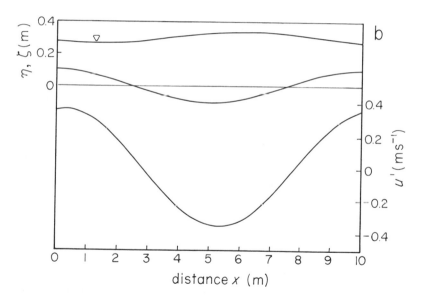

**Figure 15.15** Waveforms of velocity perturbation $u'$ and water-surface perturbation $\zeta'$ over two-dimensional sinusoidal bedforms at low wavenumber $\omega$ ($a$) and high wavenumber $\omega$ ($b$).

$(A_1^2 + A_2^2)^{1/2}/(B_1^2 + B_2^2)^{1/2}$. Here $A_1$ is the real part of $A$ (Re $A$) and $A_2$ is the imaginary part (Im $A$); likewise $B_1 = $ Re $B$ and $B_2 = $ Im $B$. The phase angles $\phi_u$ and $\phi_\zeta$ are obtained by computing the *angle* of $A/B$. This is denoted by ang $(A/B) = $ ang $A - $ ang $B$, and is given by $\tan^{-1}(A_2/A_1) - \tan^{-1}(B_2/B_1)$. (The magnitude also is referred to as the *modulus,* and the angle is referred to as the *argument.* In addition, the ratio of the amplitude of the output to the amplitude of the input, for example, $\zeta_0/\eta_0$, is referred to as the *gain*.) Using these definitions,

$$u_0 = \frac{U}{H} \frac{\beta\omega}{\sqrt{9\alpha^2 + (1 - \beta)^2\omega^2}} \eta_0 \qquad (15.119)$$

$$\zeta_0 = \frac{\sqrt{9\alpha^2 + \omega^2}}{\sqrt{9\alpha^2 + (1 - \beta)^2\omega^2}} \eta_0 \qquad (15.120)$$

$$\phi_u = -\frac{\pi}{2} - \tan^{-1}\frac{(1 - \beta)\omega}{3\alpha} \qquad (15.121)$$

$$\phi_\zeta = \tan^{-1}\frac{\omega}{3\alpha} - \tan^{-1}\frac{(1 - \beta)\omega}{3\alpha} \qquad (15.122)$$

## 15.8.4 Similitude and the logarithmic velocity law

H. Schlichting performed a set of flow experiments with artificial roughness that consisted of small spheres and hemispheres glued to flat surfaces in regular hexagonal arrangements (Figure 15.8). Based on measurements of the velocity profile above these roughened surfaces, Schlichting associated with each roughness pattern an equivalent Nikuradse roughness height $k_s$. This height may be interpreted as the roughness height of Nikuradse sand roughness that would lead to the same total drag as that measured for the artificial roughness arrangement.

Let us examine these results from another point of view. Consider the experiments involving spheres of given height $k$ with varying spatial density; each arrangement is associated with a different value of $B_s$ in the logarithmic velocity law (15.44). Further notice that $k_s$ increases with increasing spatial density, then decreases with the closest arrangement of spheres. This reflects an increasing drag with increasing sphere density; but with the closest arrangement of spheres, the surface is hydrodynamically smoother than lower spatial densities since the total exposure of individual spheres to flow decreases.

Recall that for Nikuradse sand roughness, $B_s$ takes on a constant value of 8.5 in the hydrodynamically rough regime. This result occurs because the dimensional argument embodied in (15.42), that $z_0$ is proportional to the roughness height, is based on the assumption that a condition of geometrical similitude is satisfied. By a similar argument, we may therefore anticipate that the value of $B_s$ associated with artificial roughness ought to be constant for cases where the artificial roughness is geometrically similar. Notice that two of Schlichting's experiments with spheres involve geometrically similar conditions. A plot of $B_s$ versus $u_* k/\nu$ reveals that the value of $B_s$ for these two experiments is essentially constant, but different from the value of 8.5 associated with Nikuradse sand roughness (Figure 15.7).

These results lead to an interesting implication regarding natural roughness. Namely, we cannot expect to obtain a universal value of $B_s$ for natural roughnesses, given the infinity of possible textures that natural roughnesses exhibit. Use of the logarithmic velocity law (15.44) based on a single linear dimension (for example, roughness height) to characterize the roughness is in a strict sense appropriate only for geometrically similar conditions.

## 15.9 READING

Bear, J. 1972. *Dynamics of fluids in porous media*. New York: Elsevier, 764 pp. This text provides a discussion (pp. 176–84) of attempts to generalize Darcy's law to a form that characterizes effects of inertial forces at large Reynolds numbers.

Best, J. L. 1993. On the interactions between turbulent flow structure, sediment transport and bedform development: some considerations from recent experimental research. In Clifford, N. J., French, J. R., and Hardisty, J., eds., *Turbulence: perpectives on flow and sediment transport*. Chichester: Wiley, pp. 61–92. This article provides a review of research on the interactions of turbulence structure and stream bedforms, and examines implications for the entrainment of sediment.

Clifford, N. J., French, J. R., and Hardisty, J., eds. 1993. *Turbulence: perpectives on flow and sediment transport*. Chichester: Wiley, 360 pp. This book is a collection of papers regarding laboratory and field studies of turbulence structure and its relation to bed roughness in stream channels. Several of the articles examine techniques and instrumentation for measuring fluctuating velocity components under field conditions.

Davis, J. C. 1986. *Statistics and data analysis in geology*, 2nd ed. New York: Wiley, 646 pp. This text provides a nice introductory treatment of matrix algebra, including Cramer's rule for small matrices.

Dingman, S. L. 1984. *Fluvial hydrology*. New York: W. H. Freeman and Co., 383 pp. This text contains an introductory treatment of the Darcy–Weisbach friction coefficient and its relation to other measures of flow resistance (pp. 138–49); also see Giles (1962).

Gerald, C. F. and Wheatley, P. O. 1984. *Applied numerical analysis*, 3rd ed. Reading, Mass.: Addison-Wesley, 579 pp. This text provides an applied view of matrix algebra, including the application of Cramer's rule to large matrices, and of other methods for solving systems of equations.

Giles, R. V. 1962. *Theory and problems of fluid mechanics and hydraulics*, 2nd ed. New York: McGraw-Hill, 274 pp. This edition in the Schaum's Outline Series contains a concise development of the Darcy–Weisbach coefficient base on dimensional analysis (pp. 56–57).

Ikeda, S. and Parker, G., eds. 1989. *River Meandering*. Water Resources Monograph 12, American Geophysical Union, Washington, D. C., 485 pp. This monograph contains several papers that substantially extend the linear analysis of Example Problem 15.8.2 by coupling flow with sediment transport. The idea is to apply an infinitesimal perturbation to the bed in the form of a sinusoidal waveform, then determine which wavelength grows the fastest due to interactions between the bed and flow, with the implication that this wavelength reflects that of natural bars. Papers in this monograph that use some form of this technique include those by: Fukuoka; Nelson and Smith; Seminara and Tubino; and Striksma and Crosato.

LePage, W. R. 1961. *Complex variables and the Laplace transform for engineers*. New York: McGraw-Hill (Dover edition, 1980), 474 pp. This text provides a nice introduction (Chapter 2) to complex numbers and variables, and algebraic rules of complex numbers. Its treatment of complex variables is otherwise presented at an advanced level.

Nelson, J. M., McLean, S. R., and Wolfe, S. R. 1993. Mean flow and turbulence fields over two-dimensional bed forms. *Water Resources Research* 29:3935–53. This paper is the basis for Example Problem 15.8.2.

Nowell, A. R. M. and Church, M. 1979. Turbulent flow in a depth-limited boundary layer. *Journal of Geophysical Research* 84:4816–24. This paper examines the effects of roughness density (bump spacing) on the turbulence field in a unidirectional shear flow. Vertical variations in the turbulence intensity field are characterized in terms of roughness density involving three regimes: "isolated roughness," "wake interference," and "skimming."

Schlichting, H. 1979. *Boundary-layer theory,* 7th ed. New York: McGraw-Hill, 817 pp. Part C of this classic text on boundary-layer flows covers the topic of the origin of turbulence, and Part D is given entirely to turbulent boundary layers. Students who are beginning their studies of turbulence and turbulent flows should make a point of consulting this comprehensive text. Data for Example Problem 15.8.4 are taken from section XX.g.

Smith, J. D. and McLean, S. R. 1984. A model for flow in meandering streams. *Water Resources Research* 20:1301–15. This paper presents a theoretical description of flow in meandering rivers that stands as the basis of the most comprehensive treatment to date of flow, sediment transport, and bedform dynamics in such rivers: the "Muddy Creek Experiments" (see the several articles referenced in this paper by Dietrich, Smith, and Dunne). This paper also presents an interesting treatment of the eddy viscosity distribution above a stream bed that has roughness produced by ripples and dunes, and provides a brief derivation of the equations of motion for open channel flow in curvilinear coordinates.

Tritton, D. J. 1988. *Physical fluid dynamics,* 2nd ed. Oxford: Oxford University Press, 519 pp. Chapters 18 through 21 in this text provide concise treatments of: the mechanisms of transition to turbulence; statistical descriptions of turbulence; and the structure of turbulence in boundary-layer shear flows, including the production, transport, and dissipation of turbulence energy.

Wiberg, P. L. and Smith, J. D. 1991. Velocity distribution and bed roughness in high-gradient streams. *Water Resources Research* 27:825–38. This paper presents a model for the mean velocity distribution over a stream bed with coarse sediment. The model involves partitioning the total stress into a fluid part and a part due to form drag associated with flow around particles.

# CHAPTER 16

# Thermally Driven Flows

Recall that we briefly examined the stability of thermally stratified fluids in Chapter 6. Our essential conclusion was that a condition in which hot, light fluid resides beneath cooler, heavier fluid is a potentially unstable configuration, inasmuch as gravitational forces are no longer balanced by forces associated with the vertical pressure gradient. In this regard we obtained a necessary (but insufficient) condition for instability, which stated that the magnitude of the vertical temperature gradient must exceed the adiabatic lapse rate. Further recall, however, that such a condition does not necessarily lead spontaneously to overturning (convection). It is possible for a fluid to possess a steady temperature gradient over its vertical extent, in excess of the adiabatic rate, such that the fluid remains static and merely acts like a thermally conducting solid, and we concluded that this static *conduction state* represents an unstable equilibrium. We then characterized the tendency for instability in terms of the Rayleigh number *Ra*.

Let us now extend our treatment to a description of thermally driven convection motions. We will concentrate on Rayleigh–Bénard and Hele–Shaw configurations (Chapter 6), and the buoyancy-driven flows that occur within these configurations due to steady heating and cooling of the boundaries. In such configurations, convective motions typically occupy the full region between the boundaries, and vary from steady two-dimensional rolls at small Rayleigh numbers to complex turbulent motions at large Rayleigh numbers. In addition, we will briefly consider buoyancy-driven flows that arise from localized temperature and compositional variations near a single solid boundary. Envision, for example, the roof of a magma chamber. For a sufficient contrast in temperature between the magma and country rock, melting of the magma roof occurs. If the density of the melt that is produced is less than that of the underlying magma, thermally driven convection may begin within the buoyant layer of melt, which in turn advects heat upward from the underlying hot magma to the roof. Alternatively, if the density of the melt is greater than that of the underlying magma, compositionally driven convection may occur, which in turn brings hot magma to the roof. Similar remarks apply to fluid conditions near a cool xenolith immersed within a hot magma. Such convection is particularly important for understanding the dynamics and geochemistry of magmas.

The scope of thermally driven flows is very broad. Here we will have the opportunity to only touch on this topic. In this vein we will start with free convection in purely fluid systems, and use this topic to introduce the Boussinesq approximation—a set of simplified versions of the equations of motion, energy, and state—which forms the cornerstone of formal descriptions of convection. In addition, we will consider free convection within porous geological materials.

## 16.1  BOUSSINESQ APPROXIMATION

Formal descriptions of free convection normally begin with simplified versions of the equations of motion, energy, and state, as first derived by J. Boussinesq in 1903. The Boussinesq approximation embodies two essential ideas. First, any fluctuations in density that occur with the onset of motion are produced mainly by thermal effects rather than by pressure effects. Second, all variations in fluid properties, except density, may be neglected; and variations in density may be neglected except insofar as they are coupled to the gravitational acceleration and buoyancy forces. In this section we will examine the justification of the Boussinesq approximation, then use the results to clarify how temperature variations have a dual role, thermal and mechanical, in transporting heat in free convection, and why free convection is difficult to treat theoretically. Subsequent sections contain descriptions of free convection based on laboratory experiments, and implications regarding field-scale problems.

The general case deals with compressible fluids. Herein we will follow the discussion of Spiegel and Veronis (1960), who point out that the essential objective in justifying the Boussinesq approximation is to demonstrate that the equations governing convection of an ideal gas are equivalent to those for an incompressible fluid, modified to take into account that the important temperature gradient is that in excess of the adiabatic lapse rate. Then, the approximation is subject to two conditions: First, the vertical dimension of the fluid system must be much less than any characteristic scale height (which we will define below). Second, any fluctuations in pressure and density induced by fluid motions must not exceed the total variations in these quantities in the static state. The Boussinesq approximation for an incompressible liquid is then obtained as a special case.

Consider for illustration a Rayleigh–Bénard configuration with vertical dimension $Z$ (Figure 6.4). The $x$ and $y$ coordinate axes are horizontal, and the $z$-axis is vertical in the Earth reference frame. The equations of motion are

$$\rho \frac{D\mathbf{u}}{Dt} = -\nabla p - g\rho\mathbf{k} + \mu\nabla^2\mathbf{u} + \frac{\mu}{3}\nabla(\nabla \cdot \mathbf{u}) \tag{16.1}$$

$$\frac{D\rho}{Dt} + \rho\nabla \cdot \mathbf{u} = 0 \tag{16.2}$$

where $\mathbf{k} = \langle 0, 0, 1 \rangle$ is the unit vector in the vertical direction, and the viscosity $\mu$ is assumed to be constant. Neglecting radiative energy and internal sources of heat associated with chemical reactions or phase changes, the energy equation is

$$\rho c_V \frac{DT}{Dt} + p\nabla \cdot \mathbf{u} = K_T\nabla^2 T + \Phi \tag{16.3}$$

where the specific heat $c_V$ and the thermal conductivity $K_T$ are assumed to be constant. We shall note here without proof that viscous dissipation contributes negligibly to the energy balance, so the dissipation function $\Phi$ in (16.3) may be neglected. The equation of state has the general form

$$\rho = \rho(p, T) \tag{16.4}$$

Now, let pressure $p$, temperature $T$, and density $\rho$ be represented as

$$p(x, y, z, t) = p_m + p_s(z) + p'(x, y, z, t) \tag{16.5}$$

$$T(x, y, z, t) = T_m + T_s(z) + T'(x, y, z, t) \tag{16.6}$$

$$\rho(x, y, z, t) = \rho_m + \rho_s(z) + \rho'(x, y, z, t) \tag{16.7}$$

where $p_m$, $T_m$, and $\rho_m$ denote (constant) spatially averaged values. The quantities $p_s$, $T_s$, and $\rho_s$ denote variations about the average state in absence of motion. For example, $T_s$ describes the temperature variation in the (static) conduction state, and $p_s$ and $\rho_s$ are determined by the hydrostatic equation and the equation of state, and are functions of the vertical coordinate only. The quantities $p'$, $T'$, and $\rho'$ denote fluctuations associated with fluid motion.

Now define scale heights $Z_p$, $Z_T$, and $Z_\rho$:

$$Z_p = \left| \frac{1}{p_m} \frac{dp_s}{dz} \right|^{-1} ; \qquad Z_T = \left| \frac{1}{T_m} \frac{dT_s}{dz} \right|^{-1} ; \qquad Z_\rho = \left| \frac{1}{\rho_m} \frac{d\rho_s}{dz} \right|^{-1} \tag{16.8}$$

Comparing these expressions with (2.5), we may conclude that each is a measure of the distance over which the defining state variable is vertically uniform. For example, $Z_\rho$ is a measure of the distance over which the density of the fluid varies by a fraction (of the mean density $\rho_m$) of order unity. Then, the first condition to be satisfied is that the fluid height $Z$ is much less than the smallest of the scale heights in (16.8). With regard to the Boussinesq approximation, this means that

$$Z \ll Z_\rho \tag{16.9}$$

or

$$\frac{1}{Z_\rho} \ll \frac{1}{Z} \tag{16.10}$$

Based on the definition of $Z_\rho$,

$$\left| \frac{1}{\rho_m} \frac{d\rho_s}{dz} \right| \ll \frac{1}{Z} \tag{16.11}$$

Integrating these over the height $Z$, or from the level where $\rho_s$ is a minimum to the level where $\rho_s$ is a maximum,

$$\frac{1}{\rho_m} \int_Z \left| \frac{d\rho_s}{dz} \right| dz \ll \frac{1}{Z} \int_Z dz \tag{16.12}$$

which leads to the condition that

$$\frac{\Delta\rho_s}{\rho_m} \equiv \epsilon \ll 1 \tag{16.13}$$

where $\Delta\rho_s$ is the maximum variation in $\rho_s$ over the distance $Z$. The full significance of the quantity $\epsilon$ will become clear below. Meanwhile we may think of it as a measure of the maximum acceptable error in simplifying the governing equations; that is, we will neglect terms in these equations only if they are of the same order as $\epsilon$ or smaller.

It is also necessary when considering nonlinear effects to impose an additional condition, namely that fluctuations induced by motion do not exceed variations in the static state. Thus,

$$\left| \frac{\rho'}{\rho_m} \right| \leq O(\epsilon) \tag{16.14}$$

This condition normally is satisfied when (16.13) is satisfied.

The equation of state (16.4) can be expanded as a Taylor series about the mean quantities $\rho_m$, $p_m$, and $T_m$:

$$\rho = \rho_m + \frac{\partial \rho}{\partial T}(T - T_m) + \frac{\partial \rho}{\partial p}(p - p_m) + \frac{1}{2}\frac{\partial^2 \rho}{\partial T^2}(T - T_m)^2$$

$$+ \frac{1}{2}\frac{\partial^2 \rho}{\partial T \partial p}(T - T_m)(p - p_m) + \frac{1}{2}\frac{\partial^2 \rho}{\partial p^2}(p - p_m)^2 + \dots \tag{16.15}$$

Subtracting $\rho_m$ from both sides, then dividing by $\rho_m$,

$$\frac{\rho - \rho_m}{\rho_m} = \frac{1}{\rho_m}\frac{\partial \rho}{\partial T}(T - T_m) + \frac{1}{\rho_m}\frac{\partial \rho}{\partial p}(p - p_m) + \frac{1}{2\rho_m}\frac{\partial^2 \rho}{\partial T^2}(T - T_m)^2$$

$$+ \frac{1}{2\rho_m}\frac{\partial^2 \rho}{\partial T \partial p}(T - T_m)(p - p_m) + \frac{1}{2\rho_m}\frac{\partial^2 \rho}{\partial p^2}(p - p_m)^2 + \dots \tag{16.16}$$

Recall from Chapter 4 that we defined local coefficients of thermal expansion and isothermal compressibility, $\alpha$ and $\beta$, by

$$\alpha = -\frac{1}{\rho_m}\frac{\partial \rho}{\partial T}; \qquad \beta = \frac{1}{\rho_m}\frac{\partial \rho}{\partial p} \tag{16.17}$$

where it is understood that $\alpha$ and $\beta$ are evaluated at the average density $\rho_m$. Also recall that, for an ideal gas, these are

$$\alpha = \frac{1}{T_m}; \qquad \beta = \frac{1}{p_m} \tag{16.18}$$

Using (16.17) and (16.18) it can be demonstrated that

$$\frac{1}{\rho_m}\frac{\partial^2 \rho}{\partial T \partial p} = -\frac{1}{T_m p_m} \tag{16.19}$$

Substituting (16.18) and (16.19) into (16.16),

$$\frac{\rho - \rho_m}{\rho_m} = -\frac{T - T_m}{T_m} + \frac{p - p_m}{p_m} + \left(\frac{T - T_m}{T_m}\right)^2 - \frac{1}{2}\left(\frac{T - T_m}{T_m}\right)\left(\frac{p - p_m}{p_m}\right) + \dots \tag{16.20}$$

From (16.7), $(\rho - \rho_m)/\rho_m = (\rho_s + \rho')/\rho_m$. Because $|\rho_s|$ must be less than or equal to $\Delta\rho_s$, it follows from (16.13) that $\rho_s/\rho_m$ is $O(\epsilon)$ or less. Using this conclusion together with (16.14), it follows that $(\rho - \rho_m)/\rho_m$ is $O(\epsilon)$ or less. Thus, because the left side of (16.20) is $O(\epsilon)$, the right side must be also, and we may conclude that

$$\left(\frac{T - T_m}{T_m}\right)^2 < O(\epsilon^2); \qquad \left(\frac{T - T_m}{T_m}\right)\left(\frac{p - p_m}{p_m}\right) < O(\epsilon^2) \tag{16.21}$$

Thus to order $\epsilon$,

$$\rho = \rho_m[1 - \alpha(T - T_m) + \beta(p - p_m)] \tag{16.22}$$

which is the linear equation of state introduced as (4.34) in Chapter 4.

Substituting (16.5), (16.6), and (16.7) for $p$, $T$, and $\rho$ into (16.22),

$$\rho_s + \rho' = -\rho_m\alpha(T_s + T') + \rho_m\beta(p_s + p') \tag{16.23}$$

In the static state, $\rho' = T' = p' = 0$, so

$$\rho_s = \rho_m(\beta p_s - \alpha T_s) \tag{16.24}$$

from which it also follows that

$$\rho' = \rho_m(\beta p' - \alpha T') \tag{16.25}$$

We will simplify this below.

Turning to the continuity equation (16.2), substitution of (16.7) leads to

$$\nabla \cdot \mathbf{u} = -\frac{1}{\rho_m + \rho_s + \rho'}\frac{D}{Dt}(\rho_s + \rho') \tag{16.26}$$

Factoring out $\rho_m$,

$$\nabla \cdot \mathbf{u} = -\frac{1}{\rho_m}\left(1 + \frac{\rho_s}{\rho_m} + \frac{\rho'}{\rho_m}\right)^{-1}\frac{D}{Dt}(\rho_s + \rho') \tag{16.27}$$

The first parenthetical part in (16.27) can be expanded as an infinite power series; namely,

$$\frac{1}{1 + \dfrac{\rho_s}{\rho_m} + \dfrac{\rho'}{\rho_m}} = 1 - \frac{\rho_s}{\rho_m} - \frac{\rho'}{\rho_m} + \left(\frac{\rho_s}{\rho_m}\right)^2 + \frac{2\rho_s\rho'}{\rho_m^2} + \left(\frac{\rho'}{\rho_m}\right)^2 - \ldots \tag{16.28}$$

Using this and (16.13), (16.27) becomes

$$\nabla \cdot \mathbf{u} = -\frac{\epsilon}{\Delta\rho_s}\frac{D}{Dt}(\rho_s + \rho') + \left(\frac{\epsilon}{\Delta\rho_s}\right)^2(\rho_s + \rho')\frac{D}{Dt}(\rho_s + \rho') - \ldots \tag{16.29}$$

The first term on the right side of (16.29) is $O(\epsilon)$ or smaller; the second and subsequent terms are $O(\epsilon^2)$ or smaller. We may therefore conclude that to order $\epsilon$,

$$\nabla \cdot \mathbf{u} = 0 \tag{16.30}$$

which is the first result of the Boussinesq approximation.

In absence of motion, where $\rho' = p' = 0$, substitution of (16.5) and (16.7) into the vertical component of (16.1) leads to

$$\frac{\partial p_s}{\partial z} = -g(\rho_m + \rho_s) \tag{16.31}$$

Then substituting (16.5) and (16.7) for $p$ and $\rho$ in the pressure and buoyancy terms in (16.1), recalling that $\rho_m$ and $p_m$ are constants and that $\rho_s$ is independent of $y$ and $z$, and making use of (16.30) and (16.31),

$$\rho\frac{D\mathbf{u}}{Dt} = -\nabla p' - g\rho'\mathbf{k} + \mu\nabla^2\mathbf{u} \tag{16.32}$$

After substituting (16.7) into (16.32), an expansion similar to that used to obtain (16.29) leads to

$$\frac{D\mathbf{u}}{Dt} = -\frac{1}{\rho_m}\nabla p' - g\epsilon\frac{\rho'}{\Delta\rho_s}\mathbf{k} + \nu\nabla^2\mathbf{u} \tag{16.33}$$

where $\nu = \mu/\rho_m$. Despite involving the factor $\epsilon$, the buoyancy term in this equation is retained because it is essential for evaluating fluid accelerations. Motions are caused by buoyancy forces, and are therefore driven by fluctuations in the density field. Thus, the viscous force or the inertial force, or both, must be of the same order as the buoyancy force. (We will return to this point below.) Assuming that the acceleration $\partial\mathbf{u}/\partial t$ contained in the substantive derivative on the left side of (16.33) is of the same order as the term $(g\epsilon\rho'/\Delta\rho_s)\mathbf{k}$,

$$\left|\frac{\partial\mathbf{u}}{\partial t}\right| \sim \left|g\epsilon\frac{\rho'}{\Delta\rho_s}\mathbf{k}\right| \tag{16.34}$$

which implies that

$$\frac{\partial\mathbf{u}/\partial t}{(\epsilon\rho'/\Delta\rho_s)\mathbf{k}} \sim g \tag{16.35}$$

This indicates that the gravitational acceleration $g$ in general must be much larger than $\partial\mathbf{u}/\partial t$.

Equation (16.33) can be simplified further. The vertical component of (16.33) is

$$\mathbf{u}\cdot\nabla w + \frac{\partial w}{\partial t} = -\frac{1}{\rho_m}\frac{\partial p'}{\partial z} - g\epsilon\frac{\rho'}{\Delta\rho_s} + \nu\nabla^2 w \tag{16.36}$$

Using (16.13), the pressure and buoyancy terms in (16.36) may be written as

$$-\frac{1}{\rho_m}\frac{\partial p'}{\partial z} - g\epsilon\frac{\rho'}{\Delta\rho_s} = -\frac{1}{\rho_m}\left(\frac{\partial p'}{\partial z} + \frac{g\rho_m}{p_m}p'\right) + g\frac{T'}{T_m} \tag{16.37}$$

According to the hydrostatic equation we may write $p_m = g\rho_m H$. That is, the quantity

$$H = \frac{p_m}{g\rho_m} \tag{16.38}$$

in (16.37) may be regarded as the thickness of a fluid layer with constant density $\rho_m$ whose pressure varies from $p_m$ at the bottom to zero at the top. Substituting (16.31) for $\partial p_s/\partial z$ into the expression (16.8) for $Z_p$ and rearranging,

$$Z_p = \frac{p_m}{g(\rho_m + \rho_s)} \tag{16.39}$$

Factoring $\rho_m$ from the denominator, expanding this as a power series and comparing the result with (16.38), we may conclude that $H = Z_p + O(\epsilon)$ since, from above, we obtained the result that $\rho_s/\rho_m$ is $O(\epsilon)$. Recall that we insisted that $Z \ll Z_p$. Therefore, because $p'/Z \leq \partial p'/\partial z$, we may also conclude that $p'/H = (g\rho_m/p_m)p'$ in (16.37) is negligibly small compared to $\partial p'/\partial z$, in which case (16.33) can be simplified to

$$\frac{D\mathbf{u}}{Dt} = -\frac{1}{\rho_m}\nabla p' + g\alpha T'\mathbf{k} + \nu\nabla^2\mathbf{u} \tag{16.40}$$

This is the second result of the Boussinesq approximation.

The preceding dynamical arguments lead to a simplification of (16.25). Namely, since the contribution of the pressure fluctuation $p'$ to the buoyancy term in (16.37) is small relative to the contribution of the temperature fluctuation $T'$, we may approximate (16.25) as

$$\rho' = -\rho_m \alpha T' \tag{16.41}$$

This is the third result of the Boussinesq approximation.

In absence of motion the energy equation (16.3) reduces to

$$K_T \nabla^2 T_s = 0 \tag{16.42}$$

which defines the static conduction state. After substituting (16.6) into (16.3), and using (16.42), we then obtain

$$\rho c_V \left( \mathbf{u} \cdot \nabla T + \frac{\partial T'}{\partial t} \right) + p \nabla \cdot \mathbf{u} = K_T \nabla^2 T' \tag{16.43}$$

where the dissipation function $\Phi$ has been neglected. At this point it is tempting to conclude that the term $p \nabla \cdot \mathbf{u}$ can be neglected based on the result (16.30). However, the term $p \nabla \cdot \mathbf{u}$ is of the same order as the other terms in (16.43). Dividing (16.6) by $p_m$,

$$\frac{p}{p_m} = 1 + \frac{p_s}{p_m} + \frac{p'}{p_m} \tag{16.44}$$

The quantity $p_s/p_m$ is $O(Z/H)$. (Why?) Then following the arguments leading to (16.40), we may expect that $|p'/p_m| = O(\Delta p_s/p_m)$. It follows from (16.44) that $p = p_m + O(Z/H)$. Using this result, noting that $\epsilon/\Delta \rho_s = 1/\rho_m$, then substituting (16.24), (16.41), and (16.18) into (16.29), we obtain

$$p \nabla \cdot \mathbf{u} = p_m \frac{D}{Dt} \left( \frac{T_s + T'}{T_m} - \frac{p_s}{p_m} \right) \tag{16.45}$$

where all terms of order $Z/H$ and smaller have been neglected. Now, using (16.31),

$$\frac{D}{Dt} \left( -\frac{p_s}{p_m} \right) = -\frac{1}{p_m} w \frac{\partial p_s}{\partial z} = -wg \frac{\rho_m}{p_m} - wg \frac{\rho_s}{p_m} \tag{16.46}$$

It is left as an exercise to demonstrate that the last term in (16.46) is negligibly small relative to $-wg\rho_m/p_m$, in which case (16.45) becomes

$$p \nabla \cdot \mathbf{u} = \frac{p_m}{T_m} \frac{D}{Dt} (T_s + T') + wg\rho_m \tag{16.47}$$

Substituting (16.47) into (16.43) and rearranging, we then obtain

$$\frac{DT'}{Dt} + w \left( \frac{\partial T_s}{\partial z} - \Gamma \right) = \kappa \nabla^2 T' \tag{16.48}$$

where $\Gamma = -g/c_p$ is the adiabatic lapse rate, $\kappa = K_T/\rho_m c_p$ is the thermal diffusivity, and $c_p = c_V + p_m/\rho_m T_m$ is the specific heat defined for constant pressure.

This is the final result of the Boussinesq approximation.

Let us now collect the Boussinesq equations:

$$\frac{D\mathbf{u}}{Dt} = -\frac{1}{\rho_m}\nabla p' + g\alpha T'\mathbf{k} + \nu\nabla^2\mathbf{u} \tag{16.49}$$

$$\nabla \cdot \mathbf{u} = 0 \tag{16.50}$$

$$\frac{DT'}{Dt} + w\left(\frac{\partial T_s}{\partial z} - \Gamma\right) = \kappa\nabla^2 T' \tag{16.51}$$

$$\rho_s = \rho_m(\beta p_s - \alpha T_s) \tag{16.52}$$

$$\rho' = -\rho_m\alpha T' \tag{16.53}$$

These equations, which govern convection of an ideal gas, are equivalent to those which govern convection of an incompressible fluid, but with two differences. The parenthetical part of the energy equation (16.51) describes the static temperature gradient in excess of the adiabatic gradient, and the specific heat $c_V$ is replaced by $c_p$. The validity of these equations is subject to the two conditions noted at the beginning of this section: the vertical dimension $Z$ must be much less than any of the scale heights (16.8), and fluctutations in pressure and density induced by fluid motions must not exceed the total variations in these quantities under static conditions.

Notice that (16.49) and (16.51) both involve $\mathbf{u}$ and $T$. The immediate implication is this: The velocity field is governed by the temperature field, but the temperature field in turn depends on the velocity field through advection of heat. Recall that, in contrast, the temperature field in forced convection depends on the velocity field, but the velocity field does not depend on the temperature field. The velocity field can be determined independently of temperature, then the temperature field is obtained from knowledge of the velocity field. With free convection, however, this is not possible. For this reason, theoretical descriptions of free convection are difficult to achieve.

The Boussinesq equations are simplified for the case of an incompressible liquid, and further linearized for small perturbations in the state variables and the velocity about the static conduction state, in Example Problem 16.5.1.

## 16.2 DIMENSIONLESS QUANTITIES

Experimental and theoretical descriptions of convection often involve several important dimensionless quantities. We will start by listing four of these quantities, then loosely follow discussions of Tritton (1988) and others to examine their physical significance with reference to the Boussinesq approximation.

The *Grashof number Gr* is defined as

$$Gr = \frac{g\alpha\Theta L^3}{\nu^2} \tag{16.54}$$

where $\Theta$ denotes a characteristic temperature difference and $L$ is a characteristic length scale. In the context of an experiment involving Rayleigh–Bénard convection, for example, $\Theta$ may be treated as the temperature difference $\Delta T$ between the upper and lower boundaries separated by the vertical distance $L = Z$ (Figure 6.4). The *Prandtl number Pr* is defined as

$$Pr = \frac{\nu}{\kappa} \tag{16.55}$$

which is entirely a property of the fluid. The *Rayleigh number Ra* is

$$Ra = GrPr = \frac{g\alpha\Theta L^3}{\nu\kappa} \tag{16.56}$$

and the *Nusselt number Nu* is

$$Nu = \frac{q_T L}{K_T \Theta} \tag{16.57}$$

where $q_T$ is a characteristic heat flux density.

Dimensional analysis leads to the conclusion that a condition of dynamical similitude exists for geometrically similar systems when the systems have the same values of $Gr$ and $Pr$. This means that, unlike adjusting a single quantity, the Reynolds number $Re$, for purely mechanical flows, both $Gr$ and $Pr$ must be adjusted in experiments to achieve similitude between two convecting systems. This is straightforward when the two systems involve the same fluid at the same mean temperature. For a fluid with given Prandtl number $Pr$ and coefficient of thermal expansion $\alpha$, similitude can be readily achieved by varying $\Theta$ or $L$, or both, depending on the experimental apparatus. Suppose a natural system that is to be mimicked experimentally involves a fluid with $Pr_1 = \nu_1/\kappa_1$ and $\alpha_1$ giving $Gr_1 = g\alpha_1\Theta_1 L_1^3/\nu_1^2$. Then suppose that the experiment is to involve the same fluid with $Gr_2 = g\alpha_1\Theta_2 L_2^3/\nu_1^2$. To achieve similitude then merely requires that

$$\Theta_1 L_1^3 = \Theta_2 L_2^3 \tag{16.58}$$

This suggests a strategy for adjusting $\Theta_2$ and $L_2$ depending on the fluid and characteristic length of the prototype system—so long as the mean temperatures of the prototype system and experiment are the same, or the fluid properties $\nu$, $\kappa$, and $\alpha$ are insensitive to variations in temperature, such that the Prandtl numbers and $\alpha$ are the same. Achieving similitude between systems that involve the same fluid at different mean temperatures, or different fluids altogether, is not as straightforward.

The Grashof number $Gr$ provides an indication of the relative importance of inertial and viscous forces, analogous to the Reynolds number $Re$ for purely mechanical systems; but this interpretation of $Gr$ for free convection is not as simple as that associated with $Re$. For simplicity, consider steady motion. The buoyancy force must be balanced by other terms in the momentum equation. Therefore, the inertial force or the viscous force, or both, must be of the same order as the buoyancy force. But since motion is caused by the buoyancy force, the inertial and viscous forces cannot become large relative to the buoyancy force.

Suppose that the viscous force is of the same order as the buoyancy force; that is,

$$\left| \nu \nabla^2 \mathbf{u} \right| \sim \left| g\alpha T' \mathbf{k} \right| \tag{16.59}$$

which implies that

$$\frac{\nu U}{L^2} \sim g\alpha\Theta \tag{16.60}$$

where $U$ is a characteristic fluid speed. This suggests that

$$U \sim \frac{g\alpha\Theta L^2}{\nu} \tag{16.61}$$

Now, comparing the relative magnitudes of inertial and viscous forces,

$$\frac{|\mathbf{u}\cdot\nabla\mathbf{u}|}{|\nu\nabla^2\mathbf{u}|} \sim \frac{UL}{\nu} \sim \frac{g\alpha\Theta L^3}{\nu^2} = Gr \tag{16.62}$$

which we will return to momentarily.

Consider the alternative possibility, that the inertial force is of the same order as the buoyancy force; that is,

$$|\mathbf{u}\cdot\nabla\mathbf{u}| \sim |g\alpha T'\mathbf{k}| \tag{16.63}$$

Thus,

$$\frac{U^2}{L} \sim g\alpha\Theta \tag{16.64}$$

which implies that

$$U \sim (g\alpha\Theta L)^{1/2} \tag{16.65}$$

Here we may interpret $U$ as a measure of the speed of motion arising from the temperature variation $\Theta$. Again comparing inertial and viscous forces,

$$\frac{|\mathbf{u}\cdot\nabla\mathbf{u}|}{|\nu\nabla^2\mathbf{u}|} \sim \frac{UL}{\nu} \sim \left(\frac{g\alpha\Theta L^3}{\nu^2}\right)^{1/2} = Gr^{1/2} \tag{16.66}$$

When $Gr$ is small, according to (16.62), the inertial force is negligible relative to the viscous and buoyancy forces. It is not appropriate, however, to claim that a large value of $Gr$, based on (16.62), implies a large inertial force. This would connote that the inertial force is large relative to the buoyancy force, which is in contradiction to the argument above, that the inertial force cannot become large relative to the buoyancy force. In turn, (16.66) implies that when $Gr$ is large, the viscous force is negligible relative to the inertial and buoyancy forces. But a small value of $Gr$, based on (16.66), does not imply a large viscous force, since this would connote that the viscous force is large relative to the buoyancy force. Thus, whereas $Gr$ provides some measure of the relative importance of inertial and viscous forces, a simple conclusion is not obtained from the ratio of these forces.

The Prandtl number $Pr$ is the ratio of the tendency of a fluid to diffuse momentum (and vorticity), as characterized by $\nu$, and its tendency to diffuse heat, as characterized by $\kappa$. Qualitatively, the viscosity $\nu$ governs the rate at which a velocity field involving shearing motion responds to a driving force, for example, as measured by the rate at which the boundary-layer in a conduit approaches a fully developed condition when a pressure gradient is imposed on the conduit (Example Problem 12.5.6). Conversely, it governs the rate at which such a boundary layer relaxes when the pressure gradient inducing motion is removed. Similarly, the thermal diffusivity $\kappa$ governs the rate at which a temperature field responds by conduction to an imposed temperature variation. A fluid whose Prandtl number is 1 will tend to diffuse momentum and heat at an equal rate. As a point of reference, the Prandtl number of dry air at 15°C is 0.72, and varies only slightly with temperature. The Prandtl number of water varies from 13.4 at 0°C to 1.8 at 100°C.

Consider the relative magnitudes of heat advection and heat conduction within a convecting fluid. With steady motion,

$$\frac{\left|\mathbf{u} \cdot \nabla T'\right|}{\left|\kappa \nabla^2 T'\right|} \sim \frac{UL}{\kappa} \sim \frac{UL}{\nu}\frac{\nu}{\kappa} \tag{16.67}$$

Thus for small $Gr$, using (16.62),

$$\frac{\left|\mathbf{u} \cdot \nabla T'\right|}{\left|\kappa \nabla^2 T'\right|} \sim Gr\,Pr = Ra \tag{16.68}$$

and for large $Gr$, using (16.66),

$$\frac{\left|\mathbf{u} \cdot \nabla T'\right|}{\left|\kappa \nabla^2 T'\right|} \sim Gr^{1/2}Pr = (Ra\,Pr)^{1/2} \tag{16.69}$$

The quantity $UL/\kappa$ in (16.67) is like a Peclet number, and the quantity $UL/\nu$ is like a Reynolds number. Moreover, based on (16.68), evidently the Rayleigh number $Ra$ may be interpreted as the ratio of the advection of heat to the conduction of heat. Herein the significance of the Rayleigh number, first introduced in Chapter 6, can be further clarified.

Advection of heat involves fluid motion whereas conduction occurs with or without motion. Since convective motions, and therefore heat advection, are driven by the same variations in temperature that induce conduction, temperature variations have a dual role in the transport of heat in convecting systems. One role is purely thermal, as embodied in Fourier's law, and one is mechanical, acting through the destabilizing effects of buoyancy forces. Let us therefore consider two forms of the Rayleigh number to obtain complementary interpretations of its significance. We first rewrite the Rayleigh number as

$$Ra = \frac{g\alpha\Theta L^3}{\nu\kappa} = \left(\frac{g\alpha\Theta L^2}{\nu}\right)(\rho c_p\Theta)\left(\frac{1}{K_T\dfrac{\Theta}{L}}\right) \tag{16.70}$$

The first parenthetical quantity, based on (16.61), is a measure of the speed of motion. The second is a measure of the heat stored in (or released from) a parcel of fluid that undergoes a temperature change equal to $\Theta$. Thus, at least for small $Gr$, the product of the first two parenthetical quantities in (16.70) is a measure of the rate of advective transport of stored heat. The denominator of the last parenthetical quantity is a measure of the rate of heat conduction across the fluid column. (We will examine this further below.) From this perspective the Rayleigh number characterizes the relative magnitudes of advected and conducted heat.

Let us now rewrite the Rayleigh number as

$$Ra = (g\alpha\Theta L)(\rho c_p \Theta L)\left(\frac{1}{\nu}\right)\left(\frac{1}{K_T\dfrac{\Theta}{L}}\right) \tag{16.71}$$

The first parenthetical quantity is a measure of the buoyancy force produced by the temperature variation over the fluid column, and is thus a measure of the destabilizing influence of this temperature variation. The viscosity in the third parenthetical part is a measure of the mechanical resistance to convective motion. The denominator of the last parenthetical quantity is, like above, a measure of the rate of heat conduction. Its effect is to contribute to the stability of the fluid column inasmuch as conduction of heat locally dissipates destabilizing temperature variations. Consider, for example, a Rayleigh–Bénard configuration in which the temperature of the lower boundary is suddenly increased, such that the temperature gradient at this boundary is momentarily elevated. A rapid, upward dissipation of heat by conduction in this case would decrease the gradient and thereby tend to eliminate a destabilizing thermal stratification of fluid near the boundary. Moreover, after the onset of convection, dissipation of heat from hot ascending fluid to surrounding cooler fluid tends to decrease the density contrasts that drive this motion. Herein arises the significance of the second parenthetical quantity in (16.71). This quantity is a measure of the heat storage of the column (associated with a temperature change equal to $\Theta$), and is thus a measure of the capacity of the fluid to store heat without changing temperature. In this vein, the equation governing heat diffusion,

$$\frac{\partial T}{\partial t} = \frac{K_T}{\rho c_p}\nabla^2 T \tag{16.72}$$

indicates that the effect of an increasing specific heat $c_p$ is to decrease the rate of change in temperature for a given divergence of heat flux. Thus, a large heat storage capacity tends to attenuate the rate at which temperature gradients, and therefore density contrasts, are eliminated by conduction of heat. We may therefore conclude that the numerator of the Rayleigh number (16.56) characterizes factors that tend to destabilize the fluid column, whereas the denominator characterizes factors that tend to stabilize it.

Let us rearrange the Nusselt number $Nu$ to

$$q_T = K_T\frac{\Theta}{L}Nu \tag{16.73}$$

When $Nu = 1$, this is merely Fourier's law for one-dimensional conduction of heat through a static fluid layer of thickness $L = Z$ over which a steady temperature difference $\Theta = \Delta T$ is maintained. Therefore, in the case of a Rayleigh–Bénard configuration (where $L = Z$), the Nusselt number is the ratio of the actual rate of heat transfer (by advection and conduction) to the rate that would otherwise occur in the static conduction state. Thus the onset of convection is marked by an increase in $Nu$ above unity. This condition, in fact, provides a way to determine the onset of convection in laboratory experiments. Generally, $Nu$ increases with $Gr$ after the onset of convection. Moreover, as the temperature gradient increases, the rate of heat transfer increases disproportionately faster. This indicates that convective motions become more vigorous with an increasing temperature gradient and, in turn, advective heat transport is more rapid.

## 16.3 LABORATORY EXPERIMENTS WITH RAYLEIGH–BÉNARD AND HELE–SHAW FLOWS

For reasons mentioned at the end of Section 16.1, much of our understanding of free convection derives from laboratory experiments involving Rayleigh–Bénard and Hele–Shaw configurations. A Rayleigh–Bénard configuration generally consists of fluid residing between two parallel, horizontal plane boundaries separated by a vertical distance $Z$ (Figure 6.4). Although we may envision for mathematical reasons that these boundaries extend infinitely far in the horizontal directions parallel to the $x$-axis and $y$-axis, in practice a Rayleigh–Bénard configuration must involve sidewalls. The two boundaries are heated and cooled according to specified conditions. For example, heating and refrigeration coils may be used to maintain the two boundaries at constant temperatures $T_0$ and $T_1$. Another possibility is that the upper boundary may consist of a free liquid surface that is cooled by evaporation. A Hele–Shaw configuration generally consists of two parallel, vertical plane boundaries separated by a small aperture $Y$ (Figure 16.1). Heating and cooling may involve any of the boundaries, for example, the bottom and top, or the two sides.

Several boundary conditions with respect to flow are possible. A *stress-free boundary* behaves like a surface where slip flow can occur, but tangential stresses vanish. This is a good approximation of the conditions at the free surface of a liquid, but it is an idealization in the case of a solid boundary. Nonetheless, the idea of a stress-free solid boundary is of theoretical interest. A *rigid boundary,* in contrast, is one where the no-slip condition applies, and tangential stresses may be nonzero. This realistically describes the conditions at a solid boundary. Combinations of these are possible. The combinations of *free-free* and *rigid-free* boundaries (where the lower boundary of a Rayleigh–Bénard configuration, for example, is stress-free) are of theoretical interest. The combinations of *free-rigid* and *rigid-rigid* boundaries are physically realistic. In addition, boundaries may be permeable with respect to flow; here we will consider only boundaries that are impermeable, such that the component of flow normal to the boundary is zero.

The onset of convection in a Rayleigh–Bénard configuration occurs at distinct values of the Rayleigh number depending on the boundary conditions. In the free-free case, this critical value $Ra_c = 657.5$; in the free-rigid case, $Ra_c = 1101$; and in the rigid-rigid case, $Ra_c = 1708$ (Example Problem 16.5.1). This increasing value of $Ra_c$ reflects a stabilizing influence provided by viscous friction at the

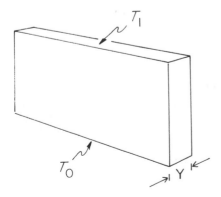

**Figure 16.1** Hele–Shaw configuration with horizontal aperture $Y$, and lower and upper boundary temperatures $T_0$ and $T_1$; $Y$ may be several mm or less in some experiments.

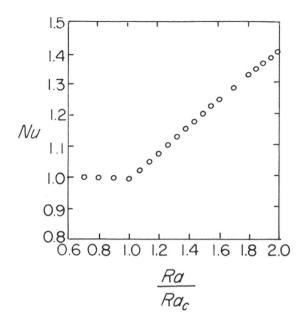

**Figure 16.2** Variation in Nusselt number $Nu$ with Rayleigh number $Ra$ illustrating conduction state ($Ra < Ra_c$, $Nu = 1$) and transition to convection ($Ra \geq Ra_c$, $Nu > 1$); adapted from Howle, Behringer, and Georgiadis (1993), reprinted with permission from *Nature*, copyright 1993 Macmillan Magazines Limited.

boundaries; that is, an increasing temperature variation is required to destabilize the fluid column when one, then both, of the boundaries are rigid.

For $Ra < Ra_c$, the static conduction state persists with increasing $Ra$ such that an increasing temperature difference between the boundaries is fully compensated by an increasing heat flow due entirely to conduction ($Nu = 1$; Figure 16.2). The onset of convection is then marked by an increase in the Nusselt number $Nu$ above unity at $Ra_c$. The steep slope of the curve relating $Nu$ to $Ra$ for $Ra > Ra_c$ reflects that as the temperature gradient increases, the rate of heat transfer, now involving both conduction and advection, increases disproportionately faster.

At Rayleigh numbers slightly greater than $Ra_c$, the flow field in Rayleigh–Bénard convection tends to become organized into regular patterns of cells. The simplest case involves two-dimensional rolls (Figure 16.3). In principle, there is no preferential alignment of roll axes in the $xy$-plane. As a consequence, rolls with different orientations can occur simultaneously in different regions of an apparatus. Nonetheless, rolls often become aligned preferentially for certain sidewall geometries. For example, roll axes will become aligned parallel to the short side of a rectangular apparatus. (This alignment probably produces the least total sidewall friction relative to other possible configurations.) Similarly, roll-like cells that develop in an apparatus with circular planform tend to bend and vary in width, such that their axes become nearly normal to the sidewalls at the roll ends.

For rigid-rigid boundaries, the horizontal wavelength of the velocity and temperature fluctuations about the static state, predicted from stability theory, is about $2Z$ (Example Problem 16.5.1). This means that sites of hot ascending fluid ought to be separated by a distance of $2Z$; that is, the width of each roll is about equal to $Z$. This indeed occurs in experiments, although the height–width ratio can be made to vary, at least momentarily, with nonuniform heating initially (see below). Thus flow sequentially alternates between clockwise and counterclockwise motion, with upward flow at one margin between neighboring rolls and downward flow at

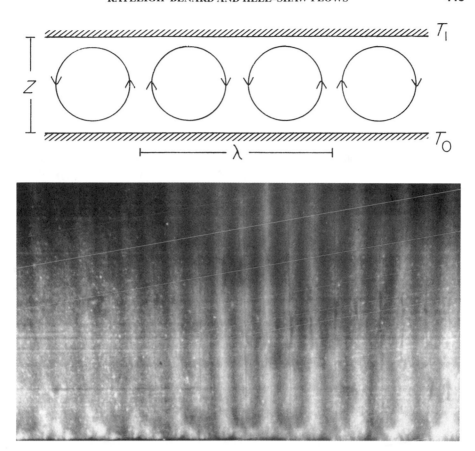

**Figure 16.3** Schematic diagram and photograph (shadowgraph) of two-dimensional roll cells; photograph courtesy of R. Krishnamurti.

the next margin (Figure 16.3). Cool fluid descends to the lower boundary, adsorbs heat by conduction as it traverses the lower boundary, then ascends when it is sufficiently buoyant. Hot ascending fluid reaches the upper boundary, loses heat by conduction as it traverses this boundary, then begins to descend again. With free-rigid boundaries, cooling at the upper boundary can occur by evaporation.

Like roll spacing, other cell patterns can be "preselected" by nonuniform heating (or cooling). An interesting experiment conducted by R. Krishnamurti (1968) to demonstrate this point involves placing a shallow vessel containing water and flow tracer on the surface of a flat heater. A paper "snowflake" with an intricate cutout pattern is placed into a slot beneath the vessel. The heater nonuniformly warms the base of the vessel through the openings in the snowflake. The snowflake is then removed after a few seconds, and a convective cell pattern that mimics the snowflake pattern momentarily develops. But this pattern generally does not persist with the uniform heating that exists once the snowflake is removed, and herein rolls take on another significance.

Stability analysis and experiments suggest that, for rigid-rigid boundaries and $Ra = Ra_c = 1708$, rolls with horizontal wavelength $2Z$ represent a stable cell configuration. (For $Ra > Ra_c$, a band of stable wavelengths close to $2Z$ exists.)

That is, if the velocity (or temperature) field of an array of roll cells is momentarily perturbed, the perturbations will decay and the cell array will tend to return to its original state. In contrast, rolls possessing a wavelength that is very different from 2Z, if perturbed, will evolve into more complex shapes that presumably are more stable. Likewise, other cell shapes (hexagons, for example) typically are unstable in the sense that, if perturbed, they tend to degenerate to more stable cell configurations. Students are urged to consult Schlüter, Lortz, and Busse (1965) regarding these points.

Nonetheless, other cell shapes occur at Rayleigh numbers slightly greater than $Ra_c$. Perhaps most familiar are hexagonal cells. (It is interesting that the hexagonal cells described by Bénard in 1900, which prompted Lord Rayleigh to systematically examine thermally driven convection, were in fact produced by surface tension effects rather than being thermally driven.) Hexagonal cells tend to develop when vertical asymmetries in fluid properties exist. Such asymmetries can occur when fluid viscosity varies strongly as a function of temperature, or when the mean temperature of the fluid varies with time. Not all polygonal shapes are probable. Only a few, including hexagons, form regular tessellated patterns. With hexagonal cells, two flow patterns are possible. One involves upward flow in the cell centers and downward flow at the margins. This usually occurs with liquids (Figure 16.4). The other involves downward flow in the cell centers and upward flow at the margins. Often, polygonal arrangements involve mixtures of cells with different numbers of sides, including hexagons.

Rayleigh–Bénard convection can exhibit a rich sequence of cell patterns and flow behaviors with steadily increasing $Ra$ above $Ra_c$. In principle, the initial cell configuration, rolls for example, depends solely on the wavelength of the temperature and velocity perturbations initially excited at the critical state associated with $Ra_c$ (Example Problem 16.5.1). Cell patterns and flow, however, are sensitive to random influences associated with the apparatus, independent of the (theoretical)

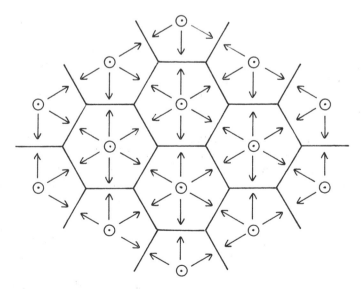

**Figure 16.4** Schematic diagram and photograph of hexagonal convection cells with upward flow in centers and downward flow at margins; photograph courtesy of R. Krishnamurti.

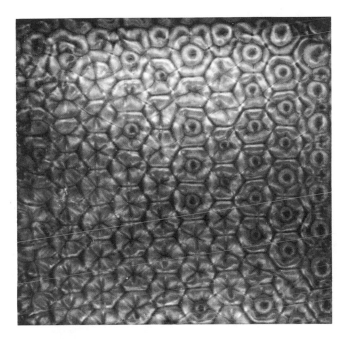

**Figure 16.4**    (*Continued*)

initial conditions at $Ra_c$. In this regard, rolls are the least sensitive to deviations from ideal conditions. That preferred cell orientations occur further suggests that nonlinear behavior (interaction among cells and sidewalls) is important. In any case, convection is characterized initially by stationary cells and steady motion (Figure 16.5). With increasing $Ra$, these may in some situations give way to more complex cell patterns that likewise involve steady motion. At sufficiently high $Ra$, steady motion ceases to occur; convection is characterized by time-dependent motions and continuously changing cell patterns, eventually giving way to fully turbulent convection.

The vertical distribution of mean temperature within turbulent convection reflects an overall mixing of the fluid associated with hot ascending plumes and cool descending plumes. The mean temperature at a position $z$ may be obtained as a time average or as a spatial average over horizontal planes. Temperature gradients close to the boundaries are steep, whereas the averaged temperature within the interior of the fluid is nearly uniform (Figure 6.4). Moreover, the gradients close to the boundaries are significantly greater than the average gradient $-\Theta/Z$ that otherwise would occur throughout the fluid column in a static conduction state. Thus, applying Fourier's law to fluid very close to the boundaries, the total heat flow through the column during turbulent convection is significantly greater than that which would occur in a static state. However, the contribution by conduction to this total heat flow is important only near the boundaries. Although heat flow by conduction occurs locally in the interior of the flow, the uniform temperature here implies that it is negligible when spatially averaged. Heat transfer in the interior is therefore dominated by advection associated with plume motions.

Similar behavior occurs with Hele–Shaw convection when a temperature difference between the top and bottom boundaries is maintained. This is essentially a Rayleigh–Bénard configuration where two opposing sidewalls are close to each

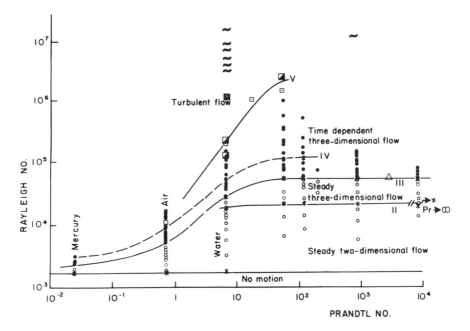

**Figure 16.5**    Regime diagram of Prandtl number $Pr$ versus Rayleigh number $Ra$ for Rayleigh–Bénard convection; courtesy of Krishnamurti.

other. Sidewall friction therefore strongly influences the behavior of the fluid. Organized convection cells occur at $Ra$ slightly above $Ra_c$, then these give way to more complex motions with increasing $Ra$.

Also of interest is the situation where the two sidewalls, separated by a distance $Y$, are maintained at temperatures $T_0$ and $T_1$ (Figure 16.6). If $T_1 < T_0$, fluid near the left sidewall is relatively warm and buoyant, and fluid near the right sidewall is cooler and denser. This induces a sense of torque, or horizontal vorticity, normal to the temperature gradient. Unlike a Rayleigh–Bénard configuration wherein a vertical temperature gradient may or may not lead to instability, this situation involving a horizontal temperature gradient is one where fluid cannot remain at rest. A static conduction state does not occur. To illustrate this, we start by writing the momentum equation (16.1) in a form consistent with a static state. That is, neglecting terms in (16.1) involving velocity, a static configuration must satisfy the condition:

$$\nabla p = -g\rho\mathbf{k} \tag{16.74}$$

which describes a balance between pressure and buoyancy forces. Upon taking the curl of both sides of (16.74), the left side vanishes according to (17.45), and therefore using (17.44) and (17.48),

$$\nabla \times (g\rho\mathbf{k}) = g(\nabla\rho \times \mathbf{k}) = 0 \tag{16.75}$$

This describes the contribution to vorticity due to the buoyancy force, which involves a density gradient. Upon expanding (16.75) it becomes apparent that this condition can be satisfied only if

$$\frac{\partial\rho}{\partial y} = 0 \tag{16.76}$$

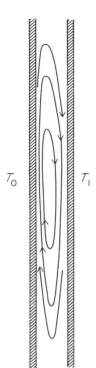

**Figure 16.6** Hele–Shaw configuration with sidewall temperatures $T_0 > T_1$ flow cells with circulation; after Elder (1965).

where $\partial \rho / \partial x$ is assumed to be zero. But because $\partial \rho / \partial y \neq 0$ due to the temperature variation across $Y$, this configuration is necessarily unstable. Convection therefore occurs spontaneously when $T_0 \neq T_1$.

At low $Ra$, circulation involving a single, vertically elongated cell with horizontal axis occurs throughout the apparatus (Figure 16.6). With increasing $Ra$, this overall circulation persists, but convection typically involves a more complex multiple-cell arrangement. Like Rayleigh–Bénard convection, the Nusselt number, now defined in terms of $Y$, increases with $Ra$.

These results of theoretical and experimental studies of convection bear on several geological phenomena. Of immediate use are values of the critical Rayleigh number, which provide a measure of the likelihood that convection will occur in a fluid system. Let us start with a simplistic example involving a tabular magma body with vertical height $Z = 100$ m, and choose conservative values of magma properties: $\rho = 2{,}700(\text{kg m}^{-3})$, $\mu = 100(\text{N s m}^{-2})$, $\alpha = 5 \times 10^{-5}(°\text{C}^{-1})$, and $\kappa = 1 \times 10^{-6}(\text{m}^2 \text{ s}^{-1})$. Assuming rigid-rigid boundaries with $Ra_c = 1708$, substitution of these values into (16.56) leads to the conclusion that convection is possible with a temperature variation $\Theta$ of less than 1 °C. (As a point of reference, the geothermal gradient varies from about 0.015 °C m$^{-1}$ to 0.075 °C m$^{-1}$, giving a temperature variation over 100 m of 1.5 °C to 7.5 °C.) We may infer that if a modest temperature variation persists over such a magma body, cooling of the magma, at least initially, is likely to be enhanced by upward transfer of heat by advection. (We will consider another part of this problem below.)

The exponent of three on $Z$ in (16.56) means that convection is sensitive to this parameter. One implication is that a small $Z$ tends to suppress convection. Thus,

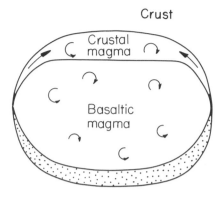

**Figure 16.7**    Schematic diagram of accumulation of buoyant melt layer next to roof of magma changer; after Kerr (1994).

a fluid-filled (horizontal) fracture with small aperture can conduct heat in a static state. Conversely, wide fractures are susceptible to convection and rapid transport of heat. This problem, however, is not entirely analogous to the Rayleigh–Bénard configuration described above, because the walls of a fluid-filled fracture in a rock are not necessarily maintained at constant temperatures. Envision a fluid-filled fracture within a rock that is conducting heat over a scale larger than the fracture due to a regional temperature gradient. Heat is delivered by conduction to the fracture wall. If convection occurs, heat is rapidly transported away from the wall by advection, which leads to cooling of the wall and suppression of the initial temperature gradient that induced convection. Similar comments apply to a Hele–Shaw configuration. Also note that, in this situation, the fracture is not a barrier to heat flow over the larger scale; in fact, convection enhances the transfer of heat across the fracture. In a similar vein, vigorous convection within a thermal geyser, enhanced by open, cavernous plumbing, brings hot water up and cool water down, and thereby tends to distribute heat evenly over the fluid column, and to contribute to dissipation of heat at the surface without eruption (Example Problem 4.10.3).

Thus far we have considered stationary boundaries that are not physically altered by the convecting fluid. Several important geological processes, however, involve fluid-boundary interactions whereby the boundary and nearby fluid may be physically and chemically altered during convection. Envision, for example, the roof of a magma chamber that contains uniformly hot magma. If the temperatures of the country rock and magma are sufficiently different, melting of the roof can occur. Suppose that the magma and country rock are compositionally different. This is the case, for example, when a basaltic magma intrudes into crustal rocks of granitic composition. If the density of the melt that is produced is less than that of the underlying magma, the melt will accumulate as a buoyant layer next to the roof (Figure 16.7). The temperature at the base of this melt layer is nearly equal to that of the underlying hot magma, whereas the temperature at the top of the layer is close to that of the country rock. If the melt layer accumulates to sufficient thickness, thermally driven convection may begin within the melt layer. This advects heat upward from the underlying hot magma to the roof, which in turn contributes to further melting. This in effect describes a Rayleigh–Bénard configuration, where the lower boundary is not rigid and the upper boundary, although rigid, is not stationary. This has been proposed as a mechanism for producing granitic melts associated with intrusions of basaltic magma into silicic country rock with lower melting temperature. Students are urged to consult Campbell and Turner (1987), and Huppert and Sparks (1988a, b), regarding these ideas.

Alternatively, if the density of the melt is greater than that of the underlying magma, compositionally driven convection may occur. This brings hot magma to the roof, which in turn contributes to further melting. Kerr (1994) uses a simple analogy to make this point: Consider ice cubes that are melting in a glass of water versus ice cubes that are melting in a glass of Scotch. The ice in the Scotch melts faster because the meltwater is denser than the surrounding alcohol-rich fluid. The dense meltwater tends to sink rapidly after being produced, and this convective overturning brings warmer fluid in contact with the ice cube. Similar remarks apply to fluid conditions near a cool xenolith immersed within a hot magma. In treating this problem, Kerr supposes that far-field conditions (magma temperature and composition) do not change during melting, that the temperature contrast between the magma and country rock is sufficiently large that compositional diffusion can be neglected relative to heat diffusion, and that thermal effects on density are negligible relative to compositional effects. Similar arguments apply to melting of the floor of a magma chamber. In particular, if the density of the melt is less than that of the magma, compositionally driven convection can occur at the floor.

The mechanisms of convection outlined above are particularly important for understanding the dynamics and geochemistry of magmas. Whereas thermally driven convection within a buoyant layer of melt near the roof of a magma chamber may produce a melt whose composition is like that of the country rock, compositionally driven convection is likely to contribute to mixing of magma and melted country rock. Students are urged to examine papers in the Reading list that bear on this topic.

## 16.4 CONVECTION IN POROUS MEDIA

Consider a Rayleigh–Bénard configuration occupied by porous material that is fully saturated with liquid, where the lower and upper boundaries are maintained at temperatures $T_0$ and $T_1$. This situation represents, for example, a horizontal stratified layer that is uniformly heated from below. When $T_1 < T_0$, cool dense fluid resides above warm buoyant fluid, so like the Rayleigh–Bénard configuration containing only a viscous fluid, this situation is potentially unstable.

Let us start with Darcy's law and the equations of continuity, energy, and state. Following the discussion of Phillips (1991), we will consider the possibility that the layer is homogeneous but transversely anisotropic such that the horizontal and vertical hydraulic conductivities are unequal (Chapter 13). Using the relation between hydraulic conductivity and permeability, (13.5), and the definition of hydraulic head, $h = z + p/\rho g$, we write Darcy's law in two parts:

$$\mathbf{q}_1 = -\frac{k_{xy}}{\mu}\nabla_1 p \tag{16.77}$$

$$q_z = -\frac{k_z}{\mu}\left(\rho g + \frac{\partial p}{\partial z}\right) \tag{16.78}$$

Here, $\mathbf{q}_1 = \langle q_x, q_y \rangle$, $k_{xy}$ and $k_z$ denote horizontal and vertical permeabilities, and

$$\nabla_1 = \frac{\partial}{\partial x}\mathbf{i} + \frac{\partial}{\partial y}\mathbf{j} \qquad (16.79)$$

For a medium that is not deforming, the continuity equation for a homogeneous liquid is

$$n\frac{\partial \rho}{\partial t} + \rho\nabla \cdot \mathbf{q} = 0 \qquad (16.80)$$

Neglecting internal sources of heat, the energy equation is

$$\rho c\mathbf{q} \cdot \nabla T + c_e\frac{\partial T}{\partial t} = K_{Te}\nabla^2 T \qquad (16.81)$$

where $c_e$ is the effective specific heat and $K_{Te}$ is the effective thermal conductivity. For an incompressible liquid, the equation of state has the linear form:

$$\rho = \rho_m[1 - \alpha(T - T_m)] \qquad (16.82)$$

Now, let the specific discharge and state variables be represented as:

$$\mathbf{q} = \mathbf{q}'(x, y, z, t); \qquad \mathbf{q}_1 = \mathbf{q}_1'(x, y, z, t); \qquad q_z = q_z'(x, y, z, t) \quad (16.83)$$

$$p = p_m + p_s(z) + p'(x, y, z, t) \qquad (16.84)$$

$$T = T_m + T_s(z) + T'(x, y, z, t) \qquad (16.85)$$

where $p_m$ and $T_m$ denote constant reference values, $p_s$ and $T_s$ denote variations about the reference state in absence of motion, and $p'$ and $T'$ denote small perturbations associated with fluid motion. In the static conduction state,

$$\frac{\partial p_s}{\partial z} = -\rho g; \qquad \frac{\partial T_s}{\partial z} = -\frac{\Theta}{Z} \qquad (16.86)$$

where $\Theta = T_0 - T_1$, and it is assumed that $T_1 < T_0$.

After substituting (16.83) through (16.85) into (16.77), (16.78), (16.80), and (16.81), order-of-magnitude arguments similar to those used in Section 16.1 are applied to expansions of the continuity and energy equations. Then using simplifications arising from (16.86), Equations (16.77), (16.78), (16.80), and (16.81) become:

$$\mathbf{q}_1' = -\frac{k_{xy}}{\mu}\nabla_1 p' \qquad (16.87)$$

$$q_z' = \frac{k_z}{\mu}\left(\rho_m g\alpha T' - \frac{\partial p'}{\partial z}\right) \qquad (16.88)$$

$$\nabla \cdot \mathbf{q}' = 0 \qquad (16.89)$$

$$\frac{c_e}{\rho_m c}\frac{\partial T'}{\partial t} + \mathbf{q}' \cdot \nabla T' - \frac{\Theta}{Z}q_z' = \kappa\nabla^2 T' \qquad (16.90)$$

Here $\kappa = K_{Te}/\rho_m c$ is the effective thermal diffusivity, and the effective specific heat $c_e = n\rho_m c + (1 - n)\rho_s c_s$, where $n$ is the porosity and $\rho_s$ and $c_s$ denote the density and specific heat of the solid phase. These equations, which govern convection of a homogeneous fluid in a porous medium, are a counterpart to the Boussinesq equations, (16.49) through (16.51). We will examine them further in Example Problem 16.5.2.

The Rayleigh number for porous media convection can be defined in terms of the horizontal permeability $k_{xy}$ as

$$Ra = \frac{g\alpha\Theta k_{xy}Z}{\nu\kappa} \qquad (16.91)$$

It also is useful to define a modified Rayleigh number:

$$\mathbb{R}a = \frac{g\alpha\Theta Z}{\nu\kappa} \frac{k_{xy}k_z}{(k_{xy}^{1/2} + k_z^{1/2})^2} \qquad (16.92)$$

Notice that this quantity is also dimensionless. The interpretation of the Rayleigh number for convection in a porous medium is the same as that for purely fluid systems. The numerator of the Rayleigh number, (16.91) or (16.92), characterizes factors that tend to destabilize the fluid column, whereas the denominator characterizes factors that tend to stabilize it. The modified Rayleigh number $\mathbb{R}a$ incorporates effects of anisotropy. With unequal horizontal and vertical permeabilities, the value of $\mathbb{R}a$ is determined more by the smaller value of the two permeabilities. With isotropic conditions, $k_{xy} = k_z = k$, and $\mathbb{R}a = Ra/4$. The modified Rayleigh number is useful in describing the speed of convective motions.

In the case of an isotropic medium, the critical Rayleigh number marking the onset of convection is about $Ra_c = 40$. However, experimentally determined values of $Ra_c$ typically are less than 40, probably due in part to uncertainty in values of the quantities in $Ra$. We will see in Example Problem 16.5.2 that the critical value $Ra_c$ increases with transverely anisotropic conditions when $k_z < k_{xy}$. The critical value $Ra_c$ also varies with boundary conditions. For example, if temperatures at both boundaries are constant, but the upper boundary is permeable and at constant pressure while the lower boundary is impermeable, then $Ra_c$ is about 27.

Like purely fluid systems, stable convection cells have the form of rolls. The height–width ratio of an individual roll cell is about one for isotropic conditions, but cells flatten when $k_{xy} > k_z$ (Example Problem 16.5.2). Cell patterns and motions become more complex with increasing $Ra$ above the critical value.

Students are urged to examine Phillips (1991), and other papers in the Reading list, regarding other factors, including spatial variations in salinity, that influence the stability of fluids and buoyancy-driven flows within porous media.

## 16.5 EXAMPLE PROBLEMS

### 16.5.1 The onset of Rayleigh–Bénard convection

It is useful to have a means to infer whether convection is likely to occur in a fluid system with specified geometry, temperature conditions, and fluid properties. For example, knowing whether convection is likely to occur within a magma chamber with specified dimensions and vertical temperature variation could provide insight regarding whether motions of crystals are governed entirely by gravitational settling, or are also influenced by convective overturning of the magma.

Here we will consider the onset of convection in a Rayleigh–Bénard configuration. Our basic objective is to obtain a critical value of the Rayleigh number, $Ra_c$, at which convection spontaneously begins. We will start with the idealized condition of stress-free boundaries, which is the case that Lord Rayleigh first examined in 1916. The more realistic cases involving a no-slip condition at one or both of the boundaries are significantly more difficult to solve. Nonetheless, our analysis of the stress-free case contains all the basic ingredients of the general problem of determining a critical value $Ra_c$, and from our analysis we will draw insight regarding how the no-slip condition influences the onset of convection.

Recall that a static conduction state, where the temperature gradient is greater than the adiabatic lapse rate, is not necessarily unstable in the sense that it leads spontaneously to convection. Rather, such a state represents an unstable equilibrium. In this situation, the problem of determining a critical value $Ra_c$ is readily treated with stability analysis. We start with the Boussinesq equations of momentum, continuity, and energy:

$$\frac{D\mathbf{u}}{Dt} = -\frac{1}{\rho_m}\nabla p' + g\alpha T'\mathbf{k} + \nu\nabla^2\mathbf{u} \tag{16.93}$$

$$\nabla \cdot \mathbf{u} = 0 \tag{16.94}$$

$$\frac{DT'}{Dt} + w\left(\frac{\partial T_s}{\partial z} - \Gamma\right) = \kappa\nabla^2 T' \tag{16.95}$$

In the static state, $\mathbf{u} = 0$; thus for an infinitesimal perturbation $\mathbf{u}'$ about the static state, $\mathbf{u} = \mathbf{u}'$. Substituting this into $D\mathbf{u}/Dt$ in (16.93) gives

$$\frac{D\mathbf{u}'}{Dt} = \mathbf{u}' \cdot \nabla\mathbf{u}' + \frac{\partial\mathbf{u}'}{\partial t} \tag{16.96}$$

Because $\mathbf{u}'$ is infinitesimal (at incipient motion), the convective term is much smaller than the local term in (16.96). Likewise, the advective term $\mathbf{u}' \cdot \nabla T'$ in (16.95) is much smaller than the local term $\partial T'/\partial t$. With an incompressible liquid the lapse rate $\Gamma$ can be neglected, and in the static state, $\partial T_s/\partial z = -(T_0 - T_1)/Z = -\Theta/Z$, where $T_0$ is the temperature of the lower boundary at $z = 0$, $T_1$ is the temperature of the upper boundary at $z = Z$, and it is assumed that $T_1 < T_0$. Using these simplifications, (16.93), (16.94), and (16.95) become

$$\frac{\partial\mathbf{u}'}{\partial t} = -\frac{1}{\rho_m}\nabla p' + g\alpha T'\mathbf{k} + \nu\nabla^2\mathbf{u}' \tag{16.97}$$

$$\nabla \cdot \mathbf{u}' = 0 \tag{16.98}$$

$$\frac{\partial T'}{\partial t} - \frac{\Theta}{Z}w' = \kappa\nabla^2 T' \tag{16.99}$$

where $\kappa = K_T/\rho_m c$. These are the fully linearized Boussinesq equations for a liquid.

It is convenient to cast (16.97), (16.98), and (16.99) in dimensionless form. Characteristic length, time, and temperature scales respectively are $Z$, $Z^2/\kappa$ and $\Theta$. Then,

$$\mathbf{x}_* = \frac{1}{Z}\mathbf{x}; \qquad t_* = \frac{\kappa}{Z^2}t; \qquad \mathbf{u}'_* = \frac{Z}{\kappa}\mathbf{u}'$$

$$T'_* = \frac{1}{\Theta}T'; \qquad p'_* = \frac{Z^2}{\rho_m \kappa^2}p' \tag{16.100}$$

Here $\mathbf{x}_* = \langle x_*, y_*, z_* \rangle = (1/Z)\langle x, y, z \rangle$ and $\mathbf{u}'_* = \langle u'_*, v'_*, w'_* \rangle = (Z/\kappa)\langle u', v', w' \rangle$. It is then left as an exercise to demonstrate, for example, that

$$\frac{\partial}{\partial x} = \frac{1}{Z}\frac{\partial}{\partial x_*}; \qquad \frac{\partial^2}{\partial x^2} = \frac{1}{Z^2}\frac{\partial^2}{\partial x_*^2}; \qquad \frac{\partial}{\partial t} = \frac{\kappa}{Z^2}\frac{\partial}{\partial t_*} \tag{16.101}$$

from which we may also demonstrate, for example, that

$$\frac{\partial T'}{\partial t} = \frac{\kappa\Theta}{Z^2}\frac{\partial T'_*}{\partial t_*}; \qquad \frac{\partial T'}{\partial x} = \frac{\Theta}{Z}\frac{\partial T'_*}{\partial x_*}; \qquad \nabla^2 T' = \frac{\Theta}{Z^2}\nabla_*^2 T'_* \tag{16.102}$$

where

$$\nabla_*^2 T'_* = \frac{\partial^2 T'_*}{\partial x_*^2} + \frac{\partial^2 T'_*}{\partial y_*^2} + \frac{\partial^2 T'_*}{\partial z_*^2} \tag{16.103}$$

Similarly evaluating other terms in (16.97), (16.98), and (16.99) then leads to

$$\frac{\partial \mathbf{u}'_*}{\partial t_*} = -\nabla_* p'_* + RaPrT'_*\mathbf{k} + Pr\nabla_*^2\mathbf{u}'_* \tag{16.104}$$

$$\nabla_* \cdot \mathbf{u}'_* = 0 \tag{16.105}$$

$$\frac{\partial T'_*}{\partial t_*} - w'_* = \nabla_*^2 T'_* \tag{16.106}$$

where

$$Ra = \frac{g\alpha\Theta Z^3}{\nu\kappa}; \qquad Pr = \frac{\nu}{\kappa} \tag{16.107}$$

The next step is to eliminate all dependent variables in (16.104) except $w'_*$. This first involves taking the curl of (16.104). Recall from Chapter 11 that the curl of the velocity vector equals the vorticity; thus curl $\mathbf{u}'_* = \nabla_* \times \mathbf{u}'_* = \boldsymbol{\omega}'_*$. Based on (17.45), the curl of the gradient of a scalar quantity equals zero; thus curl $\nabla_* p'_* = \nabla_* \times (\nabla_* p'_*) = 0$. Using (17.44) and (17.48), the curl of the scalar-vector product $T'_*\mathbf{k}$ is the same as the cross product $\nabla_* T'_* \times \mathbf{k}$, and the curl of the quantity $\nabla_*^2\mathbf{u}'_*$ is equal to $\nabla_*^2\boldsymbol{\omega}'_*$. Thus (16.104) becomes

$$\frac{\partial \boldsymbol{\omega}'_*}{\partial t_*} = Ra\,Pr(\nabla_* T'_* \times \mathbf{k}) + Pr\nabla_*^2\boldsymbol{\omega}'_* \tag{16.108}$$

In turn we take the curl of (16.108); it is left as an exercise to use formulae from Appendix 17.1 to do this, and then use (16.105) to simplify the result, to obtain:

$$\frac{\partial}{\partial t_*}\nabla_*^2\mathbf{u}'_* = Ra\,Pr\left(\nabla_*^2 T'_*\mathbf{k} - \nabla_*\frac{\partial T'_*}{\partial z_*}\right) + Pr\nabla_*^4\mathbf{u}'_* \tag{16.109}$$

Isolating the vertical component and simplifying,

$$\frac{\partial}{\partial t_*}\nabla_*^2 w'_* = RaPr\nabla_{*1}^2 T'_* + Pr\nabla_*^4 w'_* \tag{16.110}$$

where

$$\nabla^2_{*1} = \frac{\partial^2}{\partial x^2_*} + \frac{\partial^2}{\partial y^2_*} \tag{16.111}$$

Let us rearrange (16.106) and (16.110) to obtain

$$\left(\frac{\partial}{\partial t_*} - \nabla^2_*\right)T'_* = w'_* \tag{16.112}$$

$$\left(\frac{1}{Pr}\frac{\partial}{\partial t_*} - \nabla^2_*\right)\nabla^2_* w'_* = Ra\nabla^2_{*1}T'_* \tag{16.113}$$

then in turn perform the following operations on (16.112) and (16.113):

$$\nabla^2_{*1}\left(\frac{\partial}{\partial t_*} - \nabla^2_*\right)T'_* = \nabla^2_{*1}w'_* \tag{16.114}$$

$$\left(\frac{\partial}{\partial t_*} - \nabla^2_*\right)\left(\frac{1}{Pr}\frac{\partial}{\partial t_*} - \nabla^2_*\right)\nabla^2_* w'_* = Ra\left(\frac{\partial}{\partial t_*} - \nabla^2_*\right)\nabla^2_{*1}T'_* \tag{16.115}$$

Note that the order of the operations on the left side of (16.114) is unimportant; it follows that (16.115) becomes

$$\left(\frac{\partial}{\partial t_*} - \nabla^2_*\right)\left(\frac{1}{Pr}\frac{\partial}{\partial t_*} - \nabla^2_*\right)\nabla^2_* w'_* = Ra\nabla^2_{*1}w'_* \tag{16.116}$$

Let us now turn to the boundary conditions at $z_* = 0$ and $z_* = 1$. Stationary impermeable boundaries require that

$$w'_* = 0, \qquad z_* = 0, 1 \tag{16.117}$$

Moreover, a stress-free condition implies that tangential slip flow can occur, but tangential stresses must vanish, at the boundaries. Based on (12.49) and (12.50), this means that

$$\frac{\partial u'_*}{\partial z_*} + \frac{\partial w'_*}{\partial x_*} = 0, \qquad \frac{\partial v'_*}{\partial z_*} + \frac{\partial w'_*}{\partial y_*} = 0, \qquad z_* = 0, 1 \tag{16.118}$$

Based in turn on (16.117), these become

$$\frac{\partial u'_*}{\partial z_*} = \frac{\partial v'_*}{\partial z_*} = 0, \qquad z_* = 0, 1 \tag{16.119}$$

Taking the derivative of (16.105) with respect to $z_*$, and using (16.119),

$$\frac{\partial}{\partial z_*}\nabla_* \cdot \mathbf{u}'_* = \frac{\partial^2 w'_*}{\partial z^2_*} = 0, \qquad z_* = 0, 1 \tag{16.120}$$

Thus, stress-free conditions at the two boundaries are specified by

$$w'_* = \frac{\partial^2 w'_*}{\partial z^2_*} = 0, \qquad z_* = 0, 1 \tag{16.121}$$

These are referred to as *free-free* conditions. In addition we note that temperature fluctuations vanish at the boundaries, so

$$T'_* = 0, \qquad z_* = 0, 1 \tag{16.122}$$

Equations (16.112) and (16.116) are coupled partial differential equations that possess two important properties: they are linear and they have constant coefficients. This means that, like ordinary linear differential equations with constant coefficients, general solutions for the perturbed quantities $w'_*$ and $T'_*$ have exponential forms. In addition, (16.112), (16.116), (16.121), and (16.122) are symmetrical with respect to $x_*$ and $y_*$. This means that there is no preferential horizontal orientation in the problem. Thus we assume that

$$w'_* = W(z_*)f(x_*, y_*)e^{st_*}; \qquad T'_* = \theta(z_*)f(x_*, y_*)e^{st_*} \tag{16.123}$$

These are to be interpreted as follows: The distribution of perturbed values at the initial instant of disturbance may be considered an arbitrary function of the spatial coordinates $x_*$, $y_*$, and $z_*$ so long as it satisfies the boundary conditions (16.121) and (16.122). Thus, each of the perturbed quantities generally can be represented as a sum of sinusoidal functions over space in the sense of a Fourier series (or Fourier integral). Each of the functions $W(z_*)$ and $\theta(z_*)$ consists of a set of sinusoidal functions that describes how $w'_*$ and $T'_*$ vary over $z_*$, and $f(x_*, y_*)$ consists of a set of sinusoids that describes how $w'_*$ or $T'_*$ varies over $x_*$ and $y_*$. The part $e^{st_*}$ is like an amplitude of $W(z_*)f(x_*, y_*)$ or $\theta(z_*)f(x_*, y_*)$. The quantity $s$ may be complex; if its real part is positive for any of the sinusoid components, the perturbations $w'_*$ and $T'_*$ grow exponentially with time, implying that the field is unstable. If the real part of $s$ is negative, the perturbations are dampened, implying that the field is stable inasmuch as the perturbations vanish and the field returns to the static conduction state. If the real part of $s$ is zero, the field is neutrally (or marginally) stable. Our objective is to determine the value of $Ra$, that is $Ra_c$, at which $s = 0$. This describes a threshold beyond which perturbations grow, marking the onset of convection.

In (16.123), $w'_*$ and $T'_*$ are each assumed to be the product of two functions, one of which depends only on $z_*$, and one of which depends on $x_*$ and $y_*$ (and $t_*$). This permits a separation of variables. Consider (16.112), for example. Differentiating the expression for $T'_*$ in (16.123) with respect to time and space, substituting the results together with the expression for $w'_*$ in (16.123) into (16.112), then dividing all terms by $e^{st_*}$, we obtain

$$f D_z^2 \theta + \theta \nabla_{*1}^2 f - s\theta f = -Wf \tag{16.124}$$

where

$$D_z = \frac{\partial}{\partial z_*}, \qquad D_z^2 = \frac{\partial^2}{\partial z_*^2}, \ldots \tag{16.125}$$

Dividing all terms in (16.124) by $\theta f$ and rearranging,

$$\frac{1}{\theta} D_z^2 \theta - s + \frac{W}{\theta} = -\frac{1}{f} \nabla_{*1}^2 f \tag{16.126}$$

Notice that the left side of (16.126) is a function of $z_*$ only, and the right side is a function of $x_*$ and $y_*$ only. Because these are independent variables, we may conclude that each side of (16.126) is equal to the same constant, say $a^2$. (The significance of $a^2$ is examined below.) That is,

$$\frac{1}{\theta}D_z^2\theta - s + \frac{W}{\theta} = a^2 \tag{16.127}$$

$$-\frac{1}{f}\nabla_{*1}^2 f = a^2 \tag{16.128}$$

These are then rearranged to obtain

$$(D_z^2 - a^2 - s)\theta = -W \tag{16.129}$$

$$(\nabla_{*1}^2 + a^2)f = 0 \tag{16.130}$$

The latter is referred to as the *reduced wave equation* or the *Helmholtz equation*. We will return to this below.

It is left as an exercise to differentiate the expression for $w_*'$ in (16.123), substitute the results into (16.116) and (16.121), then separate variables to obtain

$$(D_z - a^2)(D_z - a^2 - s)\left(D_z - a^2 - \frac{1}{Pr}s\right)W = -a^2 Ra\, W \tag{16.131}$$

$$W = D_z^2 W = D_z^4 W = 0, \qquad z_* = 0, 1 \tag{16.132}$$

The solution to (16.131) satisfying (16.132) consists of the set of *eigenfunctions*:

$$W_j = \sin j\pi z_*, \qquad j = 1, 2, 3, \ldots \tag{16.133}$$

Note that $j\pi z_* = j\pi z/Z$. Thus when $j = 1$, $\sin \pi z/Z$ is a sinusoid with wavelength $2Z$ and wavenumber $\pi/Z$; when $j = 2$, $\sin 2\pi z/Z$ is a sinusoid with wavelength $Z$ and wavenumber $2\pi/Z$; and so on for higher *modes j*. Students who are familiar with discrete Fourier series will recognize the modes $j$ as harmonics of the function $W$. Now, whereas it may not be obvious that each of the eigenfunctions (16.133) satisfies (16.131), it should be apparent that each satisfies the boundary condition (16.132). That is, for all $j$, $\sin j\pi z_*$ equals zero at $z_* = 0$ and $z_* = 1$. Moreover, the second and fourth derivatives of $\sin j\pi z_*$, for all $j$, also equal zero at $z_* = 0$ and $z_* = 1$.

Substituting (16.133) and its derivatives into (16.131), then dividing all terms by $\sin j\pi z_*$,

$$(j^2\pi^2 + a^2)(j^2\pi^2 + a^2 + s)\left(j^2\pi^2 + a^2 + \frac{1}{Pr}s\right) = a^2 Ra \tag{16.134}$$

With marginal (neutral) stability, $s = 0$, and (16.134) reduces to

$$Ra_j = \frac{(j^2\pi^2 + a^2)^3}{a^2} \tag{16.135}$$

Here it is clear that $Ra_1 < Ra_2 < \ldots$ Thus a global minimum is associated with $Ra_1$. Setting $j = 1$, then taking the derivative $dRa_1/d(a^2)$ and setting the result to zero:

$$\frac{dRa_1}{d(a^2)} = \frac{(2a^2 - \pi^2)(\pi^2 + a^2)^2}{a^4} = 0 \tag{16.136}$$

If we insist that $a$ is a positive quantity, (16.136) is satisfied when $2a^2 = \pi^2$, or when $a = a_c = \pi/\sqrt{2}$. Substituting this result into (16.135) with $j = 1$, we obtain $Ra_c$ equal to the minimum of $Ra_1$:

$$Ra_c = \frac{27\pi^4}{4} \tag{16.137}$$

Thus a critical condition occurs at $Ra_c \approx 658$, independent of the Prandtl number $Pr$. When $Ra \leq Ra_c$, all modes $j$ are stable; when $Ra > Ra_c$, at least one mode is unstable. We may therefore infer that the onset of thermal convection with free-free boundaries occurs when $Ra \approx 658$.

Let us now return to (16.130) to briefly examine the quantities $a$ and $a_c$. Because there is no preferential horizontal orientation in this problem, it will suffice for our purpose to consider a one-dimensional form of (16.130). For example,

$$\frac{\partial^2 f}{\partial x_*^2} + a^2 f(x_*) = 0 \tag{16.138}$$

Notice that $f$ in (16.138) must be a function whose second derivative with respect to $x_*$ equals the negative of itself multiplied by $a^2$. We asserted above that $f(x_*, y_*)$ is sinusoidal. Thus, suppose that $f(x_*) = \sin 2\pi x_*$. Then $\partial^2 f/\partial x_*^2 = -(2\pi)^2 \sin 2\pi x_*$, from which we conclude that $a = 2\pi$. Rewriting $f(x_*) = \sin a x_* = \sin ax/Z = \sin 2\pi x/\lambda$, it becomes apparent that $a/Z = 2\pi/\lambda$, where $\lambda$ is an unknown horizontal wavelength. At the critical condition, $a = a_c$, whence $a_c/Z = 2\pi/\lambda$. Solving for $\lambda$ with $a_c = \pi/\sqrt{2}$, we obtain $\lambda = 2^{3/2}Z = 2.83Z$. This is the horizontal wavelength of the perturbations $w_*'$ and $T_*'$ at the onset of convection.

Two other boundary conditions are more physically meaningful than free-free boundaries: these are *rigid-rigid* and *free-rigid* boundaries. In the context of magma dynamics, for example, a rigid-rigid condition describes the situation of a tabular magma body bounded by country rock. A free-rigid condition essentially describes a liquid with free surface. Qualitatively, the effect of the no-slip condition (rigid boundary) is to increase the viscous resistance to the onset of motion; thus $Ra_c$ increases in both cases above that for free-free boundaries. For free-rigid boundaries, $Ra_c = 1101$ and $a_c = 2.682$, giving $\lambda = 2.343Z$. For rigid-rigid boundaries, $Ra_c = 1708$ and $a_c = 3.117$, giving $\lambda = 2.016Z$. These theoretical results closely match experimentally determined values of $Ra_c$, and stand as a triumph of linear stability analysis applied to the Boussinesq equations.

The result that the wavelength $\lambda = 2.016Z$ for rigid-rigid boundaries suggests that, in the case of two-dimensional rolls, the width of each roll ought to be approximately equal to its height $Z$, inasmuch as this wavelength, excited at the critical condition, persists during growth of the perturbations to finite amplitude. Indeed, this condition typically occurs in experiments; but it is possible to "preselect" different height–width ratios by initially heating the lower boundary nonuniformly. Similar comments apply to two-dimensional rolls with free-rigid boundaries.

## 16.5.2 Convection in a porous medium

Let us turn to the analogous problem of determining the critical Rayleigh number associated with the onset of convection in a porous medium. The steps required to do this are similar to those followed in the preceding Example Problem, but simpler. Here we will say more about the forms of the sinusoidal perturbations.

We start with Equations (16.87) through (16.90), neglecting the nonlinear term in the energy equation based on the assumption that perturbations are small:

$$\mathbf{q}'_1 = -\frac{k_{xy}}{\mu}\nabla_1 p' \tag{16.139}$$

$$q'_z = \frac{k_z}{\mu}\left(\rho_m g\alpha T' - \frac{\partial p'}{\partial z}\right) \tag{16.140}$$

$$\nabla \cdot \mathbf{q}' = 0 \tag{16.141}$$

$$\frac{c_e}{\rho_m c}\frac{\partial T'}{\partial t} - \frac{\Theta}{Z}q'_z = \kappa\nabla^2 T' \tag{16.142}$$

These coupled partial differential equations are linear and have constant coefficients, and we may therefore seek sinusoidal solutions of the form

$$T' = T_* e^{st} e^{i\left(\mathbf{m}\cdot\mathbf{r}+\frac{j\pi}{Z}z\right)} \tag{16.143}$$

Here $\mathbf{m}$ is a vector wavenumber defined by $\langle m/Z, n/Z\rangle$, and $\mathbf{r} = \langle x, y\rangle$. Thus $\mathbf{m}\cdot\mathbf{r} = mx/Z + ny/Z$. The quantities $m$, $n$, and $j$ are modes, and $m/Z$, $n/Z$, and $j\pi/Z$ are wavenumbers expressed in terms of the characteristic height $Z$. (We will examine these below.) The part $T_* e^{st}$ is a complex amplitude; if the real part of $s$ is positive for any of the modes, perturbations grow exponentially with time. If the real part of $s$ is negative, perturbations are dampened and the field returns to the static conduction state. If the real part of $s$ is zero, the field is marginally stable. As in the preceding Example Problem, our objective is to determine the value of $Ra$, that is $Ra_c$, at which $s = 0$.

There is no preferential horizontal orientation in this problem; for simplicity we may therefore choose the $x$-axis so that it coincides with the direction of $\mathbf{r}$. Then any dependence on $y$ vanishes and (16.139), (16.141), and (16.142) become

$$q'_x = -\frac{k_{xy}}{\mu}\frac{\partial p'}{\partial x} \tag{16.144}$$

$$\frac{\partial q'_x}{\partial x} + \frac{\partial q'_z}{\partial z} = 0 \tag{16.145}$$

$$\frac{c_e}{\rho_m c}\frac{\partial T'}{\partial t} - \frac{\Theta}{Z}q'_z = \kappa\left(\frac{\partial^2 T'}{\partial x^2} + \frac{\partial^2 T'}{\partial z^2}\right) \tag{16.146}$$

The next step is to eliminate all dependent quantities except $q'_z$ and $T'$. Differentiating (16.145) and (16.144) with respect to $z$,

$$\frac{\partial}{\partial x}\left(\frac{\partial q'_x}{\partial z}\right) = -\frac{\partial^2 q'_z}{\partial z^2} \tag{16.147}$$

$$\frac{\partial q'_x}{\partial z} = -\frac{k_{xy}}{\mu}\frac{\partial}{\partial x}\left(\frac{\partial p'}{\partial z}\right) \tag{16.148}$$

Solving (16.140) for $\partial p'/\partial z$, substituting this into (16.148) and differentiating the last term with respect to $x$, then substituting the result into (16.147) and differentiating with respect to $x$, we obtain

$$\frac{k_{xy}}{k_z}\frac{\partial^2 q_z'}{\partial x^2} + \frac{\partial^2 q_z}{\partial z^2} - \frac{g\alpha k_{xy}}{\nu}\frac{\partial^2 T'}{\partial x^2} = 0 \tag{16.149}$$

where $\nu = \mu/\rho_m$. Now we have two equations, (16.146) and (16.149), involving the two dependent variables $q_z'$ and $T'$.

Stationary impermeable boundaries require that

$$q_z' = 0, \qquad z = 0, Z \tag{16.150}$$

With $T = T_0$ at $z = 0$ and $T = T_1$ at $z = Z$, temperature fluctuations vanish at the boundaries, so

$$T' = 0, \qquad z = 0, Z \tag{16.151}$$

We now seek solutions of the form:

$$q_z' = q_* e^{st} e^{i\left(\frac{m}{Z}x + \frac{j\pi}{Z}z\right)}, \qquad j = 1, 2, 3, \ldots \tag{16.152}$$

$$T' = T_* e^{st} e^{i\left(\frac{m}{Z}x + \frac{j\pi}{Z}z\right)}, \qquad j = 1, 2, 3, \ldots \tag{16.153}$$

Let us briefly examine the quantity $m$. Consider for illustration the function $\sin j\pi z/Z$. This might characterize how the perturbation $q_z'$ varies over $z$, and is generally coincident with a horizontal perturbation of the form $\sin 2\pi x/\lambda$; but the horizontal wavelength $\lambda$ is unknown. We thus write $\sin mx/Z$, where $m = 2\pi Z/\lambda$, and note that whereas the mode $j$ is an integer, $m$ may be real. Now taking second derivatives of (16.152) and (16.153) with respect to $x$ and $z$, and taking the first derivative of (16.153) with respect to time, we obtain, for example:

$$\frac{\partial^2 T'}{\partial x^2} = -\frac{m^2}{Z^2} T_* e^{st} e^{i\left(\frac{m}{Z}x + \frac{j\pi}{Z}z\right)} \tag{16.154}$$

$$\frac{\partial^2 T'}{\partial z^2} = -\frac{j^2\pi^2}{Z^2} T_* e^{st} e^{i\left(\frac{m}{Z}x + \frac{j\pi}{Z}z\right)} \tag{16.155}$$

$$\frac{\partial T'}{\partial t} = s T_* e^{st} e^{i\left(\frac{m}{Z}x + \frac{j\pi}{Z}z\right)} \tag{16.156}$$

where we have made use of the fact that $i^2 = -1$. Substituting the results of these differentiations together with (16.152) into (16.149) and (16.146), then canceling exponential quantities,

$$-\frac{k_{xy}}{k_z}m^2 q_* - j^2\pi^2 q_* + \frac{g\alpha k_{xy}}{\nu}m^2 T_* = 0 \tag{16.157}$$

$$\frac{c_e}{\rho_m c}s T_* - \frac{\Theta}{Z}q_* + \kappa\left(\frac{m^2}{Z^2} + \frac{j^2\pi^2}{Z^2}\right)T_* = 0 \tag{16.158}$$

Solving (16.158) for $T_*$, then substituting this into (16.157) and rearranging,

$$Ra = \frac{\left(\frac{k_{xy}}{k_z}m^2 + j^2\pi^2\right)\left(\frac{Z^2}{\kappa}\frac{c_e}{\rho_m c}s + m^2 + j^2\pi^2\right)}{m^2} \tag{16.159}$$

where $Ra$ is the Rayleigh number, (16.91), defined in terms of $k_{xy}$.

With marginal stability, $s = 0$, so (16.159) becomes

$$Ra_j = \frac{\left(\dfrac{k_{xy}}{k_z}m^2 + j^2\pi^2\right)(m^2 + j^2\pi^2)}{m^2} \qquad (16.160)$$

from which it is evident that a global minimum occurs when $j = 1$. Setting $j = 1$, then taking the derivative $dRa_1/d(m^2)$ and setting the result to zero:

$$\frac{\dfrac{k_{xy}}{k_z}m^4 - \pi^4}{m^4} = 0 \qquad (16.161)$$

This is satisfied when $m^2 = \pi^2(k_{xy}/k_z)^{-1/2}$, or when $m = m_c = \pi(k_{xy}/k_z)^{-1/4}$. Substituting this result into (16.160) with $j = 1$, we obtain for $Ra_c$:

$$Ra_c = \left[1 + \left(\frac{k_{xy}}{k_z}\right)^{1/2}\right]^2 \pi^2 \qquad (16.162)$$

This critical value is given by $\mathbb{R}a_c = \pi^2$ using the modified Rayleigh number (16.92).

For isotropic conditions, $k_{xy} = k_z = k$, and the critical Rayleigh number according to (16.162) is $4\pi^2 = 39.5$, a result first obtained by E. R. Lapwood in 1948. Experimentally determined values of $Ra_c$ typically are lower than this, probably due in part to uncertainty in estimates of the quantities in $Ra$. Notice that according to (16.162) the critical value $Ra_c$ increases with transversely anisotropic conditions when $k_z < k_{xy}$, and decreases when $k_{xy} < k_z$. Using $m_c = \pi(k_{xy}/k_z)^{-1/4}$, the horizontal wavelength $\lambda = 2Z(k_{xy}/k_z)^{1/4}$. Thus, with $k_z < k_{xy}$, cells flatten by a factor of $(k_{xy}/k_z)^{1/4}$. For isotropic conditions, $\lambda = 2Z$, suggesting that the width of a roll cell is equal to its height, the same result that was obtained for purely fluid rolls.

## 16.6 READING

Campbell, I. H. and Turner, J. S. 1987. A laboratory investigation of assimilation at the top of a basaltic magma chamber. *Journal of Geology* 95:155–72. This paper, and papers by Huppert and Sparks, and Kerr listed below, examine melting of the boundaries of a magma chamber and implications for generating melts with different composition than that of the intruded magma.

Drazin, P. G. and Reid, W. H. 1981. *Hydrodynamic stability.* Cambridge: Cambridge University Press, 527 pp. This text provides a comprehensive treatment of mechanisms of instability that influence the behavior of fluids and fluid flows. The material is presented at an advanced level. Its derivation of the critical Rayleigh number for Rayleigh–Bénard convection is loosely followed in Example Problem 16.5.1.

Elder, J. W. 1965. Laminar free convection in a vertical slot. *Journal of Fluid Mechanics* 23:77–98. This and the following paper by Elder describe experiments to characterize the convective motions in a vertical slot, where the sidewalls of the slot are maintained at different temperatures.

Elder, J. W. 1965. Turbulent free convection in a vertical slot. *Journal of Fluid Mechanics* 23:99–111.

Howle, L., Behringer, R. P., and Georgiadis, J. 1993. Visualization of convective fluid flow in a porous medium. *Nature* 362: 230–32. Using conventional techniques to visualize flows in porous media normally are very difficult. This paper describes a procedure wherein a porous material is specially constructed to allow use of conventional shadowgraph techniques to visualize convective motions.

Huppert, H. E. and Sparks, R. S. J. 1988a. Melting the roof of a chamber containing a hot, turbulently convecting fluid. *Journal of Fluid Mechanics* 188: 107–31.

Huppert, H. E. and Sparks, R. S. J. 1988b. The generation of granitic magmas by intrusion of basalt into continental crust. *Journal of Petrology* 29: 599–624.

Kerr, R. C. 1994a. Melting driven by vigorous compositional convection. *Journal of Fluid Mechanics* 280: 255–85. The paper examines how melting of a solid boundary, such as the roof of a magma chamber, is driven by compositional convection, and it describes several interesting laboratory experiments to test predictions of melting rates.

Kerr, R. C. 1994b. Dissolving driven by vigorous compositional convection. *Journal of Fluid Mechanics* 280: 257–302. This is a companion paper to the one above, and emphasizes dissolution of a solid boundary.

Krishnamurti, R. 1968. Finite amplitude convection with changing mean temperature: Part 1, Theory; Part 2, An experimental test of the theory. *Journal of Fluid Mechanics* 33:445–55. These companion papers describe how vertical asymmetry in the vertical temperature profile in a Rayleigh–Bénard configuration, produced by a steady change in the mean fluid temperature, leads to a range of $Ra$ near $Ra_c$ for which hexagonal cells are stable.

Krishnamurti, R. 1973. Some further studies on the transition to turbulent convection. *Journal of Fluid Mechanics* 60: 285–303. This paper provides a context for the $Ra$–$Pr$ regime diagram illustrated in Figure 16.5.

Lapwood, E. R. 1948. Convection of a fluid in a porous medium. *Cambridge Philosophical Society Proceedings* 44: 508–21. Lapwood derives the critical Rayleigh number for the onset of convection in porous media.

Phillips, O. M. 1991. *Flow and reactions in permeable rocks*. Cambridge: Cambridge University Press, 285 pp. This text provides an alternative derivation of the critical Rayleigh number for porous media convection based on the stream function, it provides a discussion of the modified Rayleigh number, and it systematically covers other mechanisms of instability that influence flows in porous media. The text also examines how large-scale flows, including thermal convection, bear on processes such as mineralization and dolomitization.

Rayleigh, J. W. S., Lord. 1916. On convection currents in a horizontal layer of fluid, when the higher temperature is on the under side. *Philosophical Magazine* 32: 529–46. Rayleigh first describes the critical state for the onset of convection in a Rayleigh–Bénard configuration.

Schlüter, A., Lortz, D., and Busse, F. 1965. On the stability of steady finite amplitude convection. *Journal of Fluid Mechanics* 23:129–44. This paper examines the stability of different cell patterns in Rayleigh–Bénard convection and concludes that two-dimensional rolls are the only steady motions that are stable (within a certain $Ra$-$\lambda$ field) to infinitesimal perturbations.

Spiegel, E. A. and Veronis, G. 1960. On the Boussinesq approximation for a compressible fluid. *Astrophysics Journal* 131:442–47. This paper provides a rigorous justification of the Boussinesq approximation for a compressible fluid (ideal gas), and forms the basis of Section 16.1.

# CHAPTER 17

# Appendixes

## 17.1 FORMULAE IN VECTOR ANALYSIS

Definitions and formulae used at various points in the text to manipulate vectors are listed below. Additional useful formulae, including geometrical and physical interpretations complementary to those provided in this text, can be found in standard texts on vector analysis and in mathematical handbooks. The Standard Mathematical Tables published by CRC Press (Boca Raton, Florida) is a particularly handy resource, and most college-level calculus texts cover introductory vector analysis as part of the material intended for a third semester course. Appendix A in Bird, Stewart, and Lightfoot (1960) is a very good summary of vector and tensor notation presented in the context of fluid mechanics.

Section 17.1.1 begins with several basic definitions of vector quantities that generally apply to any orthogonal coordinate system. The notation for unit vectors in Cartesian coordinates, $\mathbf{i}$, $\mathbf{j}$, and $\mathbf{k}$, are used in this section, but it is understood that this notation may be directly replaced with symbols for unit vectors associated with other orthogonal coordinates. Section 17.1.2 then covers differential operations for Cartesian coordinates. Although the notation used for these differential operations in Cartesian coordinates is the same as that for other coordinate systems, the actual operations connoted by the notation are different, and must be defined separately (Appendix 17.2).

### 17.1.1 Basic definitions

Let $S$ and $T$ denote scalar functions, and let $\mathbf{U}$, $\mathbf{V}$, and $\mathbf{W}$ denote vectors. If $\mathbf{U} = \langle U_1, U_2, U_3 \rangle$, then

$$\mathbf{U} = U_1\mathbf{i} + U_2\mathbf{j} + U_3\mathbf{k} \tag{17.1}$$

where $\mathbf{i}$, $\mathbf{j}$, and $\mathbf{k}$ denote the unit vectors:

$$\mathbf{i} = \langle 1,0,0 \rangle; \qquad \mathbf{j} = \langle 0,1,0 \rangle; \qquad \mathbf{k} = \langle 0,0,1 \rangle \tag{17.2}$$

The *magnitude* of a vector $\mathbf{U}$, denoted by $|\mathbf{U}|$ or $U$, is

$$|\mathbf{U}| = U = \sqrt{U_1^2 + U_2^2 + U_3^2} \tag{17.3}$$

Further, if $\mathbf{V} = V_1\mathbf{i} + V_2\mathbf{j} + V_3\mathbf{k}$, then

$$\mathbf{U} + \mathbf{V} = \langle U_1 + V_1, U_2 + V_2, U_3 + V_3 \rangle$$

$$= (U_1 + V_1)\mathbf{i} + (U_2 + V_2)\mathbf{j} + (U_3 + V_3)\mathbf{k}$$

$$\text{(17.4)}$$

Multiplication of a scalar $S$ and a vector $\mathbf{U}$ is given by

$$S\mathbf{U} = S\langle U_1, U_2, U_3 \rangle = \langle SU_1, SU_2, SU_3 \rangle = SU_1\mathbf{i} + SU_2\mathbf{j} + SU_3\mathbf{k} \quad \text{(17.5)}$$

The *inner product* (or *dot product* or *scalar product*) of two vectors $\mathbf{U}$ and $\mathbf{V}$, denoted by $\mathbf{U} \cdot \mathbf{V}$, is given by

$$\mathbf{U} \cdot \mathbf{V} = U_1 V_1 + U_2 V_2 + U_3 V_3 = \mathbf{V} \cdot \mathbf{U} \quad \text{(17.6)}$$

The *cross product* of two vectors $\mathbf{U}$ and $\mathbf{V}$, denoted by $\mathbf{U} \times \mathbf{V}$, is given by the expansion of a third-order determinant. A second-order determinant is defined by

$$\begin{vmatrix} A_1 & A_2 \\ B_1 & B_2 \end{vmatrix} = A_1 B_2 - A_2 B_1 \quad \text{(17.7)}$$

and a third order determinant is given by

$$\begin{vmatrix} C_1 & C_2 & C_3 \\ A_1 & A_2 & A_3 \\ B_1 & B_2 & B_3 \end{vmatrix} = \begin{vmatrix} A_2 & A_3 \\ B_2 & B_3 \end{vmatrix} C_1 - \begin{vmatrix} A_1 & A_3 \\ B_1 & B_3 \end{vmatrix} C_2 + \begin{vmatrix} A_1 & A_2 \\ B_1 & B_2 \end{vmatrix} C_3 \quad \text{(17.8)}$$

where all letters represent real numbers. This operation is referred to as an *expansion by the first row*. The cross product of $\mathbf{U} \times \mathbf{V}$ is obtained by regarding $C_1$, $C_2$, and $C_3$ as the unit vectors $\mathbf{i}$, $\mathbf{j}$, and $\mathbf{k}$. Thus,

$$\mathbf{U} \times \mathbf{V} = \begin{vmatrix} \mathbf{i} & \mathbf{j} & \mathbf{k} \\ U_1 & U_2 & U_3 \\ V_1 & V_2 & V_3 \end{vmatrix} = \begin{vmatrix} U_2 & U_3 \\ V_2 & V_3 \end{vmatrix} \mathbf{i} - \begin{vmatrix} U_1 & U_3 \\ V_1 & V_3 \end{vmatrix} \mathbf{j} + \begin{vmatrix} U_1 & U_2 \\ V_1 & V_2 \end{vmatrix} \mathbf{k} \quad \text{(17.9)}$$

Applying (17.7) to each of the second-order determinants in (17.9),

$$\mathbf{U} \times \mathbf{V} = (U_2 V_3 - U_3 V_2)\mathbf{i} - (U_1 V_3 - U_3 V_1)\mathbf{j} + (U_1 V_2 - U_2 V_1)\mathbf{k} \quad \text{(17.10)}$$

If $\mathbf{U}$ is a function of the scalar $t$, then

$$\frac{d\mathbf{U}}{dt} = \frac{dU_1}{dt}\mathbf{i} + \frac{dU_2}{dt}\mathbf{j} + \frac{dU_3}{dt}\mathbf{k} \quad \text{(17.11)}$$

Several useful identities associated with the preceding definitions are listed below.

$$SU = US \quad \text{(17.12)}$$

$$\mathbf{U} + \mathbf{V} = \mathbf{V} + \mathbf{U} \quad \text{(17.13)}$$

$$(S + T)\mathbf{U} = S\mathbf{U} + T\mathbf{U} \quad \text{(17.14)}$$

$$S(\mathbf{U} + \mathbf{V}) = S\mathbf{U} + S\mathbf{V} \quad \text{(17.15)}$$

$$\mathbf{U} + (\mathbf{V} + \mathbf{W}) = (\mathbf{U} + \mathbf{V}) + \mathbf{W} = \mathbf{U} + \mathbf{V} + \mathbf{W} \quad \text{(17.16)}$$

$$(\mathbf{U} + \mathbf{V}) \cdot \mathbf{W} = \mathbf{U} \cdot \mathbf{W} + \mathbf{V} \cdot \mathbf{W} \quad \text{(17.17)}$$

$$\mathbf{U} \cdot (\mathbf{V} + \mathbf{W}) = \mathbf{U} \cdot \mathbf{V} + \mathbf{U} \cdot \mathbf{W} \tag{17.18}$$

$$\mathbf{i} \cdot \mathbf{i} = \mathbf{j} \cdot \mathbf{j} = \mathbf{k} \cdot \mathbf{k} = 1 \tag{17.19}$$

$$\mathbf{i} \cdot \mathbf{j} = \mathbf{j} \cdot \mathbf{k} = \mathbf{k} \cdot \mathbf{i} = 0 \tag{17.20}$$

$$\mathbf{U} \times \mathbf{V} = -\mathbf{V} \times \mathbf{U} \tag{17.21}$$

$$\mathbf{U} \times (\mathbf{V} + \mathbf{W}) = \mathbf{U} \times \mathbf{V} + \mathbf{U} \times \mathbf{W} \tag{17.22}$$

$$(\mathbf{U} + \mathbf{V}) \times \mathbf{W} = \mathbf{U} \times \mathbf{W} + \mathbf{V} \times \mathbf{W} \tag{17.23}$$

$$\mathbf{U} \cdot \mathbf{V} \times \mathbf{W} = \mathbf{U} \cdot (\mathbf{V} \times \mathbf{W}) = (\mathbf{U} \times \mathbf{V}) \cdot \mathbf{W} = \mathbf{V} \cdot (\mathbf{W} \times \mathbf{U}) \tag{17.24}$$

$$\mathbf{U} \times (\mathbf{V} \times \mathbf{W}) = (\mathbf{U} \cdot \mathbf{W})\mathbf{V} - (\mathbf{U} \cdot \mathbf{V})\mathbf{W} \tag{17.25}$$

$$\mathbf{i} \times \mathbf{j} = \mathbf{k}; \qquad \mathbf{j} \times \mathbf{k} = \mathbf{i}; \qquad \mathbf{k} \times \mathbf{i} = \mathbf{j} \tag{17.26}$$

$$\frac{d}{dt}(\mathbf{U} + \mathbf{V}) = \frac{d\mathbf{U}}{dt} + \frac{d\mathbf{V}}{dt} \tag{17.27}$$

$$\frac{d}{dt}(\mathbf{U} \cdot \mathbf{V}) = \frac{d\mathbf{U}}{dt} \cdot \mathbf{V} + \mathbf{U} \cdot \frac{d\mathbf{V}}{dt} \tag{17.28}$$

$$\frac{d}{dt}(\mathbf{U} \times \mathbf{V}) = \frac{d\mathbf{U}}{dt} \times \mathbf{V} + \mathbf{U} \times \frac{d\mathbf{V}}{dt} \tag{17.29}$$

## 17.1.2 Differential operations in Cartesian coordinates

For Cartesian coordinates, the *vector differential operator* in three dimensions is defined by

$$\nabla = \mathbf{i}\frac{\partial}{\partial x} + \mathbf{j}\frac{\partial}{\partial y} + \mathbf{k}\frac{\partial}{\partial z} \tag{17.30}$$

Also by definition, the differential *Laplacian operator* is

$$\nabla^2 = \frac{\partial^2}{\partial x^2} + \frac{\partial^2}{\partial y^2} + \frac{\partial^2}{\partial z^2} \tag{17.31}$$

If $S$ is a differentiable scalar function, then the *gradient* of $S$, denoted by grad $S$ or $\nabla S$, is

$$\text{grad } S = \nabla S = \frac{\partial S}{\partial x}\mathbf{i} + \frac{\partial S}{\partial y}\mathbf{j} + \frac{\partial S}{\partial z}\mathbf{k} \tag{17.32}$$

and the *Laplacian of S,* denoted by Lap $S$ or $\nabla^2 S$, is

$$\text{Lap } S = \nabla^2 S = \frac{\partial^2 S}{\partial x^2} + \frac{\partial^2 S}{\partial y^2} + \frac{\partial^2 S}{\partial z^2} \tag{17.33}$$

In Cartesian coordinates the *dyadic product* $\nabla\mathbf{U}$ is equivalent to

$$\nabla\mathbf{U} = \frac{\partial\mathbf{U}}{\partial x}\mathbf{i} + \frac{\partial\mathbf{U}}{\partial y}\mathbf{j} + \frac{\partial\mathbf{U}}{\partial z}\mathbf{k} \tag{17.34}$$

and the operation $\nabla^2\mathbf{U}$ is equivalent to

$$\nabla^2\mathbf{U} = \nabla^2 U_1\mathbf{i} + \nabla^2 U_2\mathbf{j} + \nabla^2 U_3\mathbf{k} \tag{17.35}$$

The *divergence* of a vector $\mathbf{U}$, denoted by div $\mathbf{U}$ or $\nabla \cdot \mathbf{U}$, is obtained from the formal dot product of $\nabla$ and $\mathbf{U}$:

$$\text{div } \mathbf{U} = \nabla \cdot \mathbf{U} = \frac{\partial U_1}{\partial x} + \frac{\partial U_2}{\partial y} + \frac{\partial U_3}{\partial z} \tag{17.36}$$

The *curl* of a vector $\mathbf{U}$, denoted by curl $\mathbf{U}$ or $\nabla \times \mathbf{U}$, is obtained from the formal cross product of $\nabla$ and $\mathbf{U}$:

$$\text{curl } \mathbf{U} = \nabla \times \mathbf{U} = \begin{vmatrix} \mathbf{i} & \mathbf{j} & \mathbf{k} \\ \dfrac{\partial}{\partial x} & \dfrac{\partial}{\partial y} & \dfrac{\partial}{\partial z} \\ U_1 & U_2 & U_3 \end{vmatrix} \tag{17.37}$$

Using (17.9) and (17.7),

$$\nabla \times \mathbf{U} = \left(\frac{\partial U_3}{\partial y} - \frac{\partial U_2}{\partial z}\right)\mathbf{i} + \left(\frac{\partial U_1}{\partial z} - \frac{\partial U_3}{\partial x}\right)\mathbf{j} + \left(\frac{\partial U_2}{\partial x} - \frac{\partial U_1}{\partial y}\right)\mathbf{k} \tag{17.38}$$

Several useful identities associated with the preceding definitions are listed below. Note that these identities also apply to other coordinate systems, although the operations connoted by the notation are different.

$$\nabla(S + T) = \nabla S + \nabla T \tag{17.39}$$

$$\nabla(ST) = S\nabla T + T\nabla S \tag{17.40}$$

$$\nabla \cdot (\mathbf{U} + \mathbf{V}) = \nabla \cdot \mathbf{U} + \nabla \cdot \mathbf{V} \tag{17.41}$$

$$\nabla \times (\mathbf{U} + \mathbf{V}) = \nabla \times \mathbf{U} + \nabla \times \mathbf{V} \tag{17.42}$$

$$\nabla \cdot (S\mathbf{U}) = \mathbf{U} \cdot \nabla S + S\nabla \cdot \mathbf{U} \tag{17.43}$$

$$\nabla \times (S\mathbf{U}) = \nabla S \times \mathbf{U} + S\nabla \times \mathbf{U} \tag{17.44}$$

$$\nabla \times \nabla S = 0 \tag{17.45}$$

$$\nabla \cdot \nabla \times \mathbf{U} = 0 \tag{17.46}$$

In addition, if $\mathbf{R} = x\mathbf{i} + y\mathbf{j} + z\mathbf{k}$ is a radial position vector with terminal point $(x, y, z)$, then

$$\nabla \cdot \mathbf{R} = 3; \qquad \nabla \times \mathbf{R} = 0 \tag{17.47}$$

from which it can be demonstrated that

$$\nabla \times \mathbf{i} = \nabla \times \mathbf{j} = \nabla \times \mathbf{k} = 0 \tag{17.48}$$

---

## 17.2 ORTHOGONAL CURVILINEAR COORDINATES

Various expressions introduced throughout the text in their Cartesian forms are presented in their cylindrical, spherical, and curvilinear forms in this appendix. For each coordinate system, several general identities are listed, from which specific equations can be constructed. Section 17.2.1.2 contains several examples illustrating this procedure. The first example involves a derivation of the continuity equation in cylindrical form, based on Taylor expansions about a control volume, then based on the identities listed initially. More systematic treatments of polar coordinate systems are provided in the texts by Batchelor (1967) and by Bird, Stewart, and Lightfoot (1960).

## 17.2.1 Cylindrical coordinates

Certain phenomena involve fluid motion about a single axis. Familiar examples include vortices in the forms of whirlpools, tornadoes, and dust devils. It is sometimes convenient to use cylindrical coordinates (Figure 1.2) when treating these flows. In addition, problems involving radially symmetric flows are conveniently treated using cylindrical coordinates. Groundwater flow within an isotropic porous medium in the vicinity of a well, and Poiseuille flow in a right-circular cylinder, are examples.

***17.2.1.1 Identities.*** Let $S$ denote a scalar quantity; let vectors $\mathbf{U} = \langle U_r, U_\theta, U_z \rangle$ and $\mathbf{V} = \langle V_r, V_\theta, V_z \rangle$; and let unit vectors $\mathbf{e}_r = \langle 1, 0, 0 \rangle$, $\mathbf{e}_\theta = \langle 0, 1, 0 \rangle$, and $\mathbf{e}_z = \langle 0, 0, 1 \rangle$. Then

$$\nabla S = \mathbf{e}_r \frac{\partial S}{\partial r} + \mathbf{e}_\theta \frac{1}{r} \frac{\partial S}{\partial \theta} + \mathbf{e}_z \frac{\partial S}{\partial z} \tag{17.49}$$

$$\nabla^2 S = \frac{\partial^2 S}{\partial r^2} + \frac{1}{r} \frac{\partial S}{\partial r} + \frac{1}{r^2} \frac{\partial^2 S}{\partial \theta^2} + \frac{\partial^2 S}{\partial z^2} \tag{17.50}$$

$$\nabla \cdot \mathbf{U} = \frac{\partial U_r}{\partial r} + \frac{U_r}{r} + \frac{1}{r} \frac{\partial U_\theta}{\partial \theta} + \frac{\partial U_z}{\partial z} \tag{17.51}$$

$$\mathbf{U} \cdot \nabla S = U_r \frac{\partial S}{\partial r} + \frac{U_\theta}{r} \frac{\partial S}{\partial \theta} + U_z \frac{\partial S}{\partial z} \tag{17.52}$$

$$\mathbf{U} \cdot \nabla \mathbf{V} = \mathbf{e}_r \left( \mathbf{U} \cdot \nabla V_r - \frac{U_\theta V_\theta}{r} \right) + \mathbf{e}_\theta \left( \mathbf{U} \cdot \nabla V_\theta + \frac{U_\theta V_r}{r} \right) + \mathbf{e}_z (\mathbf{U} \cdot \nabla V_z) \tag{17.53}$$

$$\nabla^2 \mathbf{V} = \mathbf{e}_r \left( \nabla^2 V_r - \frac{V_r}{r^2} - \frac{2}{r^2} \frac{\partial V_\theta}{\partial \theta} \right) + \mathbf{e}_\theta \left( \nabla^2 V_\theta + \frac{2}{r^2} \frac{\partial V_r}{\partial \theta} - \frac{V_\theta}{r^2} \right) + \mathbf{e}_z (\nabla^2 V_z)$$

$$\tag{17.54}$$

In addition, components of the rate-of-strain tensor are

$$\dot{\varepsilon}_r = \frac{\partial u_r}{\partial r}; \qquad \dot{\varepsilon}_\theta = \frac{1}{r} \frac{\partial v_\theta}{\partial \theta} + \frac{u_r}{r}; \qquad \dot{\varepsilon}_z = \frac{\partial w_z}{\partial z}$$

$$\dot{\varepsilon}_{r\theta} = \frac{1}{2} \left( \frac{1}{r} \frac{\partial u_r}{\partial \theta} + \frac{\partial v_\theta}{\partial r} - \frac{v_\theta}{r} \right); \qquad \dot{\varepsilon}_{rz} = \frac{1}{2} \left( \frac{\partial u_r}{\partial z} + \frac{\partial w_z}{\partial r} \right);$$

$$\dot{\varepsilon}_{\theta z} = \frac{1}{2} \left( \frac{1}{r} \frac{\partial w_z}{\partial \theta} + \frac{\partial v_\theta}{\partial z} \right) \tag{17.55}$$

***17.2.1.2 Example derivations.*** Consider an elementary control volume in the form of a wedge whose interior surface (closest to the $z$-axis) is at the radial coordinate position $r$ (Figure 17.1). The edges of the volume have lengths $dr$, $r \, d\theta$, $(r + dr) \, d\theta$, and $dz$. The area of the interior surface is $dA = r \, d\theta \, dz$. The mass flowing into the control volume in the direction of positive $r$ during the interval $dt$ is $\rho u_r \, dA \, dt = \rho u_r r \, d\theta \, dz \, dt$. Since the surface area $dA$ varies with $r$, the mass flowing out of the volume is to first order $\rho u_r \, dA \, dt + (\partial/\partial r)(\rho u_r \, dA) \, dr \, dt$. Note, however, that the small angle $d\theta$ and the length $dz$ contained in $dA$ are independent of $r$; these may be removed from the differential, so the mass flowing out is

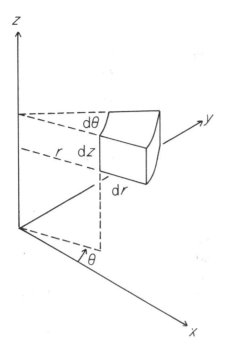

**Figure 17.1**  Definition diagram for mass flow quantities associated with an elementary control volume in cylindrical coordinates.

$[\rho r u_r + (\partial/\partial r)(\rho r u_r)\,dr]\,d\theta\,dz\,dt$. The net mass flowing out of the volume is therefore

$$\frac{\partial}{\partial r}(\rho r u_r)\,dr\,d\theta\,dz\,dt \qquad (17.56)$$

The mass flowing into the volume in the positive angular direction $\theta$ is $\rho v_\theta\,dr\,dz\,dt$. The mass flowing out of the volume is to first order $[\rho v_\theta + (\partial/\partial\theta)(\rho v_\theta)\,d\theta]\,dr\,dz\,dt$. The net mass flowing out is therefore

$$\frac{\partial}{\partial\theta}(\rho v_\theta)\,dr\,d\theta\,dz\,dt \qquad (17.57)$$

The area of each surface of the control volume normal to $z$ is $dr\,r\,d\theta$. The mass flowing into the volume in the direction of positive $z$ is $\rho w_z\,dr\,r\,d\theta\,dt$. Noting that $r$ is independent of $z$, the mass flowing out of the volume is to first order $[\rho w_z + (\partial/\partial z)(\rho w_z)\,dz]r\,dr\,d\theta\,dt$. The net mass flowing out is therefore

$$\frac{\partial}{\partial z}(\rho w_z)r\,dr\,d\theta\,dz\,dt \qquad (17.58)$$

The mass within the control volume at time $t$ is $\rho\,dr\,r\,d\theta\,dz$. The mass within the volume at time $t + dt$ is to first order $[\rho + (\partial\rho/\partial t)\,dt]r\,dr\,d\theta\,dz$. The change in mass is therefore

$$\frac{\partial\rho}{\partial t}r\,dr\,d\theta\,dz\,dt \qquad (17.59)$$

By the sign convention used, (17.56) through (17.59) must sum to zero. Forming this sum, then dividing by $r\,dr\,d\theta\,dz\,dt$,

$$\frac{1}{r}\frac{\partial}{\partial r}(\rho r u_r) + \frac{1}{r}\frac{\partial}{\partial\theta}(\rho v_\theta) + \frac{\partial}{\partial z}(\rho w_z) = -\frac{\partial\rho}{\partial t} \qquad (17.60)$$

which is the Eulerian form of the continuity equation. A similar approach involving Taylor expansions can be used to obtain the cylindrical forms of other equations.

Alternatively, let us start with the Lagrangian form of the continuity equation:

$$\frac{D\rho}{Dt} + \rho\nabla \cdot \mathbf{u} = 0 \qquad (17.61)$$

$$\mathbf{u} \cdot \nabla\rho + \frac{\partial\rho}{\partial t} + \rho\nabla \cdot \mathbf{u} = 0 \qquad (17.62)$$

Letting $S = \rho$ and $\mathbf{U} = \mathbf{u} = \langle u_r, v_\theta, w_z\rangle$, then using (17.51) and (17.52),

$$u_r\frac{\partial\rho}{\partial r} + \frac{v_\theta}{r}\frac{\partial\rho}{\partial\theta} + w_z\frac{\partial\rho}{\partial z} + \rho\frac{\partial u_r}{\partial r} + \frac{\rho u_r}{r} + \frac{\rho}{r}\frac{\partial v_\theta}{\partial\theta} + \rho\frac{\partial w_z}{\partial z} = -\frac{\partial\rho}{\partial t} \qquad (17.63)$$

which is the expanded form of (17.60).

As a second example of the use of the identities above, consider the $r$-component of the Navier–Stokes equations for incompressible fluid flow:

$$\mathbf{u} \cdot \nabla u_r + \frac{\partial u_r}{\partial t} = -\frac{1}{\rho}\nabla p - g\nabla h + \nu\nabla^2 u_r \qquad (17.64)$$

Letting $\mathbf{U} = \mathbf{V} = \mathbf{u} = \langle u_r, v_\theta, w_z\rangle$, observe that the first term in (17.64) is equivalent to the parenthetical part of the $r$-component in (17.53). Letting $S = u_r$, then using (17.52), the left side of (17.64) becomes

$$u_r\frac{\partial u_r}{\partial r} + \frac{v_\theta}{r}\frac{\partial u_r}{\partial\theta} + w_z\frac{\partial u_r}{\partial z} - \frac{v_\theta^2}{r} + \frac{\partial u_r}{\partial t} \qquad (17.65)$$

Letting $S = p$ and $S = h$, observe that each of the terms involving $p$ and $h$ in (17.64) is equivalent to the product of a constant and the $r$-component of (17.49). Using this identity, these terms become

$$-\frac{1}{\rho}\frac{\partial p}{\partial r}; \qquad -g\frac{\partial h}{\partial r} \qquad (17.66)$$

Now, with $\mathbf{V} = \mathbf{u} = \langle u_r, v_\theta, w_z\rangle$, observe that the last term in (17.64) is equivalent to the product of a constant and the parenthetical part of the $r$-component in (17.54). Letting $S = u_r$, then using (17.50), the last term in (17.64) becomes

$$\nu\left(\frac{\partial^2 u_r}{\partial r^2} + \frac{1}{r}\frac{\partial u_r}{\partial r} + \frac{1}{r^2}\frac{\partial^2 u_r}{\partial\theta^2} + \frac{\partial^2 u_r}{\partial z^2} - \frac{u_r}{r^2} - \frac{2}{r^2}\frac{\partial v_\theta}{\partial\theta}\right) \qquad (17.67)$$

Collecting (17.65), (17.66), and (17.67),

$$u_r\frac{\partial u_r}{\partial r} - \frac{v_\theta^2}{r} + \frac{v_\theta}{r}\frac{\partial u_r}{\partial\theta} + w_z\frac{\partial u_r}{\partial z} + \frac{\partial u_r}{\partial t} = -\frac{1}{\rho}\frac{\partial p}{\partial r} - g\frac{\partial h}{\partial r}$$

$$+ \nu\left(\frac{\partial^2 u_r}{\partial r^2} + \frac{1}{r}\frac{\partial u_r}{\partial r} - \frac{u_r}{r^2} + \frac{1}{r^2}\frac{\partial^2 u_r}{\partial\theta^2} - \frac{2}{r^2}\frac{\partial v_\theta}{\partial\theta} + \frac{\partial^2 u_r}{\partial z^2}\right) \qquad (17.68)$$

which is the cylindrical form of the $r$-component (17.64).

As a final example, consider the cylindrical form of the energy equation for incompressible fluid flow:

$$\rho c_V \left( \mathbf{u} \cdot \nabla T + \frac{\partial T}{\partial t} \right) = K_T \nabla^2 T + \Phi \qquad (17.69)$$

Letting $S = T$ and $\mathbf{U} = \mathbf{u} = \langle u_r, v_\theta, w_z \rangle$, then using (17.50) and (17.52), this becomes

$$\rho c_V \left( u_r \frac{\partial T}{\partial r} + \frac{v_\theta}{r} \frac{\partial T}{\partial \theta} + w_z \frac{\partial T}{\partial z} + \frac{\partial T}{\partial t} \right)$$

$$= K_T \left( \frac{\partial^2 T}{\partial r^2} + \frac{1}{r} \frac{\partial T}{\partial r} + \frac{1}{r^2} \frac{\partial^2 T}{\partial \theta^2} + \frac{\partial^2 T}{\partial z^2} \right) + \Phi \qquad (17.70)$$

where the cylindrical form of the dissipation function $\Phi = 2\mu(\varepsilon_r^2 + \varepsilon_\theta^2 + \varepsilon_z^2 + \varepsilon_{r\theta}^2 + \varepsilon_{rz}^2 + \varepsilon_{\theta_z}^2)$ is obtained by substituting the expressions in (17.55) for the strain-rate components.

### 17.2.1.3 Continuity equation. For viscous flow,

$$\frac{1}{r} \frac{\partial}{\partial r} (\rho r u_r) + \frac{1}{r} \frac{\partial}{\partial \theta} (\rho v_\theta) + \frac{\partial}{\partial z} (\rho w_z) = -\frac{\partial \rho}{\partial t} \qquad (17.71)$$

and for saturated porous-media flow,

$$\frac{1}{r} \frac{\partial}{\partial r} (\rho r q_{hr}) + \frac{1}{r} \frac{\partial}{\partial \theta} (\rho q_{h\theta}) + \frac{\partial}{\partial z} (\rho q_{hz}) = -\frac{\partial}{\partial t} (n\rho) \qquad (17.72)$$

### 17.2.1.4 Navier–Stokes equations for incompressible fluid flow. The $r$-component is

$$u_r \frac{\partial u_r}{\partial r} - \frac{v_\theta^2}{r} + \frac{v_\theta}{r} \frac{\partial u_r}{\partial \theta} + w_z \frac{\partial u_r}{\partial z} + \frac{\partial u_r}{\partial t} = -\frac{1}{\rho} \frac{\partial p}{\partial r} - g \frac{\partial h}{\partial r}$$

$$+ \nu \left( \frac{\partial^2 u_r}{\partial r^2} + \frac{1}{r} \frac{\partial u_r}{\partial r} - \frac{u_r}{r^2} + \frac{1}{r^2} \frac{\partial^2 u_r}{\partial \theta^2} - \frac{2}{r^2} \frac{\partial v_\theta}{\partial \theta} + \frac{\partial^2 u_r}{\partial z^2} \right) \qquad (17.73)$$

the $\theta$-component is

$$u_r \frac{\partial v_\theta}{\partial r} + \frac{u_r v_\theta}{r} + \frac{v_\theta}{r} \frac{\partial v_\theta}{\partial \theta} + w_z \frac{\partial v_\theta}{\partial z} + \frac{\partial v_\theta}{\partial t} = -\frac{1}{\rho r} \frac{\partial p}{\partial \theta} - \frac{g}{r} \frac{\partial h}{\partial \theta}$$

$$+ \nu \left( \frac{\partial^2 v_\theta}{\partial r^2} + \frac{1}{r} \frac{\partial v_\theta}{\partial r} - \frac{v_\theta}{r} + \frac{2}{r^2} \frac{\partial u_r}{\partial \theta} + \frac{1}{r^2} \frac{\partial^2 v_\theta}{\partial \theta^2} + \frac{\partial^2 v_\theta}{\partial z^2} \right) \qquad (17.74)$$

and the $z$-component is

$$u_r \frac{\partial w_z}{\partial u_r} + \frac{v_\theta}{r} \frac{\partial w_z}{\partial \theta} + w_z \frac{\partial w_z}{\partial z} + \frac{\partial w_z}{\partial t}$$

$$= -\frac{1}{\rho} \frac{\partial p}{\partial z} - g \frac{\partial h}{\partial z} + \nu \left( \frac{\partial^2 w_z}{\partial r^2} + \frac{1}{r} \frac{\partial w_z}{\partial r} + \frac{1}{r^2} \frac{\partial^2 w_z}{\partial \theta^2} + \frac{\partial^2 w_z}{\partial z^2} \right) \qquad (17.75)$$

## 17.2.2 Spherical coordinates

Certain flows are conveniently treated in terms of a spherical coordinate system. Diffusion of dissolved gas toward a bubble (Example Problem 8.4.4) is an example. Let $r$ denote a radial coordinate, let $\phi$ denote the angular displacement from a reference direction (the vertical direction, for example), and let $\theta$ denote the azimuthal angle about the reference direction ($\phi = 0$). The equations below are then obtained by expanding quantities about an elementary control volume, in a manner similar to that used for cylindrical coordinates.

***17.2.2.1 Identities.*** Let $S$ denote a scalar quantity; let vectors $\mathbf{U} = \langle U_r, U_\theta, U_\phi \rangle$ and $\mathbf{V} = \langle V_r, V_\theta, V_\phi \rangle$; and let unit vectors $\mathbf{e}_r = \langle 1, 0, 0 \rangle$, $\mathbf{e}_\theta = \langle 0, 1, 0 \rangle$, and $\mathbf{e}_\phi = \langle 0, 0, 1 \rangle$. Then

$$\nabla S = \mathbf{e}_r \frac{\partial S}{\partial r} + \mathbf{e}_\theta \frac{1}{r \sin \phi} \frac{\partial S}{\partial \theta} + \mathbf{e}_\phi \frac{1}{r} \frac{\partial S}{\partial \phi} \tag{17.76}$$

$$\nabla^2 S = \frac{\partial^2 S}{\partial r^2} + \frac{2}{r} \frac{\partial S}{\partial r} + \frac{1}{r^2 \sin^2 \phi} \frac{\partial^2 S}{\partial \theta^2} + \frac{1}{r^2} \frac{\partial^2 S}{\partial \phi^2} + \frac{\cos \phi}{r^2 \sin \phi} \frac{\partial S}{\partial \phi} \tag{17.77}$$

$$\nabla \cdot \mathbf{U} = \frac{\partial U_r}{\partial r} + \frac{2 U_r}{r} + \frac{1}{r \sin \phi} \frac{\partial U_\theta}{\partial \theta} + \frac{1}{r} \frac{\partial U_\phi}{\partial \phi} + \frac{\cos \phi U_\phi}{r \sin \phi} \tag{17.78}$$

$$\mathbf{U} \cdot \nabla S = U_r \frac{\partial S}{\partial r} + \frac{U_\theta}{r \sin \phi} \frac{\partial S}{\partial \theta} + \frac{U_\phi}{r} \frac{\partial S}{\partial \phi} \tag{17.79}$$

$$\mathbf{U} \cdot \nabla \mathbf{V} = \left( \mathbf{U} \cdot \nabla V_r - \frac{U_\theta V_\theta}{r} - \frac{U_\phi V_\phi}{r} \right) \mathbf{e}_r + \left( \mathbf{U} \cdot \nabla V_\theta + \frac{U_\theta V_r}{r} + \frac{\cos \phi U_\theta V_\phi}{r \sin \phi} \right) \mathbf{e}_\theta$$

$$+ \left( \mathbf{U} \cdot \nabla V_\phi + \frac{U_\phi V_r}{r} - \frac{\cos \phi U_\theta V_\theta}{r \sin \phi} \right) \mathbf{e}_\phi \tag{17.80}$$

$$\nabla^2 \mathbf{V} = \left( \nabla^2 V_r - \frac{2 V_r}{r^2} - \frac{2}{r^2 \sin \phi} \frac{\partial V_\theta}{\partial \theta} - \frac{2}{r^2} \frac{\partial V_\phi}{\partial \phi} - \frac{2 \cos \phi V_\phi}{r^2 \sin \phi} \right) \mathbf{e}_r$$

$$+ \left( \nabla^2 V_\theta - \frac{V_\theta}{r^2 \sin^2 \phi} + \frac{2}{r^2 \sin \phi} \frac{\partial V_r}{\partial \theta} + \frac{2 \cos \phi}{r^2 \sin^2 \phi} \frac{\partial V_\phi}{\partial \theta} \right) \mathbf{e}_\theta$$

$$+ \left( \nabla^2 V_\phi - \frac{V_\phi}{r^2 \sin^2 \phi} + \frac{2}{r^2} \frac{\partial V_r}{\partial \phi} - \frac{2 \cos \phi}{r^2 \sin^2 \phi} \frac{\partial V_\theta}{\partial \theta} \right) \mathbf{e}_\phi \tag{17.81}$$

In addition, components of the rate-of-strain tensor are

$$\dot{\varepsilon}_r = \frac{\partial u_r}{\partial r}; \qquad \dot{\varepsilon}_\theta = \frac{1}{r \sin \phi} \frac{\partial v_\theta}{\partial \theta} + \frac{u_r}{r} + \frac{\cos \phi w_\phi}{r \sin \phi}; \qquad \dot{\varepsilon}_\phi = \frac{1}{r} \frac{\partial w_\phi}{\partial \phi} + \frac{u_r}{r}$$

$$\dot{\varepsilon}_{r\theta} = \frac{1}{2} \left( \frac{1}{r \sin \phi} \frac{\partial u_r}{\partial \theta} + \frac{\partial v_\theta}{\partial r} - \frac{v_\theta}{r} \right); \qquad \dot{\varepsilon}_{r\phi} = \frac{1}{2} \left( \frac{1}{r} \frac{\partial u_r}{\partial \phi} + \frac{\partial w_\phi}{\partial r} - \frac{w_\phi}{r} \right)$$

$$\dot{\varepsilon}_{\theta\phi} = \frac{1}{2} \left( \frac{1}{r} \frac{\partial v_\theta}{\partial \phi} - \frac{\cos \phi v_\theta}{r \sin \phi} + \frac{1}{r \sin \phi} \frac{\partial w_\phi}{\partial \theta} \right) \tag{17.82}$$

## 17.2.2.2  Continuity equation for viscous flow.

$$\frac{\partial}{\partial r}(\rho u_r) + \frac{2\rho u_r}{r} + \frac{1}{r\sin\phi}\frac{\partial}{\partial\theta}(\rho u_\theta) + \frac{1}{r}\frac{\partial}{\partial\phi}(\rho u_\phi) + \frac{\rho\cos\phi u_\phi}{r\sin\phi} = -\frac{\partial\rho}{\partial t}$$

$$(17.83)$$

## 17.2.2.3  Navier–Stokes equations for incompressible fluid flow.   The $r$-component is

$$u_r\frac{\partial u_r}{\partial r} + \frac{u_\theta}{r\sin\phi}\frac{\partial u_r}{\partial\theta} - \frac{u_\theta^2}{r} + \frac{u_\phi}{r}\frac{\partial u_r}{\partial\phi} - \frac{u_\phi^2}{r} + \frac{\partial u_r}{\partial t} = -\frac{1}{\rho}\frac{\partial p}{\partial r} - g\frac{\partial h}{\partial r}$$

$$+\nu\left(\frac{\partial^2 u_r}{\partial r^2} + \frac{2}{r}\frac{\partial u_r}{\partial r} - \frac{2u_r}{r^2} + \frac{1}{r^2\sin^2\phi}\frac{\partial^2 u_r}{\partial\theta^2} - \frac{2}{r^2\sin\phi}\frac{\partial u_\theta}{\partial\theta}\right.$$

$$\left.+\frac{1}{r^2}\frac{\partial^2 u_r}{\partial\phi^2} + \frac{\cos\phi}{r^2\sin\phi}\frac{\partial u_r}{\partial\phi} - \frac{2}{r^2}\frac{\partial u_\phi}{\partial\phi} - \frac{2\cos\phi u_\phi}{r^2\sin\phi}\right) \qquad (17.84)$$

the $\theta$-component is

$$u_r\frac{\partial v_\theta}{\partial r} + \frac{u_r v_\theta}{r} + \frac{w_\phi}{r}\frac{\partial v_\theta}{\partial\phi} + \frac{\cos\phi v_\theta w_\phi}{r\sin\phi} + \frac{v_\theta}{r\sin\phi}\frac{\partial v_\theta}{\partial\theta} + \frac{\partial v_\theta}{\partial t}$$

$$= -\frac{1}{\rho r\sin\phi}\frac{\partial p}{\partial\theta} - \frac{g}{r\sin\phi}\frac{\partial h}{\partial\theta} + \nu\left(\frac{\partial^2 v_\theta}{\partial r^2} + \frac{2}{r}\frac{\partial v_\theta}{\partial r} - \frac{v_\theta}{r^2\sin^2\phi} + \frac{1}{r^2\sin^2\phi}\frac{\partial^2 v_\theta}{\partial\theta^2}\right.$$

$$\left.+\frac{2\cos\phi}{r^2\sin^2\phi}\frac{\partial v_\theta}{\partial\theta} + \frac{1}{r^2}\frac{\partial^2 v_\theta}{\partial\phi^2} + \frac{\cos\phi}{r^2\sin\phi}\frac{\partial v_\theta}{\partial\phi} + \frac{2}{r^2\sin\phi}\frac{\partial u_r}{\partial\theta}\right) \qquad (17.85)$$

and the $\phi$-component is

$$u_r\frac{\partial w_\phi}{\partial r} + \frac{u_r w_\phi}{r} + \frac{w_\phi}{r}\frac{\partial w_\phi}{\partial\phi} - \frac{\cos\phi v_\theta^2}{r\sin\phi} + \frac{v_\theta}{r\sin\phi}\frac{\partial w_\phi}{\partial\theta} + \frac{\partial w_\phi}{\partial t} = -\frac{1}{\rho r}\frac{\partial p}{\partial\phi} - \frac{g}{r}\frac{\partial h}{\partial\phi}$$

$$+\nu\left(\frac{\partial^2 w_\phi}{\partial r^2} + \frac{2}{r}\frac{\partial w_\phi}{\partial r} - \frac{w_\phi}{r^2\sin^2\phi} + \frac{1}{r^2\sin^2\phi}\frac{\partial^2 w_\phi}{\partial\theta^2}\right.$$

$$\left.-\frac{2\cos\phi}{r^2\sin^2\phi}\frac{\partial v_\theta}{\partial\theta} + \frac{1}{r^2}\frac{\partial^2 w_\phi}{\partial\phi^2} + \frac{\cos\phi}{r^2\sin\phi}\frac{\partial w_\phi}{\partial\phi} + \frac{2}{r^2}\frac{\partial u_r}{\partial\phi}\right) \qquad (17.86)$$

## 17.2.3  Curvilinear coordinates

Some problems involve smoothly curving boundaries whose influence on the flow field is manifest throughout it. In such problems it is sometimes convenient to choose a curvilinear coordinate system (Figure 1.2) that in some sense parallels the intrinsic lines of the boundaries. (Cylindrical and spherical coordinate systems are also curvilinear systems; the use of "curvilinear" here is merely for simplicity.) A good example is flow in a channel. In this case it is natural to choose the centerline as an intrinsic coordinate axis $s$. Let $n$ denote a cross-stream axis, positive

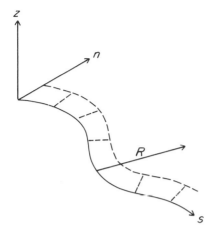

**Figure 17.2** Definition diagram for orthogonal, curvilinear coordinate system.

toward the left bank, and let $z$ denote the axis that is nearly vertical, owing to the small streamwise slope of the channel (Figure 17.2). Whereas $n$ and $z$ are orthogonal to $s$, the coordinate mesh in the $sn$-plane locally may be distorted. In addition, we may associate with each centerline coordinate position $s$ a local curvature $C$ and radius of curvature $R = 1/C$, taken as positive when the center of curvature is located in the direction of positive $n$. For any given coordinate position $(s, n, z)$, the local radius of curvature is thus $R - n$. Note that, in using this coordinate system, an important condition must be satisfied: the local radius of curvature must be larger than the channel half-width.

***17.2.3.1 Identities.*** Let $S$ denote a scalar quantity; let vectors $\mathbf{U} = \langle U_s, U_n, U_z \rangle$ and $\mathbf{V} = \langle V_s, V_n, V_z \rangle$; and let unit vectors $\mathbf{e}_s = \langle 1, 0, 0 \rangle$, $\mathbf{e}_n = \langle 0, 1, 0 \rangle$, and $\mathbf{e}_z = \langle 0, 0, 1 \rangle$. Then

$$\nabla S = \mathbf{e}_s \frac{1}{1 - Cn} \frac{\partial S}{\partial s} + \mathbf{e}_n \frac{\partial S}{\partial n} + \mathbf{e}_z \frac{\partial S}{\partial z} \tag{17.87}$$

$$\nabla^2 S = \frac{1}{(1 - Cn)^2} \frac{\partial^2 S}{\partial s^2} + \frac{n}{(1 - Cn)^3} \frac{\partial C}{\partial s} \frac{\partial S}{\partial s} + \frac{\partial^2 S}{\partial n^2} - \frac{C}{1 - Cn} \frac{\partial S}{\partial n} + \frac{\partial^2 S}{\partial z^2} \tag{17.88}$$

$$\nabla \cdot \mathbf{U} = \frac{1}{1 - Cn} \frac{\partial U_s}{\partial s} + \frac{\partial U_n}{\partial n} - \frac{C U_n}{1 - Cn} + \frac{\partial U_z}{\partial z} \tag{17.89}$$

$$\mathbf{U} \cdot \nabla S = \frac{U_s}{1 - Cn} \frac{\partial S}{\partial s} + U_n \frac{\partial S}{\partial n} + U_z \frac{\partial S}{\partial z} \tag{17.90}$$

$$\mathbf{U} \cdot \nabla \mathbf{V} = \left( \mathbf{U} \cdot \nabla V_s - \frac{C U_n V_s}{1 - Cn} \right) \mathbf{e}_s + \left( \mathbf{U} \cdot \nabla V_n + \frac{C U_s V_s}{1 - Cn} \right) \mathbf{e}_n + (\mathbf{U} \cdot \nabla V_z) \mathbf{e}_z$$

$$\tag{17.91}$$

***17.2.3.2 Continuity equation.*** For viscous flow,

$$\frac{1}{1 - Cn} \frac{\partial}{\partial s} (\rho u_s) - \frac{\rho C v_n}{1 - Cn} + \frac{\partial}{\partial n} (\rho v_n) + \frac{\partial}{\partial z} (\rho w_z) = -\frac{\partial \rho}{\partial t} \tag{17.92}$$

for turbulent, incompressible fluid flow,

$$\frac{1}{1 - Cn}\frac{\partial \overline{u}_s}{\partial s} - \frac{C\overline{v}_n}{1 - Cn} + \frac{\partial \overline{v}_n}{\partial n} + \frac{\partial \overline{w}_z}{\partial z} = 0 \tag{17.93}$$

and for saturated porous-media flow,

$$\frac{1}{1 - Cn}\frac{\partial}{\partial s}(\rho q_{hs}) - \frac{\rho C q_{hn}}{1 - Cn} + \frac{\partial}{\partial n}(\rho q_{hn}) + \frac{\partial}{\partial z}(\rho q_{hz}) = -\frac{\partial}{\partial t}(n\rho) \tag{17.94}$$

### 17.2.3.3 Momentum equations. The s-component is

$$\frac{u_s}{1 - Cn}\frac{\partial u_s}{\partial s} + v_n\frac{\partial u_s}{\partial n} + w_z\frac{\partial u_s}{\partial z} - \frac{Cu_s v_n}{1 - Cn} + \frac{\partial u_s}{\partial t} = -\frac{1}{\rho(1 - Cn)}\frac{\partial p}{\partial s} - \frac{g}{1 - Cn}\frac{\partial h}{\partial s}$$
$$-\frac{1}{\rho}\left(\frac{1}{1 - Cn}\frac{\partial \tau_{ss}}{\partial s} + \frac{\partial \tau_{ns}}{\partial n} + \frac{\partial \tau_{zs}}{\partial z} - \frac{2C\tau_{ns}}{1 - Cn}\right) \tag{17.95}$$

the n-component is

$$\frac{u_s}{1 - Cn}\frac{\partial v_n}{\partial s} + v_n\frac{\partial v_n}{\partial n} + w_z\frac{\partial v_n}{\partial z} + \frac{Cu_s^2}{1 - Cn} + \frac{\partial v_n}{\partial t} = -\frac{1}{\rho}\frac{\partial p}{\partial n} - g\frac{\partial h}{\partial n}$$
$$-\frac{1}{\rho}\left(\frac{1}{1 - Cn}\frac{\partial \tau_{sn}}{\partial s} + \frac{\partial \tau_{nn}}{\partial n} + \frac{\partial \tau_{zn}}{\partial z} + \frac{C(\tau_{ss} - \tau_{nn})}{1 - Cn}\right) \tag{17.96}$$

and the z-component is

$$\frac{u_s}{1 - Cn}\frac{\partial w_z}{\partial s} + v_n\frac{\partial w_z}{\partial n} + w_z\frac{\partial w_z}{\partial z} + \frac{\partial w_z}{\partial t} = -\frac{1}{\rho}\frac{\partial p}{\partial z} - g\frac{\partial h}{\partial z}$$
$$-\frac{1}{\rho}\left(\frac{1}{1 - Cn}\frac{\partial \tau_{sz}}{\partial s} + \frac{\partial \tau_{nz}}{\partial n} + \frac{\partial \tau_{zz}}{\partial z} - \frac{C\tau_{nz}}{1 - Cn}\right) \tag{17.97}$$

For an incompressible Newtonian fluid the constitutive equations are

$$\tau_{ss} = -2\mu\left(\frac{1}{1 - Cn}\frac{\partial u_s}{\partial s} - \frac{Cv_n}{1 - Cn}\right); \qquad \tau_{nn} = -2\mu\frac{\partial v_n}{\partial n}; \qquad \tau_{zz} = -2\mu\frac{\partial w_z}{\partial z}$$

$$\tau_{sz} = \tau_{zs} = -\mu\left(\frac{\partial u_s}{\partial z} + \frac{1}{1 - Cn}\frac{\partial w_z}{\partial s}\right); \qquad \tau_{zn} = \tau_{nz} = -\mu\left(\frac{\partial v_n}{\partial z} + \frac{\partial w_z}{\partial n}\right)$$

$$\tau_{sn} = \tau_{ns} = -\mu\left(\frac{\partial u_s}{\partial n} + \frac{1}{1 - Cn}\frac{\partial v_n}{\partial s} + \frac{Cu_s}{1 - Cn}\right) \tag{17.98}$$

For turbulent flow within a channel, where diffusion of momentum is primarily in the vertical direction (see Smith and McLean [1984]), these may be written as

$$\tau_{ss} = -2\rho\varepsilon\left(\frac{1}{1 - Cn}\frac{\partial u_s}{\partial s} - \frac{Cv_n}{1 - Cn}\right); \qquad \tau_{nn} = -2\rho\varepsilon\frac{\partial v_n}{\partial n}; \qquad \tau_{zz} = -2\rho\varepsilon\frac{\partial w_z}{\partial z}$$

$$\tau_{sz} = \tau_{zs} = -\rho\varepsilon\left(\frac{\partial u_s}{\partial z} + \frac{1}{1 - Cn}\frac{\partial w_z}{\partial s}\right); \qquad \tau_{zn} = \tau_{nz} = -\rho\varepsilon\left(\frac{\partial v_n}{\partial z} + \frac{\partial w_z}{\partial n}\right)$$

$$\tau_{sn} = \tau_{ns} = -\rho\varepsilon\left(\frac{\partial u_s}{\partial n} + \frac{1}{1 - Cn}\frac{\partial v_n}{\partial s} + \frac{Cu_s}{1 - Cn}\right) \tag{17.99}$$

where $\varepsilon$ denotes a kinematic eddy viscosity. The eddy viscosity is a function of coordinate position (Chapter 15), and therefore generally cannot be removed from the differentials when the expressions in (17.99) are substituted into the momentum equations, (17.95) through (17.97). In this regard, for fully turbulent flow the velocity components and pressure in (17.95) through (17.97) are replaced by time-averaged values, and local acceleration terms vanish to satisfy the quasi-steady condition.

## 17.3 NOTATION

| | |
|---|---|
| $a$ | acceleration component |
| $a, A$ | wave amplitude |
| $a$ | coefficient in van der Waals equation of state |
| $\mathbf{a}$ | acceleration |
| $A, B$ | specified area |
| $A$ | parameter in Glen's law |
| $A$ | eddy viscosity |
| $\mathbf{A}$ | rotation matrix |
| $A_\Psi$ | tidal amplitude |
| $b$ | half-aperture, channel half-width, channel width |
| $b$ | coefficient in van der Waals equation of state |
| $B, B_s$ | coefficient in logarithmic velocity law |
| $\mathbf{B}$ | rotation matrix |
| $\mathbf{B}$ | body force |
| $\mathbf{B}_V$ | body force per unit volume |
| $c$ | sample covariance |
| $c$ | specific heat capacity |
| $c$ | wave speed |
| $c, c_s$ | speed of sound |
| $c$ | molecular speed |
| $c$ | solute mass concentration |
| $c$ | specific moisture capacity |
| $c_e$ | effective specific heat |
| $c_p$ | specific heat capacity at constant pressure |
| $c_V$ | specific heat capacity at constant volume |
| $C$ | curvature |
| $C$ | molar specific heat capacity |
| $\mathbf{C}$ | rotation matrix |
| $C_D$ | coefficient of drag |
| $C_p$ | molar specific heat capacity at constant pressure |
| $C_V$ | molar specific heat capacity at constant volume |
| $d$ | grain diameter |
| $D$ | fractal dimension |
| $D$ | soil-water diffusivity |
| $D$ | coefficient of mechanical dispersion |
| $D$ | sediment particle diameter |
| $\mathbf{D}$ | rotation matrix |
| $D_c, D_l$ | coefficient of mass diffusion |
| $D_c^*$ | apparent coefficient of mass diffusion |
| $D^*$ | coefficient of hydrodynamic dispersion |
| $D_L^*$ | coefficient of longitudinal dispersion |
| $D_T^*$ | coefficient of transverse dispersion |
| $\mathbf{D}'$ | local dispersion coefficient tensor |

|   |   |
|---|---|
| $\mathbf{D}^*$ | dispersion coefficient tensor |
| $\mathbf{e}$ | unit vector in curvilinear coordinate systems |
| $E$ | modulus of elasticity |
| $E$ | evaporation rate |
| $E$ | total energy |
| $f_s$ | discrete series |
| $F, \mathbf{F}$ | force, force per unit weight |
| $F_D$ | drag force |
| $Fr$ | Froude number |
| $F_V$ | force per unit volume |
| $g$ | scalar acceleration due to gravity |
| $\mathbf{g}$ | acceleration due to gravity |
| $G$ | specific gravity |
| $Gr$ | Grashof number |
| $h$ | hydraulic head |
| $h$ | vertical coordinate position |
| $h$ | height of liquid in capillary tube (capillary rise) |
| $h$ | fluid depth, channel depth, dike height |
| $H$ | heat energy |
| $i$ | imaginary number defined by $i^2 = -1$ |
| $\mathbf{i}$ | unit vector parallel to $x$-axis |
| $I$ | moment of inertia |
| $\mathbf{I}$ | identity matrix |
| $j$ | Fourier mode |
| $\mathbf{j}$ | unit vector parallel to $y$-axis |
| $k, k_h$ | intrinsic permeability |
| $k, k_B$ | Boltzmann's constant |
| $\mathbf{k}$ | unit vector parallel to $z$-axis |
| $k_g$ | constant in Henry's law |
| $k_n$ | empirical coefficient relating $D_c^*$ and $D_c$ |
| $k_s$ | Nikuradse sand roughness height |
| $k_0, k_1$ | constant in logarithmic velocity law |
| $K, K_h$ | hydraulic conductivity |
| $\mathbf{K}$ | hydraulic conductivity tensor |
| $\mathbf{K}'$ | local hydraulic conductivity tensor |
| $Kn$ | Knudsen number |
| $K_T$ | thermal conductivity |
| $K_{Te}$ | effective thermal conductivity |
| $K_u$ | unsaturated hydraulic conductivity |
| $l$ | mean free path |
| $l$ | scaling factor |
| $l$ | foraminifer spine length |
| $l, l_0$ | Prandtl's mixing length |
| $L$ | characteristic length |
| $L$ | latent heat |
| $L_0$ | characteristic length |
| $L_0$ | scale of turbulence |
| $m$ | mass |
| $m$ | exponent in Glen's law |
| $m$ | Fourier mode |
| $M$ | molecular mass |
| $M$ | Mach number |
| $n$ | transverse coordinate in curvilinear system |
| $n$ | porosity |
| $n$ | number of moles in ideal gas law and van der Waals equation of state |

| | |
|---|---|
| $n$ | number of foraminifer spines |
| $N$ | sample size |
| $N$ | number of molecules |
| $N$ | momentum, mass |
| $Nu$ | Nusselt number |
| $p$ | fluid pressure, thermodynamic pressure |
| $p$ | pore size |
| $p_b, P_b$ | pressure in bubble |
| $p_b$ | bubbling pressure |
| $p_c$ | capillary pressure |
| $p_T$ | stagnation pressure |
| $p_0$ | atmospheric pressure |
| $p_0$ | reference pressure |
| $p'$ | excess-fluid pressure |
| $P$ | partial pressure |
| $P$ | pore volume |
| $P$ | force component per unit volume due to pressure gradient |
| $\mathbf{P}$ | force per unit mass due to pressure gradient |
| $Pe$ | Peclet number |
| $Pe_c, Pe_c*$ | Peclet number for solute transport |
| $Pe_T$ | Peclet number for heat transport |
| $Pr$ | Prandtl number |
| $q_c, \mathbf{q}_c$ | solute mass flux density |
| $q, q_h$ | specific discharge component |
| $\mathbf{q}, \mathbf{q}_h$ | specific discharge |
| $q_T, \mathbf{q}_T$ | heat flux density |
| $Q$ | volumetric discharge |
| $Q$ | heat energy |
| $Q_h$ | volumetric flux |
| $Q_T$ | heat flux |
| $r$ | correlation coefficient, autocorrelation |
| $r$ | radial coordinate |
| $r$ | radius; molecular radius, foraminifer spine radius |
| $r$ | linear similarity ratio |
| $r$ | longitudinal strain rate of glacier |
| $\mathbf{r}$ | position vector |
| $r_b$ | bubble radius |
| $R$ | radius, radius of curvature |
| $R$ | pore radius |
| $R$ | well radius |
| $R$ | specific gas constant |
| $R$ | recharge rate to unconfined aquifer |
| $\mathbf{R}$ | radial position vector |
| $Ra, \mathbb{R}a$ | Rayleigh number |
| $Ra_c$ | critical Rayleigh number |
| $R_c$ | capillary radius |
| $R_c$ | rate of heat generation per unit volume due to chemical reaction |
| $Re, \mathbb{R}e$ | Reynolds number |
| $R_0$ | universal gas constant |
| $s$ | arbitrary coordinate |
| $s$ | curvilinear coordinate |
| $s$ | complex exponent in stability analysis of Rayleigh–Bénard convection |
| $s^2$ | sample variance |
| $S$ | arbitrary scalar quantity |
| $S$ | characteristic length |

| | |
|---|---|
| $S$ | entropy |
| $S$ | aquifer storativity |
| $S_m$ | shear modulus |
| $\mathbf{S}$ | surface force |
| $\mathbf{S}_V$ | surface force per unit volume |
| $t$ | time |
| $t_0$ | time constant |
| $t_0$ | wave period |
| $t_0, t_*$ | characteristic time interval |
| $T$ | temperature |
| $T$ | time interval |
| $T$ | aquifer transmissivity |
| $T$ | shear force |
| $T_c$ | temperature of country rock |
| $T_m$ | magma temperature |
| $T_w$ | temperature at magma-country rock interface |
| $T_0$ | characteristic time interval |
| $u$ | fluid velocity component parallel to $x$-axis |
| $\mathbf{u}$ | fluid velocity |
| $u_c$ | dispersive velocity component |
| $u_m$ | molecular velocity component parallel to $x$-axis, diffusive velocity component |
| $u_r$ | fluid velocity component parallel to $r$-axis |
| $u_s$ | fluid velocity component parallel to $s$-axis |
| $u_s$ | velocity component of solute molecule |
| $\mathbf{u}_m$ | molecular velocity |
| $\mathbf{u}_s$ | solid matrix velocity |
| $u_0$ | reference velocity component |
| $u_0$ | average velocity component, depth-averaged velocity component |
| $u_*$ | characteristic velocity component |
| $u_*$ | shear velocity |
| $U$ | maximum velocity, free-stream velocity |
| $U$ | internal energy |
| $\mathbf{U}$ | arbitrary vector |
| $U_c$ | autocovariance of dispersive velocity |
| $U_m$ | autocovariance of diffusive velocity |
| $v$ | fluid velocity component parallel to $y$-axis, fluid speed |
| $v_m$ | molecular velocity component parallel to $y$-axis |
| $v_m$ | mean molecular speed |
| $v_n$ | fluid velocity component parallel to $n$-axis |
| $v_\theta$ | fluid velocity component parallel to $\theta$-axis |
| $v_0$ | depth-averaged velocity component |
| $V$ | volume, partial volume |
| $\mathbf{V}$ | arbitrary vector |
| $V_b$ | bubble volume |
| $V_g$ | gas volume |
| $V_l$ | liquid volume |
| $V_p$ | pore volume |
| $V_T$ | total volume |
| $V_0$ | reference volume |
| $V_*$ | characteristic volume, representative elementary volume |
| $w$ | fluid velocity component parallel to $z$-axis |
| $w_m$ | molecular velocity component parallel to $z$-axis |
| $w, w_s$ | settling velocity |
| $w_z$ | fluid velocity component parallel to $z$-axis |

| | |
|---|---|
| $w_\phi$ | fluid velocity component parallel to $\phi$-axis |
| $w_*$ | characteristic velocity component |
| $W$ | weight |
| $W$ | work |
| $W_V$ | work per unit volume |
| $\mathbf{W}$ | arbitrary vector |
| $x$ | Cartesian coordinate |
| $X$ | control volume dimension |
| $y$ | Cartesian coordinate |
| $y_0$ | critical depth associated with Bingham flow |
| $Y$ | control volume dimension |
| $z$ | Cartesian coordinate |
| $z$ | potential energy head |
| $z_0$ | vertical coordinate axis |
| $z_0$ | zero velocity position in logarithmic velocity law |
| $z_*$ | average fluid depth |
| $Z$ | control volume dimension |
| $Z$ | vertical fluid thickness in Rayleigh–Bénard configuration |
| $Z$ | characteristic scale height |
| $\alpha$ | coefficient of thermal expansion |
| $\alpha$ | porous medium compressibility |
| $\alpha$ | arbitrary angle |
| $\alpha_d$ | dispersivity |
| $\beta$ | fluid compressibility |
| $\gamma$ | specific weight |
| $\gamma$ | covariance, autocovariance |
| $\gamma$ | ratio equal to $c_p/c_V$ |
| $\gamma$ | coefficient relating fluctuating and mean dispersive velocities |
| $\gamma$ | static-confined tidal efficiency |
| $\Gamma$ | adiabatic lapse rate |
| $\Gamma$ | autocorrelation of dispersive velocity coefficient $\gamma$ |
| $\Gamma$ | friction coefficient in open-channel flow |
| $\Gamma_a$ | environmental lapse rate |
| $\delta$ | characteristic distance associated with dispersion |
| $\delta$ | boundary layer thickness |
| $\delta$ | viscous sublayer thickness |
| $\delta'$ | turbulent boundary-layer thickness |
| $\delta^*$ | correlation length of dispersion |
| $\varepsilon$ | kinematic eddy viscosity |
| $\varepsilon$ | strain |
| $\dot{\varepsilon}$ | strain rate, effective strain rate |
| $\dot{\varepsilon}$ | volumetric dilation rate |
| $\varepsilon_R$ | dimensionless kinematic eddy viscosity |
| $\zeta$ | vorticity component about $z$-axis |
| $\zeta$ | water-surface coordinate |
| $\eta$ | vorticity component about $y$-axis |
| $\eta$ | channel bed-surface coordinate |
| $\eta$ | exponent in Corey–Brooks model of unsaturated hydraulic conductivity |
| $\theta$ | angular coordinate |
| $\theta$ | arbitrary angle |
| $\theta$ | moisture content |
| $\theta_e$ | effective saturation |
| $\theta_r$ | residual saturation |
| $\theta_s$ | degree of saturation |

| | |
|---|---|
| $\theta_{sr}$ | residual degree of saturation |
| $\Theta$ | characteristic temperature difference |
| $\kappa$ | von Kármán constant |
| $\kappa, \kappa_e$ | effective thermal diffusivity |
| $\kappa_c$ | molecular diffusivity |
| $\kappa_h$ | hydraulic diffusivity |
| $\kappa, \kappa_T$ | thermal diffusivity |
| $\Lambda$ | linear spacing between bedrock bumps |
| $\lambda$ | length of bedrock bump |
| $\lambda$ | wave length |
| $\lambda$ | coefficient relating effective strain rate and effective shear stress |
| $\lambda$ | exponent in Corey–Brooks model of characteristic moisture curve |
| $\lambda_*$ | characteristic horizontal distance |
| $\mu$ | population mean |
| $\mu$ | dynamic viscosity |
| $\mu_b$ | bulk viscosity |
| $\nu$ | kinematic viscosity |
| $\xi$ | vorticity component about $x$-axis |
| $\xi$ | sum of mean squared velocity fluctuations; square of turbulence intensity |
| $\Xi$ | sum of squared mean velocity components |
| $\Pi$ | Buckingham pi term |
| $\rho$ | mass density |
| $\rho_b$ | bulk density |
| $\rho_c$ | density of country rock |
| $\rho_e$ | effective density |
| $\rho_m$ | magma density |
| $\rho_p$ | protoplasm density |
| $\rho_s$ | density of solid |
| $\rho_N$ | molecular density |
| $\rho_0$ | reference density |
| $\sigma$ | space lag, space interval |
| $\sigma$ | coefficient of surface tension |
| $\sigma$ | normal stress |
| $\sigma$ | total stress |
| $\sigma'$ | Reynolds stress |
| $\sigma_e$ | effective stress |
| $\sigma_0$ | mechanical pressure |
| $\sigma^2$ | population variance |
| $\tau$ | time lag, time interval |
| $\tau$ | torque |
| $\tau$ | shear stress |
| $\tau$ | surface force component per unit volume |
| $\tau$ | effective shear stress |
| $\mathrm{T}$ | total stress (sum of viscous and Reynolds stress) |
| $\tau'$ | Reynolds stress |
| $\tau_b$ | average bed stress |
| $\tau_c$ | characteristic time scale of randomization of solute motion |
| $\tau_d$ | characteristic time scale of molecular diffusion |
| $\tau_f$ | characteristic time scale of solute motion in pore |
| $\tau_0$ | yield stress |
| $\tau_0$ | average deviatoric normal stress |
| $\tau_0$ | boundary stress |
| $\boldsymbol{\tau}$ | stress tensor |
| $\boldsymbol{\tau}'$ | Reynolds stress tensor |

| $\phi$ | arbitrary angle |
|---|---|
| $\phi$ | angular coordinate |
| $\phi$ | phase shift |
| $\Phi$ | Hubbert's potential |
| $\Phi$ | dissipation function |
| $\Phi_d$ | direct dissipation function |
| $\Phi_t$ | turbulent dissipation function |
| $\psi$ | wetting angle |
| $\psi$ | pressure head |
| $\psi_b, \psi_{cb}$ | bubbling head |
| $\psi_c$ | capillary pressure head |
| $\omega$ | angular frequency |
| $\omega$ | angular acceleration component |
| $\omega$ | wavenumber |
| $\boldsymbol{\omega}$ | vorticity |
| $\omega_x$ | angular velocity component about $x$-axis |
| $\omega_y$ | angular velocity component about $y$-axis |
| $\omega_z$ | angular velocity component about $z$-axis |

# References

Acheson, D. J. 1990. *Elementary fluid dynamics*. Oxford: Oxford University Press, 397 pp.

Anderson, D. A., Tannehill, J. C., and Pletcher, R. H. 1984. *Computational fluid mechanics and heat transfer*. New York: McGraw-Hill, 599 pp.

Anderson, M. P. and Woessner, W. W. 1992. *Applied groundwater modeling: simulation of flow and advective transport*. San Diego: Academic Press, 381 pp.

Bagnold, R. A. 1954. Experiments on a gravity-free dispersion of large solid spheres in a Newtonian fluid under shear. *Proceedings of the Royal Society of London* Series A, 255:49–63.

Barker, R. W. 1960. Taxonomic Notes on the Species Figured by H. B. Brady in his Report on the Foraminifera Dredged by H. M. S. *Challenger* During the Years 1873–1876. Society of Economic Paleontologists and Mineralogists, Special Publication No. 9, Tulsa, Okla., 238 pp.

Batchelor, G. K. 1967. *An introduction to fluid dynamics*. Cambridge: Cambridge University Press, 615 pp.

Bear, J. 1972. *Dynamics of fluids in porous media*. New York: Elsevier (reprinted 1988 by Dover, New York), 764 pp.

Benoit, A. T. 1992. Predicting unsaturated hydraulic conductivity of coarse unconsolidated sand. M.S. thesis, Florida State University, Tallahassee, 109 pp.

Best, J. L. 1993. On the interactions between turbulent flow structure, sediment transport and bedform development: some considerations from recent experimental research. In Clifford, N. J., French, J. R., and Hardisty, J., eds., *Turbulence: perspectives on flow and sediment transport*. Chichester: Wiley, pp. 61–92.

Bethke, C. M. and Corbet, T. F. 1988. Linear and nonlinear solutions for one-dimensional compaction flow in sedimentary basins. *Water Resources Research* 24:461–67.

Bird, R. B., Stewart, W. E., and Lightfoot, E. N. 1960. *Transport phenomena*. New York: Wiley, 780 pp.

Box, G. E. P. and Jenkins, G. M. 1976. *Time series analysis: forecasting and control*. San Francisco: Holden-Day, 575 pp.

Bras, R. L. 1990. *Hydrology: an introduction to hydrologic science*. Reading, Mass.: Addison-Wesley, 643 pp.

Bredehoeft, J. D. and Hanshaw, B. B. 1968. On the maintenance of anomalous fluid pressures: I. thick sedimentary sequences. *Geological Society of America Bulletin* 79: 1097–106.

Bredehoeft, J. D. and Papadopulos, I. S. 1965. Rates of vertical groundwater movement estimated from the earth's thermal profile. *Water Resources Research* 1:325–28.

Bredehoeft, J. D., Roeloffs, E. A., and Riley, F. S. 1987. Dipping into the well to predict earthquakes. *Geotimes* 32:16–19.

Brooks, R. H. and Corey, A. T. 1964. Hydraulic properties of porous media. Hydrology Paper 3, Colorado State University, Fort Collins, 27 pp.

Brown, R. A. 1991. *Fluid mechanics of the atmosphere.* San Diego: Academic Press, 486 pp.

Bruce, P. M. and Huppert, H. E. 1990. Solidification and melting along dykes by the laminar flow of basaltic magma. In Ryan, M. P., ed., *Magma transport and storage.* Chichester: Wiley, pp. 87–101.

Burdine, N. T. 1953. Relative permeability calculations from pore-size distribution data. *Transactions of the American Institute of Mineralogy, Metallurgy and Petroleum Engineering* 198:71–78.

Campbell, I. H. and Turner, J. S. 1987. A laboratory investigation of assimilation at the top of a basaltic magma chamber. *Journal of Geology* 95:155–72.

Carslaw, H. S. and Jaeger, J. C. 1959. *Conduction of heat in solids,* 2nd ed. Oxford: Oxford University Press, 510 pp.

Chu, S. and Sposito, G. 1980. A derivation of the macroscopic solute transport equation for homogeneous, saturated, porous media. *Water Resources Research* 16:542–46.

Clark, S. P., Jr., ed. 1966a. Handbook of Physical Constants. Geological Society of America, Memoir 97, New York, 587 pp.

Clark, S. P., Jr., 1966b. Thermal conductivity. In Clark, S. P., Jr., ed., Handbook of Physical Constants. Geological Society of America, Memoir 97, New York, pp. 459–82.

Clauser, C. 1992. Permeability of crystalline rocks. *EOS, Transactions,* American Geophysical Union, 73:233–38.

Clifford, N. J., French, J. R., and Hardisty, J., eds. 1993. *Turbulence: perspectives on flow and sediment transport.* Chichester: Wiley, 360 pp.

Cooper, H. H., Bredehoeft, J. D., and Papadopulos, I. S. 1967. Response of a finite diameter well to an instantaneous charge of water. *Water Resources Research* 3:263–69.

Cox, K. G., Bell, J. D., and Pankhurst, R. J. 1979. *The interpretation of igneous rocks.* London: Allen & Unwin, 450 pp.

Davis, J. C. 1986. *Statistics and data analysis in geology,* 2nd ed. New York: Wiley, 646 pp.

DeFay, R., Prigogine, I., Bellemans, A., and Everett, D. H. 1966. *Surface tension and adsorption.* New York: Wiley, 432 pp.

Dingman, S. L. 1984. *Fluvial hydrology.* New York: W. H. Freeman and Co., 383 pp.

Domenico, P. A. and Schwartz, F. W. 1990. *Physical and chemical hydrogeology.* New York: Wiley, 824 pp.

Donnelly, R. J., Herman, R., and Prigogine, I., eds. 1966. *Non-equilibrium thermodynamics, variational techniques and stability.* Chicago: University of Chicago Press, 313 pp.

Drake, J. M. and Klafter, J. 1990. Dynamics of confined molecular systems. *Physics Today* 43(5):46–55.

Drazin, P. G. and Reid, W. H. 1981. *Hydrodynamic stability.* Cambridge: Cambridge University Press, 527 pp.

Dritschel, D. G. and Legras, B. 1993. Modeling oceanic and atmospheric vortices. *Physics Today* 46(3):44–51.

Elder, J. W. 1965a. Laminar free convection in a vertical slot. *Journal of Fluid Mechanics* 23:77–98.

Elder, J. W. 1965b. Turbulent free convection in a vertical slot. *Journal of Fluid Mechanics* 23:99–111.

Freeze, R. A. and Cherry, J. A. 1979. *Groundwater.* Englewood Cliffs, N. J.: Prentice-Hall, 604 pp.

Frisch, U. and Orszag, S. A. 1990. Turbulence: challenges for theory and experiment. *Physics Today* 43(1):24–32.

Furbish, D. J. 1987. Conditions for geometric similarity of coarse stream-bed roughness. *Mathematical Geology* 19:291–307.

Gelhar, L. W., Mantoglou, A., Welty, C., and Rehfeldt, K. R. 1985. A review of field-scale physical solute transport processes in saturated and unsaturated porous media. Electric Power Research Institute EPRI EA-4190 Project 2485-5, 116 pp.

Gerald, C. F. and Wheatley, P. O. 1984. *Applied numerical analysis,* 3rd ed. Reading, Mass.: Addison-Wesley, 579 pp.

Gibson, R. E. 1958. The progress of consolidation in a clay layer increasing in thickness with time. *Géotechnique* 8:171–82.

Giles, R. V. 1962. *Theory and problems of fluid mechanics and hydraulics,* 2nd ed. Schaum's Outline Series. New York: McGraw-Hill, 274 pp.

Goldstein, H. 1980. *Classical mechanics,* 2nd ed. Reading, Mass.: Addison-Wesley, 672 pp.

Granick, S. 1991. Motions and relaxations of confined liquids. *Science* 253:1374–79.

Grishanin, K. V. 1979. *Dynamics of alluvial flows* (in Russian). Leningrad: Gidrometeoizdat, 311 pp.

de Groot, S. R. and Mazur, P. 1984. *Non-equilibrium thermodynamics.* New York: Dover, 510 pp.

Guggenheim, E. A. 1959. *Thermodynamics,* 4th ed. Amsterdam: North-Holland, 476 pp.

Häcker, V. 1908. Tiefseeradiolarien. Wiss. Ergebn. Deutsch. Tiefsee Exp. *Valdivia* pp. 417–706.

Hallet, B. 1976. The effect of subglacial chemical processes on glacier sliding. *Journal of Glaciology* 17:209–21.

Hargraves, R. B., ed. 1980. *Physics of magmatic processes.* Princeton, N.J.: Princeton University Press.

Hess, H. H. and Poldervaart, A., eds. 1968. *Basalt: the Poldervaart treatise on rocks of basaltic composition,* volumes 1 and 2. New York: Wiley, 862 pp.

Hillel, D. 1971. *Soil and water: physical principles and processes.* New York: Academic Press, 288 pp.

Hinze, J. O. 1959. *Turbulence: an introduction to its mechanism and theory.* New York: McGraw-Hill, 586 pp.

Howle, L., Behringer, R. P., and Georgiadis, J. 1993. Visualization of convective fluid flow in a porous medium. *Nature* 362:230–32.

Hsieh, P. A., Bredehoeft, J. D., and Farr, J. M. 1987. Determination of aquifer transmissivity from phase analysis of Earth-tide fluctuations of water levels in artesian wells. *Water Resources Research* 23:1824–32.

Hubbert, M. K. 1940. The theory of groundwater motion. *Journal of Geology* 48: 785–944.

Huppert, H. E. 1986. The intrusion of fluid mechanics into geology. *Journal of Fluid Mechanics* 173:557–94.

Huppert, H. E. and Sparks, R. S. J. 1988a. Melting the roof of a chamber containing a hot, turbulently convecting fluid. *Journal of Fluid Mechanics* 188:107–31.

Huppert, H. E. and Sparks, R. S. J. 1988b. The generation of granitic magmas by intrusion of basalt into continental crust. *Journal of Petrology* 29:599–624.

Ikeda, S. and Parker, G., eds. 1989. River meandering. American Geophysical Union, Water Resources Monograph 12, Washington, D. C., 485 pp.

Ingebritsen, S. E. and Rojstaczer, S. A. 1993. Controls on geyser periodicity. *Science* 262:889–92.

Jacob, C. E. 1940. On the flow of water in an elastic artesian aquifer. *Transactions of the American Geophysical Union* 22:574–86.

Jaeger, J. C. and Cook, N. G. W. 1976. *Fundamentals of rock mechanics,* 2nd ed. New York: Halsted Press, 585 pp.

Javandel, I., Doughty, C., and Tsang, C. F. 1984. Groundwater Transport: Handbook of Mathematical Models. Water Resources Monograph 10, American Geophysical Union, Washington, D. C., 228 pp.

Jenkins, G. M. and Watts, D. G. 1968. *Spectral analysis and its applications.* San Francisco: Holden-Day, 525 pp.

Johnson, A. M. 1970. *Physical processes in geology.* San Francisco: Freeman, Cooper & Co., 577 pp.

Karniadakis, G. E. and Orszag, S. A. 1993. Nodes, modes and flow codes. *Physics Today* 46(3):34–42.

Kerr, R. C. 1994a. Melting driven by vigorous compositional convection. *Journal of Fluid Mechanics* 280:255–85.

Kerr, R. C. 1994b. Dissolving driven by vigorous compositional convection. *Journal of Fluid Mechanics* 280:287–302.

Kieffer, S. W. 1977. Sound speed in liquid-gas mixtures: water-air and water-steam. *Journal of Geophysical Research* 82:2895–904.

Kieffer, S. W. 1981. Blast dynamics at Mount St Helens on 18 May 1980. *Nature* 291: 568–70.

Kieffer, S. W. 1984. Seismicity at Old Faithful geyser: an isolated source of geothermal noise and possible analogue of volcanic seismicity. *Journal of Volcanology and Geothermal Research* 22:59–95.

Kieffer, S. W. and Sturtevant, B. 1984. Laboratory studies of volcanic jets. *Journal of Geophysical Research* 89(B10):8253–68.

Kluitenberg, G. A. 1966. Application of the thermodynamics of irreversible processes to continuum mechanics. In Donnelly, R. J., Herman, R., and Prigogine, I., eds., *Nonequilibrium thermodynamics, variational techniques and stability.* Chicago: University of Chicago Press, pp. 91–99.

Klute, A., ed. 1986. Methods of Soil Analysis, Part 1: Physical and Mineralogical Methods, 2nd ed. American Society of Agronomy and Soil Science Society of America, Agronomy Series No. 9, Madison, Wis., 1188 pp.

Komar, P. D. 1972. Mechanical interactions of phenocrysts and flow differentiation of igneous dikes and sills. *Geological Society of America Bulletin* 83:973–88.

Krishnamurti, R. 1968a. Finite amplitude convection with changing mean temperature: Part 1, theory. *Journal of Fluid Mechanics* 33:445–55.

Krishnamurti, R. 1968b. Finite amplitude convection with changing mean temperature: Part 2, an experimental test of the theory. *Journal of Fluid Mechanics* 33:457–63.

Krishnamurti, R. 1973. Some further studies on the transition to turbulent convection. *Journal of Fluid Mechanics* 60:285–303.

Kushiro, I. 1980. Viscosity, density, and structure of silicate melts at high pressure, and their petrological applications. In Hargraves, R. B., ed., Physics of magmatic processes. Princeton, N.J.: Princeton University Press, pp. 93–120.

Langseth, M. G. and Moore, J. C. 1990. Introduction to special section on the role of fluids in sediment accretion, deformation, diagenesis, and metamorphism in subduction zones. *Journal of Geophysical Research* 95:8737–41.

Lapwood, E. R. 1948. Convection of a fluid in a porous medium. *Cambridge Philosophical Society Proceedings* 44:508–21.

LePage, W. R. 1961. *Complex variables and the Laplace transform for engineers.* New York: McGraw-Hill (Dover edition, 1980), 475 pp.

Levich, V. G. 1962. *Physicochemical hydrodynamics.* Englewood Cliffs, N. J.: Prentice-Hall, 700 pp.

Ligrani, P. M. and Moffat, R. J. 1986. Structure of transitionally rough and fully rough turbulent boundary layers. *Journal of Fluid Mechanics* 162:69–98.

Long, J. C. S. and Billaux, D. M. 1987. From field data to fracture network modeling: an example incorporating spatial structure. *Water Resources Research* 23:1201–16.

Loper, D. E. 1992. A nonequilibrium theory of a slurry. *Continuum Mechanics and Thermodynamics* 4:213–45.

Marsh, B. D. 1981. On the crystallinity, probability of occurrence and rheology of lava and magma. *Contributions in Mineralogy and Petrology* 78:85–98.

de Marsily, G. 1986. *Quantitative hydrogeology*. London: Academic Press, 440 pp.

May, M. 1991. Aerial defense tactics of flying insects. *American Scientist* 79:316–28.

Meyer, R. E. 1971. *Introduction to mathematical fluid dynamics*. New York: Dover, 185 pp.

Middleton, G. V. and Southard, J. B. 1984. Mechanics of Sediment Motion. Lecture Notes for Short Course No. 3, Society of Economic Paleontologists and Mineralogists, Tulsa, Okla., 401 pp.

Middleton, G. V. and Wilcock, P. R. 1994. *Mechanics in the earth and environmental sciences*. Cambridge: Cambridge University Press, 459 pp.

Miller, F., Jr. 1977. *College physics,* 4th ed. New York: Harcourt Brace Jovanovich, 836 pp.

Nafe, J. E. and Drake, C. L. 1968. Physical properties of rocks of basaltic composition. In Hess, H. H. and Poldervaart, A., eds., *Basalt: the Poldervaart treatise on rocks of Basaltic composition,* Volume 2. New York: Wiley, pp. 483–502.

Nakayama, Y., Woods, W. A., and Clark, D. G., eds. 1988. *Visualized flow: fluid motion in basic and engineering situations.* Oxford: Pergamon, 137 pp.

Nelson, J. M., McLean, S. R., and Wolfe, S. R. 1993. Mean flow and turbulence fields over two-dimensional bed forms. *Water Resources Research* 29:3935–53.

Nelson, J. M. and Smith, J. D. 1989. Evolution and stability of erodible channel beds. In Ikeda, S. and Parker, G., eds., River Meandering. American Geophysical Union, Water Resources Monograph 12, Washington, D. C., pp. 321–77.

Neuzil, C. E. 1986. Groundwater flow in low-permeability environments. *Water Resources Research* 22:1163–95.

Neuzil, C. E. 1993. Low fluid pressure within the Pierre shale: a transient response to erosion. *Water Resources Research* 29:2007–20.

Nielson, J. E. and Nakata, J. K. 1994. Mantle origin and flow sorting of megacryst-xenolith inclusions in mafic dikes of Black Canyon, Arizona. U. S. Geological Survey Professional Paper 1541, Washington, D. C., 41 pp.

Nikuradse, J. 1933. Strömungsgesetze in rauhen Rohren. Forschung auf dem Gebiete des Ingenieur-Wesens, 361, Berlin.

Nowell, A. R. M. and Church, M. 1979. Turbulent flow in a depth-limited boundary layer. *Journal of Geophysical Research* 84:4816–24.

Nye, J. F. 1957. The distribution of stress and velocity in glaciers and ice-sheets. *Proceedings of the Royal Society of London* Series A, 239:113–33.

Oke, T. R. 1978. *Boundary layer climates*. London: Methuen, 372 pp.

Orszag, S. A. and Zabusky, N. J. 1993. High-performance computing and physics. *Physics Today* 46(3):22–23.

Page, A. L., Miller, R. H., and Keeney, D. R., eds. 1986. Methods of Soil Analysis, Part 2: Chemical and Microbiological Properties, 2nd ed. American Society of Agronomy and Soil Science Society of America, Agronomy Series No. 9, Madison, Wis., 1159 pp.

Paterson, W. S. B. 1981. *The physics of glaciers,* 2nd ed. Oxford: Pergamon, 380 pp.

Phillips, O. M. 1991. *Flow and reactions in permeable rocks*. Cambridge: Cambridge University Press, 285 pp.

Prandtl, L. 1905. Über Flüssigkeitsbewegung bei sehr kleiner Reibung. Verhandlungen des III Internationalen Mathematiker-Kongresses, Heidelberg, 1904, Leipzig, pp. 484–91.

Prandtl, L. and Tietjens, O. G. 1957. *Fundamentals of hydro- and aeromechanics*. New York: Dover, 270 pp.

Priestley, M. B. 1981. *Spectral analysis and time series*. London: Academic Press, 940 pp.

Proussevitch, A. A., Sahagian, D. L., and Anderson, A. T. 1993. Dynamics of diffusive bubble growth in magmas: isothermal case. *Journal of Geophysical Research* 98:22, 283–307.

Ray, R. J., Krantz, W. B., Caine, T. N., and Gunn, R. D. 1983. A model for sorted patterned-ground regularity. *Journal of Glaciology* 29:317–37.

Rayleigh, J. W. S., Lord. 1916. On convection currents in a horizontal layer of fluid, when the higher temperature is on the under side. *Philosophical Magazine* 32:529–46.

Reynolds, O. 1883. On the experimental investigation of the circumstances which determine whether the motion of water shall be direct or sinuous, and the law of resistance in parallel channels. *Philosophical Transactions of the Royal Society* 174:935–82.

Rieu, M. and Sposito, G. 1991a. Fractal fragmentation, soil porosity, and soil water properties: I. Theory. *Soil Science Society of America Journal* 55:1231–38.

Rieu, M. and Sposito, G. 1991b. Fractal fragmentation, soil porosity, and soil water properties: II. Applications. *Soil Science Society of America Journal* 55:1239–44.

Roeloffs, E. A., Burford, S. S., Riley, F. S., and Records, A. W. 1989. Hydrologic effects on water level changes associated with episodic fault creep near Parkfield, California. *Journal of Geophysical Research* 94:12,387–402.

Rojstaczer, S. 1988. Determination of fluid flow properties from the response of water levels in wells to atmospheric loading. *Water Resources Research* 24:1927–38.

Rojstaczer, S. and Agnew, D. C. 1989. The influence of formation material properties on the response of water levels in wells to Earth tides and atmospheric loading. *Journal of Geophysical Research* 94:12,403–11.

Rose, H. E. 1945. An investigation into the laws of flow of fluids through beds of granular material. *Proceedings of the Institution of Mechanical Engineers* 153:141–48.

Ryan, M. P., ed. 1990. *Magma transport and storage*. Chichester: Wiley, 420 pp.

Sahagian, D. L. and Proussevitch, A. A. 1992. Bubbles in volcanic systems. *Nature* 359(6395):485.

Schlichting, H. 1979. *Boundary-layer theory,* 7th ed. New York: McGraw-Hill, 817 pp.

Schlüter, A., Lortz, D., and Busse, F. 1965. On the stability of steady finite amplitude convection. *Journal of Fluid Mechanics* 23:129–44.

Shafer, N. E. and Zare, R. N. 1991. Through a beer glass darkly. *Physics Today* 44(10): 48–52.

Shaw, H. R., Wright, T. L., Peck, D. L., and Okamura, R. 1968. The viscosity of basaltic magma: analysis of field measurements in Makaopuki Lava Lake, Hawaii. *American Journal of Science* 266:225–64.

Smith, J. D. and McLean, S. R. 1984. A model for flow in meandering streams. *Water Resources Research* 20:1301–15.

Sparks, R. S. J. 1978. The dynamics of bubble formation and growth in magmas: a review and analysis. *Journal of Volcanology and Geothermal Research* 3:1–38.

Spera, F. J. 1980. Aspects of magma transport. In Hargraves, R. B., ed., *Physics of magmatic processes*. Princeton, N. J.: Princeton University Press, pp. 265–323.

Spiegal, E. A. and Veronis, G. 1960. On the Boussinesq approximation for a compressible fluid. *Astrophysical Journal* 131:442–47.

Steinberg, G. S., Steinberg, A. S., and Merzhanov, A. G. 1989a. Fluid mechanism of pressure growth in volcanic (magmatic) systems. *Modern Geology* 13:257–65.

Steinberg, G. S., Steinberg, A. S., and Merzhanov, A. G. 1989b. Fluid mechanism of pressure growth and the seismic regime of volcanoes prior to eruption. *Modern Geology* 13:267–74.

Steinberg, G. S., Steinberg, A. S., and Merzhanov, A. G. 1989c. Fluid mechanism of pressure rise in volcanic (magmatic) systems with mass exchange. *Modern Geology* 13: 275–81.

Stokes, G. G. 1851. On the effect of internal friction of fluids on the motion of pendulums. *Transactions of the Cambridge Philosophical Society* 9(Part II):8–106.

Struiksma, N. and Crosato, A. 1989. Analysis of a 2-D bed topography model for rivers. In Ikeda, S. and Parker, G., eds., River Meandering. American Geophysical Union, Water Resources Monograph 12, Washington, D. C., 153–80.

Takahashi, K. and Bé, A. W. H. 1984. Planktonic foraminifera: factors controlling sinking speeds. *Deep-Sea Research* 31:1477–1500.

Taylor, G. I. 1935. Statistical theory of turbulence: Parts 1–4. *Proceedings of the Royal Society* Series A, 151:421–78.

Tennekes, H. and Lumley, J. L. 1972. *A first course in turbulence.* Cambridge, Mass.: MIT Press, 300 pp.

Terzaghi, K. 1925. Erdbaumechanic auf Bodenphysikalischer Grundlage. Franz Deuticke, Vienna.

Theis, C. V. 1935. The relation between the lowering of the piezometric surface and rate and duration of discharge of a well using groundwater storage. *Transactions of the American Geophysical Union* 2:519–24.

Tolkien, J. R. R. 1937. *The hobbit.* London: Allen & Unwin.

Tolkien, J. R. R. 1954. *The lord of the rings* (trilogy). London: Allen & Unwin.

Tolman, R. C. 1979. *The principles of statistical mechanics.* New York: Dover, 661 pp.

Tritton, D. J. 1988. *Physical fluid dynamics,* 2nd ed. Oxford: Oxford University Press, 519 pp.

Tsang, Y. W. 1984. The effect of tortuosity on fluid flow through a single fracture. *Water Resources Research* 20:1209–15.

Turcotte, D. L. 1990. On the role of laminar and turbulent flow in buoyancy driven magma fractures. In Ryan, M. P., ed., *Magma transport and storage.* Chichester: Wiley, pp. 103–12.

Turner, F. J. and Weiss, L. E. 1963. *Structural analysis of metamorphic tectonites.* New York: McGraw-Hill, 545 pp.

U. S. Geological Survey. 1977. National Handbook of Recommended Methods for Water-Data Acquisition. Reston, Va.

van Genuchten, M. Th. and Alves, W. J. 1982. Analytical solutions of the one-dimensional convective-dispersive solute transport equation. U. S. Department of Agriculture Technical Bulletin 1661, 149 pp.

Vogel, S. 1981. *Life in moving fluids: the physical biology of flow.* Boston: Willard Grant Press, 352 pp.

Wang, H. F. and Anderson, M. P. 1982. *Introduction to groundwater modeling.* San Francisco: W. H. Freeman, 237 pp.

Weast, R. C., ed. 1988. *Chemical Rubber Company handbook of chemistry and physics.* Boca Raton, Fla.: CRC Press.

Weertman, J. 1957. On the sliding of glaciers. *Journal of Glaciology* 3:33–38.

Whipple, K. X. and Dunne, T. 1992. The influence of debris-flow rheology on fan morphology, Owens Valley, California. *Geological Society of America Bulletin* 105:887–900.

Wiberg, P. L. and Smith, J. D. 1991. Velocity distribution and bed roughness in high-gradient streams. *Water Resources Research* 27:825–38.

Williams, J. and Elder, S. A. 1989. *Fluid physics for oceanographers and physicists*. Oxford: Pergamon, 300 pp.

Young, H. D. 1964. *Fundamentals of mechanics and heat*. New York: McGraw-Hill, 638 pp.

Zabusky, N. J., Silver, D., Pelz, R., and Vizgroup '93. 1993. Visiometrics, juxtaposition and modeling. *Physics Today* 46(3):24–31.

Zemansky, M. W. 1981. *Temperatures very high and very low*. New York: Dover, 127 pp.

# Index